ADVANCES IN CHEMICAL ENGINEERING

Volume 11

CONTRIBUTORS TO THIS VOLUME

DEE H. BARKER
JEAN-CLAUDE CHARPENTIER
SHINTARO FURUSAKI
YONEICHI IKEDA
C. R. MITRA
TERUKATSU MIYAUCHI
SHIGEHARU MOROOKA
MICHAEL STAMATOUDIS
LAWRENCE L. TAVLARIDES

ADVANCES IN
CHEMICAL ENGINEERING

Edited by

THOMAS B. DREW

Department of Chemical Engineering
Massachusetts Institute of Technology
Cambridge, Massachusetts

GILES R. COKELET

Department of Chemical Engineering
University of Rochester
Rochester, New York

JOHN W. HOOPES, JR.

ICI Americas, Inc.
Wilmington, Delaware

THEODORE VERMEULEN

Department of Chemical Engineering
University of California
Berkeley, California

Volume 11
1981

ACADEMIC PRESS

A Subsidiary of Harcourt Brace Jovanovich, Publishers

New York   London   Toronto   Sydney   San Francisco

COPYRIGHT © 1981, BY ACADEMIC PRESS, INC.
ALL RIGHTS RESERVED.
NO PART OF THIS PUBLICATION MAY BE REPRODUCED OR
TRANSMITTED IN ANY FORM OR BY ANY MEANS, ELECTRONIC
OR MECHANICAL, INCLUDING PHOTOCOPY, RECORDING, OR ANY
INFORMATION STORAGE AND RETRIEVAL SYSTEM, WITHOUT
PERMISSION IN WRITING FROM THE PUBLISHER.

ACADEMIC PRESS, INC.
111 Fifth Avenue, New York, New York 10003

United Kingdom Edition published by
ACADEMIC PRESS, INC. (LONDON) LTD.
24/28 Oval Road, London NW1 7DX

LIBRARY OF CONGRESS CATALOG CARD NUMBER: 56–6600

ISBN 0–12–008511–9

PRINTED IN THE UNITED STATES OF AMERICA

81 82 83 84   9 8 7 6 5 4 3 2 1

CONTENTS

CONTRIBUTORS TO VOLUME 11	vii
PREFACE	ix
CONTENTS OF PREVIOUS VOLUMES	xi

Mass-Transfer Rates in Gas–Liquid Absorbers and Reactors
JEAN-CLAUDE CHARPENTIER

I.	Introduction	2
II.	Mass Transfer in Gas–Liquid Reactors	2
III.	Measurement of Interfacial Areas and Mass-Transfer Coefficients	35
IV.	Mass-Transfer Coefficients and Interfacial Areas in Absorber Scale-Up	67
V.	Simulation of Absorbers in Laboratory-Scale Apparatus Using Chemical Reaction	114
VI.	Nomenclature	123
	References	125
	Note Added in Proof	133

The Indian Chemical Industry—Its Development and Needs
DEE H. BARKER AND C. R. MITRA

I.	Summary	136
II.	Development to 1800	137
III.	Development from 1800 to 1947	142
IV.	Structure of the Chemical Industry	147
V.	Development since Independence	154
VI.	Awards for Progress	184
VII.	Professional Societies and Education	187
VIII.	Needs of the Chemical Industry	189
	References	195

The Analysis of Interphase Reactions and Mass Transfer in Liquid–Liquid Dispersions
LAWRENCE L. TAVLARIDES AND MICHAEL STAMATOUDIS

I.	Introduction	200
II.	Flow Field in Agitated Dispersions	200

III.	Behavior of Liquid–Liquid Dispersions	207
IV.	Measurements and Analysis of the Properties of the Dispersion	221
V.	Mathematical Models for Mass Transfer with Reaction in Liquid–Liquid Dispersions	233
VI.	Conclusions	262
	Nomenclature	263
	References	266

Transport Phenomena and Reaction in Fluidized Catalyst Beds

TERUKATSU MIYAUCHI, SHINTARO FURUSAKI, SHIGEHARU MOROOKA, AND YONEICHI IKEDA

I.	Introduction	276
II.	Flow Properties of Fluid Beds	285
III.	Turbulent-Flow Phenomena in Bubble Columns and Fluidized Catalyst Beds	310
IV.	Longitudinal Dispersion Phenomena as Derived from Flow Properties	330
V.	Bubble Phenomena in Relation to Bed Performance	340
VI.	Heat and Mass Transfer in Fluidized Catalyst Beds	360
VII.	The Successive Contact Mechanism for Catalytic Reaction	381
VIII.	Further Properties of the Successive Contact Mechanism	402
IX.	Nonisothermal Effect on the Bed Performance	413
X.	Discussion and Summary	425
	Nomenclature	432
	References	437
	Note Added in Proof	448

INDEX ... 449

CONTRIBUTORS TO VOLUME 11

Numbers in parentheses indicate the pages on which the authors' contributions begin.

DEE H. BARKER* (135), *Birla Institute of Technology and Science, Pilani, Rajasthan, India*

JEAN-CLAUDE CHARPENTIER (1), *Laboratoire des Sciences du Génie Chimique, Centre National de la Recherche Scientifique, Ecole Nationale Supérieure des Industries Chimiques, 54042 Nancy, France*

SHINTARO FURUSAKI (275), *Department of Chemical Engineering, University of Tokyo, Bunkyo-ku, Hongo, Tokyo 113, Japan*

YONEICHI IKEDA (275), *Fluidization Engineering Laboratory, Tokyo, Japan*

C. R. MITRA (135), *Birla Institute of Technology and Science, Pilani, Rajasthan, India*

TERUKATSU MIYAUCHI (275), *Department of Chemical Engineering, University of Tokyo, Bunkyo-ku, Hongo, Tokyo 113, Japan*

SHIGEHARU MOROOKA (275), *Department of Applied Chemistry, Kyushu University, Hakozaki, Higashi-ku, Fukuoka 812, Japan*

MICHAEL STAMATOUDIS† (199), *Department of Chemical Engineering, Illinois Institute of Technology, Chicago, Illinois 60616*

LAWRENCE L. TAVLARIDES (199), *Department of Chemical Engineering, Illinois Institute of Technology, Chicago, Illinois 60616*

---

* Present address: Chemical Engineering Department, Brigham Young University, Provo, Utah 84601.
† Present address: Department of Chemical Engineering, University of Thessaloniki, Thessaloniki, Greece.

PREFACE

With the publication of the eleventh volume of *Advances in Chemical Engineering* the Board of Editors takes pleasure in piping aboard a new and able captain: James Wei, the Warren Kendall Lewis Professor of Chemical Engineering at the Massachusetts Institute of Technology. The senior members of the present Board have served as pilots since 1956; it is time for them to rest upon their oars. They are most happy that the publishers have accepted as the new Editor in Chief the man they have recommended.

Professor Wei will announce the new Board. We wish them well and feel sure that they will enjoy the voyage as much as we have.

Thomas B. Drew
John W. Hoopes, Jr.
Giles R. Cokelet
Theodore Vermeulen

CONTENTS OF PREVIOUS VOLUMES

Volume 1

Boiling of Liquids
J. W. Westwater
Non-Newtonian Technology: Fluid Mechanics, Mixing, and Heat Transfer
A. B. Metzner
Theory of Diffusion
R. Byron Bird
Turbulence in Thermal and Material Transport
J. B. Opfell and B. H. Sage
Mechanically Aided Liquid Extraction
Robert E. Treybal
The Automatic Computer in the Control and Planning of Manufacturing Operations
Robert W. Schrage
Ionizing Radiation Applied to Chemical Processes and to Food and Drug Processing
Ernest J. Henley and Nathaniel F. Barr
AUTHOR INDEX—SUBJECT INDEX

Volume 2

Boiling of Liquids
J. W. Westwater
Automatic Process Control
Ernest F. Johnson
Treatment and Disposal of Wastes in Nuclear Chemical Technology
Bernard Manowitz
High Vacuum Technology
George A. Sofer and Harold C. Weingartner
Separation by Adsorption Methods
Theodore Vermeulen

Mixing of Solids
Sherman S. Weidenbaum
AUTHOR INDEX—SUBJECT INDEX

Volume 3

Crystallization from Solution
C. S. Grove, Jr., Robert V. Jelinek, and Herbert M. Schoen
High Temperature Technology
F. Alan Ferguson and Russell C. Phillips
Mixing and Agitation
Daniel Hyman
Design of Packed Catalytic Reactors
John Beek
Optimization Methods
Douglass J. Wilde
AUTHOR INDEX—SUBJECT INDEX

Volume 4

Mass-Transfer and Interfacial Phenomena
J. T. Davies
Drop Phenomena Affecting Liquid Extraction
R. C. Kintner
Patterns of Flow in Chemical Process Vessels
Octave Levenspiel and Kenneth B. Bischoff
Properties of Cocurrent Gas–Liquid Flow
Donald S. Scott
A General Program for Computing Multistage Vapor–Liquid Processes
D. N. Hanson and G. F. Somerville
AUTHOR INDEX—SUBJECT INDEX

Volume 5

Flame Processes—Theoretical and Experimental
J. F. Wehner
Bifunctional Catalysts
J. H. Sinfelt
Heat Conduction or Diffusion with Change of Phase
S. G. Bankoff
The Flow of Liquids in Thin Films
George D. Fulford

Segregation in Liquid–Liquid Dispersions and Its Effect on Chemical Reactions
K. Rietema
AUTHOR INDEX—SUBJECT INDEX

Volume 6

Diffusion-Controlled Bubble Growth
S. G. Bankoff
Evaporative Convection
John C. Berg, Andreas Acrivos, and Michel Boudart
Dynamics of Microbial Cell Populations
H. M. Tsuchiya, A. G. Fredrickson, and R. Aris
Direct Contact Heat Transfer between Immiscible Liquids
Samuel Sideman
Hydrodynamic Resistance of Particles at Small Reynolds Numbers
Howard Brenner
AUTHOR INDEX —SUBJECT INDEX

Volume 7

Ignition and Combustion of Solid Rocket Propellants
Robert S. Brown, Ralph Anderson, and Larry J. Shannon
Gas–Liquid-Particle Operations in Chemical Reaction Engineering
Knud Østergaard
Thermodynamics of Fluid-Phase Equilibria at High Pressures
J. M. Prausnitz
The Burn-Out Phenomenon in Forced-Convection Boiling
Robert V. Macbeth
Gas–Liquid Dispersions
William Resnick and Benjamin Gal-Or
AUTHOR INDEX—SUBJECT INDEX

Volume 8

Electrostatic Phenomena with Particulates
C. E. Lapple
Mathematical Modeling of Chemical Reactions
J. R. Kittrell
Decomposition Procedures for the Solving of Large-Scale Systems
W. P. Ledet and D. M. Himmelblau

The Formation of Bubbles and Drops
  R. Kumar and N. R. Kuloor
AUTHOR INDEX—SUBJECT INDEX

Volume 9

Hydrometallurgy
  Renato G. Bautista
Dynamics of Spouted Beds
  Kishan B. Mathur and Norman Epstein
Recent Advances in the Computation of Turbulent Flows
  W. C. Reynolds
Drying of Solid Particles and Sheets
  R. E. Peck and D. T. Wasan
AUTHOR INDEX—SUBJECT INDEX

Volume 10

Heat Transfer in Tubular Fluid–Fluid Systems
  G. E. O'Connor and T. W. F. Russell
Balling and Granulation
  P. C. Kapur
Pipeline Network Design and Synthesis
  Richard S. H. Mah and Mordechai Shacham
Mass-Transfer Measurements by the Limiting-Current Technique
  J. Robert Selman and Charles W. Tobias
AUTHOR INDEX—SUBJECT INDEX

ADVANCES IN CHEMICAL ENGINEERING

Volume 11

MASS-TRANSFER RATES IN GAS–LIQUID ABSORBERS AND REACTORS

Jean-Claude Charpentier

Laboratoire des Sciences du Génie Chimique
Centre National de la Recherche Scientifique
Ecole Nationale Supérieure des Industries Chimiques
Nancy, France

I.	Introduction	2
II.	Mass Transfer in Gas–Liquid Reactors	2
	A. Physical Absorption	3
	B. Mass Transfer with Chemical Reaction	7
	C. Solubility and Diffusivity of Gases in Liquids	20
III.	Measurement of Interfacial Areas and Mass-Transfer Coefficients	35
	A. Physical Methods	36
	B. Chemical Methods	40
	C. Apparatus for Determining the Physicochemical Parameters	49
	D. Mass-Transfer Rates with Chemical and Purely Physical Processes	65
IV.	Mass-Transfer Coefficients and Interfacial Areas in Absorber. Scale-Up	67
	A. Packed Columns	67
	B. Plate Columns	87
	C. Bubble Columns	90
	D. Tube Reactors	93
	E. Spray Towers	94
	F. Mechanically Agitated Bubble Reactors	97
	G. Jet Reactors	107
	H. Conclusions Regarding Scale-Up Problems	112
V.	Simulation of Absorbers in Laboratory-Scale Apparatus Using Chemical Reaction	114
	A. Criteria for Simulating an Industrial Absorber	115
	B. Practical Selection of a Simulative Laboratory Model	116
	C. Example: Stirred Cell Simulating a Packed Column	119
	D. Applications	121
VI.	Nomenclature	123
	References	125
	Note Added in Proof	133

I. Introduction

Gas–liquid reactions and absorptions widely used in four main fields of the chemical industry are:

(1) *Liquid-phase processes:* oxidation, hydrogenation, sulfonation, nitration, halogenation, alkylation, sulfation, polycondensation, and polymerization (B5, B6, P14).

(2) *Gas scrubbing:* CO_2, H_2S, CO, SO_2, NO, NO_2, N_xO_y, HF, SiF_4, Cl_2, P_2O_5, and hydrocarbons (very often to combat air pollution).

(3) *Manufacture of pure products:* H_2SO_4, HNO_3, $BaCO_3$, $BaCl_2$, adipic acid, nitrates, phosphates, and so on.

(4) *Biological systems:* fermentation, oxidation of sludges, production of proteins from hydrocarbons, and biological oxidations.

At the heart of these processes is the absorber or the reactor of a particular configuration best suited to the chemical absorption or reaction being carried out. Its selection, design, sizing, and performance depend on the hydrodynamics and axial dispersion, mass and heat transfer, and reaction kinetics.

Two books deal almost exclusively with the subject of mass transfer with chemical reaction, the admirably clear expositions of Astarita (A6) and Danckwerts (D2). Since then a flood of theoretical and experimental work has been reported on gas absorption and related separations. The principal object of this chapter is to present techniques, results, and opinions published mainly during the last 6 or 7 years on mass-transfer coefficients and interfacial areas in most types of absorbers and reactors. This necessitates some review of mass transfer with and without chemical reaction in the first section, and comments about the simulation of industrial reactors by laboratory-scale apparatus in the concluding section. Although many gas–liquid reactions are accompanied by a rise in temperature that may be great enough to affect the rate of gas absorption, our attention here is confined to cases where the rise in temperature does not affect the absorption rate. This latter topic (treated by references B20, T10, S3, T3, V5) could justify another complete chapter.

II. Mass Transfer in Gas–Liquid Reactors

In this section we consider the rate of absorption of gases into liquids that are agitated so that dissolved gas is transported from the interfacial surface to the interior by convective motion. The next section, based on this one, treats chemical methods for determining interfacial areas and mass-transfer coefficients in agitated gas–liquid reactors.

When a soluble gas is mixed with an insoluble one, it must first diffuse through the latter to reach the interfacial surface. As a result, the partial pressure of the soluble gas at the interface is generally less than in the bulk of the gas. The liquid in which the soluble gas is absorbed may be agitated in different ways:

(1) The liquid may flow in a turbulent layer over an inclined or vertical surface (wetted columns, Votator). Discontinuities in the surface may cause periodic mixing of the liquid layer during its flow (packed columns).

(2) The gas may be sparged through the liquid as a stream or cloud of bubbles (bubble columns, plate columns, and sparged vessel).

(3) The liquid may be stirred by a mechanically driven agitator.

(4) The liquid may be sprayed through the gas as drops or jets (spray columns, jets, ejector reactor, and venturi).

To study gas–liquid mass-transfer phenomena it is convenient to consider, first of all, steady-state situations in which the composition of the gas and the liquid are statistically constant when averaged over time in a specified region, such as a short, vertical slice of a tubular column or the entire volume of a single-compartment agitated vessel. In such a situation, diffusion, convection, and reaction often proceed simultaneously, and the convective movement of both gas and liquid is ill defined. Nevertheless, useful predictions have been developed for describing the behavior of complicated systems, using highly simplified models that simulate the situation for practical purposes without introducing a large number of parameters. The procedure differs depending on whether physical or chemical absorption is involved.

A. Physical Absorption

If we consider first physical absorption, whereby gas dissolves in the liquid without reacting, the basic representation for transport of the solute gas is based upon the concept of additivity of a gas-phase resistance and a liquid-phase resistance, assuming that the interfacial resistance can be neglected. The rate of absorption is then

$$\Phi = \varphi a = k_G a(p - p_i) = k_L a(C_A^* - C_{A0}) \tag{1}$$

Here a is the interfacial area between gas and liquid per unit volume of the system, φ is the average rate of transfer of gas per unit area, p and p_i are the partial pressures of soluble gas in the bulk gas and at the interface, C_A^* is the concentration of dissolved gas corresponding to equilibrium with p_i, and C_{A0} is the average concentration of dissolved gas in the bulk liquid.

k_G is the "true" gas-side mass-transfer coefficient. It is possible to refer to the gas film resistance, which implies a stagnant film of gas of finite thickness δ_G across which the soluble gas is transported by molecular diffusion alone (while the bulk of the gas has a uniform composition).

k_L is the physical liquid-side mass-transfer coefficient that applies in the absence of chemical reaction. Absorption is physical when the dissolved gas molecules cannot be divided into reacted and unreacted categories. For liquid-side processes, three models based on three distinct hypothesis about the behavior of the liquid lead to different expressions for k_L.

1. The Film Model

This model assumes a stagnant film of thickness δ_L at the surface of the liquid next to the gas, while the rest of the liquid below the film boundary is kept uniform in composition by turbulent agitation (W5). Convection is assumed absent in the liquid film, within which the transport of the soluble gas takes place by molecular diffusion alone. The concentration in the film falls from C_A^* at the interface to C_{A0} at the inner edge of the film, that is, in the bulk liquid. This simple model leads to

$$\varphi = (D_A/\delta_L)(C_A^* - C_{A0}), \qquad k_L = D_A/\delta_L \qquad (2)$$

where D_A is the diffusivity of the dissolved gas in the liquid.

The hydrodynamic properties of the gas–liquid system are taken into account by the film thickness δ_L, which depends on the geometry, physical properties, and liquid agitation. Though this model is not very realistic (experimentally k_L is often found to vary as $D_A^{0.5}$), it has the advantage of simplicity. Predictions based on it, especially when a chemical reaction occurs, are usually similar to those based on more sophisticated models.

2. Surface-Renewal Models

Here we assume the periodic replacement of elements of liquid at the interface by liquid from the interior, of mean bulk composition. While the element of liquid is at the interface and is exposed to the gas, it absorbs gas as if it were a stagnant layer of infinite depth. The rate of absorption is a function of the time of exposure of the element, being rapid initially and decreasing with time. Such a replacement of liquid at the interface by fresh liquid of the bulk composition can be brought about in a reactor by turbulent circulation of the liquid, or in a laminar flow system by flow discontinuity. The form of the surface-renewal model originally proposed by Higbie (H6) assumes that every element of the surface is exposed to the gas for the same length of time θ before being replaced by liquid of the bulk composition. During this time, every element of the liquid absorbs the same amount Q of gas per unit area, and the average rate of absorption is Q/θ. The relations between θ and k_L are derived from the equations for physical absorption into quiescent liquids (D1):

$$Q = 2(C_A^* - C_{A0})(D_A \theta/\pi)^{1/2} \tag{3}$$

$$\varphi = Q/\theta = 2(C_A^* - C_{A0})(D_A/\pi\theta)^{1/2} \tag{4}$$

$$k_L = 2(D_A/\pi\theta)^{1/2} \tag{5}$$

The hydrodynamic properties of the system are accounted for by the exposure time θ; thus k_L is defined here as the mean value of the mass-transfer coefficient during the time interval from $t = 0$ up to $t = \theta$. Equation (5) shows that the liquid-side mass-transfer coefficient is proportional to the square root of the diffusivity, in accord with experimental evidence.

Higbie's assumption of equal times of exposure may apply, for example, for the absorption of gas in the movement of separate gas bubbles, in swarms of rigid liquid droplets, in jets of liquid, and in laminar liquid films (see Section III,B). However, the uniform-time model is unrealistic for industrial gas–liquid contactors such as packed columns, plate columns, and mechanically agitated vessels. Danckwerts' model (D1) supposes instead that the probability of an element of an interface being replaced by a fresh eddy is independent of the length of time for which it has been exposed. This leads to a stationary distribution of surface ages. Instead of the exposure time θ, the rate of surface renewal s is introduced. Writing that the probability that the interface element will disappear in the time between t and $t + dt$ equals the probability that this element will still be there at time t, $s\,dt$, leads to the following relation between s and k_L:

$$\varphi = (D_A s)^{1/2} = k_L(C_A^* - C_{A0}) \tag{6}$$

with

$$k_L = (D_A s)^{1/2} \tag{7}$$

Here again the hydrodynamic properties of the system are accounted for by a single parameter s, the surface renewal frequency; $1/s$ may be regarded as the average life of interface elements. Equation (7) shows also that k_L is proportional to the square root of the diffusivity. Practical use of Danckwerts' model is restricted to cases where something is known a priori about the renewal frequency (as in gas absorption in packed columns with the liquid flowing over the packing elements being completely mixed at each point of contact between two packing elements).

3. *Comments on Use of the Models*

The physical significance given to k_L is different in each of these three models, as can be seen in Fig. 1. However, they lead to the same predic-

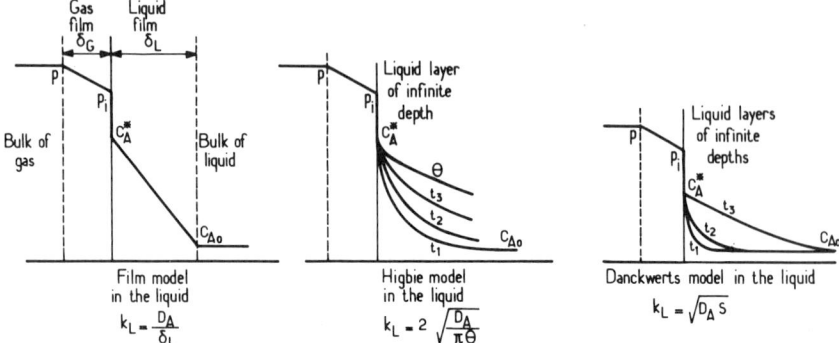

FIG. 1. Mass-transfer models for the liquid side.

tion concerning the effect of the driving force $C_A^* - C_{A0}$ in the liquid phase on the average rate of transfer φ. The models of Higbie and Danckwerts predict that k_L is proportional to $(D_A)^{1/2}$, which agrees well with many experimental observations, while the film model predicts that k_L varies as D_A. But it is difficult to make an accurate test of this dependence, because the diffusivities of solute gases are difficult to determine (see Section II,C).

Apart from this, each model contains an empirical parameter: effective film thickness δ_L, effective exposure time θ, and effective rate of surface renewal s. Consequently, when predicting the average rate of purely physical absorption for large-scale conditions, one cannot expect better results from refined models than from the empirical mass-transfer coefficient first defined by Whitman (W5).

In fact the three models can be regarded as interchangeable for many purposes, and it is mainly a question of convenience which of the three is used. As explained by Danckwerts (D2), in certain cases the mathematical solution may be found analytically for one model, while for another it must be computed numerically. Computations relating to the film model are the simplest, since they involve only ordinary differential equations. Analytical expressions for average absorption rate φ derived from the Higbie model are generally more complicated than those from the Danckwerts model. Also computed values of φ for the film and Higbie models are generally almost equal, except insofar as the diffusivities of dissolved reactants are very different from those of the dissolved gas.

To describe gas absorption with simultaneous chemical reaction, the film theory will be used for purposes of illustration, because the predictions based on the three models are quite similar (except in regard to the effect of solute gas and reactant diffusivities on the rate of absorption).

When necessary, results of surface-renewal theories will be presented simultaneously, as these models may be applied directly to determine interfacial area and mass-transfer coefficients in the laboratory apparatus considered in Sections III,B and V.

B. Mass Transfer with Chemical Reaction

The type of chemical system that has received the most attention is the one in which the dissolved gas (component A) undergoes an irreversible second-order reaction with a reactant (component B) dissolved in the liquid. For the present, the gas will be taken as consisting of pure A, so that complications arising from gas film resistance can be avoided. The stoichiometry of the reaction is represented by

$$A + zB \xrightarrow{k_2} \text{products}$$

with the rate equations

$$r_A = k_2 C_A C_B, \qquad r_B = z r_A \tag{8}$$

If steady-state conditions are assumed in the film, material balances for the two are

$$D_A(d^2 C_A/dx^2) - k_2 C_A C_B = 0 \tag{9}$$

$$D_B(d^2 C_B/dx^2) - z k_2 C_A C_B = 0 \tag{10}$$

Here x is the distance in the liquid from the interface, and D_A and D_B are the diffusivities of A and B in the liquid.

Boundary conditions at the gas–liquid interface are

$$C_A = C_A^* \qquad \text{at} \quad x = 0 \tag{11}$$

$$dC_B/dx = 0 \qquad \text{at} \quad x = 0 \tag{12}$$

At the inner edge of the liquid film ($x = \delta_L$), the boundary condition for component B is

$$C_B = C_{B0}, \qquad x = \delta_L$$

To obtain the boundary condition for component A, it must be indicated that some amount of A reacts within the film while the rest is transferred across the film and reacts in the bulk of the liquid. If a is the specific interfacial area (per unit volume of reactor space) and β is the liquid holdup, the volume of this bulk of liquid may be written as $(\beta/a) - \delta_L$, and the boundary condition for component A will thus be

$$-D_A(dC_A/dx)_{x=\delta_L} = k_2 C_{A0} C_{B0}(\beta/a - \delta_L) \tag{13}$$

Typical forms of the concentration profiles are shown in Fig. 2.

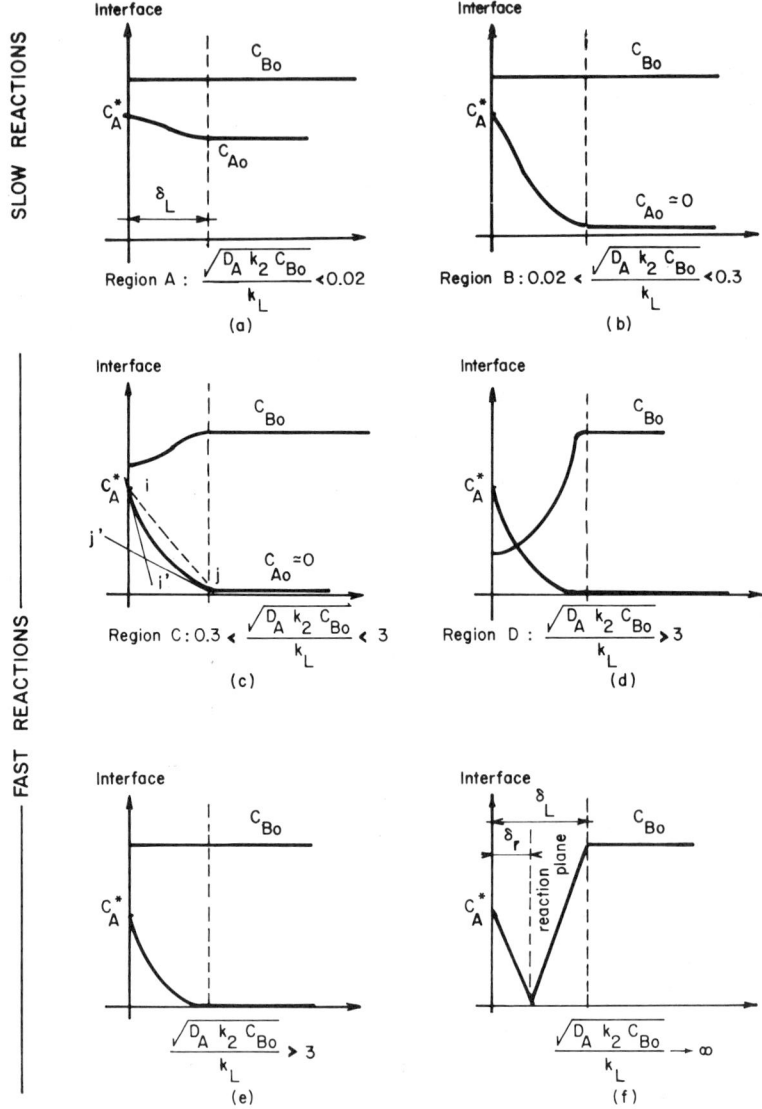

FIG. 2. Liquid-phase concentration profiles for mass transfer with chemical reaction: film theory.

A complete analytical solution of Eqs. (9)–(13) is not possible, but an approximate set of numerical and analytical solutions may be computed for part of the range of variables (D2). The results of these solutions are discussed in terms of an enhancement factor E (or reaction factor) defined by

$$\varphi = E k_L C_A^* \qquad (14)$$

This may be compared with Eq. (2) which applies in the absence of a reaction. When greater than 1, E represents the ratio of the average rate of absorption into an agitated liquid, in the presence of a reaction, to the average rate of pure physical absorption with a zero concentration of component A in the bulk liquid.

A graphical plot of the complete solution of the set of equations is presented in Fig. 3 in which the enhancement factor E is plotted as a function of the dimensionless parameter (Hatta number)

$$\text{Ha} = (D_A k_2 C_{B0})^{1/2}/k_L \qquad (15)$$

of the concentration–diffusion parameter

$$Z_D = (D_B/zD_A)(C_{B0}/C_A^*) \qquad (16)$$

and of the ratio between the volume of liquid associated with unit inter-

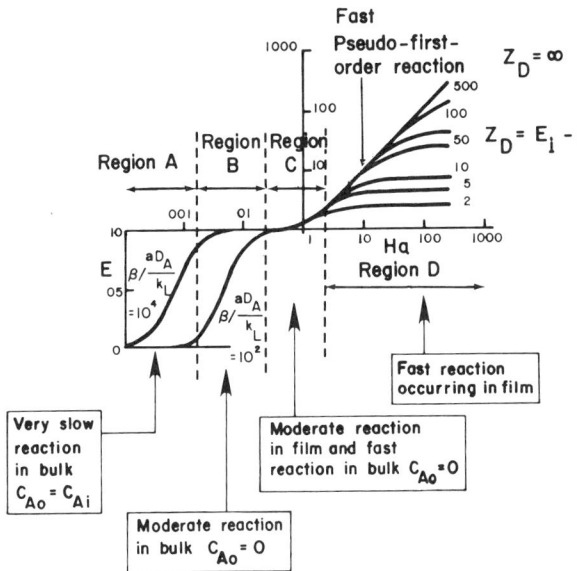

FIG. 3. Enhancement factor for second-order reaction, plotted against Hatta number (C17). Ordinate scale is linear below 1, logarithmic above. Region A: very slow reaction in bulk; $C_{A0} \approx C_{Ai}$; $\beta K_L/aD_A = 10^4$. Region B: moderate reaction in bulk; $C_{A0} \approx 0$; $\beta k_L/aD_A = 10^2$. Region C: moderate reaction in film; fast reaction in bulk; $C_{A0} = 0$. Region D: fast reaction in film; pseudo-first-order curves are asymptotic to $E_i = Z_D + 1$.

face area and the film thickness parameter

$$(\beta/a)(k_L/D_A) \tag{17}$$

The physical significance of the regions covered by Fig. 3 is best appreciated by considering the corresponding profiles. This figure is very important for the choice of the type of gas–liquid reactor or absorber, as the value of the Hatta number provides an important indication of whether a large specific interfacial area a or a large liquid holdup β is required for a particular reaction of rate constant k_2. Four regions are considered in the following paragraphs.

1. *Very Slow Reaction in Bulk Liquid:* Ha < 0.02 (*Region A*)

No reaction occurs in the film; mass transfer is used to keep the bulk concentration of component A (C_{A0}) close to the saturation value C_A^* (Fig. 2a), and sufficient interfacial area is necessary for this purpose. A high liquid holdup is the more important requirement; the use of a bubble column would be suitable. If $C_{A0} \simeq C_A^*$, then the rate of transformation of component A by unit volume of reactor is

$$R_A = k_2 C_A^* C_{B0} \beta \tag{18}$$

Near the limit between regions A and B in Fig. 3, where actually no reaction occurs in the film, the concentration of component A in the bulk of the liquid C_{A0} is determined by indicating that the amount of A transported across the film reacts in the bulk:

$$\Phi = \varphi a = k_L a(C_A^* - C_{A0}) = R_A = k_2 C_{A0} C_{B0} \beta \tag{19}$$

Hence

$$C_{A0} = C_A^*[1 + (\beta k_2 C_{B0}/k_L a)]^{-1} \tag{19a}$$

and

$$\Phi = \varphi a = k_L a C_A^*(1 + k_L a/\beta k_2 C_{B0})^{-1} = C_A^*/[(k_L a)^{-1} + (\beta k_2 C_{B0})^{-1}] \tag{20}$$

Equation (19) signifies that, in the case of a very slow chemical reaction ($C_{A0} \simeq C_A^*$), the rate of the transport process is completely determined by the rate of the chemical reaction. As seen in Eq. (19a), the condition for a very slow reaction is that

$$\beta k_2 C_{B0}/k_L a \ll 1 \tag{21}$$

2. *Slow Reaction in Bulk Liquid:* 0.02 < Ha < 0.3 (*Region B*)

In this case an appreciable amount of the absorbed gas reacts before leaving the reactor, but a negligible proportion of component A reacts in the diffusion film. The process is essentially one of physical absorption followed by reaction in the bulk liquid. The governing equations are again Eqs. (19) and (19a). The condition now, so that the reaction will be sufficiently fast to hold C_{A0} close to zero in the bulk liquid (Fig. 2b), is deduced from Eq. (19a):

$$\beta k_2 C_{B0}/k_L a \gg 1 \qquad (22)$$

Moreover, the liquid holdup must be large. When the condition

$$\beta k_L/aD_A > 10^2 \quad \text{or} \quad aD_A/\beta k_L < 10^{-2} \qquad (23)$$

is not verified, the enhancement factor E is substantially smaller than 1 in the part of region B (Fig. 3) where Ha < 0.1. So, for the important situation of a slow reaction with $C_{A0} = 0$ and $E = 1$, that is, when conditions (22) and (23) are fulfilled,

$$\Phi = \varphi a = k_L a(C_A^* - 0) = k_L a C_A^* = R_A \qquad (24)$$

thus both interfacial area and liquid holdup should be high; the use of mechanically agitated tank would be suitable. Equation (24) means that the rate of the transport process is completely determined by the mass transport across the liquid film. The mass flux through the interface is proportional to $k_L a$. As seen in Section III,B, this result is the basis for using chemical reaction to measure $k_L a$ directly from the rate of absorption when C_A^* is known, whatever the degree of mixing of the liquid phase.

3. *Moderately Fast Reaction:* $0.3 <$ Ha < 3 *(Region C)*

In this case, the reaction is fast enough for a substantial amount of component A to react in the film rather than be transferred unreacted to the bulk liquid where, because of fast reaction, C_{A0} is very low (Fig. 2c). When the reaction undergone by the dissolved gas was slow, in the sense of Sections II,B,1 and 2, the rate at which unreacted component A diffused out of the film into the bulk liquid was the same as the rate at which it diffused across the film from the interface. The concentration profile of A in the diffusion film was a straight line (*ij*).

Now as a substantial amount of component A reacts in the film, the profile becomes curved, so that the concentration gradient at the surface (*ii'*) is greater than that at the inner edge of the film (*jj'*). The ratio of the rate of absorption to the rate of transfer of unreacted gas into the bulk liquid is the ratio of the slopes of *ii'* and *jj'*. The enhancement factor E is the ratio of the slopes of *ii'* and *ij*. So, if E is appreciably greater than 1, the reaction has an appreciable effect on the rate of absorption, which is given by Eq. (25) in the next section. In this case, the surface area of the interface begins to outweigh the volume of the reaction phase in controlling the total conversion rate.

4. Fast Reaction in the Diffusion Film: Ha > 3 (Region D)

The reaction is fast and occurs completely in the liquid film during the transport of component A (Fig. 2d). The concentration of A in the bulk liquid is virtually zero. Thus

$$\Phi = \varphi a = E k_L a C_A^* \tag{25}$$

Therefore, for such reactions, the rate of absorption will be large if the interfacial area is large. So a high interfacial area is required in the reactor, but the liquid holdup is not important; the use of a packed column or a plate column would be suitable.

Van Krevelen and Hoftijzer (V4) have computed an approximate set of solutions indicating that the enhancement factor can be expressed as a function of the Hatta number [Eq. (15)] and the concentration–diffusion parameter [Eq. (16)]. The relation between E, Ha, and $Z_D = E_i - 1$ (given in Fig. 3, region D, or in detail in Fig. 4) is expressed mathematically by

$$E = \frac{\text{Ha}[(E_i - E)/(E_i - 1)]^{1/2}}{\tanh\{\text{Ha}[(E_i - E)/(E_i - 1)]^{1/2}\}} \tag{26}$$

where E_i is the enhancement factor corresponding to an instantaneous reaction. For a given value of E_i (or $Z_D + 1$), an increase in Ha brings about an increase in E until a limiting value is approached where $E = E_i$:

$$E_i = 1 + (D_B/zD_A)(C_{B0}/C_A^*) \tag{27}$$

Several limiting types of behavior may be identified in Fig. 4.

  a. *Fast Pseudo-First-Order:* $3 < \text{Ha} < E_i/2$. If the concentration of component B in the bulk liquid is much greater than C_A^* (see Fig. 2e), the kinetics of the reaction becomes pseudo-first order ($k_1 = k_2 C_{B0}$). Equation (9) with boundary condition (11) and with $C_A = C_{A0}$ at $x = \delta_L$ yields

$$C_A = \frac{1}{\sinh \text{Ha}} \left[C_{A0} \sinh x \left(\frac{k_2 C_{B0}}{D_A}\right)^{1/2} + C_A^* \sinh\left(\frac{D_A}{k_L} - x\right)\left(\frac{k_2 C_{B0}}{D_A}\right)^{1/2} \right]$$

The average absorption rate is

$$\varphi = -D_A \left(\frac{dC_A}{dx}\right)_{x=0} = k_L \left(C_A^* - \frac{C_{A0}}{\cosh \text{Ha}}\right) \frac{\text{Ha}}{\tanh \text{Ha}} \tag{28}$$

If Ha $< E_i/2$, the point representing the enhancement factor falls very close to the limiting diagonal in Fig. 4. Physically this means that reactant B diffuses toward the surface fast enough to prevent the reaction from causing any significant depletion there, so that C_{B0} is kept virtually constant. The local rate of reaction of dissolved gas is $k_2 C_{B0} C_A$.

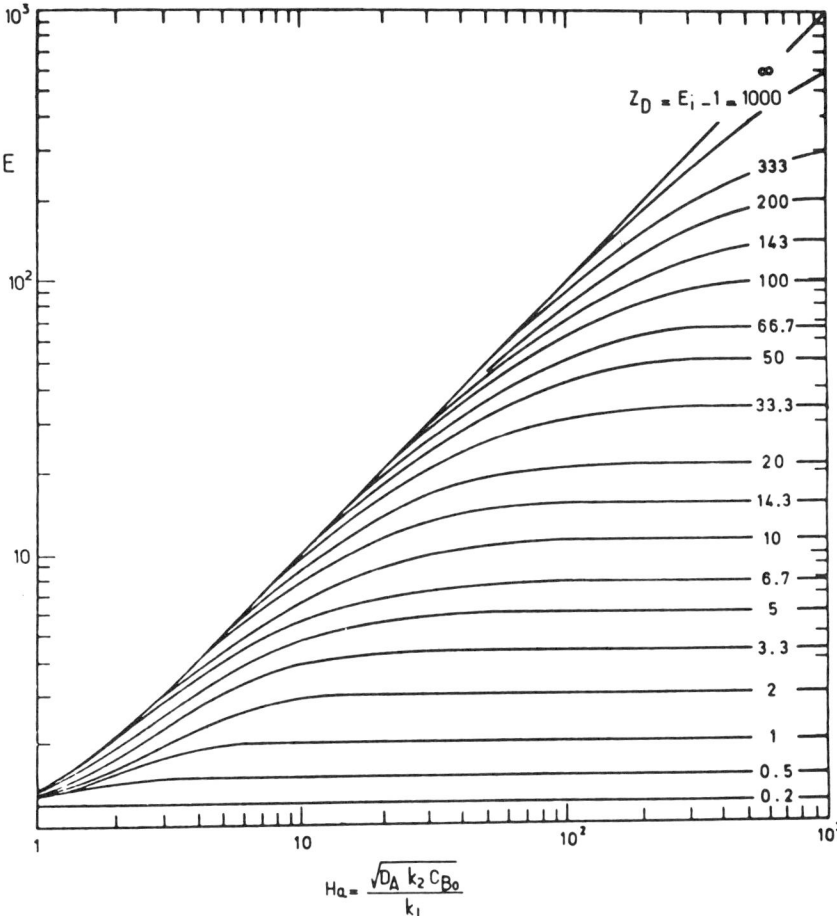

FIG. 4. Enhancement factor for second-order reaction for Hatta number greater than 3 (Ha > 3) (V4).

As also Ha > 3, the dissolved gas all reacts in the diffusion film and none diffuses unreacted into the bulk liquid ($C_{A0} = 0$). The concentration profile is sketched in Fig. 2e. Thus

$$\varphi = k_L C_A^* (\text{Ha}/\tanh \text{Ha}) \simeq \text{Ha}\, k_L C_A^* \qquad (29)$$

and, as can be seen from Fig. 4, to a close approximation,

$$E = \text{Ha} \qquad (30)$$

that is,

$$\varphi = C_A^*(D_A k_2 C_{B0})^{1/2} \tag{31}$$

$$\Phi = \varphi a = a C_A^*(D_A k_2 C_{B0})^{1/2} \tag{31a}$$

This corresponds to absorption with fast pseudo-first-order kinetics. Thus the film thickness, or the value of k_L, is irrelevant and does not appear in the expression for the average rate of absorption φ or for the rate of absorption per unit volume of reactor Φ. This important case will be the basis of a chemical method for measuring the interfacial area directly from the rate of absorption when $C_A^*(D_A k_2 C_{B0})^{1/2}$ is known.

b. *Instantaneous Reaction*, $\text{Ha} > 10 E_i$. It may be seen from Fig. 4 that, for a given value of E_i (that is, for given values of z, D_A, D_B, C_A^*, and C_{B0}) when $\text{Ha} > 10E_i$, an increase in the Hatta number leads to a limiting value of $E = E_i$. The reaction rate constant is high, the concentration of reactant B is substantially less than the solubility of the gas, or the physical mass-transfer rate is low; the dissolved gas A reacts instantaneously with component B. There is a plane beneath the interface at a distance δ_r where the concentration of both components is zero and the rate of reaction is equal to the rate at which the two components can diffuse to the reaction plane. The actual kinetics of the reaction is then immaterial. The concentration profile is similar to that shown in Fig. 2f. The reactant B is depleted in the neighborhood of the interface, to the extent that the rate of reaction is determined by diffusion alone.

The rate at which both components reach the reaction plane is

$$zD_A[(C_A^* - 0)/\delta_r] = D_B[(C_{B0} - 0)/(\delta_L - \delta_r)]$$

Therefore, from the average rate of absorption,

$$\varphi = D_A \left(\frac{C_A^* - 0}{\delta_r}\right) = \frac{D_A C_A^*}{\delta_L}\left(1 + \frac{D_B}{zD_A}\frac{C_{B0}}{C_A^*}\right)$$

that is,

$$\varphi = k_L C_A^*\left(1 + \frac{D_B}{zD_A}\frac{C_{B0}}{C_A^*}\right) = k_L C_A^* E_i \tag{32}$$

and

$$\Phi = \varphi a = k_L a C_A^* E_i \tag{32a}$$

This important case will be the basis of a chemical method for measuring either $k_G a$, for solute gas mixed with insoluble gas (with the reaction plane at the interface) or $k_L a$ (when $C_A^* \ll C_{B0}$).

5. Surface-Renewal Models

Transient absorption of gas, followed by irreversible second-order reaction, has been studied extensively by Brian *et al.* (B26) and Danckwerts (D1). The average rate of absorption for a contact time θ is also given in terms of the enhancement factor

$$E = (Q/2C_A^*)(\pi/D_A\theta)^{1/2} \tag{33}$$

where Q is the amount of gas absorbed per unit surface area.

According to the Higbie model, these values of E can be used to find the average rate of absorption into an agitated liquid:

$$\varphi = E k_L C_A^*$$

and can be expressed as a function of Hatta number Ha' and an instantaneous enhancement factor E_i' defined as

$$\text{Ha}' = \left(\frac{\pi}{4} k_2 C_{B0} \theta\right)^{1/2}, \quad E_i' = \left(\frac{D_A}{D_B}\right)^{1/2} + \left(\frac{D_B}{D_A}\right)^{1/2} \frac{C_{B0}}{zC_A^*} = E_i \left(\frac{D_A}{D_B}\right)^{1/2} \tag{34}$$

The result is approximated well by Eq. (26), using Ha = Ha' and $E_i = E_i'$.

So the computed values of E for the Higbie and the film models are almost equal for given values of Ha or Ha' and E_i or E_i'. Therefore the same diagram (that is, Fig. 4, showing E as a function of Ha with E_i as a parameter), with the results of Sections II,B,1–4, can be used for either with sufficient accuracy.

Turning to Danckwerts' model, we observe that irreversible second-order reactions are not conveniently dealt with except where the reaction is pseudo-first-order based on component A. In this case (D2), it is found for

$$\left(1 + \frac{k_2 C_{B0}}{s}\right)^{1/2} = \left(1 + \frac{D_A k_2 C_{B0}}{k_L^2}\right)^{1/2}$$

$$= (1 + \text{Ha}^2)^{1/2} \ll \left(\frac{D_A}{D_B}\right)^{1/2} + \left(\frac{D_B}{D_A}\right)^{1/2} \frac{C_{B0}}{zC_A^*} \tag{35}$$

that the average rate of absorption is

$$\varphi = k_L C_A^* \left(1 + \frac{k_2 C_{B0}}{s}\right)^{1/2}$$

$$= k_L C_A^* \left(1 + \frac{D_A k_2 C_{B0}}{k_L^2}\right)^{1/2} = k_L C_A^* (1 + \text{Ha}^2)^{1/2} \tag{36}$$

whence

$$\Phi = \varphi a = k_L a C_A^* (1 + \text{Ha}^2)^{1/2} \qquad (36a)$$

If $\text{Ha} \gg 1$ (rapid pseudo-first-order), $\varphi = k_L C_A^* \text{Ha}$ and $E = \text{Ha}$.

Therefore, from a comparison with Eq. (29), the three models give nearly comparable results in this case. Also, when $D_A = D_B$, the instantaneous enhancement factor E_i has the same form for the film and the surface-renewal models (D2). Recently De Coursey (D10) derived an approximate solution for the Danckwerts model, given in the next section, which can help considerably when this model is used for design. Therefore, even though analytical expressions for the average rate of absorption based on the three models look very different, nevertheless the three will give the same value of the enhancement factor to within a few percent for all values of Ha between 0.1 and ∞.

6. *Other Correlations for an Irreversible Second-Order Reaction*

As explained in Section II,B,4, Van Krevelen and Hoftijzer (V4) computed approximated solutions for the film model, expressed by Eq. (26). It is interesting to note that other correlations giving implicit or explicit dependence between the enhancement factor E and the parameters Ha and E_i have been developed. Hikita and Asai (H7) have proposed an implicit expression for the Higbie model:

$$E = \left[\frac{\text{Ha}[(E_i - E)/(E_i - 1)]^{1/2} + \tfrac{1}{8}\pi}{\text{Ha}[(E_i - E)/(E_i - 1)]^{1/2}} \right] \text{erf}\left(\frac{4}{\pi} \text{Ha}^2 \frac{E_i - E}{E_i - 1} \right)^{1/2}$$
$$+ \frac{1}{2} \exp\left(-\frac{4}{\pi} \text{Ha}^2 \frac{E_i - E}{E_i - 1} \right)$$

There are also explicit solutions. Thus Porter (P9) has suggested:

$$E = 1 + (E_i - 1)\{1 - \exp[-(\text{Ha} - 1)/(E_i - 1)]\}$$

and Kishinevskii *et al.* (K6):

$$E = 1 + (\text{Ha}/\alpha)[1 - \exp(-0.65\,\text{Ha}\,\sqrt{\alpha})]$$

with

$$\alpha = \frac{\text{Ha}}{E_i - 1} + \exp\left[\frac{0.68}{\text{Ha}} - \frac{0.45\,\text{Ha}}{E_i - 1} \right]$$

De Coursey (D10), using the Danckwerts model, has given

$$E = -\frac{\text{Ha}^2}{2(E_i - 1)} + \left[\frac{\text{Ha}^4}{4(E_i - 1)^2} + \frac{E_i\,\text{Ha}^2}{E_i - 1} + 1 \right]^{1/2}$$

and recently Baldi and Sicardi (B2) have modified Porter's equation:

$$E = 1 + (E_i - 1)\left[1 - \exp\left(-\frac{(1 + \text{Ha}^2)^{1/2} - 1}{E_i - 1}\right)\right]$$

All these equations fit Fig. 4 and lead to very close numerical values.

7. Other Types of Reactions

Van Krevelen and Hoftijzer (V4) developed Eq. (26) or Fig. 4 originally only for irreversible second-order reactions that are first-order in each reactant. Later it was applied successfully to other situations. Brian et al. (B27, B28) generalized it for other mass-transfer models, and also for reactions of general order n, and showed that it agreed reasonably with numerical solutions. Hikita and Asai and Onda et al. developed the approximate application to gas absorption accompanied by an (m,n)-th-order irreversible chemical reaction (H7), $(m,n)-(p,q)$-th reversible (O4), consecutive (O5), and parallel (O6). Again, the approximate solutions agree with numerical and analytical solutions within a few percent. Onda et al. (O7) have developed this approximate treatment for other orders using the surface-renewal models, with agreement also within a few percent. Very useful tables for approximate solutions, based on the three models for many types of kinetics, published by Onda et al. (O7), are recommended for reactor design.

All these studies lead to an analytical implicit relation or a graphical plot of the enhancement factor E as a function of the concentration–diffusion parameter (or E_i) and Hatta number. In the case of an irreversible (m,n)-th-order reaction, where

$$r_A = k_{mn} C_A^m C_B^n$$

Hikita and Asai (H7) have shown that, approximately,

$$E = \frac{\text{Ha}[(E_i - E)/(E_i - 1)]^{n/2}}{\tanh\{\text{Ha}[(E_i - E)/(E_i - 1)]^{n/2}\}} \tag{37}$$

where the Hatta number is defined as

$$\text{Ha} = (1/k_L)\{[2/(m + 1)]k_{mn} D_A C_A^{*m-1} C_{B0}^n\}^{1/2} \tag{38}$$

and E_i is given by Eq. (27). Thus knowledge of Ha and E_i, with use of the Van Krevelen–Hoftijzer diagram, will lead to the value of E.

When the concentration of component B has negligible variation throughout the liquid phase (that is, when $\text{Ha} \ll E_i$), the reaction is then pseudo-mth-order with $k_m = k_{mn} C_{B0}^n$. Thus Hikita and Asai have also shown that Eq. (29), valid for $m = 1$ (with $C_{A0} = 0$),

$$E = \text{Ha}/(\tanh \text{Ha}) \tag{39}$$

can be used for the cases $m = 0, 2,$ and 3 with results correct to within 6%. The expression

$$E = (1 + \text{Ha}^2)^{1/2} \tag{40}$$

also gives results correct to within 8% for $m > 1$. Furthermore, when $1 \ll \text{Ha} \ll E_i$, it develops that $E = \text{Ha}$, and the average rate of absorption is then

$$\varphi = Ek_L C_A^* = \{[2/(m + 1)]k_{mn} D_A C_A^{*m+1} C_{B0}^n\}^{1/2} \tag{41}$$

This is the fast pseudo-mth-order regime, where the average rate of absorption is independent of k_L.

8. *Mass Transfer with Gas-Side Resistance*

As explained in Section II,A, when a soluble gas is mixed with an insoluble gas, it must diffuse through the latter to reach the interface. It is usual to refer to a gas film resistance. This implies a stagnant film of gas across which the soluble gas is transferred by molecular diffusion from the bulk gas with partial pressure p to the interface where the partial pressure is p_i. If the component B has negligible vapor pressure, the reaction will proceed only in the liquid phase.

If we now adopt the film model to describe the phenomena occurring on both liquid and gas sides of the interface, we arrive at the two-film model. The soluble gas is being transported across both films by a steady-state process, and the rate of absorption is expressed by

$$\Phi = \varphi a = k_G a(p - p_i) = Ek_L a(C_A^* - C_{A0}) \tag{42}$$

For a system obeying Henry's law for equilibrium at the interface ($p_i = HC_A^*$),

$$\Phi = \varphi a = \frac{a(p - HC_{A0})}{(k_G)^{-1} + (H/Ek_L)} = \frac{a[(p/H) - C_{A0}]}{(Ek_L)^{-1} + (Hk_G)^{-1}} \tag{43}$$

or

$$\Phi = \varphi a = K_G a(p - HC_{A0}) = EK_L(p/H - C_{A0}) \tag{44}$$

where

$$(K_G)^{-1} = (k_G)^{-1} + H/Ek_L, \quad (K_L)^{-1} = (Hk_G)^{-1} + (Ek_L)^{-1}$$

Thus the overall resistance, $(K_G)^{-1}$ or $(K_L)^{-1}$, is the sum of the resistances referring to the two films. As already explained, in reality the situation is more complicated, with Ek_L and k_G likely to vary from point to point. However, for practical engineering purposes, we treat the quantities k_G and Ek_L as if they had uniform values on all parts of the surface.

In the overall picture, different expressions are proposed for the rate of gas absorption with chemical reaction, depending on the forms of the enhancement factor E corresponding to different kinetic regimes, going from reaction-controlling to mass transfer-controlling. Typical cases are:

(1) Slow irreversible second-order reaction in the bulk liquid, with $C_{A0} \neq 0$ (Section II,B,2):

$$\Phi = \varphi a = \frac{pa}{\dfrac{1}{k_G} + \dfrac{H}{k_L} + \dfrac{Ha}{k_2 C_{B0} \beta}} \tag{45}$$

Here the two films and the main body of liquid act as resistances in series, and the interfacial area and liquid holdup (or the surface/volume ratio) enter this expression.

(2) Pseudo-mth-order reaction in the liquid phase, $C_{A0} = 0$ (Section II,B,7):

$$E = \text{Ha} = (1/k_L)\{[2/(m+1)]k_{mn} D_A C_A^{*m-1} C_{B0}^n\}^{1/2}$$

$$\Phi = \varphi a = \frac{p}{\dfrac{1}{k_G a} + \dfrac{H}{a\{[2/(m+1)]k_{mn} D_A C_A^{*m-1} C_{B0}^n\}^{1/2}}} \tag{46}$$

Here the two films only act as resistances in series; the film thickness (or a/β) does not enter into this expression, since component A disappears within the film and does not penetrate the main body of liquid.

(3) Instantaneous irreversible reaction (Section II,B,4,b):

$$\Phi = \varphi a = \frac{p + (H D_B C_{B0}/z D_A)}{(k_G a)^{-1} + (H/k_L a)} \tag{47}$$

Equations (45)–(47) are used intensively in the chemical method of measuring mass-transfer parameters. Intermediate rates with more complex reaction kinetics are discussed in texts devoted to this subject (A6, D2).

9. Conclusions: Film Conversion Parameter; Significance of the Hatta Number

In using the film model in the absence of gas-phase resistance to tell whether the reaction is slow, fast, or intermediate, it is interesting to focus on a unit surface of the gas–liquid interface and to define a film conversion parameter:

$$M = \frac{\text{maximum possible conversion in film}}{\text{maximum diffusion transport through film}}$$

$$= \frac{k_2 C_A^* C_{B0} \delta_L}{D_A[(C_A^* - 0)/\delta_L]} = \frac{k_2 C_{B0} \delta_L^2}{D_A} = \frac{D_A k_2 C_{B0}}{k_L^2} = \text{Ha}^2 \qquad (48)$$

If $\text{Ha}^2 \gg 1$, all reaction occurs within the film, and the surface area is the controlling factor. On the contrary, if $\text{Ha}^2 \ll 1$, no reaction occurs within the film, and the bulk volume controls the rate with no benefit from an increase in interfacial area. Therefore, as seen in previous sections, the Hatta number is actually the criterion for whether the reaction occurs completely in the bulk of the liquid ($\text{Ha} < 0.3$) necessitating a large volume of liquid, completely in the boundary layer ($\text{Ha} > 3$) necessitating contacting devices that create a large interfacial area, or in both the bulk and the film necessitating both a large volume of liquid and a large interfacial area. Table XVIII presents typical data for a and β in various gas–liquid contacting devices.

Mass transfer and kinetics are both taken into consideration in the Hatta number. In a reactor designed for a particular reaction with a rate constant k_{mn}, the value of Ha is a significant indicator of whether a large specific interfacial area or a large liquid holdup is needed. Conversely, in a given designed reactor, it indicates the extent to which reaction kinetics on the one hand, or the mass-transfer coefficient and the interfacial area on the other, will control the overall rate. These questions are fully explored in Section III.

C. Solubility and Diffusivity of Gases in Liquids

The design of equipment for diffusional separation of gas–liquid mixtures is determined by two major considerations: the distribution of components between phases in a state of thermodynamic equilibrium (solubility of gases in a liquid) and the rate at which mass transfer occurs under the prevailing conditions (liquid diffusivity and chemical reaction). Therefore solubilities and diffusivities of gases in liquids are practically always required for design. Obtaining the data is often a challenging problem, so wide is the range of solutes and solvents the chemical engineer or researcher may encounter.

Although theory on solubility and diffusivity has not yet progressed to the point where purely theoretical predictions are possible, many strides have been made in this direction. We give here only the most representative correlations and references. The reader should always keep in mind that such relationships are to be used only in the absence of experimental data and after careful study of the conditions of validity.

1. *Solubility of Gases in Liquids*

Different approaches have been made, depending upon whether the solutions are electrolytic or nonelectrolytic.

The main sources for gas solubility data are Linke and Seidell's handbook (L21, inorganic compounds) and Seidell and Linke's handbook (S13, inorganic and organic compounds) which do not contain thermodynamic data; Appendix 3 in Hildebrand *et al.*'s excellent book (H9); Wilhem and Battino's paper (W6), which surveyed the literature through the end of 1970; and more recent papers by Prausnitz *et al.* (P15, C18, C9), Battino *et al.* (W7, F3, B34, C15), Hildebrand and Lamoreaux (H10), and Fleury and Hayduck (F4). Battino and Clever (B10) give a comprehensive review of the physicochemical aspects of gas solubilities in liquids, including even molten salts and liquid metals; they also provide many references to solubility determinations:

a. *Nonelectrolytes.* A popular way to express the solubility of a gas in a liquid is to use the Ostwald coefficient:

$$\text{Os} = \frac{V_A^*}{V_B} = \frac{\text{volume of gas absorbed at temperature } T}{\text{volume of absorbing liquid at temperature } T} \quad (49)$$

Another is to use the Henry constant (K_{H0} or H) in the relation

$$p_A = K_{H0} X_A^* = H C_A^* \quad (50)$$

where p_A is the partial pressure of the gas over the solution in equilibrium with the mole fraction X_A^* (also called the mole fraction gas solubility) or with the concentration C_A^* of the solute in solution.

If ideal-gas behavior is assumed, the mole fraction of the dissolved gas is

$$X_A^* = [(RT/\text{Os} \cdot p_A V_B^0) + 1]^{-1} \quad (51)$$

where R is the gas constant, T is the Kelvin temperature, and V_B^0 is the molar volume of the pure solvent.

Equation (50) is the low-pressure approximation of the general thermodynamic equation

$$f_A = K_H X_A^* \quad (52)$$

Because f_A becomes equal to p_A at low pressures,

$$K_{H0} = \lim_{X_A^* \to 0} K_H \quad (53)$$

where f_A is the fugacity of the gas solute, measured in the gas phase. K_H generally increases with an increase in temperature, while the solubility of the gas generally decreases. In an ideal solution, which is the case for very few gas–liquid systems, K_H is constant for a given solute–solvent pair at a given temperature. When the gas-phase mole fraction Y_A is used instead of the partial pressure p_A, the equilibrium constant is K_H/P, and the equilibrium term then depends also on the total pressure P.

When temperature and pressure are constant, the equilibrium values corresponding to a constant K_H are expressed graphically by a straight "equilibrium line" passing through the origin with a slope of K_{H0} or H.

Thermodynamics remind us that Henry's law is applicable only over a restricted range for dilute solutions, which means in practice that it is a limiting law for sparingly soluble gases and for readily soluble gases at low concentrations. For the limiting cases, values of K_{H0} and H for aqueous solutions of different gases can be found for instance in references such as Perry's handbook (P5, p. 3-96) and Ramm's book (R1). For higher concentrations, the solubility is generally lower than that given by Henry's law. (Gas nonideality overrides the change in solution nonideality.) In such cases K_{H0} or H depends on the composition of the liquid, and the equilibrium line becomes an equilibrium curve.

The limiting Henry constant can be predicted from nonideal solution behavior through the relation

$$f_A = \gamma_A P_{SA} X_A^* \quad \text{or} \quad K_H = \gamma_A P_{SA}$$

where P_{SA} is the extrapolated vapor pressure of component A, and γ_A is its activity coefficient (here, at infinite dilution). In accordance with the theory of regular solutions (H9), this activity coefficient can be calculated from the relation

$$\ln \gamma_A = \ln[(X_A^*)_{\text{ideal}}/(X_A^*)_{\text{real}}] = (V_A^0/RT)(\delta_A - \delta_B)^2 \tag{54}$$

Here V_A^0 is the partial molar volume of the solute gas, and δ_A and δ_B are the solubility parameter values of the solute and the bulk liquid. The solubility parameter is also cohesive-energy density, a measure of the forces between the molecules given by

$$\delta = (\Delta E/V^0)^{1/2}$$

where ΔE is the molar energy of vaporization, and V^0 is the molar volume.

V_A^0 is often assumed as the value for a liquid at 25°C, but it may also be considered linearly variable with T/T_c (R5). When water is the solvent, V_A^0 is determined approximately as the molar volume of the gaseous component at its atmospheric boiling point if the given temperature is above the critical.

Values of δ_A, δ_B, and V_B^0 can be obtained either from Reid and Sherwood's book (R5) or for 59 compounds from the article by Wilhem and Battino (W6) in which solubility data for 16 gases in 39 solvents at 25°C are presented (also included in this paper are data for several fluorine-containing gases). A detailed analysis of the cohesive energies of 31 hydrocarbons of different families enabled Maffiolo et al. (M5) to propose a simple correlation for predicting the cohesive energy of any liquid hydrocarbon from its molar volume and the polarizabilities and Van der

Waals volume of the group forming it. Also, in the case of hydrocarbons, it is important to consult the set of equations derived by Meisner (M13) to calculate molar volumes from carbon number and temperature for homologous unbranched hydrocarbons in the liquid state.

If the size of the molecules of the gas solute differs greatly from that of the molecules of the solvent, the following equation may be used:

$$\ln \gamma_A = (V_A^0/RT)(\delta_A - \delta_B)^2 + \ln(V_A^0/V_B^0) + (1 - V_A^0/V_B^0) \quad (55)$$

Equations (54) and (55) have proved to be remarkably satisfactory for solutions where both solvent and solute are nonpolar, and for slightly polar solvents. A more thorough discussion of these equations, with references to gas–liquid systems, is given by Battino and Clever (B10). Moreover, Hildebrand and Lamoreaux (H10) recently published an interesting empirical linear relationship and a graphical plot allowing calculation of the solubility of any inert gas except H_2, in any nonpolar solvent except fluorochemicals, from the energy of vaporization of the gas at its boiling point and the solvent cohesive-energy density. Complementary and alternate results were then produced by Fleury and Hayduck (F4), which even apply to perfluorinated gases and solvents.

For gas solutes in large-molecule solvents, Chappelow and Prausnitz (C9), have recently proposed a modified equation for γ_A in which both the size and shape of the molecules are taken into consideration. They provide accurate gas solubility data for 26 hydrocarbon gas–liquid systems in the temperature range 25°–200°C in the vicinity of 1 atm pressure. Complementary information on the solubility of organic solutes in polymers has also been given recently by Maloney and Prausnitz (M6) and by Stiel and Harnish (S39).

The temperature dependence of solubility has often been used to obtain the molar heat of solution ΔH, the difference between the partial molar enthalpy of a solute at infinite dilution in a solvent, and the enthalpy of a solute as an ideal gas at the same temperature. In this case, plots of $\ln X_A^*$ against $1/T$ are usually linear to within the accuracy of measurement, indicating a temperature-independent ΔH which can be calculated from the slope. However, for several binary systems, a plot of gas solubility versus termperature goes through a minimum. For instance, by measuring the solubilities of hydrogen, methane, and ethane in hexadecane, bicyclohexyl, and diphenylmethane, respectively, in the range 25°–200°C, Cukor and Prausnitz (C18) have observed that the solubility of hydrogen rises and that of ethane falls with increasing temperature, while that of methane goes through a minimum near 150°C. It is likely that a minimum may be observed in low-pressure solubility for any gas in any nonpolar solvent, provided the range of temperature is wide enough (P15, C9).

The foregoing laws are obeyed to only moderate pressures, and considerable deviation from the experimental results occurs if the pressure is high, even if X_A^* is small. Battino and Clever (B10) have provided a term to fit high-pressure gas solubility in water, methanol, and hydrocarbons generally, and hydrogen solubility in binary and ternary mixed hydrocarbon solvents. Also, in addition to the effect of total pressure, Orentlicher and Prausnitz (O10) have added a term taking into account the effect of composition. The combined equation is

$$\ln \frac{f_A}{X_A^*} = \ln K_H + \frac{V_A^0(P - P_{SB})}{RT} + \frac{A(X_B^2 - 1)}{RT} \qquad (56)$$

where P_{SB} is the solvent saturation vapor pressure, P is the total pressure, and K_H is the Henry constant related to the partial molar fugacity of the solute at the solvent saturation pressure; A is an empirical coefficient corresponding usually to $V_A^0(\delta_A - \delta_B)^2$ in Eq. (54) and depending only on temperature for a given solute–solvent system. A correlation essentially identical to Eq. (56) has been proposed by Kritchevsky and Ilniskaya (K11) to fit the solubility of mixtures of gases. For gas solubilities in mixed nonelectrolyte solvents, Battino and Clever (B10) have presented numerous references for different systems, and O'Connell and Prausnitz (O1) have proposed an equation for a gas dissolved in any number of miscible solvents.

Atmospheric pressure solubilities of many gases in water range from about 0.000007 mole fraction for helium to about 0.3 for ammonia at 25°C. The correlations given above for nonpolar solvents are not satisfactory for predicting gas solubilities in polar associating solvents such as water and alcohols. Using solubilities of 20 gases (Fig. 5) in water, methanol, ethanol, butanol, acetone, methyl acetate, acetic acid, ethylene glycol, and chlorobenzene, selected from the available literature, Hayduck and Laudie (H4) developed a systematic correlation in terms of association or hydrogen bonding. Strong hydrogen bonding appears to have the effect of excluding solute molecules. Hydrogen bonding factors were defined by Hayduck and Laudie (H4) for each gas, based on the ratio of its actual solubility to its ideal solubility. The hydrogen-bonding factors for gases in water are related to those in the primary normal alcohols. Similarly, hydrogen-bonding factors in acetone are related to those in methyl acetate, acetic acid, and ethylene glycol. Except where the dissolved gas reacts chemically with the solvent, gas solubilities can be predicted from relations between hydrogen-bonding factors in the solvents, provided the gas solubility is known in at least one of them. As for nonpolar solvents, solubilities of all gases in water tend toward a constant molar concentration at the solvent critical temperature (Fig. 5), confirming a similar trend for other associated solvents. This makes it possible to estimate the tem-

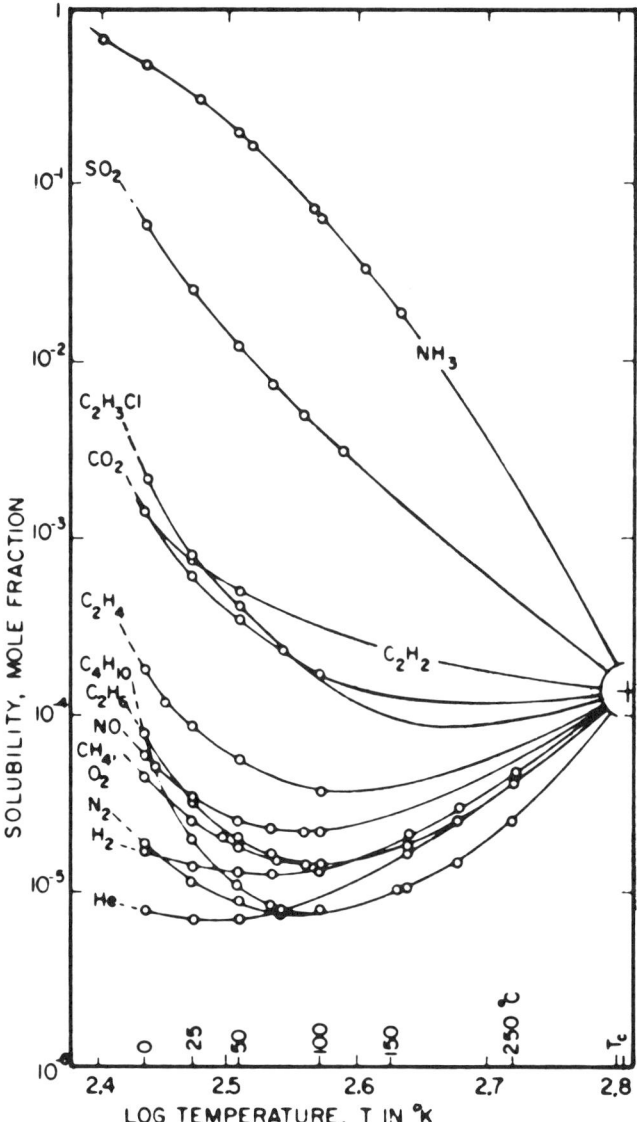

FIG. 5. Solubility of gases in water (H4).

perature coefficient of solubility for any gas in any one solvent from data for other gases in this solvent (H4).

For associating solutions, a comprehensive and credible model has been presented by Tamir and Wisniak (T2) for the case of binary mixtures containing two associating species and undergoing heterodimerization in the vapor phase (between two fatty acids, for example). This interesting model has not yet been generalized.

An interesting empirical correlation for nonpolar gas solubilities in mixed aqueous alcohol solutions (such as methanol, ethanol, and isopropanol) has been also recently presented by Tokunaga (T7) in terms of an excess quantity of gas solubility, defined as

$$\ln K = \ln(Os)_m - \Phi \ln(Os)_a - (1 - \Phi) \ln(Os)_w$$
$$= \Phi(1 - \Phi)(A + B\Phi + C\Phi^2 + D\Phi^3) \quad (57)$$

where Φ is the alcohol volume fraction in the liquid phase and $(Os)_m$, $(Os)_a$, and $(Os)_w$ are the Ostwald coefficients for the solute gas in aqueous alcohol solution, alcohol, and water, respectively. The empirical coefficients seem to depend only on the nature of the alcohol and temperature regardless of the kind of solute gas, for example, O_2, CO_2, N_2, or even CH_4 (T7, T8).

Last, for several systems, a linear relationship is found to exist between log Os and solvent surface tension σ for a gas in a series of solvents at a constant temperature (B10).

b. *Electrolytes.* The solubility of electrolyte solutions can be estimated by the empirical method of Van Krevelen and Hoftijzer (V4), which relates solubility in solution to that in pure solvent at the same temperature:

$$\log_{10}(H/H_B) = hI \quad (58)$$

where H_B is the value in pure solvent and I is the ionic strength of the solution:

$$I = \tfrac{1}{2} \sum C_i Z_i^2$$

where C_i is the concentration of ions of valency Z_i. The salting coefficient h is the sum of contributions referring to the species of gas h_G and to the species of positive and negative ions present:

$$h = h_G + h_+ + h_-$$

Van Krevelen and Hoftijzer (V4), Barrett (B7), and Onda *et al.* (O8) have evaluated h for various species. The values presented in Tables I and II, which extend to ionic strengths of 2 M and greater, are characteristic of industrial systems. The effect of pressure on h_G values is small up to 200 atm (O8).

More recently, Tiepel and Gubbins (T4, T5) have presented a theory for salt effects on gas solubility based on perturbation theory for mixtures.

TABLE I

Salting Coefficients h_+ and h_- for Inorganic Ions

Ion	h_+ (liters/gm mole)	Ion	h_- (liters/gm mole)
H^+	-0.1110	Cl^-	0.3416
Na^+	-0.0183	Br^-	0.3310
K^+	-0.0362	I^-	0.3124
NH_4^+	-0.0737	NO_3^-	0.3230
Sr^+	-0.0445	OH^-	0.3875
Li^+	-0.0416	CNS^-	0.2612
Cs^+	-0.0584	HSO_3^-	0.3869
Rb^+	-0.0449	HS^-	0.3718
Mg^{2+}	-0.0568	HCO_3^-	0.4286
Zn^{2+}	-0.0590	CO_3^{2-}	0.3754
Ca^{2+}	-0.0547	SO_4^{2-}	0.3446
Ba^{2+}	-0.0473	SO_3^{2-}	0.3275
Mn^{2+}	-0.0624	PO_4^{3-}	0.3265
Fe^{2+}	-0.0602	$C_6H_5O^-$	0.4084
Co^{2+}	-0.0534	MnO_4^-	0.2600
Ni^{2+}	-0.0520		
Cd^{2+}	-0.0062		
Cr^{3+}	-0.0986		

This theory satisfactorily predicts the heat of solution and the concentration and pressure dependence of activity coefficients of nonpolar solutes as a salting-out effect, when the molecules and ions are not very large and chemical association between ions and solute does not occur. When the

TABLE II

Salting Coefficients h_G for Various Gases[a]

T (°C)	H_2	O_2	CO_2	N_2O	H_2S	NH_3	C_2H_2	C_2H_4	SO_2
0	—	-0.1653	-0.2110	—	—	—	—	—	—
10	-0.2170	—	—	-0.2156	—	—	—	—	—
15	-0.2197	-0.1786	-0.2222	-0.2118	—	—	-0.2124	-0.2003	—
20	-0.2132	-0.1771	—	-0.2128	—	—	—	—	—
25	-0.2115	-0.1892	-0.2277	-0.2141	-0.2551	-0.2394	-0.2240	-0.1951	-0.3154
35	—	—	—	—	—	—	—	—	-0.3122
40	—	—	-0.2327	-0.2179	—	—	—	—	—

T (°C)	N_2	He	Ne	Ar	Kr	NO
25	-0.1904	-0.2222	-0.2240	-0.1866	-0.1762	-0.1825

[a] Values are in liters/gm mole.

ions are large or when there are association forces, salting-in occurs, and the theory gives undependable results. Tiepel and Gubbins give data complementary to those of Tables I and II for salting coefficients for CH_4, C_2H_6, SF_6, and C_6H_6 in LiCl, NaCl, and KI.

In mixed electrolytes it may be assumed that H is given by an expression of the form

$$\log_{10}(H/H_B) = h_1 I_1 + h_2 I_2 + \cdots \quad (59)$$

where I_1 is the ionic strength attributable to species 1 of electrolyte and h_1 is a coefficient characteristic of that electrolyte. This additive rule has been confirmed by Onda et al. (O9) for salting-out parameters for the absorption of CO_2, C_2H_4, and C_2H_2 in three mixed-salt solutions.

Finally, it is important to remember that the properties of solutions may change when gas reacts with the solution. In such cases the given equations may be still applied to the concentration of the unreacted gas solute, provided one accounts for the ongoing reaction.

2. *Diffusivity in Liquids*

In contrast to the situation for gases, there are no satisfactory theoretical methods for predicting diffusivities in liquid systems. Different approaches are needed, depending on whether the solutions are electrolytic or nonelectrolytic. Most studies have been devoted to the estimation of diffusion coefficients in very dilute solutions. However, some papers report substantial variations with increasing concentrations of the diffusing solute. The theories and the experimental methods available for estimating diffusivities in liquids are well reviewed by Kamal and Canjar (K1), Nienow (N8), Bretsznajder (B24), Tyn (T12), Dullien et al. (E4, G6), and Simons and Ponter (S29).

a. *Nonelectrolytes.* Several correlations for dilute solutions are presented in Table III with the average errors given by Reid and Sherwood (R5), Skelland (S31), and the just-cited authors. In this table, A is the solute (dissolved gas or reactant), B is the solvent, D_A is the diffusion coefficient in cm²/sec, μ_{AB} and μ_B are the solution and the solvent viscosities in cP, T is the absolute temperature in °K, M_B is the molecular weight of the solvent, and V_{mA} and V_{mB} are the solute and solvent molecular volumes at the normal boiling point in cm³/gm mole. The molecular volumes of some simple substances are given in Table IV. Molecular volumes of complex substances are estimated by adding up the contributions of the atoms in the molecule according to Table V and also, in the case of ring compounds, subtracting according to Table VI.

For estimating diffusivity, a convenient relationship is that of Wilke and

TABLE III

RELATIONSHIPS FOR DIFFUSIVITY IN VERY DILUTE BINARY SOLUTIONS OF NONELECTROLYTES

Equation number	Equation	Average error (%)				Reference
		Organic solvents	Water as solvent	Water as solute	Overall (by authors)	
(1)	$\dfrac{D_A \mu_{AB}}{T} = 7.4 \times 10^{-8} \dfrac{(xM_B)^{1/2}}{V_{mA}^{0.6}}$	27	11	Up to 250	10	Wilke and Chang (W8)
(2)	$\dfrac{D_A \mu_{AB}}{T} = 8.2 \times 10^{-8} \dfrac{1}{V_{mA}^{1/3}} \left[1 + (3V_{mB}/V_{mA})^{2/3} \right]$	25	11	Up to 250	—	Scheibel (S11)
(3)	$\dfrac{D_A \mu_B}{T} = 5.4 \times 10^{-8} \left(\dfrac{M_B^{1/2} \Delta E_B^{1/3}}{V_{mA}^{0.5} \Delta E_A^{0.3}} \right)^{0.93} \times \left(\dfrac{T}{\mu_B} \right)^{0.07}$	26	12	12	13	Sitaraman et al. (S30)
(4)	$\dfrac{D_A \mu_B}{T} = 10 \times 10^{-8} \dfrac{M_B^{1/2}}{(V_{mA} V_{mB})^{1/3}}$ for $\dfrac{V_{mB}}{V_{mA}} < 1.5$	15	9	—	13.5	Reddy and Doraiswamy (R4)
(5)	$\dfrac{D_A \mu_B}{T} = 8.5 \times 10^{-8} \dfrac{M_B^{1/2}}{(V_{mA} V_{mB})^{1/3}}$ for $\dfrac{V_{mA}}{V_{mB}} > 1.5$	18	—	26	18	Reddy and Doraiswamy (R4)
(6)	$\dfrac{D_A \mu_B}{T} = 8.5 \times 10^{-8} \dfrac{\left[1.40(V_{mB}/V_{mA})^{1/3} + (V_{mB}/V_{mA}) \right]}{V_{mB}^{1/3}}$	16	—	—	—	Lusis and Ratcliff (L25)
(7)	$D_A = 14 \times 10^{-5} \mu_{wT}^{-1.1} \times V_{mA}^{-0.6}$	—	—	—	5	Othmer and Thakar (O11)

TABLE IV

MOLECULAR VOLUME V_m OF SIMPLE SUBSTANCES (P5)[a]

Substance	V_m	Substance	V_m	Substance	V_m
Air	29.9	COS	51.5	N_2	31.2
Br_2	53.2	H_2	14.3	NH_3	25.8
Cl_2	48.4	H_2O	18.9	NO	23.6
CO	30.7	H_2S	32.9	N_2O	36.4
CO_2	34	He	16	O_2	25.6
		I_2	71.5	SO_2	44.8

[a] Values are in cm^3/gm mole.

Chang (W8) based on the Stokes–Einstein equation. In this formula the association parameter x allows for differences in solvent behavior: $x = 2.6$ for water, 1.9 for methanol, 1.5 for ethanol, and 1.0 for benzene, ether, heptane, and other unassociated solvents. The average error for systems surveyed by the authors was about 10%; the relationship cannot be used when a complex is formed between solute and solvent. For amyl alcohol,

TABLE V

ATOMIC CONTRIBUTIONS TO MOLECULAR VOLUMES OF COMPLEX SUBSTANCES (P5)

Atom	cm^3/gm mole	Atom	cm^3/gm mole	Atom	cm^3/gm mole	Atom	cm^3/gm mole
As	30.5	F	8.7	P	27	Sn	42.3
Bi	48	Ge	34.5	Pb	48.3	Ti	35.7
Br	27	H	3.7	S	25.6	V	32
C	14.8	Hg	19	Sb	34.2	Zn	20.4
Cr	27.4	I	37	Si	32		

Atom	cm^3/gm mole	Atom	cm^3/gm mole
Cl, terminal	21.6	N, tertiary amine	10.8
Cl, medial	24.6	O (usual)	7.4
N, double-bonded	15.6	O, methyl esters and ethers	9.1
N, triple-bonded	16.2	O, ethyl esters and ethers	9.9
N, primary amine	10.5	O, higher esters and ethers	11
N, secondary amine	12	O, acids	12
		O, joined to S, P, N	8.3

TABLE VI

Deduction of Molecular Volumes for Rings, (P5)

Ring	V_m (cm³/gm mole)
Three-membered	6
Four-membered	8.5
Five-membered	11.5
Six-membered	15
Naphthalene	30
Anthracene	47.5

isobutanol, ethylene glycol, and glycerol, Akgerman and Gainer (A2) have recently approximated the associated parameter by the relation

$$x = \left(\frac{\Delta E_{\text{vap}} \text{ of hydrogen-bonded substance}}{\Delta E_{\text{vap}} \text{ of homolog}} \right)^{0.6} \tag{60}$$

The homolog is defined by substituting a $-CH_3$ group for the $-OH$ group. The uncertainty in assigning values to x has resulted in efforts to eliminate this factor. Scheibel (S11) and Lusis and Ratcliff (L25) have proposed equations introducing the ratio of the molar volumes of solvent and solute. Sitaraman et al. (S30) and Reddy and Doraiswamy (R4) replaced x by functions of the latent heats of vaporization at the normal boiling temperature (in calories per gram) or of the molar volumes. Until these relations have been fully verified, our recommendation relative to Table III (particularly for the liquid reactant) is to use Eq. (2) or (4) when applicable, or Eq. (6) for diffusion in organic solvents (S31), or Eq. (7) of this table when water is the solvent. The monograph of Kuong (K14) expedites calculation of Eq. (2). These relationships have been confined to the temperature range 15°–25°C. In the range 25–90°C, it is judicious to use a linear relationship between $\ln D_A$ and $1/T$ and between $\ln D_A$ and $\ln \mu$ (R13).

For small solute gas molecules such as helium and H_2, or for systems involving solvents having viscosities greater than 3 cP, we suggest using the relationship proposed by Akgerman and Gainer (A2):

$$\frac{D_A \mu_B}{T} = \frac{k_B}{6} \left(\frac{V_{mB}}{V_{mA}} \right)^{1/6} \left(\frac{N}{V_{mB}} \right)^{1/3} \left(\frac{M_B}{M_A} \right)^{1/2} \exp\left(\frac{E_{\mu B} - E_{DAB}}{RT} \right) \tag{61}$$

with

$$E_{\mu B} - E_{DAB} = E_{BB}^j[1 - (E_{AA}^j/E_{BB}^j)^{1/(\xi_A+1)}]$$

$$E_{AA}^j = 5875.3 M_A^{-0.186}$$

$$E_{BB}^j = -\left[R \ln\left(\frac{\mu_{B2}}{\mu_{B1}}\right) + \frac{R}{2} \ln\left(\frac{T_1}{T_2}\right)\right] \frac{T_1 T_2}{T_2 - T_1}$$

$$\xi_A = 6(V_{mA}/V_{mB})^{1/6}$$

Equation (61) is less easy to handle than Wilke and Chang's equation, but it uses only the physical properties of the constituents involved and is claimed to work equally well with associated and nonassociated systems and with water. Forty-nine gas–liquid systems have been tested in the range 0°–40°C.

Moreover the diffusivity D_A^m for dilute solute A in mixed solvents B and C may be determined by the equation of Leffler and Cullinan (L7):

$$D_A^m \mu_{ABC} = (D_{AB} \mu_B)^{X_B}(D_{AC} \mu_C)^{X_C} \qquad (61a)$$

where μ_{ABC} is the viscosity of the solvent mixture B and C, D_{AB} and D_{AC} are binary diffusivities estimated by one of the correlations in Table III, and X_B, and X_C are the mole fractions of B and C.

In thermodynamically nonideal solutions, the effect of concentration on diffusion coefficient $(D_A)_{conc}$ can be estimated in terms of the activity coefficient of the solute γ_A, the binary diffusivities D_A and D_B in very dilute solutions, and the mole fractions X_A and X_B of A and B (V8):

$$(D_A)_{conc} = (D_A)^{X_B}(D_B)^{X_A}\{1 + [(d \ln \gamma_A)/(d \ln X_A)]\} \qquad (61b)$$

This relationship has been modified by Leffler and Cullinan (L8) to account for the viscosities of the solution and the pure components. The effect of pressure can usually be neglected.

Before using one of the above relationships, it is judicious to check whether the gas–liquid system of interest is included in the tabulation of experimental diffusivities prepared by Johnson and Babb (J8) or in more recent but scattered tables presented by Davies *et al.* (D8), Dim *et al.* (D15), Tyn (T12), Akgerman and Gainer (A2), Perry (P5, p. 3-224), and De Kee and Laudie (D11).

b. *Electrolytes.* Diffusion of a dissolved ionized electrolyte involves the diffusion of both cations and anions which, because of their smaller size, diffuse more rapidly than the undissociated molecules. Both types of ions diffuse at the same rate, so that electrical neutrality of any solution is preserved.

On the assumption of complete dissociation in a solution containing two species of ions, the diffusion coefficients of electrolytes can be predicted very accurately at infinite dilution using the equation

$$D_A = 8.931 \times 10^{-10} T(\lambda_+^0 \lambda_-^0/\Lambda^0)[(Z_+ + Z_-)/Z_+ Z_-] \qquad (62)$$

where D_A is the combined diffusivity of electrolyte in cm²/sec, Z_- and Z_+ are the absolute values of the anion and cation valences, $\Lambda^0(=\lambda_+^0 = \lambda_-^0)$ is the electrolyte conductance at infinite dilution in mho cm²/gm equivalent, λ_-^0 and λ_+^0 are the anionic and cationic conductances at infinite dilution in mho cm²/gm equivalent, λ_-^0 and λ_+^0 are the anionic and cationic conductances at infinite dilution in mho cm²/gm equivalent, and T is the temperature in K. Useful tabulations of ionic conductances at infinite dilution in water are given in Perry (P5, p. 3-235) at 25°C and in Bretsznajder (B24, pp. 386–7) at other temperatures (0.18° and 100°C).

Diffusion coefficients at higher concentrations (up to $2N$) may be evaluated from the relationship proposed by Gordon (G12):

$$(D_A)_{conc} = \frac{D_A}{C_{BS} \bar{V}_B} \frac{\mu_B}{\mu_{AB}} \left(1 + \frac{m \, \partial \ln \gamma_\pm}{\partial m}\right) \tag{63}$$

where D_A is the diffusivity at infinite dilution in cm²/sec, C_{BS} is the gm moles of solvent/cm³ of solution, $\bar{V}_B$ is the partial molar volume of solvent in solution in cm³/gm mole, μ_B is the viscosity of solvent in P, m is the molarity in moles/1000 gm solvent, and $\gamma_\pm$ is the mean molar activity coefficient (G10).

For the general case, where the solution contains more than two species of ions, Vinograd and McBain (V9) have proposed the relations

$$Z_+\varphi_+ = -8.931 \cdot 10^{-10} \times T \frac{\lambda_+}{Z_+} \left(G_+ - Z_+C_+ \frac{\sum \lambda_+ G_+/Z_+ - \sum \lambda_- G_-/Z_-}{\sum \lambda_+ C_+ + \sum \lambda_- C_-}\right)$$

$$Z_-\varphi_- = -8.931 \cdot 10^{-10} \times T \frac{\lambda_-}{Z_-} \left(G_- + Z_-C_- \frac{\sum \lambda_+ G_+/Z_+ - \sum \lambda_- G_-/Z_-}{\sum \lambda_+ C_+ + \sum \lambda_- C_-}\right)$$

(63a)

where Z_+ and Z_- are the absolute values of the cation and anion valences, φ_+ and φ_- are the diffusion fluxes of cation and anion in gm equivalents/cm² sec, λ_+ and λ_- are the cationic and anionic conductances in mho cm²/gm equivalent, C_+ and C_- are the concentrations in gm equivalents/cm³, and G_+ and G_- are the concentration gradients in the direction of diffusion in gm equivalents/cm³. These relationships have been used judiciously by several authors. Where a gas is absorbed and reaction with an ion in solution is second-order and irreversible, the various ionic species having different intrinsic diffusivities, Brian *et al.* (B27) have presented values of the enhancement factor significantly different from the behavior for un-ionized molecules. The system studied was a generalization of acetic acid vapor with NaOH solution. Similarly Sherwood and Wei (S25) have considered absorption of HCl by a solution of NaOH, in which the reaction is instaneous and irreversible and reactants and products are ionized. Their results were recalculated by Danckwerts (D2, p. 143) to give the enhancement factor E as a function of

$$C_{B0}/C_A^* = C_{OH^-}/C_{H^+}^*$$

with C_{OH^-}/C_{bNa^+} as a parameter, where C_{bNa^+} is the concentration of Na^+ in the bulk liquid, as shown in Fig. 6. The line labeled "molecular diffusion" is calculated for a hypothetical reaction between the molecules HCl and NaOH, with D_{HCl} and D_{NaOH} calculated by Eq. (62). The other lines allow for the fact that the reaction involves ions and account for different amounts of NaCl present in the bulk solution. The line labeled "$C_{OH^-}/C_{bNa^+} = 0$" corresponds to a very large excess of Na^+ and Cl^- concentrations relative to OH^-. The enhancement factors calculated for molecular diffusion are much lower than those for the more realistic basis of ions diffusing. The same considerations have been used by Nijsing et al. (N10, N11) for the absorption of CO_2 by solutions of KOH, NaOH, and LiOH, in which the irreversible reaction between CO_2 and OH^- is either rapid pseudo-first-order or instaneous, depending on the gas–liquid contactor used (laminar jet or falling film).

In most investigations where gas absorption takes place into a liquid containing several species of ions, the investigators have ignored the need to use ionic diffusion coefficients. The relationships presented above should, of course, be used only in the absence of experimental data. A comprehensive tabulation of experimentally measured diffusivities for electrolytes has been presented by Robinson and Stokes (R12).

In fact, for quick practical calculations, the formula of Wilke and Chang (W8) is often sufficient. It has been verified with many systems, both electrolyte and nonelectrolyte. For water at 10°–60°C, Shrier (S26) has

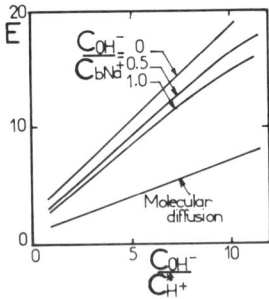

FIG. 6. Enhancement factor for absorption of HCl by aqueous hydroxide. Adapted from Sherwood and Wei (S25) by Danckwerts (D2).

examined the experimental results of Wise and Houghton (W9) on 10 slightly soluble gases (H_2, O_2, N_2, air, helium, argon, CH_4, C_2H_6, C_3H_8, n-C_4H_{10}) and has found that the Wilke and Chang formula gives adequate predicted values except for H_2 and helium. From an examination of literature data for diffusivity in aqueous solutions at 25°C, Hayduk and Laudie (H5) have suggested that a revised association parameter for water, $x = 2.26$ (instead of 2.6), improves the accuracy of the Wilke and Chang equation. So it is only for small solute gas molecules such as H_2 and helium, for viscous solutions ($\mu_L > 3$ cP), and for highly nonideal systems such as gas into an aqueous alcohol mixture that the theory developed by Akgerman and Gainer (A2) is required.

In the case of CO_2 absorption, often used to determine interfacial parameters in gas–liquid reactors, Ratcliff and Holdcroft (R2) have found that the diffusivities in aqueous solutions of chlorides, nitrates, and sulfates of sodium and magnesium vary roughly as $\mu^{-0.64}$ (rather than as μ^{-1}, in the Wilke and Chang formula). Nijsing *et al.* (N11) and Danckwerts and Alper (D7) found the diffusivity of CO_2 in sulfates of sodium and magnesium, and in Na_2CO_3, to vary as $\mu^{-0.85 \text{ to} -0.82}$. However, all these results were obtained for small variations in viscosities.

Such difficulties of knowing the solubility or the diffusivity of gas with good accuracy are overcome by using a laboratory model, such as a laminar-jet or a wetted-wall column (Section III,C), for determination of the interfacial parameters. Only a mathematical combination of solubility and diffusivity such as $C_A^* \sqrt{D_A}$ is necessary, not the separate values of C_A^* and D_A. But for industrial design, separate knowledge of these two parameters is often necessary.

[Note that Takeuchi *et al.* (T1) have recently published results for the diffusivity and solubility of CO_2 in nonelectrolytic liquids simultaneously measured in a diaphragm cell. The results satisfy the two-film models of Whitman and show good agreement with others in the literature.]

III. Measurement of Interfacial Areas and Mass-Transfer Coefficients

Gas holdup, interfacial area, and mass-transfer coefficients are the main variables determining the mass-transfer rates in gas–liquid contacting devices. The methods used to measure these parameters can be classified into two categories: local measurements with physical techniques such as light scattering or reflection, photography, or electric conductivity methods, and global measurements with chemical techniques. Each method has its advantages and its drawbacks.

A. Physical Methods

Physical measurements can be made of gas holdup α, bubble size, and specific surface area a' in gas–liquid dispersions, as usually encountered in bubble columns, plate columns, mechanically agitated tanks, and spray towers. Any two of these interfacial parameters are sufficient to define all three, since they are interrelated:

$$a' = 6\alpha/d_{SM} \tag{64}$$

where d_{SM} is the volume surface mean diameter or Sauter mean diameter:

$$d_{SM} = \sum_i n_i d_b^3 \bigg/ \sum_i n_i d_b^2 \tag{65}$$

d_b is the diameter of a single bubble or drop, and n_i is the number of bubbles or drops of diameter d_b.

1. Gas Holdup

The gas holdup α is determined directly by measuring the height of the aerated liquid (Z_a) and that of the clear liquid without aeration (Z). The average gas holdup is then

$$\alpha = (Z_a - Z)/Z_a \tag{66}$$

This method, used for plate columns (G2), bubble columns (F2, M7, H13, Y4, A3, E3), and mechanically agitated tanks (R7, V3, Y3, M16), is rapid but only accurate to ± 15–20%, especially when waves or foam occur on top of the dispersion.

A more accurate, manometric, technique has been used by Reith *et al.* (R10) and by Burgess and Calderbank (B32). The gas holdup in the dispersion is computed from measurements of the clear liquid height in the dispersion at successive manometer taps on the side of the froth-containing vessel. If x_i and x_{i-1} are the clear liquid heights recorded on the manometer at vertical positions Z_i and Z_{i-1} above the column floor, then the gas holdup at the midlevel $(i, i-1)$ is given by

$$\alpha = 1 - [(x_i - Z_i) - (x_{i-1} - Z_{i-1})]/(Z_i - Z_{i-1}) \tag{67}$$

Linek and Mayrhoferova (L16) have used electric conductivity to measure the dispersion height, based on locating the surface elevation at certain selected points by means of an electrically conductive probe. The height is determined by the vertical position of the probe tip for which the integrated time of contact is exactly half the period of measurement. The accuracy of the measured value of the total surface elevation is claimed by the authors to be ±0.2 mm, and the gas holdup is then calculated from the total surface elevation and the cross section of the reactor.

The γ-ray transmission technique has been used to determine the point holdup of gas in mechanically agitated tanks by Calderbank (C1); in plate columns by Calderbank and Rennie (C4), Vinokur and Dil'man (V10), Macmillan (M3), and Bernard and Sargent (B17); and in packed columns by Eisenklam and Ford (E2), Hwa and Beckmann (H14), and Saada (S1). The application of γ-ray absorption to holdup measurements is based on the relationship

$$\ln(I_0/I) = \lambda \rho s \qquad (68)$$

where I_0/I is the intensity ratio between the incident beam of radiation (I_0) and the transmitted beam (I), λ is the mass absorption coefficient, values of which have been published for most atoms (C7), ρ is the density to be measured (simply related to gas holdup), and s is the thickness of the absorbing medium.

Thus Calderbank (C1) used a cesium-137 source in conjunction with a scintillation counter and scaler. Traverses of the reactor were made, and readings taken of the γ-rays transmitted through the empty reactor, the reactor filled with liquid, and the reactor containing the dispersion. The point gas holdup was then calculated by

$$\alpha = \frac{\log(t_2/t_1)}{\log(t_0/t_1)} \qquad (68a)$$

where t_0, t_1, and t_2 are the times for a fixed number of counts, respectively, for the reactor empty, filled with liquid, and containing the dispersion.

2. *Bubble Size*

The Sauter mean particle size of dispersion [Eq. (65)] is evaluated directly by a statistical analysis of high-speed flash photomicrographs when the dispersion is dynamically maintained. Photographs are taken through the wall of the transparent reactor (C4, T9, V3, V6, P11, R16, V12, A3, A5) or in the interior of the reactor with the aid of an intrascope (V3).

To avoid any wall effect or perturbation effect that may occur with these methods, a sampling apparatus for the photographic technique has been proposed by Kawecki *et al.* (K5). Bubbles are extracted from the tank containing the dispersion by means of a tube connected to a small square-section column through which a continuous flow of liquid and bubbles rises. The flow rate is chosen high enough so that differences in the free-rise velocity of the bubbles do not affect the mean residence time of the bubbles in the column. The bubbles in the column are then photographed, and their diameters are determined from the negatives by enlargement or by projection on a screen. The photographic technique for measurement of bubble size most often gives local values; it may miss a small number of large bubbles and is best used in the case of small gas holdup, that is, low gas velocity.

An alternative sampling method has been used by Todtenhauft (T6), in which bubbles extracted from the dispersion are photographed through a calibrated capillary tube. A conductivity probe used by Ross and Curl for droplets should also prove suitable for sizing bubbles (R19).

3. *Interfacial Area*

The local interfacial contact area is determined directly by light transmission and reflection techniques. In the light transmission technique, a parallel beam of light is passed through the dispersion and a photocell is placed some distance from it. Light scattered by the bubbles passes outside the photocell and is lost, while the unscattered part of the incident parallel beam is recorded by the photocell at the extremity of an internally blackened tube (V6). Calderbank (C1) showed that, for scattering bubbles, which are large in comparison with the wavelength of light, the scattering cross section was equal to its projected area. Furthermore, the total interfacial area per unit volume of the dispersion equals four times the projected area per unit volume, giving the equation

$$\ln(I_0/I) = a'l/4 = \ln(t/t_0) \qquad (69)$$

where l is the optical pathlength. By connecting the photocell to a light quantity meter and electric timer, it is possible to measure the times for a given quantity of light to be received by the photocell, when the light passes through the liquid (t_0) and through the dispersion (t). This technique holds for values of $a'l < 25$ and for bubble diameters larger than 50 μm.

With optically dense dispersions such as occur in plate columns, the transmission method may fail because of intense multiple scattering. In such cases a reflectivity probe may be used (C5), where the optical reflectivity or backward scattered light from the dispersion is measured. The specific interfacial area is calculated by

$$(R_\infty/R) - 1 = 46.5/a' \qquad (70)$$

where R is the intensity of light back-reflected by the dispersion and R_∞ is the intensity of light reflected back when a' is infinite (as obtained by extrapolation of a plot of $1/R$ against $1/a'$ to $1/a' = 0$). For a given dispersion, R_∞ is a function of the refractive index ratio (C5).

4. Complementing Physical Methods by Chemical Methods

The specific surface area of contact for mass transfer in a gas–liquid dispersion (or in any type of gas–liquid reactor) is defined as the interfacial area of all the bubbles or drops (or phase elements such as films or rivulets) within a volume element divided by the volume of the element. It is necessary to distinguish between the overall specific contact area S for the whole reactor with volume V_R and the local specific contact area S_i for a small volume element ΔV_i. In practice ΔV_i is directly determined by physical methods. The main difficulty in determining overall specific area from local specific areas is that S_i varies strongly with the location of ΔV_i in the reactor—a consequence of variations in local gas holdup and in the local Sauter mean diameter [Eq. (64)]. So there is a need for a direct determination of overall interfacial area, over the entire reactor, which is possible with use of the chemical technique.

At the outset, we recognize that a technique that measures overall values cannot be used without the restrictions that arise from the results observed with physical methods. For example, the chemical method can hardly be used with fast-coalescing systems, since the presence of a chemical compound may well reduce the coalescence rates. In fast-coalescing systems, as observed with physical methods, the wide variation of specific contact area at different locations in the reactor negates the meaning of an average value. In fact, physical and chemical techniques should be used simultaneously to identify more fully the phenomena that occur in gas–liquid reactors. While chemical methods provide overall values of interfacial area that are immediately usable for design, we must also know the variations in the local interfacial parameters (α, d_{SM}) within the reactor in order to deal competently with scale-up. These complementary data, measured by physical methods, should be obtained from local simultaneous measurements of two of the three interfacial parameters as discussed above.

The electroresistivity probe, recently proposed by Burgess and Calderbank (B32, B33) for the measurement of bubble properties in bubble dispersions, is a very promising apparatus. A three-dimensional resistivity probe with five channels was designed in order to sense the bubble approach angle, as well as to measure bubble size and velocity in sieve tray froths. This probe system accepts only bubbles whose location and direction coincide with the vertical probe axis, the discrimination function being achieved with the aid of an on-line computer which receives signals from five channels communicating with the probe array. Gas holdup, gas-flow specific interfacial area, and even gas and liquid-side mass-transfer efficiencies have been calculated directly from the local measured distributions of bubble size and velocity. The derived values of the dispersion parameter for the air-water system have been found to be in excellent agreement with independently observed and previously published data.

This promising method has revealed an interesting result: The interfacial areas reported compare very favorably with those computed from experimental global measurements using liquid-phase-controlled chemical gas absorption but are lower than those measured using photography. Whereas photography through the container wall appears to truncate data above equivalent diameters of about 15 mm, the probe used by Burgess and Calderbank only deletes data below a diameter of 4 mm. So the difference between the interfacial values measured by the chemical and the photographic methods may be due to a small number fraction of large bubbles dominating the interfacial area parameters of an assembly of small and large bubbles, as in a sieve tray, and to their apparent inadvertent omission by photography.

This is a good example of the limitation of the physical methods, where they provide one datum within a local volume (d_{SM}) and the other datum information α for the entire reactor volume.

B. Chemical Methods

Chemical methods for determining gas-liquid interfacial areas and mass-transfer coefficients have been intensively developed for the last 10 years. The principles of these methods are deduced from the results presented in Section III,B,2: A gas A is absorbed into a liquid where it undergoes a reaction with a dissolved reactant B:

$$A + zB \xrightarrow{k} \text{products}$$

By choosing a reactant having a suitable solubility and concentration along with an adequate rate of reaction, either the mass-transfer coefficients or the interfacial area (or both) can be deduced from the overall rate of absorption. Generally a steady flow of each phase through the reactor is assumed.

1. Determination of $k_L a$

If the resistance to transfer of component A is entirely in the liquid phase ($k_L a \ll H k_G a$) and no chemical reaction occurs, the rate of absorption per unit volume of gas-liquid mixture is as stated initially:

$$\Phi = \varphi a = k_L a (C_A^* - C_{A0})$$

where C_A^* and C_{A0} are the concentrations, respectively, at the interface and in the bulk liquid.

The value of $k_L a$ can, in certain circumstances, be determined by purely physical experiments in the reactor. For instance, $k_L a$ may be evaluated from the observed total rate of absorption in the case of a piston-like countercurrent flow of the two phases, where C_A^* becomes a known function of C_{A0}, or where the gas and the liquid are well stirred, so that C_A^* and C_{A0} are the same at all points.

However, there are two possible difficulties in determining $k_L a$ by physical absorption. First, the flow patterns and residence-time distributions of the phases may be undetermined (C_{A0} is not a known function of C_A^*), hence the value of $k_L a$ cannot be deduced from Φ. Second, in efficient contacting devices the gas and liquid may approach equilibrium quite closely. The determination of $k_L a$ depends on the difference between the actual and the equilibrium extent and necessitates extremely accurate measurement of the flow rates.

a. *Irreversible Slow Reaction.* The problem of approaching the liquid saturation can be avoided by using a solution that reacts with the dissolved gas in the slow-reaction regime (Section II,B,2)—a reaction too slow to affect the rate of absorption directly but not too slow to reduce the bulk concentration of dissolved gas to effectively zero. If the contemplated reaction is irreversible and second-order (first-order with respect to both A and B), the rate is

$$\Phi = \varphi a = k_L a C_A^* = R_A = k_2 C_A^* C_{B0} \beta \tag{24}$$

The conditions to be satisfied are

$$k_L a \ll \beta k_2 C_{B0} \tag{22}$$

with

$$a D_A / \beta k_L \ll 1 \quad \text{and} \quad D_A k_2 C_{B0} / k_L^2 = \text{Ha}^2 \ll 1 \tag{23}$$

Thus the rate of absorption Φ is the same at all points in the reactor, and the residence-time distribution of the liquid is irrelevant.

It is possible similarly to use a reaction that is mth-order with respect to component A and nth-order with respect to component B, the local rate being thus $k_{mn} C_A^m C_B^n$. The condition for $C_{A0} = 0$ to be zero is

$$k_L a \ll \beta k_{mn} C_A^{*m-1} C_{B0}^n \tag{22a}$$

and, for no reaction in the film,

$$\frac{2}{m+1} \frac{D_A k_{mn} C_A^{*m-1} C_{B0}^n}{k_L^2} \ll 1 \tag{23a}$$

It is not always easy to satisfy both conditions (22) or (22a) and (23) or (23a) simultaneously. Thus, if $k_{mn} C_{B0}^n$ is made large enough to satisfy condition (22a) ($C_{A0} = 0$), it may become too large to satisfy condition (23a) (no reaction in the film). When condition (23a) is satisfied but condition (22a) is not, Eq. (24) may be written

$$\Phi = \varphi a = k_R a C_A^* \tag{71}$$

with

$$(k_R a)^{-1} = (k_L a)^{-1} + (k_{mn} C_{A0}^{m-1} C_{B0}^n \beta)^{-1} \tag{72}$$

If $k_{mn} C_{B0}^n$ is varied, keeping $k_L a$ constant, a plot of $1/k_R a$ against $1/k_{mn} C_{B0}^n$ will be a straight line of slope $1/C_{A0}^{m-1} \beta$ with intercept $1/k_L a$. This offers a method of determining $k_L a$ when it is not possible to satisfy condition (22a).

Some suitable chemical systems for determining $k_L a$ in the slow-reaction regime are presented in Table VII.

b. *Determination of $k_L a$ with an Irreversible Instantaneous Chemical Reaction.* Another quite different method of using a chemically reacting system to determine $k_L a$ involves an irreversible instantaneous reaction where the rate of absorption (Section II,B,4,b) is

$$\Phi = \varphi a = k_L a C_A^* E_i = k_L a C_A^* [1 + (D_B/zD_A)(C_{B0}/C_A^*)] \tag{32a}$$

The condition to be satisfied is $Ha > 10 E_i$. If, in addition,

$$C_{B0} \ll C_A^* \tag{73}$$

then the rate of absorption is

$$\Phi = \varphi a = k_L a (C_{B0}/z)(D_B/D_A) \tag{74}$$

TABLE VII

CHEMICAL SYSTEMS USED TO DETERMINE $k_L a$ IN THE SLOW-REACTION REGIME

Solute gas A	Reactant B	Catalyst in the absorbent	Reference
CO_2	K_2CO_3 + $HKCO_3$	NaClO	D4, D5
O_2 diluted with air	CuCl	—	J4
O_2 diluted with air	Na_2SO_3	$CuSO_4$	D14, W3 Y1, A7
O_2 diluted with air	Na_2SO_3	$CoSO_4$	B11, S37 L13, L17 R7, O3
O_2	Glucose	Glucose oxidase	L19

TABLE VIII
Chemical Systems Used to Determine $k_L a$ in the Instantaneous Reaction Regime

Solute gas A	Reactant B	Reference
NH_3	H_2SO_4	S22
SO_2, Cl_2, HCl	NaOH	S22
H_2S, HCl, CO_2	Amines	S22
O_2 diluted with air	$Na_2S_2O_4$	J5

independent of the gas-phase concentration of component A. It is also independent of the gas-phase residence-time distribution. In practice Eq. (32a) can be used when $E_i > 4$. Some suitable chemical systems for determination of $k_L a$ in the instantaneous regime are given in Table VIII.

2. Determination of a

When the reaction between components A and B in the liquid phase is mth-order in A and nth order in B, it has been seen (Section II,B,7) that under certain circumstances the concentration of component B is everywhere the same as in the bulk solution ($k_{mn} C_{B0}^n$ is constant), and the reaction is said to be rapid pseudo-mth-order in A. This condition arises if

$$3 < \text{Ha} = \frac{\{[2/(m+1)]k_{mn} D_A C_A^{*m-1} C_{B0}^n\}^{1/2}}{k_L} \ll E_i \quad (75)$$

the rate of absorption

$$\Phi = \varphi a = E k_L a C_A^* = \text{Ha} \cdot k_L a C_A^* = a \left(\frac{2}{m+1} k_{mn} D_A C_A^{*m+1} C_{B0}^n \right)^{1/2} \quad (76)$$

is independent of k_L, that is, of the hydrodynamic conditions. Thus, provided that the average rate of absorption

$$\varphi = \{[2/(m+1)]k_{mn} D_A C_A^{*m+1} C_{B0}^n\}^{1/2} \quad (77)$$

is known, and provided that C_A^* and C_{B0} each have effectively the same value in all parts of the system, the interfacial area is determined directly by measurement of the rate of absorption Φ.

It is not always necessary to know the kinetics of the reaction to determine φ. Indeed the values of φ may be measured by absorbing the component A into the same solution in some laboratory apparatus with a known interfacial area (Section III,C,1), the agitation speed or the flow rate being varied to confirm that φ is really independent of k_L and β. In reactors where the residence-time distribution of the gas phase is unknown, it is sometimes possible to choose conditions corresponding to a high gas flow rate and a low value of φ, so that there is practically no change in the partial pressure of the gas. Moreover, if component A is diluted by a carrier gas, it must be confirmed that the gas-side resistance is negligible. Some suitably aqueous, viscous, and organic chemical systems for the determination of a in the pseudo-mth-order regime are presented in Table IX.

3. *Determination of $k_G a$*

If the resistance to transfer of component A is entirely in the gas phase ($Hk_G a \ll k_L a$), as has been stated (Section II,A), in the absence of a chemical reaction the rate of absorption per unit volume of a gas–liquid system is

$$\Phi = \varphi a = k_G a (p - p_i)$$

where p and p_i are the pressure, respectively, in the bulk gas and at the interface. In certain cases the value of $k_G a$ can be determined by purely physical experiments in the reactor, for instance in the case of piston-like countercurrent gas–liquid flow or when both phases are well stirred. However, the rate of absorption depends on the residence-time distributions in both phases, which may be undetermined. Normally, in addition, an appreciable resistance on the liquid side must be taken into account.

The liquid-side resistance can be eliminated and the rate of absorption can be made independent of the liquid-side residence-time distribution by using a solution that reacts instantaneously and irreversibly with the dissolved gas and thus eliminates the back pressure. Therefore (Section II,B,4,b)

$$\Phi = \varphi a = k_G a(p - p_i) = k_L a C_A^* \left(1 + \frac{D_B}{zD_A} \frac{C_{B0}}{C_A^*} \right) = k_L a C_A^* E_i \quad (78)$$

The condition for satisfying is $Ha > 10 E_i$; in this case, dissolved gas A reacts instaneously with component B in a plane beneath the interface at which the concentration of both components is effectively zero. If the reaction plane is moved to the interface, the surface reaction provides the relation

$$p_i = HC_A^* = \frac{H[k_G ap - k_L a(D_B/zD_A)C_{B0}]}{Hk_G + k_L} = 0 \quad (79)$$

where

$$\Phi = k_G ap = k_L a(D_B/zD_A)C_{B0} \quad (80)$$

TABLE IX

CHEMICAL SYSTEMS USED FOR THE DETERMINATION OF a IN RAPID PSEUDO-mTH-ORDER REGIME

Solute gas A	Reactant B	Catalyst	References
CO_2 diluted with air	$Na_2CO_3 + HNaCO_3$, $K_2CO_3 + HKCO_3$	As $(OH)_2$ O^-, ClO^-	D4, R11, S17 S19, R3
	Aqueous or organic solutions of amine	—	D4, S18
	LiOH–NaOH, KOH–Ba(OH)$_2$	—	R11, N10, D3
	Na_2S	—	S22
COS diluted with air	Aqueous solutions of amine	—	S18
O_2 in air	$Na_2S_2O_3$	—	J5
O_2 diluted with air	Na_2SO_3	$CoSO_4$, $CuSO_4$	D13, W3, Y1 A7, B7, S37 L13, L17, R7 O3
Isobutylene in C_4 fraction or air	H_2SO_4	—	G5
CO_2 diluted with air	Cyclohexylamine in toluene or xylene containing 10% isopropanolamine	—	
	Cyclohexylamine in cyclohexanol	—	S35
	Monoethanolamine in aqueous di- or polyethylene glycol	—	
CO_2 and O_2 diluted with air	Aqueous cuprous amine complex solution		
O_2 diluted with air or N_2	Propionaldehyde	Manganese propionate	G18
O_2 in air	C_{10} trialkylaluminum dissolved in organic solvents	—	B20
Cl_2	p-Cresol dissolved in dichlorobenze	—	P1
H_2	Edible oil	Ziegler–Natta	G1
Desorption of isoamylene into N_2	n-Heptane, toluene	—	S4

In this situation, the dissolved component B reaches the interface by diffusion through the liquid as fast as the gaseous component A reaches it by diffusion through the gas; the transfer process is controlled by diffusion in both phases.

If at all points in the reactor the condition is satisfied that

$$k_G ap - k_L a(D_B/zD_A)C_{B0} < 0 \tag{79a}$$

the rate of absorption will remain equal to $k_G p$ but the interfacial concentration of component B will be greater than zero. In this case, the absorption process is entirely by the transport of component A across the gas film:

$$\Phi = k_G ap \tag{80a}$$

This forms the basis of methods for measuring $k_G a$. If, at both entrance and exit, low partial pressures of the soluble gas p_e and p_s are measured for a reactor of length h, $k_G a$ is calculated by

$$k_G a = (G_m/Ph) \ln(p_e/p_s) \tag{80b}$$

where G_m is the superficial molar mass flow rate of the insoluble gas in gm moles/cm² sec and P is the total pressure in atm.

In all cases, the calculation of $k_G a$ should be based on analysis of the gas stream, as a small error in analysis of the liquid stream can lead to large errors in $k_G a$. Some suitable chemical systems for determination of $k_G a$ in the instantaneous regime are given in Table X.

4. *Simultaneous Measurement of Mass-Transfer Coefficients and Interfacial Area*

In certain cases, significant mass-transfer resistances may occur in both phases, requiring knowledge of both $k_G a$ and $k_L a$. Moreover, for the purpose of calculating the effect of chemical reaction on the rate of absorption of a gas, it is generally necessary to know parameters k_L and a separately. Not only in the two-film model but also in the surface-renewal model, the separate mass-transfer parameters can be determined by chemical methods with a pseudo-first-order regime or an instantaneous regime.

Thus the condition for satisfying the pseudo-first-order behavior in component A, using Danckwerts' model (Section II,B,5), is

$$(1 + \text{Ha}^2)^{1/2} < (D_A/D_B)^{1/2} + (D_B/D_A)^{1/2}(C_{B0}/zC_A^*) = E_i(D_A/D_B)^{1/2} \tag{35}$$

With mass-transfer resistance in both phases, Eq. (35) gives

$$\Phi = \varphi a = p\left(\frac{1}{k_G a} + \frac{H}{k_L a(1 + \text{Ha}^2)^{1/2}}\right)^{-1} \tag{81}$$

TABLE X

CHEMICAL SYSTEMS USED TO DETERMINE $k_G a$ IN INSTANTANEOUS AND SURFACE REGIMES

Solute gas A	Insoluble gas diluent	Reactant B	References
SO_2 or Cl_2	Air, Freon 12, Freon 22, Freon 114	NaOH	V7
NH_3	Air, Freon 12, Freon 22, Freon 114	H_2SO_4	V7, S27
Triethylamine	Air, Freon 12, Freon 22, Freon 114	H_2SO_4	V7
I_2	Air	NaOH	L25
SO_2	Air	Na_2SO_3	P12
Propylene, CO	Air	Cuprous amine complex solution	S36

Several important cases are now presented, depending on whether k_L and a or k_G and a are needed.

a. *Determination of k_L and a.* When the gas-phase resistance is negligible, the rate of absorption is given by Eq. (81) with $k_G a = \infty$:

$$\Phi = \varphi a = (p/H)k_L a(1 + \text{Ha}^2)^{1/2} = aC_A^*(D_A k_2 C_{B0} + k_L^2)^{1/2} \qquad (82)$$

If the rate of absorption is measured with different values of $k_2 C_{B0}$ and the hydrodynamic conditions remain constant, a plot of Φ^2 against $k_2 C_{B0}$ will give a straight line with slope $D_A a^2 C_A^{*2}$ and intercept $(k_L a C_A^*)^2$. This is called the Danckwerts plot. If C_A^* and D_A are known, both $k_L a$ and a can be determined. A similar result is obtained in using the double-film model to interpret an irreversible $(m,n\text{th})$-order, when the condition $\text{Ha} \ll E_i$ is satisfied but not the condition $\text{Ha} > 3$ [Section II,B,7, Eq. (40)].

In using the Danckwerts plot, the physical properties of the system should not vary as $k_2 C_{B0}$ is changed. For this reason it is convenient to use a catalytic reaction and to change k_2 by adding small amounts of catalyst. If the catalyst is sufficiently active, the reaction rate can be varied over a wide range without much alteration in the concentration of the solution. The rate of absorption is thus

$$\Phi = \varphi a = aC_A^*(D_A k_c C_{\text{cat}} + k_L^2)^{1/2} \qquad (82a)$$

The concentration C_{cat} of the catalyst can be varied, and a and k_L determined by plotting Φ^2 against C_{cat}. Some suitable chemical systems for the determination of k_L and a with the use of the Danckwerts plot are presented in Table XI.

TABLE XI

CHEMICAL SYSTEMS FOR DETERMINATION OF k_L AND a, WITH USE OF DANCKWERTS PLOT

Solute gas A	Reactant B	Catalyst	References
CO_2 diluted with air	$HNaCO_3 + Na_2CO_3$	Arsenite	R3, M9
CO_2 diluted with air	$HNaCO_3 + Na_2CO_3$	Hypochlorite	D5, M4
O_2 in air	CuCl	—	J5, L23
O_2 in air	Na_2SO_3	$CoSO_4$	D13

It is interesting to know that the hydrodynamic conditions (that is, k_L) may be influenced by the chemical reaction, causing a change in a and k_L with a change in the reaction rate. Thus information complementary to that given by the Danckwerts plot can be provided when the rates of simultaneous chemisorption of one gas (giving a with a pseudo-mth-order reaction) and of physical absorption or desorption of another (giving k_L or $k_L a$) are determined experimentally (G9, L15, R14, B11, P13).

b. *Simultaneous Determination of k_G and a.* Even if the gas-side resistance is not negligible, it is still possible to determine k_G and a by use of a fast irreversible pseudo-mth-order reaction (Ha $\gg$ 1 and $(1 + Ha^2)^{1/2}$ = Ha). Then the rate of absorption expressed by Eq. (81) becomes

$$\Phi = \varphi a = p[(k_G a)^{-1} + H/(k_L a \cdot Ha)^{-1}]^{-1} \quad (83)$$

whence

$$\frac{p}{\Phi} = \frac{1}{a}\left(\frac{1}{k_G} + \frac{H}{k_L \cdot Ha}\right)$$

$$= \frac{1}{a}\left[\frac{1}{k_G} + \frac{H}{\{[2/(m+1)]k_{mn}D_A C_A^{*m-1}C_{B0}^n\}^{0.5}}\right] \quad (84)$$

Thus, if $k_2 C_{B0}$ or $k_{mn} C_{B0}^n$ is varied, a plot of p/Φ against $H/\sqrt{D_A k_A C_{B0}}$ or against $H/\{[2/(m+1)]k_{mn}D_A C_A^{*m-1}C_{B0}^n\}^{0.5}$ will give a straight line of intercept $1/k_G a$ and of slope $1/a$, so that a and k_G can be calculated simultaneously. A convenient system for this method is the absorption of dilute CO_2 into an aqueous solution of NaOH or amines (S22, W1).

c. *Simultaneous Measurement of $k_G a$, $k_L a$, and a.* When there is a mass-transfer resistance in both phases, it is possible, first, to measure $k_G a$ in the reactor by use of an irreversible instantaneous surface chemical reaction. Thus the values of $k_G a$ with the actual gas–liquid system will be calculated in assuming that this parameter varies as the power 0.5 of the solute gas diffusivity. Then measurement of the rate of absorption given by Eq. (81) will yield the values of $k_L a$ (S23).

Similarly, separate mass-transfer parameters can be determined from several measurements with a simple chemical system. In the case of an irreversible pseudo-first-order reaction, Eq. (81) may be written

$$\Phi = \varphi a = p \left(\frac{1}{k_G a} + \frac{H}{a(D_A k_2 C_{B0} + k_L^2)^{1/2}} \right)^{-1} \qquad (84a)$$

If it is assumed that $D_A k_2 C_{B0}$ and Φ are variable, whereas the other parameters p, H, k_L, and k_G are kept constant, differentiation of Eq. (84a) will lead to

$$\left(\frac{d(k_2 D_A C_{B0})}{d\Phi} \frac{\Phi^2}{2 C_A^*} \right)^{2/3} = a^{2/3} D_A k_2 C_{B0} + a^{2/3} k_L^2 \qquad (85)$$

Values of the differential term are obtained from the curves $k_2 D_A C_{B0}$ versus Φ. Thus a plot of the left side of Eq. (85) against $D_A k_2 C_{B0}$ will yield the values of a and k_L. Then the obtained values are used in Eq. (84a) where graphical analysis will give k_G. This chemical method for the simultaneous measurement of k_L, a, and k_G has been applied by Bartholomai *et al.* (H1, B9) to an agitated tank and a plate column.

It is also theoretically possible to measure these parameters simultaneously, using a similar sequence for interpretation of Eq. (47), that is, using an irreversible instantaneous reaction (S35).

C. Apparatus for Determining the Physicochemical Parameters

It was seen in Section III,B that chemical determination of the interfacial mass-transfer parameters required knowledge of

(1) Kinetics of the chosen chemical gas–liquid system
(2) Diffusivities of the soluble gas or gases, and the reactant or reactants, in the liquid
(3) Solubility of the gas or gases

Eventually these parameters are combined in the terms

$$C_A^* \sqrt{D_A} \quad \text{and} \quad \{[2/(m+1)] k_{mn} C_{B0}^n C_A^{*m+1}\}^{1/2}$$

The physicochemical parameters are determined in a laboratory apparatus that allows careful control of gas–liquid contact time and of the interfacial area. The recommended technique and its application to the oxidation of aqueous Na_2SO_3 solution will now be discussed.

1. *Determination of the Kinetic Regime*

We again consider the irreversible second-order reaction

$$A + zB \xrightarrow{k_2} \text{products}$$

Eight kinetic regimes are identified in terms of the two-film theory, as seen in Section II,B. The interface concentration profiles and the corresponding absorption rate forms are shown in Fig. 7. The noticeable differences in form of the rate equations aid in identifying the separate regimes. As explained by Levenspiel and Godfrey (L11), Table XII shows which factors affect the rate in each regime: + indicates that a change in a particular variable affects the rate; − indicates that it does not; and ? indicates a probable effect, but that the defining rate equations are unavailable. Going further, the schematic of Table XIII shows how to identify the kinetic regime for a given gas–liquid reaction from a series of systematic experiments where the variables are changed independently (L2, L11). Very often the corresponding rate is a function of the physicochemical parameters that must therefore be determined.

2. *Experiments and Interpretation*

The physicochemical parameters are determined in the apparatus of Fig. 8 by a dynamic method. A jet or film of liquid moves continuously through the gas to which it is exposed for a known time interval lying

TABLE XII

Factors Affecting the Overall Rate of Absorption–Reaction (L11)

Variable	Symbol	Kinetic regime							
		A	B	C	D	E	F	G	H
Concentration of reactant B in the bulk of the liquid	C_{B0}	+	−	+	+	?	?	+	+
Partial pressure of component A in the bulk of the gas	p	+	+	+	+	?	?	+	+
Interfacial area	a	+	+	+	+	+	+	+	−
Liquid holdup	β	−	−	−	−	+	+	+	+
Liquid-side mass-transfer coefficient	k_L	+	−	+	−	?	?	+	−
Gas-side mass-transfer coefficient	k_G	+	+	+	+	?	?	+	−
Rate constant for second-order chemical reaction	k_2	−	−	+	+	?	?	+	+

Regime	Interface concentration profiles	Rate equations	
Instantaneous	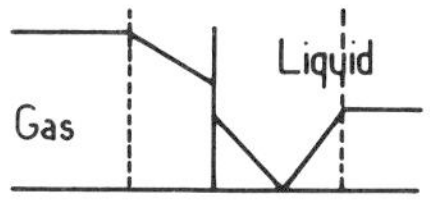	$\varphi = \dfrac{\dfrac{p}{He} + \dfrac{D_B}{D_A}\dfrac{C_{Bo}}{z}}{\dfrac{1}{k_L} + \dfrac{1}{He\,k_G}}$	A
Instantaneous and surface	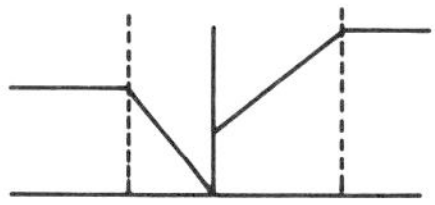	$\varphi = k_G\, p$	B
Rapid	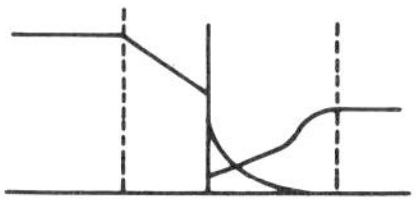	$\varphi = \dfrac{p}{\dfrac{1}{k_G} + \dfrac{He}{E\,k_L}}$	C
Rapid pseudo 1st or mth order	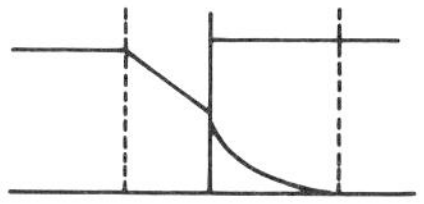	$\varphi = \dfrac{p}{\dfrac{1}{k_G} + \dfrac{He}{\sqrt{D_A\,k_2\,C_{Bo}}}}$	D
Intermediate	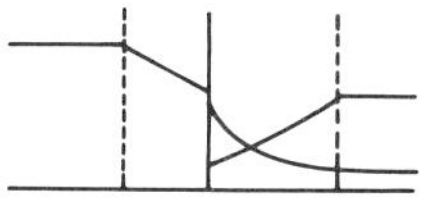	No exact general expression developed	E
Intermediate	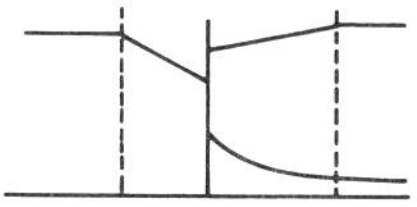	No exact general expression developed	F
Slow diffusional process	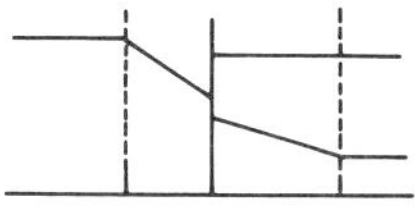	$\varphi = \dfrac{p}{\dfrac{1}{k_G} + \dfrac{He}{k_L} + \dfrac{He\,a}{\beta\,k_2\,C_{Bo}}}$	G
Very slow chemical process in the bulk of the liquid	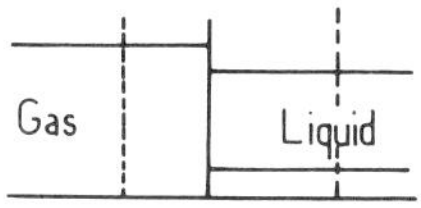	$R_A = k_2\, C_A^{*}\, C_{Bo}\, \beta$	H

FIG. 7. Interface concentration profiles for the eight distinct kinetic regimes for mass transfer with reaction.

between 10^{-4} and 10 sec. In these experiments, the interfacial area A_m is known, and movement of the liquid is carefully controlled so that it can be considered quiescent during its passage through the gas. During the contact time θ each element of liquid absorbs the same amount $Q(\theta)$ of gas per

TABLE XIII

Organigram for Identification of the Kinetic Regimes (L11)

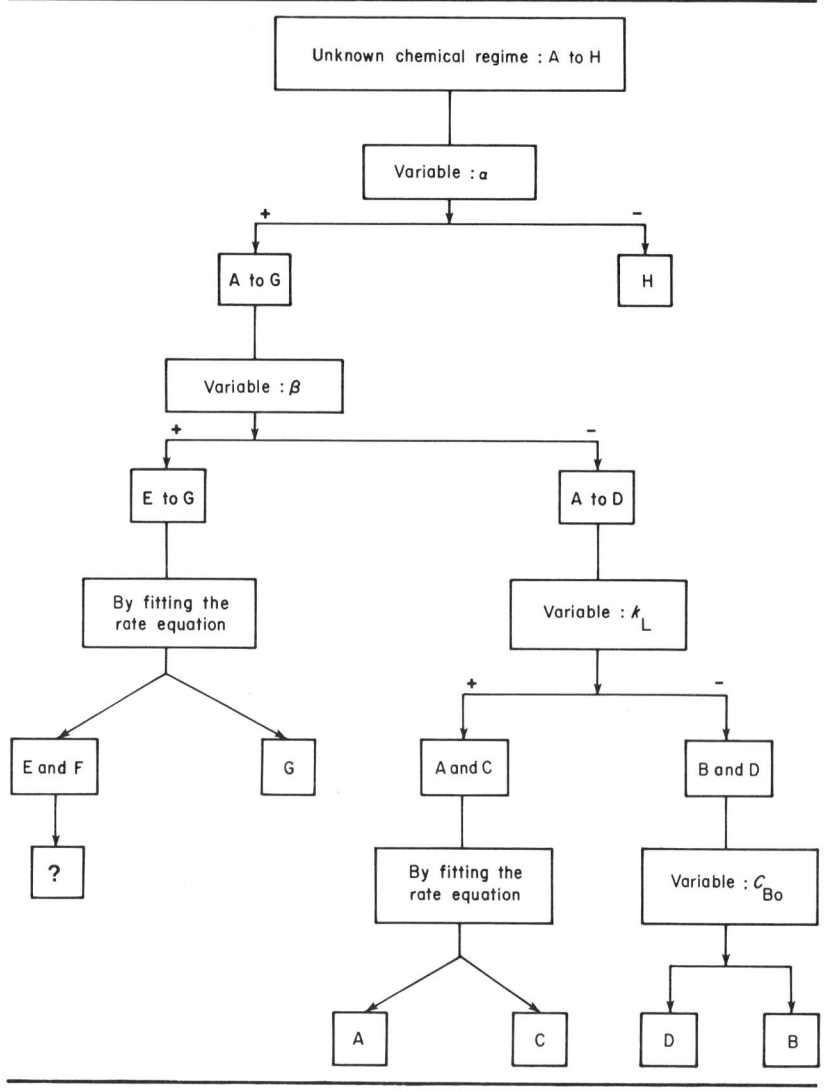

unit area as though it were stagnant and infinitely deep, as in the Higbie model. The different values of the contact time θ are generally obtained by varying the liquid volumetric flow rate Q_L and the length of the liquid film or jet h. Indeed the contact time is defined as

TYPE	LAMINAR JET	CYLINDRICAL WETTED WALL	CONIC WETTED WALL	SPHERICAL WETTED WALL	STRING OF DISKS	ROTATING DRUM	STIRRED VESSEL	STIRRED VESSEL
SCHEME								
k_L cm/sec	0.016 –0.16	$3.6 \cdot 10^{-3}$ –0.016	$5 \cdot 10^{-3}$ –0.011	$5 \cdot 10^{-3}$ –0.016	$3.6 \cdot 10^{-3}$ –0.016	0.016 –0.36	$1.6 \cdot 10^{-3}$ –0.02	$2 \cdot 10^{-3}$ –0.02
CONTACT TIMES	10^{-3}–10^{-1} sec	10^{-1}–2 sec	0.2–1 sec	0.1–1 sec	10^{-1}–2 sec	$2 \cdot 10^{-4}$–10 sec	0.06–10 sec	0.08–10 sec
INTERFACIAL AREA	0.3–10 cm² high precision	10–100 cm² high precision	80 cm² high precision	10–40 cm² high precision	30–360 cm² moderate precision	diameter 10cm length 12cm high precision	80 cm² good precision	diameter 10cm length 15cm 2–30% open

FIG. 8. Principal types of laboratory equipment.

$$\theta = h/u_s \tag{86}$$

where u_s is the velocity at the surface, which is a function of Q_L and the geometry of the equipment (diameter of the jet or film, angle of the cone, and so on). Expressions of θ for different types of equipment are given in Table XIV. If $Q(\theta)$ is the amount of gas absorbed per unit interfacial area during the contact time θ, the average rate of absorption during this time is $Q(\theta)/\theta$. Since the total area exposed in the laboratory equipment is A_m, the measured rate of absorption $\Phi(\theta)$ into the film is related to $Q(\theta)$ by

$$\Phi(\theta)/A_m = Q(\theta)/\theta = \varphi \tag{87}$$

The absorption rate $\Phi(\theta)$ (in gm moles/sec or cm³/sec of component A), is measured experimentally, and $Q(\theta)/\theta$ is calculated for different kinetic regimes from the Higbie theory. The contact time θ, calculated from Eq. (86), can be altered by altering Q_L and the geometric parameters of the equipment. Thus, by carrying out experiments with the same chemical systems and with the same kinetic regime used to determine mass-transfer parameters in the industrial gas–liquid reactor, $\Phi(\theta)$ can be determined as a function of θ in the laboratory equipment and the variations of $Q(\theta) = \theta\Phi(\theta)/A_m$ for different values of θ serve to determine the parameters.

For the case of purely physical absorption into liquid initially gas-free ($C_{A0} = 0$), the amount of component A absorbed per unit area during the time θ is (Section III,A,2)

$$Q(\theta) = 2C_A^*(D_A\theta/\pi)^{1/2} \tag{88}$$

It follows from Eq. (87) that

$$\Phi(\theta) = [Q(\theta)/\theta]A_m = 2A_m C_A^*(D_A/\pi\theta)^{1/2} = \gamma(\theta)C_A^*\sqrt{D_A} \tag{89}$$

TABLE XIV
Characteristic Parameters of Laboratory Equipment[a]

Equipment	θ	$\gamma(\theta)$
Laminar jet	$\dfrac{\pi d^2 h}{4 Q_L}$	$4(Q_L h)^{1/2}$
Cylindrical wetted wall	$\dfrac{2h}{3}\left(\dfrac{3\mu_L}{\rho_L g}\right)^{1/3}\left(\dfrac{\pi d}{Q_L}\right)^{2/3}$	$d(6h)^{1/2}\left(\dfrac{Q_L}{d}\right)^{1/3}\left(\dfrac{\pi \rho_L g}{3\mu_L}\right)^{1/6}$
Conic wetted wall	$\dfrac{4\pi R^2}{5 \sin\alpha}\left(\dfrac{3\mu_L}{2\pi \rho g R \cos\alpha}\right)^{1/3}\dfrac{1}{Q_L^{2/3}}\left(1+\dfrac{\sin\alpha}{R}l\right)^{5/3}-1$	$\dfrac{2(\pi l^2 \sin\alpha + 2\pi R l)}{\sqrt{\pi}}\dfrac{Q_L^{1/3}}{\sqrt{\delta}}$ $\delta = \left[\dfrac{4\pi R^2}{5\sin\alpha}\left(\dfrac{3\mu_L}{2\pi\rho g R \cos\alpha}\right)\right]^{1/3}\left[\left(1+\dfrac{\sin\alpha}{R}l\right)^{5/3}-1\right]$
Spherical wetted wall	$\left(\dfrac{8}{4.5}\right)^2 \pi R_s^{5/6}\left(\dfrac{2\pi\rho_L g}{3\mu_L}\right)^{-1/6} Q_L^{-1/3}$	$4.5\left(\dfrac{2\pi\rho_L g}{3\mu_L}\right)^{1/6} Q_L^{1/3} R_s^{7/6}$
Rotating drum	$\dfrac{Z_r}{u}$	$4 Q_L^{1/2}\left(\dfrac{W_r Z_r}{2\pi \delta_r}\right)^{1/2}$

[a] α, angle of the conic wetted wall; l, generating line of the cone; R, radius of the basis of the cone; R_s, radius of the spherical wetted wall; W_r, belt width of the rotating drum; Z_r, exposed belt length; δ_r, film thickness; θ, contact time.

where $\gamma(\theta)$ is a parameter characteristic of each apparatus. Values of $\gamma(\theta)$ for several types of equipment are shown in Table XIV. A plot of measured $\Phi(\theta)$ against $\gamma(\theta)$ at various liquid flow rates and various film or jet lengths should give a straight line with slope $C_A^* \sqrt{D_A}$ through the origin. If the solubility of the gaseous species is known or determined separately (Section II,C,1) its diffusivity may be determined.

For the case of absorption accompanied by chemical reaction, $Q(\theta)$ is deduced from solution of the equation

$$D_A(\partial^2 C_A/\partial x^2) - r(x, \theta) = \partial C_A/\partial \theta$$

where $r(x, \theta)$ is the rate per unit volume of liquid at which the reaction consumes the solute gas at time θ and at distance x from the interface. Analytical or numerical solutions of the diffusion reaction equations are available for a number of kinetic regimes. For an irreversible first- or pseudo-first-order reaction with

$$\text{Ha}' = [(\pi/4)k_2 C_{B0}\theta]^{1/2} \ll (D_A/D_B)^{1/2} + (C_{B0}/zC_A^*)(D_B/D_A)^{1/2} = E_i \quad (90)$$

Danckwerts (D2) has proposed the following approximate solutions:

$$Q(\theta) = C_A^*(D_A k_2 C_{B0})^{1/2}[\theta + (2k_2 C_{B0})^{-1}] \quad (91)$$

to within 3% when $k_2 C_{B0}\theta > 2$, and

$$Q(\theta) = 2C_A^*(D_A \theta/\pi)^{1/2}[1 + \tfrac{1}{3}(k_2 C_{B0}\theta)] \quad (92)$$

to within 5% when $k_2 C_{B0}\theta < \tfrac{1}{2}$. Thus experimental determination of the amount $\Phi(\theta)$ of gas absorbed up to time θ leads to the following results. For long contact times ($\theta > 2/k_2 C_{B0}$), a plot of $Q(\theta) = \theta\Phi(\theta)/A_m$ against θ will give a straight line of slope $C_A^*(D_A k_2 C_{B0})^{1/2}$ and intercept $\tfrac{1}{2}C_A^*(D_A/k_2 C_{B0})^{1/2}$, and the slope/intercept ratio will give $k_2 C_{B0}$. For short contact times ($\theta < 1/2k_2 C_{B0}$), a plot of $Q(\theta)/\sqrt{\theta}$ against θ will give a straight line of slope $\tfrac{2}{3}[(k_2 C_{B0})]C_A^*(D_A/\pi)^{1/2}$ and intercept $2C_A^*(D_A/\pi)^{1/2}$.

In principle, both k_2 and $C_A^*\sqrt{D_A}$ can be estimated in either case. In practice, the measurements at long contact times give more accurate values of $C_A^*(D_A k_2)^{1/2}$, while those at short contact times give more accurate values of $C_A^*\sqrt{D_A}$(D2).

If, in addition to Eq. (90), the following condition is satisfied,

$$\text{Ha}' > 3 \quad \text{or} \quad k_2 C_{B0}\theta > 12 \quad (93)$$

then, within 5%,

$$Q(\theta) = \theta C_A^*(D_A k_2 C_{B0})^{1/2} \quad (94)$$

This is the fast pseudo-first-order case, where the rate of absorption $Q(\theta)/\theta$ is the same at all points on the surface and therefore independent of the hydrodynamics. Similar considerations apply to the fast psuedo-mth-order reaction of component A, in which case (D2)

$$Q(\theta) = \theta\{[2/(m + 1)]D_A k_{mn} C_A^{*m+1} C_{B0}^n\}^{1/2} \tag{94a}$$

Under these circumstances, measurements of $Q(\theta)$ from experimental $\Phi(\theta)$, through Eq. (87), do not lead to separate values of $C_A^* \sqrt{D_A}$ and k_2 or of $C_A^{*(m+1)/2} \sqrt{D_A}$ and k_{mn}. Nevertheless they lead to the value of the grouped parameters $C_A^* (D_A k_2 C_{B0})^{1/2}$ or $[2/(m + 1)]D_A k_{mn} C_A^{*m+1} C_{B0}^n$ sufficient for determination of interfacial area in the gas–liquid reactor [Eq. (76)] without detailed knowledge of each physicochemical parameter.

Last, for an irreversible instantaneous reaction (that is, when $Ha' \gg E_i$), the diffusion reaction equations governing this case have been given by Danckwerts (D2). The solution is

$$Q(\theta) = 2E_i C_A^* \left(\frac{D_A \theta}{\pi}\right)^{1/2}$$

$$= 2C_A^* \left(\frac{D_A \theta}{\pi}\right)^{1/2} \left[\left(\frac{D_A}{D_B}\right)^{1/2} + \frac{C_{B0}}{zC_A^*}\left(\frac{D_B}{D_A}\right)^{1/2}\right] \tag{95}$$

In this case, in addition to the quantities C_A^*, D_A, and C_{B0} as in the pseudo-first-order case, the diffusivity D_B of the reactant B is also involved. Any one of these quantities can be deduced from the experimental $\Phi(\theta)$ if the values of all the others are known or estimated. An interesting possibility is the reduction of C_A^* to a value much less than C_{B0}, in which case $(D_B/D_A)(C_{B0}/zC_A^*)$ may be much more than 1. The diffusivity D_B can then be determined without the same uncertainties as for C_A^* and D_A (S17). Thus the quantity of component A absorbed per unit area in the laboratory equipment in time θ is

$$Q(\theta) = 2C_A^* \left(\frac{D_A \theta}{\pi}\right)^{1/2} \left(\frac{D_A}{D_B}\right)^{1/2} \left(1 + \frac{D_B}{D_A}\frac{C_{B0}}{zC_A^*}\right) = \frac{2C_{B0}}{z}\left(\frac{D_B \theta}{\pi}\right)^{1/2} \tag{96}$$

3. *Use of Sodium Sulfite Oxidation*

The absorption of oxygen into sulfite solutions is very often used to measure interfacial areas in pilot plant or industrial reactors, as reviewed by Reith and Beek (R9) and Laurent *et al.* (L4). To illustrate the strategy outlined in the previous sections, a study of this reaction by Laurent (L2) will be discussed now in relation to the experimental conditions suitable for determining interfacial area in gas–liquid reactors. The reaction is

$$O_2 + 2Na_2SO_3 \xrightarrow[k_c]{CoSO_4} 2Na_2SO_4$$

with the rate

$$r = k_c C_A^m C_B^n C_C^q = k_c C_{O_2}^m C_{Na_2SO_3}^n C_{CoSO_4}^q$$

To use the technique proposed earlier in this section for the kinetic regime the kinetic parameters are investigated with the following experimental conditions:

$$0.2 \leq p_{O_2} \leq 0.8\text{--}1 \text{ atm}, \quad 0.4 < C_{Na_2SO_3} \leq 0.8 \text{ gm moles/liter}$$

$$10^{-4} \leq C_{CoSO_4} \leq 1\text{--}2 \times 10^{-3} \text{ gm moles/liter},$$

$$C_{Na_2SO_3} + C_{Na_2SO_4} = \text{const.} \tag{97}$$

$$20 \leq t \leq 50°\text{C}, \quad 7.5 \leq \text{pH} \leq 8.5$$

Three types of laboratory equipment have been used (L2): a cylindrical wetted wall ($d = 1.6$ cm, $h = 10$ cm), a conic wetted wall ($\alpha = 14°, l = 14$ cm, $2R_s = 3.8$ cm), and a stirred vessel ($D = 10.1$ cm, $V_R = 1$ liter, $V_L = 0.5$ liter).

a. *Determination of the Kinetic Regime.* The influence of interfacial area A_m is given in Table XVA. For the same operating conditions, the volumetric absorption rate of oxygen Φ'_{O_2} (in cm³/sec) varies proportionally with the interfacial area in the three equipment types. Thus this reaction can not be considered a very slow chemical process (regime H, Table XIII; Fig. 7).

The influence of liquid holdup β is shown in Table XVB. For any given operating condition, the average volumetric rate of absorption of oxygen

TABLE XVA

DETERMINATION OF THE KINETIC REGIME FOR THE OXIDATION OF AQUEOUS Na_2SO_3: VARIATION IN INTERFACIAL AREA[a] A_m (L2)

Laboratory apparatus	Interfacial area A_m (cm²)	Φ'_{O_2} (cm³/sec)
Cylindrical wetted wall	49.8	0.195
Conic wetted wall	74	0.296
Stirred cell	78	0.308

[a] $t = 20°\text{C}; p_{O_2} = 1$ atm; pH $= 8.5; C_{Na_2SO_3} = 0.8$ gm moles/liter; $C_{CoSO_4} = 0.1$ gm/liter.

TABLE XVB

DETERMINATION OF THE KINETIC REGIME FOR THE OXIDATION OF AQUEOUS Na_2SO_3: VARIATION IN LIQUID HOLDUP[a] $\beta = V_L/V_R$

Laboratory apparatus	$V_L = \beta V_R$ (cm³)	φ'_{O_2} (cm³/cm² sec)
Cylindrical wetted wall	1.5	2×10^{-3}
Conic wetted wall	1	2×10^{-3}
Stirred cell	500	2×10^{-3}
	1000	2×10^{-3}

[a] $t = 20°C$; $p_{O_2} = 1$ atm; pH = 7.5; $C_{Na_2SO_3} = 0.8$ gm/liter; $C_{CoSO_4} = 0.1$ gm/liter.

φ'_{O_2} (in cm³/cm² sec) is independent of the liquid holdup. So the kinetic regimes E, F, and G (Table XIII; Fig. 7) can be eliminated.

The influence of the liquid-side mass-transfer parameter k_L, a function of apparatus hydrodynamics, is obtained from the different geometries of the three apparatuses (and also by varying the stirring speed N_L in the stirred cell) and from the liquid flow rate Q_L in each apparatus. The variations in $\varphi' = \varphi'_{O_2}$ with Q_L for the wetted-film equipment, and with N_L for the stirred cell, are given in Fig. 9. In a certain range of these latter variables (that is, in a defined range of contact times for the three equipment types), values of φ' are independent of k_L. Therefore, in this range of variables, kinetic regimes A and C (Table XIII; Fig. 7) can be eliminated.

The influence of the reactant (Na_2SO_3) concentration C_{B0} is also given in Fig. 9 as an example for the cylindrical wetted wall. Since φ' is influenced by C_{B0}, the kinetic regime B (Table XIII; Fig. 7) can be eliminated and the reaction can be considered a rapid pseudo-mth-order type (regime D, Table XIII; Fig. 7). This deductive sequence has to be repeated for the different operating conditions previously given.

b. *Determination of Partial Orders of Reaction.* If the gas-phase resistance is negligible, the rate equation corresponding to kinetic regime D is, within 5%,

$$\Phi = \varphi a = a\{[2/(m+1)]D_A k_{mn} C_A^{*m+1} C_{B0}^n C_{cat}^q\}^{1/2} \qquad (98)$$

That is, for any laboratory equipment, from Eqs. (87) and (94a), the specific equation for sulfite oxidation is

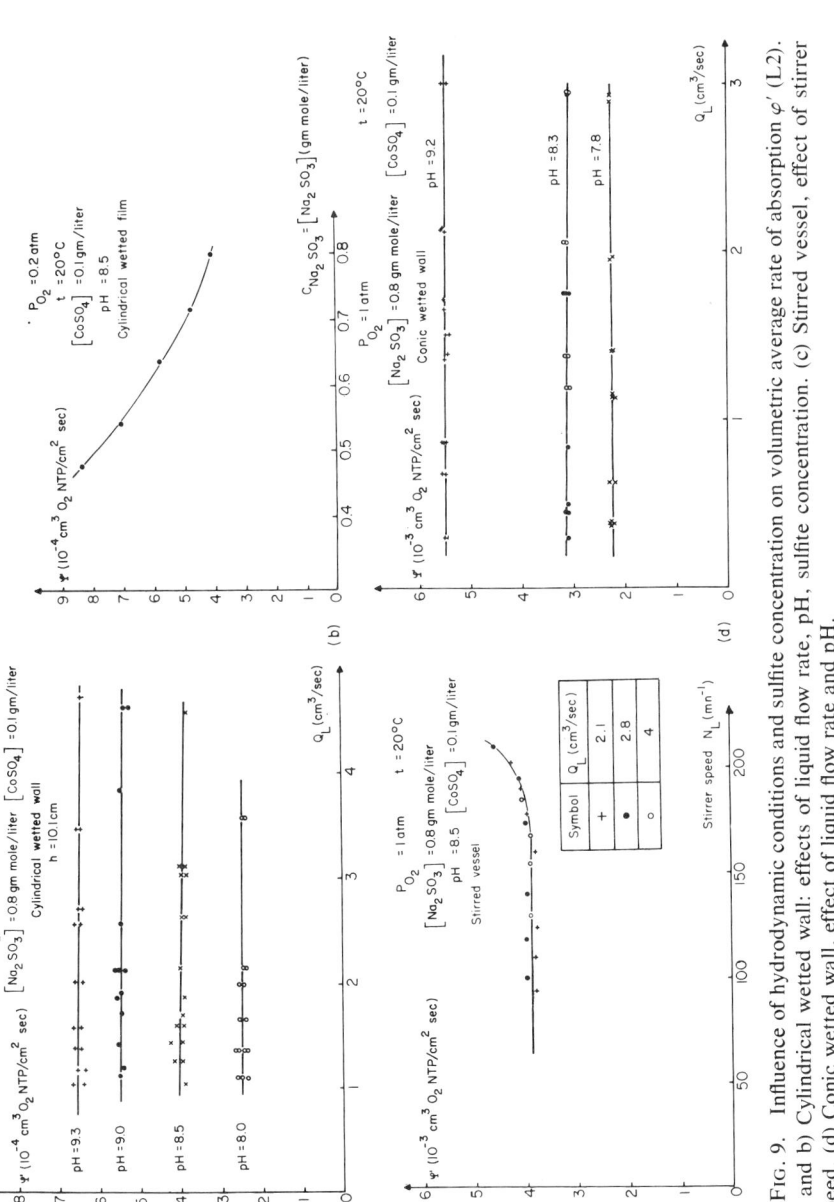

FIG. 9. Influence of hydrodynamic conditions and sulfite concentration on volumetric average rate of absorption φ' (L2). (a and b) Cylindrical wetted wall: effects of liquid flow rate, pH, sulfite concentration. (c) Stirred vessel, effect of stirrer speed. (d) Conic wetted wall, effect of liquid flow rate and pH.

$$\Phi(\theta) = A_m \frac{Q(\theta)}{\theta} = \varphi A_m$$

$$= A_m \left(\frac{2}{m+1} D_{O_2} k_{mn} C_{O_2}^{*m+1} C_{Na_2SO_3}^n C_{CoSO_4}^q \right)^{1/2} \quad (99)$$

We remember that these equations are valid if $3 < \text{Ha} \ll E_i$. To determine the interfacial area in a gas–liquid reactor, it is necessary to know the partial orders of reaction m, n, and q in the range of the operating conditions. Values of these orders from the literature, given in Table XVI, show large discrepancies for closely similar experimental conditions. For example, the value of n may be 1 or 2, or may vary continuously from 1 to 2.

However, in kinetic regime D, the variations in the average rate of absorption with the concentration of each component should conform to Eq. (99), and the experimental values measured in equipment where the interfacial area A_m is known should lead to the values of the partial orders. Examples of such variation in the average volumetric rate of absorption of oxygen φ' (here in cm³/cm² sec), as a function of $p_{O_2} = HC_{O_2}^*$, in the three laboratory equipment types are shown in Fig. 10a. The reaction is second-order in oxygen, as indicated by the slope of the straight-line plots of $\log \varphi'$ versus $\log p_{O_2}$, is $\frac{1}{2}(m + 1) = 1.5$ for p_{O_2} in the range 0.2–0.8 atm. It was observed for $p_{O_2} = 0.8$–1 atm and above that φ' increased less than would be expected for $m = 2$ (L2, A4). In practice, most of the experiments in pilot plants or in industrial reactors are carried out with air at atmospheric pressure, and under these conditions the order in oxygen appears to be $m = 2$.

The influence of the sulfite concentration on φ' is seen in Fig. 10b. As previously shown, the specific absorption rate decreases as the Na_2SO_3 concentration increases (curve a). However, experiments carried out keeping the initial sulfite plus sulfate concentration constant by the addition of Na_2SO_4 show a less marked decrease (curve b). More careful analysis reveals that the specific rate of absorption is independent of sulfite concentration when the experiments are carried out keeping a constant total sulfite plus sulfate concentration. This is demonstrated by recycling a solution partially oxidized into sulfate through the equipment (curve c). This interpretation may explain the different values of n published in the literature.

The residual decrease in curve b may perhaps be caused by variations in ionic strength, by impurities whose influence increases with sulfite concentration (L17), or by reduction of the solubility of oxygen at high concentrations of sulfite (W3). In the experiments corresponding to curve b, the ionic strength I is constant; the difference from curve a may be due to

TABLE XVI

REPORTED ORDERS OF REACTION FOR OXIDATION OF AQUEOUS SULFITE SOLUTIONS

$C_{Na_2SO_3}$ (gm mole/liter)	$p_{O_2} = He\ C^*_{O_2}$ (atm)	C_{CoSO_4} (gm mole/liter)	Temp. (°C)	pH	n	m	q	Ref.
0–0.08	0.2 and 1	0–7×10^{-7}	20	8.0–9.2	1	1	1	Y1
0.00375–1	0.1–1	10^{-3}–10^{-4}	—	—	1 if $[Na_2SO_3] < 0.06$, 0 if $[Na_2SO_3] \approx 0.25$, <0 if $0.25 < [Na_2SO_3] < 1$	0	—	A7
25–100 gm/liter	1	10^{-6}–10^{-3}	15–33	7.5–8.5	0 if $50 < [Na_2SO_3] < 100$	1	1	D13
0.4–0.8	0.1–1	4×10^{-4}	30	7.5–8.5	−1	2	1	W3
						Decreasing order if $p_{O_2} \geq 1$ atm if $[Na_2SO_3] = 0.4$ gm mole/liter		
100 gm/liter	0.2	10^{-3}	30	—	—	—	—	W4
0.25–0.795	0.2–1	0–10^{-3}	15–35	7.8–9.2	0	1 if $p_{O_2} > 0.9$ atm; 2 if $p_{O_2} < 0.9$ atm	1	L13, L16, L17
0.3–1	0.2–1	5×10^{-6}–10^{-4}	30	8.5	Variable order	Continuous variation from 2 to 1	—	O3
50–100 gm/liter	0.2–1	2×10^{-5}–2×10^{-3}	15–60	7.5–8.0	0	2	1	D2, L18, R8
0.4–0.8	0.2–1	3×10^{-6}–3×10^{-3}	15–60	7.5–8.5	0	2	1	R9, L4, B23, N15
0.4–0.8	0.1–0.5	$\leq 10^{-4}$	20	7.5–8.0	0	2	—	P2
0.8	0.3–7	10^{-4}–10^{-3}	30	8.3	0	2 if $p_{O_2} < 0.8$ atm; 1 if $p_{O_2} > 1.1$ atm	1	A4

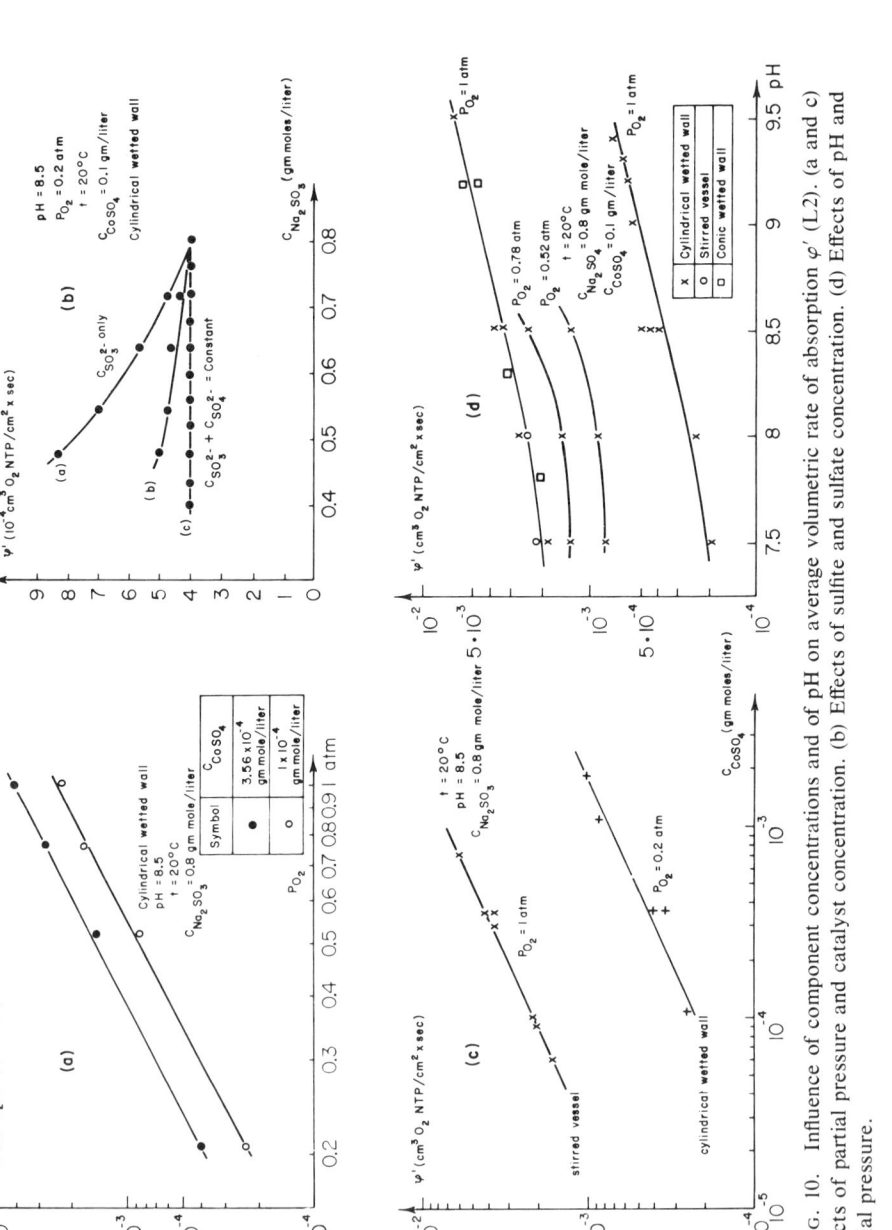

FIG. 10. Influence of component concentrations and of pH on average volumetric rate of absorption φ' (L2). (a and c) Effects of partial pressure and catalyst concentration. (b) Effects of sulfite and sulfate concentration. (d) Effects of pH and partial pressure.

a different solubility and diffusivity of O_2 in the solution. In fact, curve c represents the actual variations in the sulfite oxidation system under actual operation of an industrial reactor for which, with fixed feed conditions, the sulfite concentration, produced sulfate, and impurities are all maintained constant during a run. Under such experimental conditions, the average rate of absorption is independent of the sulfite concentration in the range 0.4–0.8 M, giving $n = 0$.

The influence of the catalyst concentration on φ' is shown in Fig. 10c. The reaction order in catalyst is unity, as indicated by the slope of 0.5 or the slope of the straight lines of log φ' versus log C_{CoSO_4}, for cobalt concentrations in the range 1×10^{-4} to $2 \times 10^{-3} M$. Any cobalt concentration higher than these will form a solid precipitate with the sulfite. A gradual increase in the volumetric average rate of absorption with pH is illustrated in Fig. 10d.

The partial orders for the defined operating conditions are sufficiently accurate for industrial practice, because only the rate of absorption per unit interfacial area needs to be known (without the rate constant and the activation energy). However, it is interesting to compare the values of these parameters as obtained by different authors to show the errors that could be made when different qualities of products are used. When m, n, q, and the diffusivity and the solubility of O_2 are all known, the second-order reaction rate constant k_{mn} ($=k_2$) and its variations with pH and

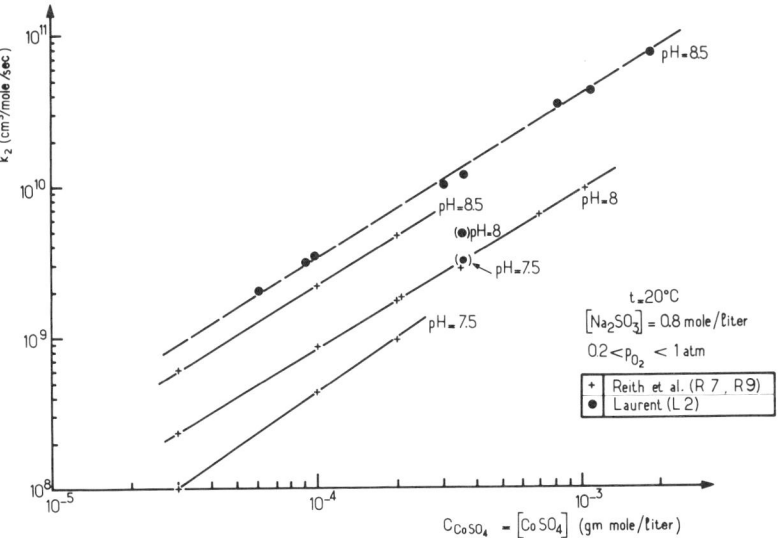

FIG. 11. Second-order reaction rate constant k_2 as a function of cobalt concentration and pH.

temperature can be determined. The dependence of k_2 on cobalt concentration and on pH is shown in Fig. 11, based on the work of Reith and Beek (R7, R9) and of Laurent (L2), whose rates are systematically higher. Linek and Mayrhoferova (L17) measured a second-order reaction rate in O_2 but did not interpret their data in terms of a second-order rate constant.

The Arrhenius relation, shown by a plot of log k_2 versus $1/T$ for different values of pH, gives straight lines. From their slope, values of E_a reported by different authors are given in Table XVII. They show a large divergence and again indicate a strong influence of the purity of the solution.

Thus the oxidation of aqueous Na_2SO_3 solutions with $CoSO_4$ as a catalyst proves to be a convenient model reaction for determining interfacial area in gas–liquid reactors. The kinetics of the reaction is not simple; many variables influence the reaction rate but, provided the range of cobalt and sulfite concentrations, pH values, and temperatures previously indicated is satisfied, the reaction is zero-order in sulfite, first-order in cobalt, and second-order in O_2. The specific rate of absorption is

$$\varphi = C_{O_2}^{*3/2} (\tfrac{2}{3} D_{O_2} k_2)^{1/2}$$

if the following conditions are verified:

$$3 < (D_{O_2} k_2 C_{O_2}^*)^{1/2}/k_L \ll 1 + (D_{Na_2SO_3}/D_O)(C_{Na_2SO_3}/2C_{O_2}^*)$$

Under these conditions, the interfacial area in a given reactor can be found by measuring the rate of absorption of O_2 with the same partial pressure of O_2, the same absorbent, and the same temperature and pH in laboratory equipment such as a wetted falling film, laminar jet, or stirred vessel in which the interfacial area A_m is known. It is emphasized that

TABLE XVII

Reported Activation Energy E_a for Oxidation of Na_2SO_3[a]

	pH			
t (°C)	7.5	8	8.5	Reference
15–60	18,421	13,875	12,918	R7
15–60	—	18,800	—	R9
15–33	12,000	12,000	12,000	D13
15–35	12,000–13,000	12,000–13,000	12,000–13,000	L17
30	—	10,526	—	W3
20–50	12,250	12,000	12,250	L2

[a] Values are in cal/gm mole.

exactly the same reactants must be used in both types of equipment, so that the influence of the catalyst and of trace impurities will be the same. Impurities such as traces of oil in the air, or from contact of the solution with metal, should be avoided. Phenomena such as foaming (reflecting the quality of the components used) should be carefully controlled. The literature suggests that use of copper as a catalyst (instead of cobalt) leads to poor results, perhaps even to inhibition.

Compared with the more common model reaction of CO_2 absorption, sulfite oxidation is moderated by the low solubility of O_2. It can be used when gas-phase conversion is very fast, as in a stirred tank. The results are typical of ionic solutions, whose properties may differ from those of other process liquids.

The equation just given becomes inaccurate at O_2 pressures above about 0.8 atm, because the reaction shifts from second-order to first-order as the pressure increases above this value. Alper (A4) has observed first-order dependence on O_2 pressure in the range from near 1 to 7 atm.

D. Mass-Transfer Rates with Chemical and Purely Physical Processes

To characterize the gas-liquid mass-transfer performance of a reactor, $k_L a$, $k_G a$, and a can be determined as described above. The question then arises how the parameters so evaluated can be extrapolated to other operating conditions, for example, from chemical absorption to physical absorption, or vaporization. In other words, is the value of k_L or a for the hydrodynamic conditions of chemical absorption the same as for the hydrodynamic conditions of physical absorption?

In a packed column, some zones of liquid in the packing are almost motionless, becoming saturated by the absorbing gas during physical absorption, hence almost ineffective to mass transfer. When the absorbing capacity of the liquid is increased by a chemical reactant, these zones may still be effective, and measurement of the amount of gas absorbed may give the impression that k_L and a have greater values. Also, in a mechanically agitated reactor, the effective interfacial area in a reacting system may not equal the area in a physical absorption or desorption, where the absorbing capacity of the liquid is not increased. In the case of absorption with a fast chemical reaction, the mass-transfer coefficient is independent of the hydrodynamics and equal at every point in the vessel, and the interfacial areas in all parts of the agitated vessel contribute equally to mass transfer. But in the case of physical desorption or absorption, the mass-transfer coefficient can have quite different values, for example, around the agitator and far away from it, or similarly for any gas-liquid reactor that exhibits hydrodynamic inhomogeneity; and the interfacial regions in different parts of the reactor do not contribute equally to the mass transfer.

Thus assumption of the same value for interfacial area in physical and chemical absorption leads to uncertainty, especially if the mass transfer coefficient is deduced from $k_L a$ measured by physical absorption or desorption and from a in chemical absorption. The effective interfacial area in the case of a fast-reaction system where the absorbing capacity is increased by a chemical reactant is substantially larger than the effective interfacial area for physical absorption or desorption, as pointed out by Joosten and Danckwerts (J10). These authors introduced a correction factor γ, the ratio between the increase in liquid absorption capacity and the increase in mass transfer due to chemical reaction:

$$\gamma = \frac{1 + (C_{B0}/zC_A^*)}{E}$$

Experimental results showed that, for physical absorption ($\gamma = 1$) and absorption with instantaneous chemical reaction ($\gamma \simeq 1$), the effective areas were the same; while for $\gamma \gg 1$, the interfacial area increased. The different effective areas found by Ratcliff et al. (R3), Joosten and Danckwerts (J10), and Laurent (L2), using similar packing and similar absorbing solutions, can be explained by the different γ values in the different cases.

The different values of $k_L a$ depending on the type of mass-transfer process (vaporization, chemical, or physical absorption) are due not only to variations of the liquid areas involved in mass transfer operation (J10, P16), but also to variations in the local mass-transfer coefficients within these zones (B2, B3, P13), for example, those due to the effect of interfacial turbulence which may accompany chemical absorption (L15).

The technique of simultaneous absorption with fast pseudo-mth-order reaction and physical absorption or desorption concurrently, used by Wilke and Robinson (R14), Linek (L15), and Beenacker and Van Swaaij (B11), is certainly a promising effort to understand the whole complex problem of transport in gas–liquid reactors, since it provides simultaneous measurement of $k_L a$ and a. But still it may leave some doubt as to a value of k_L, which can be changed by the occurrence of chemical reaction. As discussed by Prasher (P13), it will be even more promising to conduct such simultaneous experiments in a regime where both hydrodynamics and reaction have comparable effects.

As explained in Section III,A,4, a perfect knowledge of mass-transfer data theoretically necessitates simultaneous measurements of mass-transfer parameters by two methods: local physical measurement of a, k_L, and $k_L a$; and global chemical or physical measurement of a, k_L, and $k_L a$ to follow the effect of point-to-point variations. It is important to note that adequate techniques now exist, and the coming years should provide the data needed to clarify the picture.

IV. Mass-Transfer Coefficients and Interfacial Areas in Absorber. Scale-up

The choice of a suitable reactor for gas–liquid reaction or absorption is very often a question of matching the reaction kinetics with the capabilities of the proposed reactor. The specific interfacial area a, liquid holdup β, and mass-transfer coefficients k_L and k_G (or $k_L a$ and $k_L a$) are the most significant characteristics of a reactor. A synthesis of published values of the mass-transfer parameters will be given in this section.

Our objective here is to try to answer the following questions: For a proposed type of gas–liquid contactor compatible with the properties and flow rates of the phases and with the reaction type, what are the likely values of the specific interfacial area and the gas and liquid mass-transfer coefficients by which the contact performance can be predicted? And what is the expected accuracy of these values? Table XVIII gives typical values of these parameters in typical contactors shown in Fig. 12 for fluids with properties not very different from those of air and water (especially, liquid viscosity under 5 cP where the liquid is nonfoaming). Because this review is especially concerned with the chemical method of determining these parameters, experimental data obtained by this method will be given in subsequent tables and figures.

A. Packed Columns

Packed columns are used conventionally to obtain a low pressure drop or low liquid holdup when there is practically no heat to remove or supply or when the gas or the liquid is corrosive. They are not used when solids are present in the feed or are formed in the reaction. Although packed columns or reactors can be operated cocurrently, their operation is usually countercurrent. In particular, countercurrent use is preferred when a higher concentration driving force is needed, that is, for distillation or for most physical absorption. However, when irreversible reaction occurs between dissolved gases and the absorbent, the mean concentration driving force is the same for both modes of operation. In this case the capacity of cocurrent columns is not limited by flooding, and at any given flow rates

TABLE XVIII
Mass-Transfer Coefficients and Effective Interfacial Areas in Gas–Liquid Reactors

Type of reactor	β (% gas–liquid volume)	k_G (gm moles/cm² sec atm) × 10^4	k_L (cm/sec) × 10^2	a (cm²/cm³ reactor)	$k_L a$ (sec^{-1}) × 10^2
Packed columns					
Countercurrent	2–25	0.03–2	0.4–2	0.1–3.5	0.04–7
Cocurrent	2–95	0.1–3	0.4–6	0.1–17	0.04–102
Plate columns					
Bubble cap	10–95	0.5–2	1–5	1–4	1–20
Sieve plates	10–95	0.5–6	1–20	1–2	1–40
Bubble columns	60–98	0.5–2	1–4	0.5–6	0.5–24
Packed bubble columns	60–98	0.5–2	1–4	0.5–3	0.5–12
Tube reactors					
Horizontal and coiled	5–95	0.5–4	1–10	0.5–7	0.5–70
Vertical	5–95	0.5–8	2–5	1–20	2–100
Spray columns	2–20	0.5–2	0.7–1.5	0.1–1	0.07–1.5
Mechanically agitated bubble reactors	20–95	—	0.3–4	1–20	0.3–80
Submerged and plunging jet	94–99	—	0.15–0.5	0.2–1.2	0.03–0.6
Hydrocyclone	70–93	—	10–30	0.2–0.5	2–15
Ejector reactor	—	—	—	1–20	—
Venturi	5–30	2–10	5–10	1.6–25	8–25

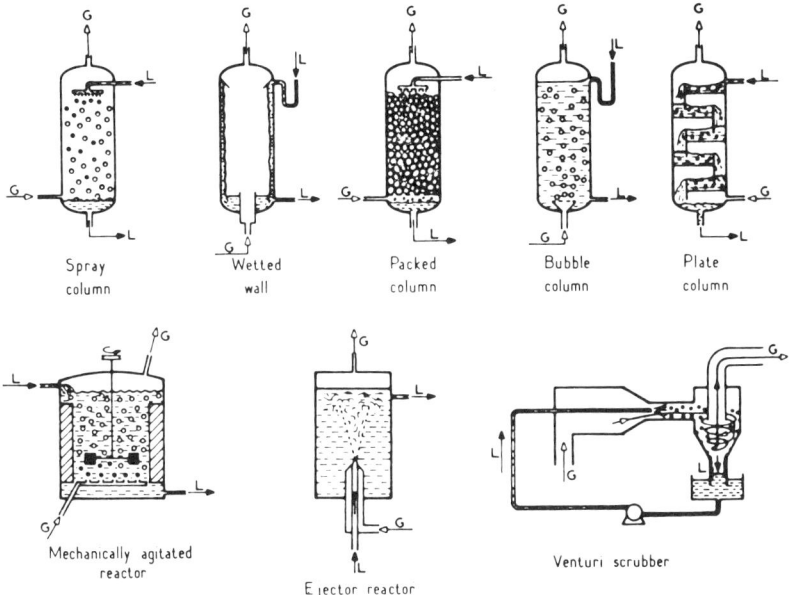

FIG. 12. Principal types of industrial equipment.

of gas and liquid the pressure drop in a cocurrent column is less. Also, in some three-phase reactors, with packing serving as a catalyst, it is advantageous to use cocurrent operation (S9, C10).

1. *Countercurrent Packed Columns*

A great number of published values of interfacial area per unit packed volume a and true liquid-side mass-transfer coefficient k_L for different packings have been compiled in Fig. 13. Also shown are the variations in a and k_L with superficial liquid velocity u_L. The compilation covers columns operating in the trickle-flow regime, that is, with gas and liquid rates insufficient to reach the loading state. For the chemical gas–liquid systems covered (mostly aqueous solutions of electrolytes), the interfacial area is independent of the chemical system to within ±20% (J5, V7, D6; (curves E, L, N). The interfacial area depends on the type and size of the packing and is independent of the column height when reported per unit packed volume (D6). For any type of packing, a decreases when the size of the packing increases. In large-scale columns, Eckert (E1) has suggested that 2-in. packings are the most economical. The interfacial area is independent of the gas superficial velocity u_G when the column operates below the loading condition (D14, B13). Also, the interfacial area depends on the column diameter (curves D, E, J, K). For a given value of packing diameter, Pall rings give the highest interfacial area.

For a given packing shape, plastic materials often offer the smallest interfacial area. However, plastics are now much used as material for packings, their main advantages being low specific weight, resistance to breakage, easy shaping, low cost, corrosion resistance, and temperature stability to as high as 130°C. Their main drawback is their contact behavior with the process liquid; they may withstand the liquid phase but not be wetted by it, or they may be well wetted and also be swollen and softened by it. To overcome these drawbacks, an interesting study has been reported by Linek *et al.* (L20), using 15-mm Raschig rings made of ceramic, polyethylene, and polypropylene. It was found that the effective interfacial area of plastic packings was about 40% of that of geometrically similar ceramic packing. The area of the plastic packing was then increased about 2.5 times (Fig. 14) by chemical oxidation [with H_2SO_4, $K_2Cr_2O_7$, and $Cr_2(SO_4)_3$ at about 90°C] to form a hydrophilic layer. Such surface treatment is inexpensive but provides significant improvement in mass transfer.

When the results of Fig. 13 do not apply, a suitable equation may be used. Among several formulas proposed for correlating interfacial areas (O2, J1, M17), the most general one is probably the relationship of Onda *et al.* (O2)

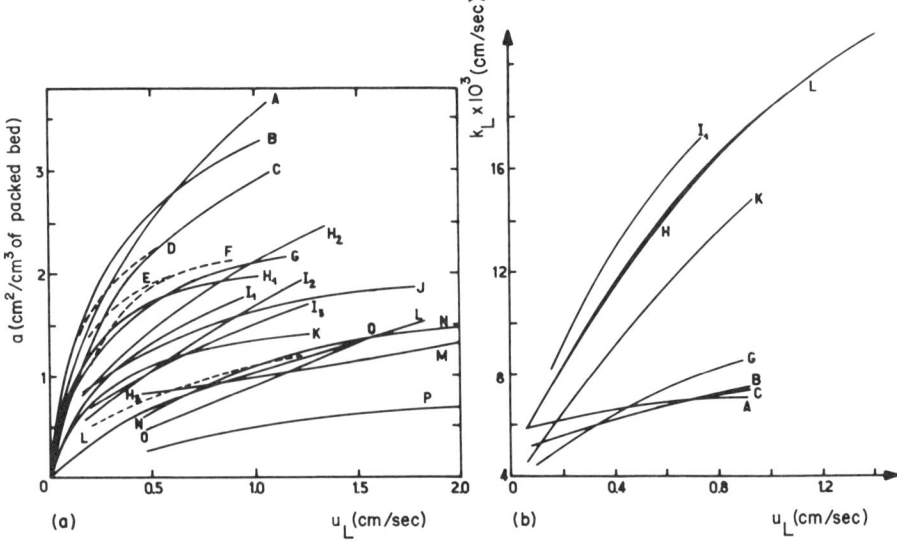

FIG. 13. Interfacial areas (a) and true liquid-side mass-transfer coefficients (b) in countercurrent packed columns. (See tabulation on p. 71.)

TABULATION FOR FIGURE 13

Curve	Packing Type	Nominal diam. (in.)	Temp. (°C)	Column diam. D	No. of particules by unit of packed vol.	Packing surface area (cm^{-1})	Chemical systems
A	Ceramic Intalox saddles	1/2	25	4 in.	640,000	4.7	Absorption CO_2
B	Ceramic Pall rings	1/2	25	4 in.	360,000	4.2	Absorption CO_2
C	Steel Pall rings	5/8	25	6 in.	220,000	3.5	Absorption CO_2
D	Ceramic Raschig rings	3/8	30	10 and 20 cm	1,070,000		CO_2–NaOH
E	Ceramic Raschig rings	3/8	30	4.37 cm	980,000		9 different systems
F	Ceramic Intalox saddles	1/2	30	10 cm	385,000		CO_2–NaOH
G	Ceramic Raschig rings	1/2	25	4 in.	370,000	3.8	CO_2–HNaCO$_3$–Na$_2$CO$_3$–H$_2$AsO$_2$
H$_1$	Ceramic Pall rings	1	25	9 in.	49,000	2.2	Absorption CO_2
H$_2$	Inox Steel Pall rings	1	25	20 cm	49,000	2.0	CO_2–NaOH
							Air–dithionite
H$_3$	Polypropylene Pall rings	1	25	20 cm	51,000	2.0	CO_2–NaOH
I$_1$	Ceramic Intalox saddles	1	25	9 in.	84,000	2.5	Absorption CO_2
I$_2$	Ceramic Intalox saddles	1	25	20 cm	75,300	2.5	CO_2–NaOH
							CO_2–DEA
							Air–dithionite
I$_3$	Polypropylene Intalox saddles	1	25	20 cm	53,500	2.0	CO_2–NaOH
J	Ceramic Raschig rings	1	25	12 in.		2.03	O_2–Na$_2$SO$_3$–Co^{2+}
K	Ceramic Raschig rings	1	25	9 in.	48,000	1.8	Absorption CO_2
	Ceramic Raschig rings	1	25	20 cm	50,600	1.9	Air + dithionite
	PVC Raschig rings	1	25	20 cm	51,400	1.9	Air + dithionite
L	Ceramic Raschig rings	1½	20	18 in.	14,000	1.3	CO_2–NaOH–CO_2–HKCO$_3$–K$_2$CO$_3$–ClO$^-$
M	Ceramic Intalox saddles	1½	11–34	0.5 m	21,000	1.6	O_2–Na$_2$SO$_3$–Co^{2+}
N	Ceramic Raschig rings	1½	11–34	0.5 m	13,000	1.3	O_2–Na$_2$SO$_3$–Co^{2+}
O	Polypropylene Pall rings	1½	11–34	0.5 m	15,500	1.3	O_2–Na$_2$SO$_3$–Co^{2+}
P	Polypropylene Intalox saddles	2	11–34	0.5 m	6200	1.1	O_2–Na$_2$SO$_3$–Co^{2+}

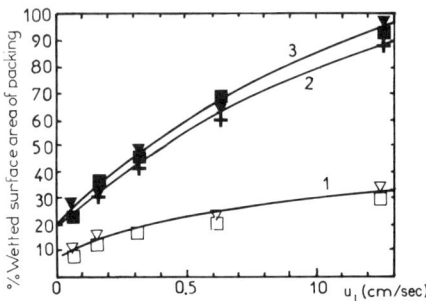

FIG. 14. Wetted surface area of packing as a function of superficial liquid velocity (L20): (1) plastic; (2) ceramic; (3) plastic coated with a hydrophilic layer (corresponding to solid symbols).

$$\frac{a}{a_c} = \frac{a_W}{a_c} = 1 - \exp\left[-1.45\left(\frac{\sigma_c}{\sigma}\right)^{0.75}\left(\frac{L}{a_c\mu_L}\right)^{0.1}\left(\frac{L^2 a_c}{\rho_L^2 g}\right)^{-0.05}\left(\frac{L^2}{\rho_L \sigma a_c}\right)^{0.2}\right] \quad (100)$$

where a_W is the wetted area per unit of packed volume, a_g is the specific surface area of the packing, a_c is the total dry area of the packing per unit of packed volume [$a_c = a_g(1 - \epsilon)$], L is the mass superficial liquid flow rate, ϵ is the void fraction, ρ_L and μ_L are the density and viscosity of liquid, σ is the surface tension of liquid and, σ_c is the critical surface tension of liquid for a particular packing material. This equation correlates results within a maximum uncertainty of ±20%, except in the case of Pall rings where it is conservative, the reason being that here the match of interfacial area with wetted area does not fit. The construction of the Pall rings disperses part of the liquid as small droplets not taken into account, and this effect may double the values of a_W.

An alternate correlation has been proposed by Puranik and Vogelpohl (P16) for the values of effective interfacial area a_V (during vaporization), a_{AC} (absorption with reaction), a_{AP} (without reaction), and corresponding values of the wetted surface area a_W. The effective interfacial area is then divided into static and dynamic areas ($a_{st} + a_{dyn}$) for vaporization and absorption with chemical reaction. For absorption without reaction, the effective interfacial area is only the dynamic area. Thus

$$a/a_c = (a_{st} + a_{dyn})/a_c = a_{AC}/a_c = a_W/a_c = a_V/a_c$$

$$a/a_c = 1.05(L/a_c\mu_L)^{0.041}(L^2/\rho_L\sigma a_c)^{0.133}(\sigma/\sigma_c)^{-0.182} \quad (101)$$

$$a_{st}/a_c = (a - a_{AP})/a_c = 0.229 - 0.091 \ln(\rho_L g/\sigma a_c^2) \quad (102)$$

Equations (101) and (102) have been used to correlate a large body of experimental data, again within a maximum error of ±20%. The range of

TABLE XIX

RANGE OF VARIABLES AND PHYSICAL PROPERTIES FOR EQS. (100)–(102)

Variable	Range
$L = \rho_L u_L$	0.25–12 kg/m sec
μ_L	0.5–13 cP
σ	25–75 dyn/cm
ρ_L	800–1900 kg/m³
d	10–37.5 mm
σ/σ_c	0.3–1.3
a/a_c	0.08–0.8

variables and the physical properties covered by Eqs. (100)–(102) are given in Table XIX.

The true liquid-side mass-transfer coefficients k_L all lie between 4×10^{-3} and 2×10^{-2} cm/sec (Fig. 13) and may be considered independent of u_G. The most representative relationship for the liquid-side mass-transfer coefficient, accurate within a range of $\pm 20\%$, is probably the one contributed by Mohunta et al. (M17):

$$k_L a = 25 \times 10^{-4} \left(\frac{g\rho_L}{a_c \mu_L}\right)^{0.66} \left(\frac{g^2 \rho_L}{\mu_L}\right)^{1/9} \left(\frac{\mu_L L^3 a_c^3}{g^2 \rho_L^4}\right)^{0.25} \left(\frac{\mu_L}{\rho_L D_L}\right)^{-0.5} \quad (\text{sec}^{-1}) \tag{103}$$

The range of variables and physical properties tested with this equation is given in Table XX.

Most of the reported mass-transfer data are confined to aqueous systems. It is interesting to underline the recent work of Sridharan and

TABLE XX

RANGE OF VARIABLES AND PHYSICAL PROPERTIES FOR EQ. (103)

Variable	Range
L	0.1–42 kg/m² sec
G	0.015–1.22 kg/m² sec
μ_L	0.7–1.5 cP
$\mu_L/\rho_L D_L$	140–1030
d	6–50 mm
D	6–50 cm

Sharma (S35), including the use of organic solvents (such as toluene, xylene, diethylene glycol, and polyethylene glycol) for the measurement of a and k_L by the chemical method. In each case, the reaction between CO_2 and selected amines is employed to determine a. For example, values of interfacial areas obtained in a reaction of CO_2 with cyclohexylamine in xylene plus 10% isopropanol, in a 10-cm-i.d. column packed with 0.5 in. ceramic Raschig rings, are reported in Fig. 15. A comparison with the values for aqueous systems shows a 50% improvement attributable to the lower surface tension of xylene ($\sigma_L = 26.8$ dyn/cm).

Data on $k_L a$ and a obtained by absorption of CO_2 and of propylene in cuprous amine solutions (containing complexed NH_3) are reported in the same figure (V7).

Finally, the true gas-side mass-transfer k_G values as measured by the chemical method (V7, M10) may be correlated with the use of dimensional analysis (D2), within a range of $\pm 30\%$, by

$$k_G P/G = (C/M)(a_c d)^{-1.7}(Gd/\mu_G)^{-0.3}(\mu_G/\rho_G D_G)^{-0.5} \quad (104)$$

where P is the total pressure in atm, M is the gas molecular weight in gm/mole, G is the mass superficial gas flow rate in gm/cm² sec, D_G is the solute gas diffusivity, $C = 2.3$ for $d < 1.5$ cm and $C = 5.23$ for $d > 1.5$ cm, and ρ_G and μ_G are the density and viscosity of gas. Additional results

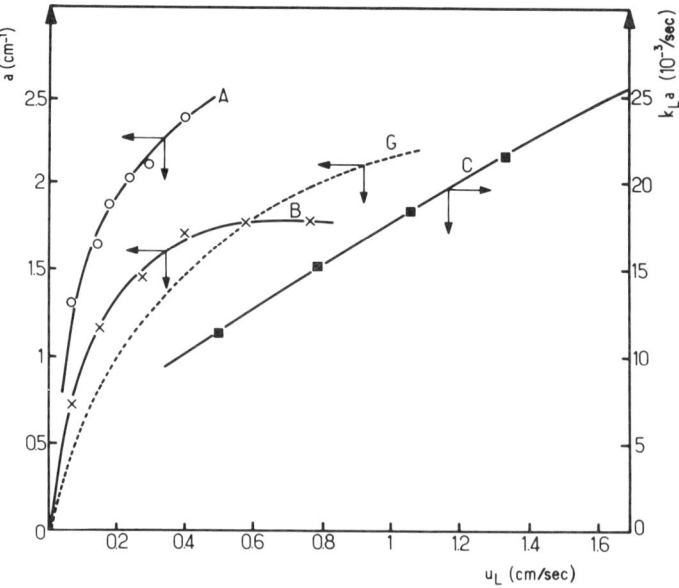

FIG. 15. Mass-transfer data for CO_2 in an amine, alkali, or NH_3 solution (S35).

TABULATION FOR FIGURE 15

Curve	Packing d (in.)	Packing ε	Column D (cm)	Column Z (cm)	Gas–liquid system	u_G (cm/sec)	Ref.
A	½	0.52	10	61	CO_2–cyclohexylamine in xylene + 10% isopropanol	40	S35
G	½		10		CO_2–aqueous $NaHCO_3$ + Na_2CO_3 with H_2AsO_3 as a catalyst		D4
B	⅜	0.64	5	27	CO_2–aqueous cuprous amine–NH_3 solution	25	S35
C	⅜	0.64	5	27	propylene—aqueous cuprous amine–NH_3 solution	4	S35

complementary to the above have been developed by Fellinger and are included in Perry's handbook (P5, pp. 18–41).

2. *Packed Bubble Columns (Liquid Flow Negligible)*

As shown in Section IV,C, bubble columns are frequently operated as absorbers and reactors because of their low cost, simplicity of operation, and the ease with which the liquid residence time can be varied. However, they involve the disadvantage of severe gas back-mixing and bubble coalescence phenomena, which can be reduced substantially by packing the column. Packed bubble columns show 15–100% improvement in effective interfacial area a_d and liquid-side mass-transfer coefficients $(k_L a)_d$, based on void volume, over those for empty bubble columns under otherwise similar conditions (M8, M9). However, when mass-transfer results are based on total column volume (that is, the actual design volume), the improvement over an empty bubble column is less (at times the interfacial area in a packed bubble column is actually smaller), because of the substantial part of the column volume occupied by solid. To avoid this effect while keeping the advantages of packed bubble columns, it is desirable to use packings with high porosities, such as screen packings (V13, C14, A1) for which the interfacial areas may be two to four times as great as in unpacked vessels.

Sahay and Sharma (S2) have reported a detailed study for 10- to 38-cm-i.d. columns with packings of different sizes and shape (ceramic, plastic, metal). The sparger design for introduction of gas is unimportant in the range of superficial gas velocities covered, 10–25 cm/sec. Also, the packing type and size have no effect on a_d and give a maximum variation of about 50% in $(k_L a)_d$ for any given packing size.

Both a_d and $(k_L a)_d$ vary as the 0.5 power of the superficial gas velocity (Fig. 16), which indicates that k_L is independent of u_G. In Fig. 16, one

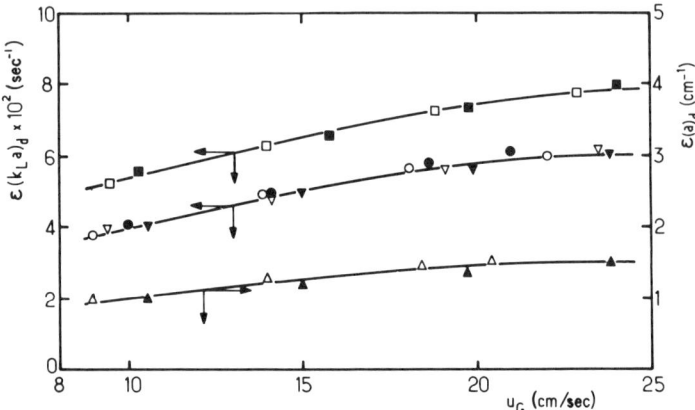

FIG. 16. Scale-up of packed bubble column (S2). Solid symbol indicates the longer column in each pair. Lower curve, areas: air–dithionite + TCP, 1-in. ceramic Intalox saddles, column heights 20 cm ($\epsilon = 0.75$) and 38.5 cm ($\epsilon = 0.77$). Middle curve, $k_L a$: CO_2–Na_2CO_3 + $NaHCO_3$ + TCP, 1-in. stainless steel Pall rings, heights 10 cm ($\epsilon = 0.84$) and 20 cm ($\epsilon = 0.90$) also air–CuCl + NaCl + HCl + TCP, 1-in. ceramic Raschig rings heights 20 cm ($\epsilon = 0.73$) and 38.5 cm ($\epsilon = 0.79$). Upper curve, $k_L a$: air–CuCl + HCl + TCP, 1-in. ceramic Intalox saddles, heights 20 cm ($\epsilon = 0.77$) and 38.5 cm ($\epsilon = 0.73$).

curve applies to tricresyl phosphate, (TCP) employed as an antifoaming agent. This behavior has been confirmed by Pexidr and Charpentier (P6) for low values of u_G (<5 cm/sec) and for a greater range of liquid velocity (up to 1.75 cm/sec) in a 10-cm-i.d. column packed with 6.4- or 10-mm Raschig rings.

3. Cocurrent Packed Columns

As explained above, cocurrent gas–liquid flow in packed beds, packing being either catalytic or inert, is advantageously employed in the petroleum and chemical industries. Successful modeling of mass transfer in packed-bed reactors requires careful study of the three-phase hydrodynamics—fluid flow patterns, pressure drops, and liquid holdup.

Because cocurrent flow is not bounded by the phenomenon of flooding, it offers a greater range of hydrodynamic patterns, which must be specified before considering the mass-transfer behavior.

a. *Hydrodynamics.* In the case of cocurrent downflow, the packed-bed reactor works in two main regimes: first, the *trickle-flow regime* in which for initially a zero gas rate the liquid phase trickles over the packing in a network of films, rivulets, and drops adjacent to a stagnant continuous gas phase; and second, the *single-phase liquid regime* in which for initially a zero gas rate the liquid phase fills the packing voids. When the liquid mass flow rate L is kept constant and the mass gas flow rate G is started and increased, the following flow patterns are observed.

The initial trickle flow of liquid ($L < 20$ kg/m² sec) through beds of spheres, beads, and pellets, as the gas flow increases, gives way to an alternate gas-rich or liquid-rich flow downward through the column (pulse flow) and finally to a turbulent state which appears to involve a continuous gas phase with part of the liquid suspended as a mist and with the other part covering the packing as a film (spray flow). For small values of L ($L < 2$–5 kg/m² sec), there is not enough liquid to wet the whole packing surface, and pulse flow caused by the liquid obstructing the gas flow is not encountered. When the liquid phase foams during gas flow, two new flow regimes arise, foaming flow and foaming, pulsing flow, which appear between trickling flow and pulsing flow.

Flow patterns, and transitions from one pattern to another as the flow rates change, have been described by several authors. This behavior is summarized in Fig. 17 for various gases, with water and different foaming

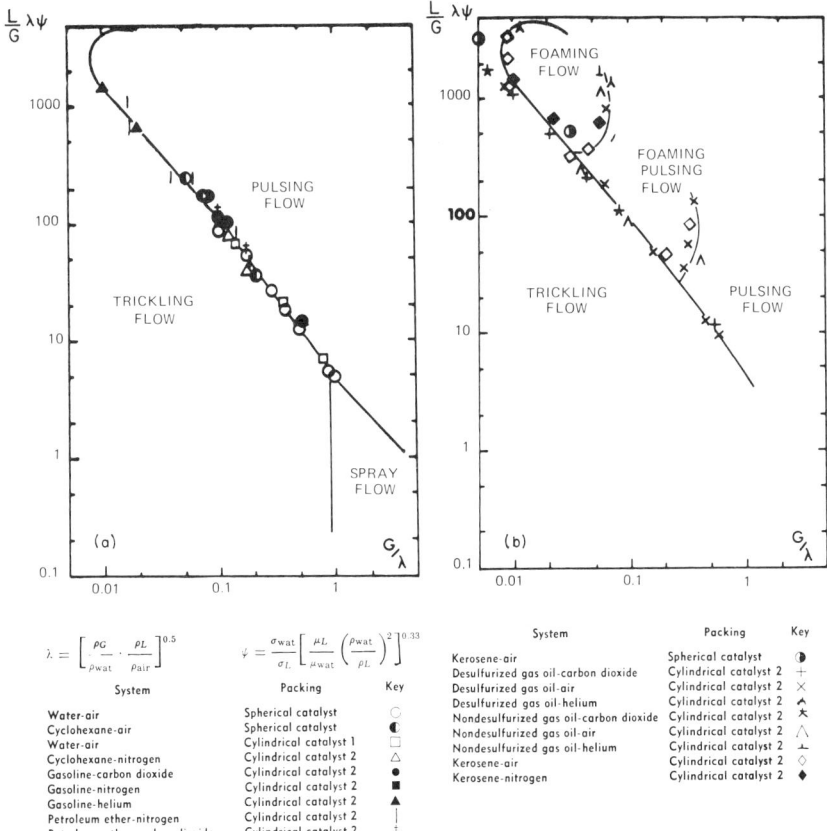

FIG. 17. Flow pattern diagrams in gas–liquid downward flow in packed beds (C12). (a) Nonfoamable liquid, (b) foamable liquid.

and nonfoaming hydrocarbons, in beds of 3-mm glass spheres, 3-mm alumina spheres, and 1.8 × 6 mm and 1.5 × 5 mm alumina pellets (C12). Diagrams are given in the literature (T11, B12, S6) for other shapes and sizes of packings, mostly confined to the water–air system.

Starting from a liquid-full system at constant liquid flow ($L > 20$ kg/m^2 sec downward, or at any L upward), the following patterns are observed as the gas flow increases: first, bubble flow, in which small bubbles appear unbroken in the continuous liquid phase; then, distorted bubble flow in which the bubbles begin to coalesce and to surround several packing elements; then pulse flow; and finally spray flow.

These different flow patterns provide several different gas–liquid geometric configurations. Hence the mass-transfer properties in downward cocurrent flow are related to gas and liquid energy dissipation rates and depend on pressure drop ΔP_{LG} and liquid holdup β. There are several correlations for predicting these parameters, the most famous being proposed by Larkins *et al.* (L1) for cocurrent bubble-packed reactors ($\pm 20\%$):

$$\log[\Delta P_{LG}/(\Delta P_L + \Delta P_G)] = 0.416/[(\log X)^2 + 0.666] \quad (105)$$

$$\log \beta = -0.774 + 0.525 \log X - 0.109 (\log X)^2 \quad (105a)$$

where $X = (\Delta P_L/\Delta P_G)^{1/2}$, for values of X between 0.05 and 30. ΔP_L and ΔP_G are the pressure drops that would exist if the liquid and the gas were to flow separately at the same rates as those in two-phase flow. For design, values for ΔP_L and ΔP_G may be obtained by an Ergun-type relationship:

$$\frac{f}{2} = \frac{\Delta P}{Z} \frac{\epsilon^3}{\rho u^2 a_g (1-\epsilon)} = h_K \frac{\mu a_g (1-\epsilon)}{\rho u} + h_B \quad (106)$$

where a_g is the specific surface area of the packing, with different values of h_K and h_B for different packings (C13, S7, R6). However, a quick experimental measurement of h_K and h_B for the packing to be used will lead to better values (C12). Note that Larkin's relationship does not apply to trickling liquid.

For air–water flows through beds of spheres, Sato *et al.* (S7) proposed two related expressions valid over nearly the same range of X and intended to include the trickle-flow regime:

$$\log \frac{\Delta P_{LG}}{\Delta P_L + \Delta P_G} = \frac{0.70}{[\log(X/1.2)]^2 + 1.00} \quad (107)$$

and

$$(\Delta P_{LG}/\Delta P_L)^{0.5} = 1.30 + 1.85 X^{0.85} \quad (108)$$

Charpentier *et al.* (C13) have compared the pressure drop ΔP_{LG} in two-phase flow with that in single-phase flow of each phase (trickle flow or single-phase flow of liquid when the gas rate is zero, and single-phase flow of gas with no liquid in the packed bed). This comparison was made in terms of a nondimensional energy ratio, accurate only to ±50%, and the relationship proposed is only valid for foaming, pulsing, and spray flow. A more complete review of pressure drop and liquid holdup correlations for gas–liquid downflow, with foaming and nonfoaming liquids, has been published by Midoux et al. (M15).

For a quick first approximation in the case of trickle flow, pressure loss may be expressed by an Ergun-type relationship in which the porosity ϵ is replaced by $\epsilon(1 - \beta)$, the total liquid extraparticle holdup β being calculated either by the correlation proposed by Sato et al. for spheres (S7) of diameter 2 mm or larger:

$$\beta = 0.42[a'_g(1 - \epsilon)]^{0.33} X^{0.22} \tag{109}$$

with a'_g in mm^{-1}; or by the correlation proposed by Charpentier et al. (B1, C12) for Raschig rings, spheres, and pellets with diameters of more than 2 mm:

$$\log \beta = P + Q \log X' + R(\log X')^2 \tag{110}$$

with

$$0.05 < X' = (L/G)^{1/2}[(\rho_G g_c)^{-1}(\Delta P_G/Z) + 1]^{-1/2} < 100$$

ΔP_G is in N/m². Also, $P = -0.570$, $Q = 0.165$, $R = -0.095$ for Raschig rings; $P = -0.280$, $Q = 0.175$, $R = -0.047$ for spherical packings; and $P = -0.363$, $Q = 0.168$, $R = -0.043$ for pellets. These equations underestimate β for particles below 2 mm in diameter.

Two points should be emphasized: First, the previous relationships for pressure drop and holdup are not valid for saddles and Pall rings, but data for these packings in specific situations are found in the literature (R6, D16, W2). Second, as a rough approximation, the equations for downward flow in a packed bubble reactor are still valid for upward flow. [The upflow data for pellets and spheres from Turpin and Huntington (T11) and Sato et al. (S8) will be useful for specific cases.]

b. *Gas–Liquid Mass Transfer.* Numerous published studies are reported in Table XXI.

The liquid-phase mass-transfer coefficient $k_L a$ is affected by both gas and liquid rates. The following correlation is recommended by Reiss (R6) and Satterfield (S9) for downflow systems:

$$k_L a = 0.0173 \, E_L^{0.5} [D_A/(2.4 \times 10^{-9})]^{0.5} \tag{111}$$

where E_L is an energy dissipation term in W/m³ for liquid flow evaluated as $u_L \Delta P_{LG}/Z$, u_L is the superficial liquid velocity in m/sec, and D_A is the diffusivity of solute gas in the liquid in m²/sec. If the viscosity of the liquid differs much from that of water, a complementary correction should be

TABLE XXI
Experimental Data Available in Literature for Cocurrent Gas–Liquid Downward Flow in Packed Bed

Reference	Packing d(mm)	Packing ε	Column D(cm)	Column Z(m)	Gas	Solute	Liquid	Superficial velocity (10^{-2} m/sec) u_L	Superficial velocity (10^{-2} m/sec) u_G	$\frac{\Delta P}{Z}$	β	$k_L a$ (sec^{-1})	Relationships (MKS units)
Reiss (1967)	12.5^a 25^a 76^a 25^b	0.72–0.78 0.79–0.88 0.79 0.68	7.5–10 41	0.5	Air Air	NH_3 O_2	H_2O H_2O	4–30	60–450	+	+	0.1–3	$k_G a = 2.0 + 0.1 E_G^{0.66}$ $k_L a = 0.0143 E_L^{0.5}$ $E_G = \left(\frac{\Delta P}{Z}\right)_{LG} u_G, E_L = \left(\frac{\Delta P}{Z}\right)_{LG} u_L$
Gianetto et al. (1970, 1973)	6^a 6^b 6^a 6^a	0.41 0.59 0.50 0.71	8	0.2 0.4 0.6	Air Air Air	CO_2 NH_3 O_2	NaOH (0.5–2 N) Na_2SO_4 (1 N) NaOH (2 N)	0.08–4.4 0.25–4.3 0.25–4.3	40–230 40–220 40–220	+	–	0.011–1.6 0.009–0.9 0.012–1.2 0.007–0.6	$\dfrac{a}{a_x(1-\epsilon)} = 0.27\left[\left(\dfrac{\Delta P}{Z}\right)_{LG}\dfrac{\epsilon}{a_x(1-\epsilon)}\right]^{0.5}$ $\dfrac{k_G \epsilon}{u_G} = 0.035\left[\dfrac{(\Delta P/Z)_{LG}}{\Psi(\rho_G u_G^2 + \rho_L u_L^2)}\right]$ $\dfrac{k_L \epsilon}{u_L} = 0.03\left\{\left[\dfrac{\epsilon(\Delta P/Z)_{LG}}{a_x(1-\epsilon)\rho_L u_L^2}\right]^{0.07} - 1\right\}$
Nagel et al. (1972)	5^a 5^a	0.41 0.64	11	3.2	Air Air	CO_2	Na_2SO_3	0.3–3 0.3–3	1.5–33 1.5–60	– +	– –	—	Experimental curves: $\dfrac{a}{a_x(1-\epsilon)}$ versus u_G for u_L = const
Ufford and Perona (1973)	6.3^a 12.7^a 19^b	— — —	10	0.25 0.30	Air	CO_2	H_2O	0.4–8.3	4.3–12	–	–	0.006–0.13	$k_L a = 13.5 a_t u_L^{0.75}$
Sato et al. (1974)	2.6^c 5.6^c	0.37 0.39											$k_L a = 2.96 a_t^{0.93} u_L^{0.42}$ $k_L a = 2.73 a_t^{0.82} u_L^{0.46}$ $\dfrac{k_L a}{\epsilon} = 0.074\left(\dfrac{u_L}{\epsilon}\right)^{0.56} E_L^{0.31}$

Size	ε		Gas	Solute	Liquid	u_L	u_G		Range	Correlation
8^c	0.41		N_2	O_2	H_2O	1–20	5–100	+	0.02–3	$\frac{a}{\epsilon} \propto (1-\beta)E_L^{1/4}$
12^c			N_2	CO_2	NaOH (0.1–1 N)	1–10	1–100	+		$E_L = \left(\frac{\Delta P}{Z}\right)_{LG} \left(\frac{u_L}{\epsilon \beta}\right)$

Shende and Sharma (1974)

Size	ε		Gas	Solute	Liquid	u_L	u_G		Range	Correlation
5^c	—		Air	CO_2	NaOH					Experimental curves:
10^a	—	0.84								a versus u_G and u_L
12.5^c	—									
16^c	—	0.90	Air	O_2	$S_2O_4Na_2$	0.1–3.5	50–300	—		$k_G a = A u_G^m u_L^n$
16^d	0.93		N_2	O_2	$S_2O_4Na_2$				—	$k_G a$ in gm mole/cm³ sec⁻¹ atm⁻¹
		1.21								

	$10^3 A$	m	n
25.4^b	1.64	0.64	0.38
25.4^d	1.57	0.65	0.40
16^d	0.61	0.87	0.34

Size	ε		Gas	Solute	Liquid	u_L	u_G		Range	Correlation
25.4^a	0.74									
25.4^b	0.70–0.92									
25.4^d	0.90–0.94	20	Air	SO_2	NaOH					$\frac{k_L a}{D_A} = 4440 \left(\frac{\rho_L u_L}{\mu_L}\right)^{0.40} \left(\frac{\mu_L}{\rho_L D_L}\right)^{0.5}$

Goto and Smith (1975)

Size	ε		Gas	Solute	Liquid	u_L	u_G		Range
4.1^c	0.37	0.15	O_2	O_2	H_2O	0.05–0.5	0–0.8	+	0.002–0.00
2.9^c	0.44		N_2	O_2	H_2O				0.005–0.01

Ruether et al. (1975)

Size	ε		Gas	Solute	Liquid	u_L	u_G		
6.3^c	0.45	0.76	Air	—	H_2O	1.2–4	27–55	—	—

$k_L a / D_A = 9080 \rho_L u_L / \mu_L)^{0.41} (\mu_L/\rho_L D_L)^{0.5}$

$k_L a \left(\frac{\mu_L}{\rho_L D_L}\right)^{0.5}$ vs E_L for pulsing flow only

---

^a Raschig rings.
^b Saddles.
^c Spheres.
^d Pall rings.
^e Pellets.

applied (see Section II,C). Reiss's relationship is based mainly on pulse and spray flow ($E_L > 80$ W/m³).

For lower gas and liquid rates, corresponding to trickle flow of liquid, the reported values of $k_L a$ (varying from 0.01 to 0.1 sec⁻¹) are smaller than predicted by Eq. (111) (U2, G13). Comparison of the different experimental data has led Charpentier (C10) to propose as a first approximation (Fig. 18).

$$k_L a = 0.0011 E_L [D_A/(2.4 \times 10^{-9})] \qquad (112)$$

for $5 < E_L < 100$ W/m³, with a complementary correction for viscosity when needed. For smaller values of u_L and u_G, $\Delta P/Z$ is only a few centimeters of water per meter of packing, and the energy dissipation term is irrelevant. In this case, a mean value of 0.008 sec⁻¹ applies for packing with $d > 2$ mm. For more precise results, see the experimental data of Goto et al. (G13, G14).

Finally, at very high flow rates of both gas and liquid, $k_L a$ may exceed 1 sec⁻¹, a level difficult to attain in other types of gas–liquid contactors such as bubble columns or agitated vessels. In contrast to this, when the liquid is trickling over the packing, $k_L a$ values are of the same magnitude as those obtained countercurrently under the same working conditions.

The gas-phase mass-transfer coefficient $k_G a$ is also affected by both gas and liquid flow rates. Extensive results have been reported by Reiss (R6) for 12.5-, 25-, and 76-mm polyethylene Raschig rings and 25-mm Intalox saddles; by Gianetto et al. (G8) for 6-mm spheres, Berl saddles, and glass and ceramic Raschig rings; and by Shende and Sharma (S24) for 25-mm

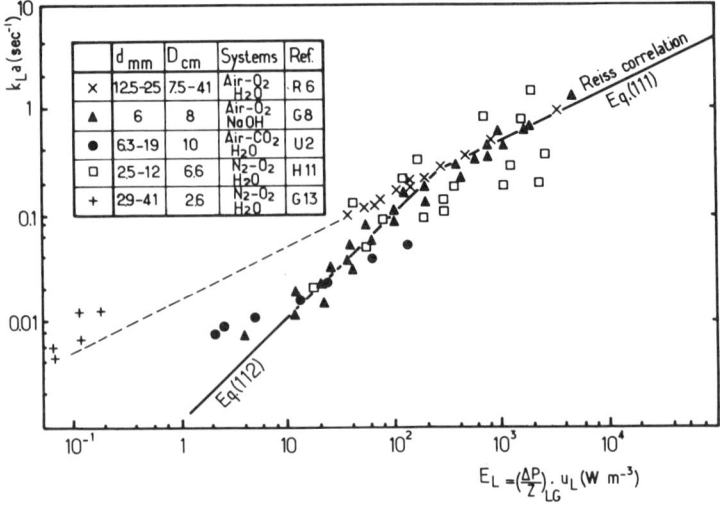

FIG. 18. Correlations for $k_L a$ values for cocurrent downward flow in packed beds (C10).

ceramic Intalox saddles and polypropylene Pall rings and 16-mm stainless-steel Pall rings. In these different studies, the value of $k_G a$ vary between 2 and 70 sec^{-1} for liquid rates ranging from 0.5 to 30 cm/sec and gas rates from 40 to 450 cm/sec. However, discrepancies exist in representing the results. For packings not studied by the previous workers, Reiss's relationship will be a first approximation for any flow pattern except for trickle flow of liquid:

$$k_G a = 2.0 + 0.1 E_G^{0.66} \tag{112a}$$

where E_G is an energy dissipation term in W/m^3 for gas flow evaluated as $u_G \, \Delta P_{LG}/Z$. As in the trickle-flow regime, the $k_G a$ values for concurrent flow are higher than those for countercurrent flow, when comparable; therefore use of the correlations for countercurrent flow will give conservative results.

The effective interfacial area a increases with either gas or liquid superficial velocity, in contrast to countercurrent operation where effective area up to the loading point is independent of superficial gas velocity. The values of effective area for different packings vary substantially. As shown by Shende and Sharma (S24), polypropylene Pall rings perform better than polypropylene Intalox saddles, and stainless steel Pall rings for each given size of packing are still better. The areas for cocurrent flow are much larger than with countercurrent flow, and also much larger than the geometric packing area $a_g(1 - \epsilon)$ at high liquid and gas flow rates. For pulse and spray flow (G7) where much of the liquid phase is in droplet form, there exists a relationship between the area a and the prsssure gradient gravity (excluding gravity effects) for 6-mm packings:

$$a/a_g(1 - \epsilon) = 0.25\{(\Delta P/Z)_{LG}[\epsilon/a_g(1 - \epsilon)]\}^{0.5} \tag{113}$$

Unfortunately most other authors have not given pressure drop data to match their interfacial area data. To use flow pattern diagrams (C13, B12, S6, C12) and to calculate pressure drop from energy relationships (see Section IV,A,3,a) Charpentier (C10) has shown that Eq. (113) fits the experimental results for other packing shapes and sizes (Fig. 19) for both pulse flow and spray flow.

However, Eq. (113) fails completely for trickle flow of liquid. In this case, the data fit the equation

$$a/a_g(1 - \epsilon(= 0.25\{(\Delta P/Z)_{LG}[\epsilon/a_g(1 - \epsilon)]\}^{+.5} \tag{114}$$

for

$$(\Delta P/Z)_{LG}[\epsilon/a_g(1 - \epsilon)] < 12 \quad \text{N/m}^2$$

Equation (114) is proposed mainly for sphere and pellet packings with $\epsilon < 0.50$.

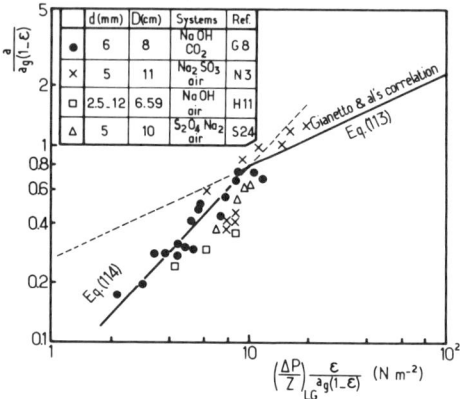

FIG. 19. Correlations for a values for cocurrent downward flow in packed beds (C10).

c. *Mass-Transfer Rates for Upward Cocurrent Flow.* A few mass-transfer results for upward cocurrent flow have been published recently (S1, S32, S34, B22). For 6-mm packings, Gianetto et al. (G8) have found that upflow values of $k_L a$ average 100% greater than downflow values, in the pulse-flow and spray-flow regimes given by Eq. (111), because gravitational forces lead to a higher liquid holdup and a higher pressure drop. In these regimes, interfacial areas may be three times as large as the geometric packing area. Both k_L and a increase with u_L and u_G and correlate with the pressure drop. Unfortunately it is difficult to generalize these results; pressure drop correlations, needed as input, depend upon knowledge of the flow regimes for which data are lacking. Moreover such correlations are not valid for the bubble flow regime, where experimental values of $k_L a$ lie between 0.08 and 0.4 sec^{-1} (S32, S34).

Nevertheless, as a first approximation for design, use of Eq. (111) is recommended with the coefficient doubled for a gas–liquid reaction in pulse or spray flow over inert packing. For the bubble flow regime, with $G > 0.01$ kg/m^2 sec, it is conservative to assume $k_L = 0.15$ sec^{-1} for any gas–liquid reaction. For the dependence of effective interfacial area, despite the lack of a general correlation, it is judicious to consider that this area will vary with the 0.5 power of superficial gas velocity regardless of packing size and type, column diameter, and liquid superficial velocity.

d. *Liquid-to-Solid Mass Transfer.* Where both the gas absorption and particle resistances are significant, the overall transfer coefficient should take into account the particle external mass-transfer coefficient $k_S a$. Four recent studies on solid–liquid mass transfer under two-phase downflow conditions are summarized in Table XXII. Sato et al. (S5) reported k_S values from the benzoic acid–water system for 5.5- and 12.2-mm-diameter solids in gas and liquid flow rate ranges corresponding to trickle, pulse, and dispersed bubble flow. At low gas flows ($G < 0.1$ kg/m² sec), the results seem to agree with an extrapolation of Van Krevelen and Krekel's correlation for filmlike flow:

$$\frac{k_S}{a_g(1-\epsilon)D_L} = 1.8\left[\frac{\rho_L u_L}{a_g(1-\epsilon)\mu_L}\right]^{0.5}\left(\frac{\mu_L}{\rho_L D_L}\right)^{0.33} \quad (115)$$

Actually Sato *et al.* expressed their particle mass-transfer coefficients in terms of an enhancement factor representing the ratio of k_S with two-phase flow to k_S at the same liquid flow rate in single-phase flow. For pulsing and dispersed bubble flow this enhancement factor was found to be inversely proportional to liquid holdup β, which in turn is a function of the two-phase parameter X or X' (see Section IV,A,3,a). For comparison, the data for single-phase liquid flow are best represented by an equation of the same form as Eq. (115) but with a constant of 0.8.

Values of k_S for 6.3-mm-diameter benzoic acid spheres in water, measured by use of a fluorescent dye (rhodamine B) introduced as a 1.0 wt % solution in molten benzoic acid before casting the spheres, have been reported by Ruether *et al.* (L9). Their results suggest that k_S is practically insensitive to radial position. The data, lying in the pulse flow regime, agree well with the Sato correlation. However, Ruether *et al.* have proposed an alternate representation with a term for power dissipation per unit mass of liquid holdup (see Table XXII) that requires the two-phase pressure drop and liquid holdup as inputs. The results of Sylvester and Pitayagulsarn (S40) for 3.2-mm pellets of benzoic acid, although higher than in the two previous studies by a factor between 4 and 12, seem still in reasonably good agreement because of different particle shape and size.

Goto and Smith (G13) measured mass transfer from particle to liquid for short beds of β-naphthol spheres and pellets. Runs were made with smaller particle diameters (0.54 and 2.41 mm) and smaller flow rates than in the three previous studies. Not all the external surface is effective for mass transfer in trickle beds containing very small packings, and the effect of particle size on $k_S a$ appears to change for particles smaller than about 2 mm. Goto and Smith proposed two equations specific for a particular size and shape of packing (Table XXII); these relationships are independent of the gas flow rate because that rate is so low.

To sum up, the particle external mass-transfer coefficients for particle diameters larger than 3 mm may be determined with the relationships proposed by Sato *et al.* or Ruether *et al.* for $G > 0.01$ kg/m² sec; that is, when the gas–liquid interaction is significant. For trickle flow, Eq. (115) with a constant of 0.8 will give reasonable values. For particle diameters smaller than 3 mm, the specific relations proposed by Goto (G13) should be used.

TABLE

REPORTED EXPERIMENTAL DATA ON LIQUID-TO-SOLID MASS TRANSFER

Reference	Packing		Column		Solute	Liquid
	d (10^{-3} m)	ϵ	D (10^{-2} m)	Z (m)		
Sato et al. (1972)	5.5[c]	0.41[c]	6.6	0.05	Benzoic acid	H_2O
	12.2[c]	0.45[c]	6.6	0.20	Benzoic acid	H_2O
Goto and Smith (1975)	0.54[d]	0.52[d]	2.6	0.020	β-naphtol	H_2O
	2.4[d]	0.48[d]	2.6	0.026	β-naphthol	H_2O
Ruether et al. (1975)	6.3[c]	0.45[c]	7.6	0.20	Benzoic acid with rhodamine B as dye	H_2O
Sylvester and Pitayagulsarm (1975)	3.2[e]	—	15	0.026	Benzoic acid	H_2O

[a] a is calculated by assuming that the particles have a spherical diameter d; $a = a_c = 6(1 - \epsilon)/d = a_g(1 - \epsilon)$.

[b] $Re = \rho_L u_L / a\mu_L$; $Sc = \mu_L / \rho_L D_L$.

Note that all these data concern experiments in the absence of reaction and in conditions where it can be assumed that all surfaces of the packing are wetted (dissolution of solid particles). However, when a reaction is occurring, mass transfer may be constrained to a region close to the pore openings on the outer surface of the particles and hence the effective surface area for mass transfer could be considerably less than the external surface of the catalyst, leading to lower values $k_s a$. More studies are needed on porous catalysts at conditions where external transport is important.

XXII

FOR COCURRENT GAS-LIQUID DOWNWARD FLOW IN PACKED BEDS

Superficial velocity (10^{-2} m/sec)		k_S (10^{-2} m/sec)	$k_S a$ (sec^{-1})^a	Relationships (international units)^b
u_L	u_G			
1–20	1–100	0.003–0.05 ($t = 15$–$27°C$)	0.02–0.16	$\dfrac{k_S}{aD_L} = \dfrac{1.05}{\beta} Re^{0.5} Sc^{0.33}$
1–20	1–100	0.002–0.03 ($t = 15$–$27°C$)	0.006–0.05	
0.05–0.5	0–0.8	0.008–0.04 ($t = 25°C$)	0.04–0.2	$\dfrac{k_S}{aD_L} = 69.9 \times 10^3 \times a^{-1.33} Re^{0.67} Sc^{0.33}$ for 2.4-mm particles
0.05–0.5	0–0.8	0.001–0.003 ($t = 25°C$)	0.01–0.03	$\dfrac{k_S}{aD_L} = 34.1 \times 10^3 \times a^{-1.44} Re^{0.56} Sc^{0.33}$ for 2.4-mm particles
1.2–4	27–55	0.007–0.012 ($t = 24°C$)	0.04–0.07	$k_S = 0.2 Sc^{-0.66} \left(\dfrac{E_L' \mu_L}{\rho_L}\right)^{0.25}$ $E_L' = \left(-\dfrac{\Delta P}{Z}\right)_{LG} \dfrac{u_L}{\beta \rho_L}$
0.3–1.8	18–72	0.016–0.07	0.01–0.07	$k_S = 0.13 \times 10^{-4} \times G^{0.38} \left[L^{-0.78}\left(1 - \dfrac{e^y}{1 + e^y}\right)\right]$ $y = \dfrac{L}{8}$, L and G in lb/hr ft^{-2}

^{c} Spheres of benzoic acid.
^{d} β-Naphthol and naphthalene pellets and particles.
^{e} Benzoic acid pellets.

B. Plate Columns

Plate columns are used for operations requiring a large number of transfer units, high pressure, high gas flow rates and low liquid flow rates, when it is necessary to supply or to remove heat, when solids are present in the liquid (or gas), and when the diameter is greater than 70 cm. They have the ability to handle large variations in gas and liquid flow rates. Mass-transfer data will be presented here for the most common designs—bubble-cap plates and sieve plates.

1. Bubble-Cap Plates

For systematic study of several gas–liquid chemical reactions using a laboratory model bubble-cap column, Sharma *et al.* (S23) have shown that the presence of electrolytes, size of caps, type of slots, ionic strength, liquid viscosity, and presence of solids do not affect the mass-transfer rates. These rates chiefly depend on the gas and liquid flow rates (S23, M2). The influence of the superficial gas flow rate on k_L and k_G is indicated in Fig. 20. Interfacial area a'' per unit area of plate (or per unit floor area) for plate diameters varying between 0.15 and 1.20 m have been grouped in Fig. 21, which with the following correlations (S23) can be used to scale up bubble-cap plates up to 2 or 3 m in diameter:

$$k_G a' = 6.2 u_G^{0.75} S^{-0.67} D_G^{0.5}, \quad k_L a' = 7 u_G^{0.75} S^{-0.67} D_L^{0.5}$$

$$k_G = 11.5 u_G^{0.25} S^{-0.5} D_G^{0.5}, \quad a' = 0.54 u_G^{0.5} S^{0.83}$$

$$k_L = 13 u_G^{0.25} S^{-0.5} D_L^{0.5}$$

$k_G a'$ and $k_L a'$ are in sec^{-1}, k_G is in cm/sec, and a'' is in cm^2/cm^2; S is the submergence in cm, defined as the height of the bubble travel and measured from halfway up the slot height to the top of the dispersion; and a' is the interfacial area per unit volume of dispersion in cm^2/cm^3 ($a'' = a' \times S$).

2. Sieve Plates

On a perforated plate the liquid-side mass-transfer coefficient $k_L a$ and gas-side mass-transfer coefficient $k_G a$, based on the column volume, vary

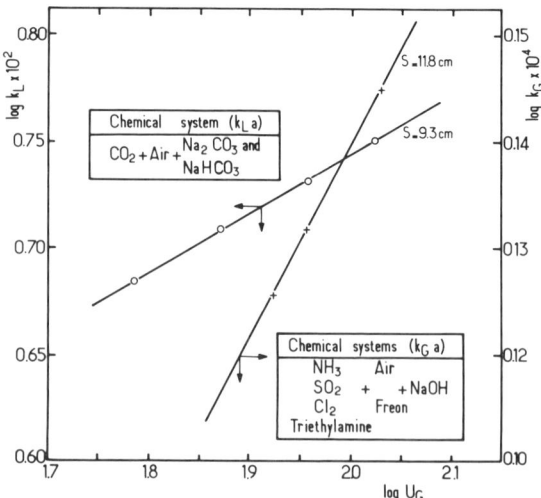

FIG. 20. Effect of superficial gas velocity on k_L and k_G in 22.5-cm model bubble-cap plate column (S23).

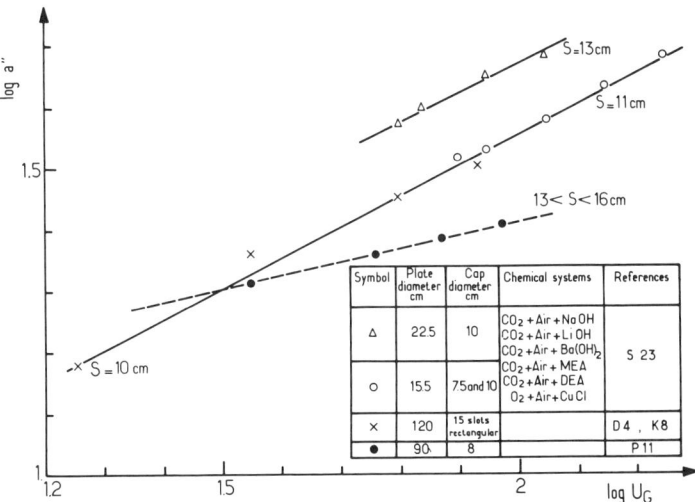

FIG. 21. Influence of u_G on the interfacial area per unit floor area in a bubble-cap plate column (L3).

linearly with the dispersion height. The true liquid- and gas-side mass-transfer coefficients k_L and k_G first increase with the dispersion height and then go through a maximum and decrease slightly (S23, S33, R17). Sharma and Gupta (S20) attribute this to different behavior of the density of dispersion and the average bubble size with increase in gas flow rate, which leads to a phase inversion point. These authors correlate their experimental data for 10-cm-i.d. perforated plates without downcomers by the following expressions:

$$k_G a = 2.6 \times 10^{-4} F^{-1.75} L^{0.6} u_G^{1.2}, \qquad k_L a = 4.2 \times 10^{-4} F^{-2.2} L^{0.6} u_G^{1.2}$$

$k_G a$ is in moles/sec atm cm^3, $k_L a$ is in sec^{-1}, u_G is in m/sec, L is in kg/m^2 hr, and F is the percentage free area of the plate (14–30%).

We note that $k_G a$ and $k_L a$ are nearly independent of the free area of the plate if the gas and liquid velocities are based on perforated area rather than column cross-sectional area. Moreover the values of the volumetric gas and liquid-side mass-transfer coefficients based on disperson volume, $k_G a'$ and $k_L a'$, are each practically constant whatever the gas and liquid mass flow rate. Perforated plates, especially those with a high free area, can handle relatively higher liquid and gas flow rates and provide higher values of overall mass-transfer coefficients than at corresponding flow rates through bubble-cap plate columns. However, discrepancies exist between different reported values of k_L (S20, B9, P3, H3) because of varying ionic strength and the presence of solids and antifoaming agents. Therefore the formulas proposed by different authors should be carefully studied before use.

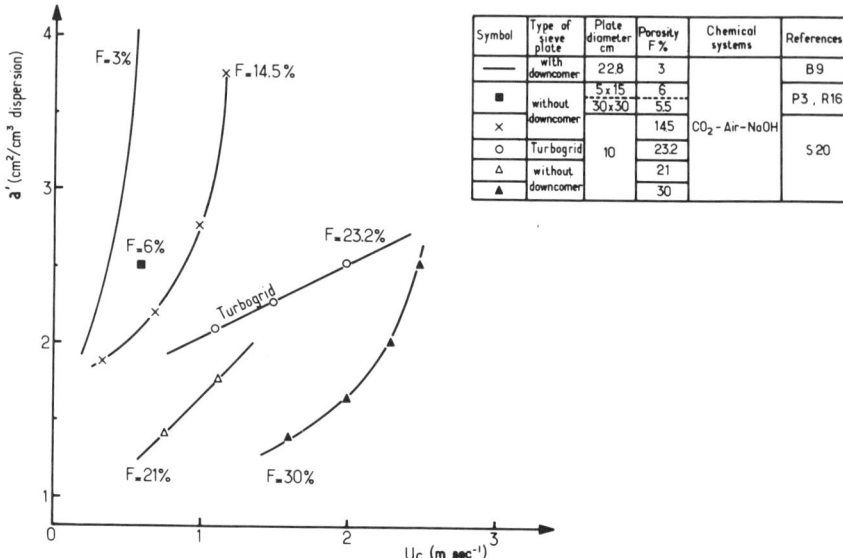

FIG. 22. Variation in interfacial area per unit volume of dispersion with u_G for sieve plates (L3).

The data on interfacial area are more homogeneous. Interfacial area a' based on dispersion volume is in the range of 2–2.5 cm²/cm³ for dispersion heights of 8–16 cm; it is not influenced much by the liquid flow rate, especially for sieve plates without downcomers. Values of a' increase with increasing gas velocity, percentage free area, ionic strength, and liquid viscosity (S20, S33, R17, P8, B33). Laurent and Charpentier (L3) have regrouped the literature data for plates with and without downcomers, and for turbogrid plates, in Fig. 22, which applies for a mean value of the dispersion height of 12 cm.

In conclusion, reliable values of interfacial area per unit floor area (N12) are given by the equation

$$a'' = 30 G^{0.5} \rho_G^{-0.25} \tag{116}$$

where a'' is in m²/m², G is in kg/m² sec, and ρ_G is in kg/m³.

C. Bubble Columns

Bubble columns where a gas is dispersed through a deep pool of liquid are commonly used in industry as absorbers, strippers, or reactors when a large liquid holdup, large liquid residence time, or large heat transfer is needed. They may be operated either countercurrently, cocurrently, or semibatch. Other advantages of bubble columns are the absence of moving parts, minimum maintenance, small floor space, ability to handle solids, relatively low cost, large interfacial area, and large mass-transfer coefficient.

The principal disadvantages of bubble columns are a large extent of liquid-phase back-mixing, a high pressure drop of the gas due to the high static head of liquid, and a decrease in the specific interfacial area for length/diameter ratios greater than 12–15 because of coalescence. Coalescence may be minimized by inserting fixed or fluidized packings, grids, or perforated plates (K2), or by pulsation.

A systematic study of mass transfer in bubble columns by Mashelkar and Sharma (M8, M9, S23) is summarized in Fig. 23. Increasing the superficial gas velocity increases the gas holdup α, the volumetric mass-transfer coefficients, and the interfacial area per unit volume of dispersion, but not the true mass-transfer coefficients. Correlations proposed for k_L seem too specific to be extended to practical systems (H13, F1, A3). Sharma and Mashelkar (S21) found good agreement between their experimental values of k_G and the values from Geddes' stagnant sphere model equation (G3):

$$k_G = -\frac{d_B}{6t_c} \ln\left\{ \frac{6}{\pi^2} \sum_1^\infty \frac{1}{n^2} \exp\left(-\frac{D_G n^2 \pi^2 t_c}{(d_B/2)^2}\right) \right\} \quad (116a)$$

in using the first term of that series ($t_c = Z/v_B$ where v_B is the velocity of the bubble rising through a height Z). The interfacial area (G4) is increased by an increase in viscosity or in temperature, the presence of solids, a decrease in surface tension, or the presence of electrolytes. However, the physical properties of the gas appear to have no effect.

As bubble column hydrodynamics depends on the physicochemical properties of the gas–liquid system, a generalized correlation for k_L and a' is not attempted. However, small-scale experiments with the system of interest will allow scale-up on the basis of equal superficial velocity of the gas. So the data in Fig. 23 or of specific experiments can be used, noting that $k_G a'$, $k_L a'$, and a' vary as $u_G^{0.8}$ (K2, D9).

The preceding discussion refers to mass-transfer rates based on the unit volume of dispersion. But the interfacial area per unit volume of liquid a''' is often required. To convert a' to a''', one must know the gas holdup α, with $a' = (1 - \alpha) a'''$. A correlation covering a wide range of column dimensions, flow conditions, and system properties has been developed by Hughmark (H13) and modified by Mashelkar (M8):

$$\alpha = [u_G/(30 + 2u_G)][(1/\rho_L)(72/\sigma_L)]^{0.33} \quad (117)$$

This equation is valid for $0.6 < \rho_L < 1.3$ gm/cm^3, $0.9 < \mu_L < 150$ cP, and $25 < \sigma_L < 76$ dyn/cm, for bubble diameters greater than 0.1 cm and for column diameters up to 1.5 m. However, electrolytes or foaming systems can give holdup values 30% higher than those of nonelectrolyte systems

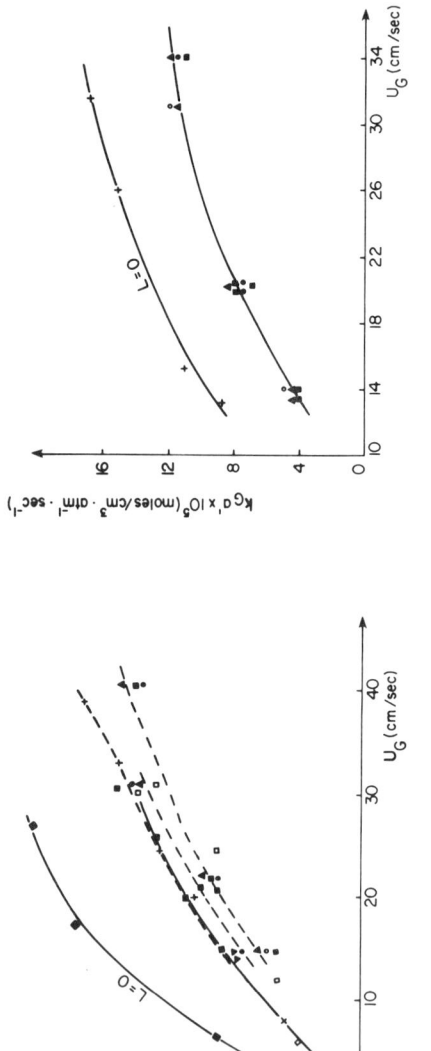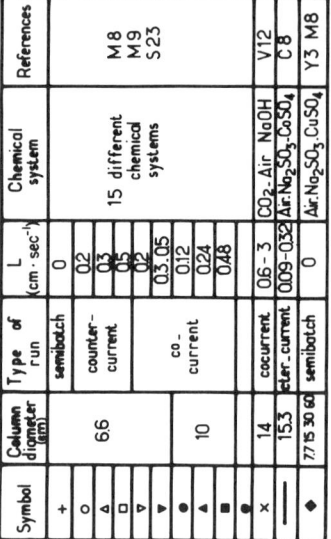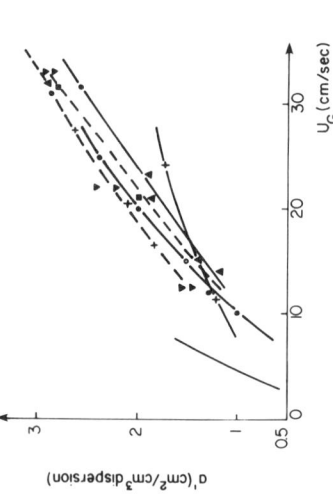

FIG. 23. Variation in mass-transfer data per unit volume of dispersion with superficial gas velocity in bubble columns (L3).

predicted by Eq. (117), as shown by Akita and Yoshida (A3). Equation (117) applies for a stationary liquid. Although the superficial liquid velocity u_L does not have a large effect, it can easily be taken into account. For example, in the case of cocurrent flow, α from Eq. (117) is corrected to the true holdup α' by the relation $u_G/\alpha = u_G/\alpha' - u_L/(1 - \alpha')$ or by the relations given by Nicklin et al. (N7a) where the terminal velocity is taken into account. An alternate, and general correlation, that has been tested in industrial equipment for the oxidation of toluene, cyclohexane, and alipharic acids under pressure, has been proposed by Van Dierendonck et al. (V3):

$$\alpha = 1.2(\mu_L u_G/\sigma_L)^{3/4} (\rho_L \sigma_L^3/g\mu_L^4)^{1/8}$$

for $\alpha < 0.45$; $3 < u_G < 40$ cm/sec; $0 < u_L < 2$ cm/sec; $0.8 < \rho_L < 1.3$ gm/cm³; $0.5 < \mu_L < 5$ cP; $20 < \sigma_L < 75$ dyn/cm; and $D > 15$ cm.

D. Tube Reactors

Reactors slightly different from bubble columns are the cocurrent gas–liquid horizontal and vertical pipeline reactors and the coiled reactor, with diameters generally less than 5 cm. These tube reactors are advantageous for compactness, high flow rates of both gas and liquid, and wide range of operating temperature and pressure. An interesting feature is the variety of two-phase flow regimes—bubble flow, plug flow, slug flow, froth flow, annular flow, and spray flow. Design is complicated by the fact that different flow regimes exhibit different mass-transfer efficiencies. Mass-transfer rates increase as the gas and liquid flow rates increase, that is, when the flow regime varies from bubble flow to annular and annular spray flow, where values of a can be found to be 10 cm²/cm³ and $k_L a$ to be 0.7 sec⁻¹. Several general studies on mass transfer have been published by Scott and Hayduck (S12), Jepsen (J3), and Gregory and Scott (G17) for horizontal reactors; and by Banerjee et al. (B4), Jepsen (J3), and Kulic and Rhodes (K12) for coil reactors. As a first approximation for design, for any two-phase flow except bubble flow and plug flow in horizontal and coiled reactors, $k_L a$ depends strongly on the energy dissipation $\epsilon' = (\Delta P/Z)_{LG}(u_L + u_G)$. Using this idea, Jepsen has given the following correlations:

$$k_L a = 3.5 D_A^{0.5} \sigma_L^{0.5} \mu_L^{0.05} D^{0.68} \epsilon'^{0.40} \quad \text{for} \quad \epsilon' < 0.05 \text{ atm/sec} \quad (118)$$

$$k_L a = 18.7 D_A^{0.5} \sigma_L^{0.5} \mu_L^{0.05} \epsilon'^{0.80} \quad \text{for} \quad \epsilon' \geq 0.05 \text{ atm/sec} \quad (118a)$$

where $k_L a$ is in sec⁻¹, $\Delta P_{LG}/Z$ is in atm/m, $0.5 < D < 4$ in., $0.6 < \mu_L < 27$ cP, $24 < \sigma_L < 74$ dyn/cm, and $0.1 < D_A < 5.10^{-5}$ cm²/sec. Gregory and Scott (G17) and Kulic and Rhodes (K12) have proposed other coefficients for specific cases. The true liquid-side mass-transfer coefficient is well represented by the modified Banerjee model:

$$k_L = D_A^{0.5}[(\Delta P/Z)_{LG} \times (D\alpha^{0.5}/4\mu_L n)]^{0.5} \qquad (119)$$

where the factor n has been defined by Gregory and Scott (G16):

$$n = 10^9/(\text{Re}_{GL})^{1.7} = 10^9 \times [(1-\alpha)\mu_L/D\rho_L u_L]^{1.7} \qquad (120)$$

These correlations require both pressure-drop and gas-holdup values which may be predicted by the Lockhart–Martinelli correlation (L22) or the Hughmark correlation (H13).

For bubble flow and plug flow in horizontal tubes, the following correlation is proposed by Scott and Hayduk (S12) and Shah and Sharma (S16):

$$k_L a = (2.3 \times 10^{-4})u_L[u_G/(u_G + u_L)]^{0.78}\sigma_L^{0.5}\mu_L^{0.09}D_A^{0.4}D^{-1.9} \qquad (121)$$

with the same units as in previous equations, except that D is in cm.

The literature on measurement of mass transfer in vertical tubular reactors is very sparse. Kasturi and Stepanek (K3, K4) have presented data for a, $k_L a$, and $k_G a$ measured under identical conditions in the case of annular flow, annular spray flow, and slug flow. For the aqueous systems used (CO_2, air, NaOH) they have proposed the following correlation for the interfacial area: $a = 0.23[(1-\alpha)/Q_L](\Delta P/Z)_{LG}^{1.1}$ where Q_L is in cm^3/sec and $\Delta P/Z$ is in N/m^3. Correlations for true liquid-side and gas-side mass-transfer coefficients by the same authors are difficult to generalize, as viscosity and surface tension were not varied.

Values of $k_G a$ have been measured by Golding and Mah (G11) for absorption of dilute hexane into paraffin oils ($\sigma_L = 36$ dyn/cm, $40 < \mu_L < 140$ cP), in vertical slug flow.

However, no general correlation is yet available for $k_L a$ and $k_G a$ in vertical tubular reactors when the two-phase flow regime is different from bubble flow. So for design, scale-up should be based on laboratory data for mass-transfer coefficients and on the ratio of energy dissipation terms ϵ', as in the method defined by Jepsen (J3). For any tubular reactor, great care must be taken with the distributor design and with the size of the inlet section so as to minimize the entrance effects.

E. SPRAY TOWERS

Spray towers are of particular interest because of their ability to handle corrosive and solid-laden fluids when only one or at most two theoretical stages of contact are required or when the gas pressure drop must be kept to a minimum. They are also used for exothermic chemical reactions or for the absorption of highly soluble gases, where large volumes of liquid must flow through the column to avoid an excessive temperature rise. The disadvantages of spray towers lie in the power consumption needed for pressuring the absorbing liquid through spray nozzles, in the height necessary to achieve one theoretical stage of absorption, and in the need for installing mist eliminators.

The large number of variables, variety in methods of operation, and differences in mechanical details such as spray nozzle design have led to scattered mass-transfer data (K9, R1). Mass-transfer correlations for single drops or even clouds of drops do not apply because normal production of the sprays introduces effects that cannot be interpreted quantitatively. In order to obtain good contact and avoid bypassing of upflowing gas, the spray must cover the entire tower cross section. Part of the liquid spray impinges on the tower wall and trickles down as a film, which reduces the mass-transfer efficiency that would be predicted from drop-type contact. Also, as no spray nozzle produces perfectly uniform drops, drop coalescence occurs as the drops fall through the tower. Mehta and Sharma (M11) are the only ones to report comprehensive research into the effect of commercial nozzles, column height, gas and liquid flow rates, and physical properties of the liquid on spray column performance, with diameters up to 0.4 m. These studies provide the following conclusions useful for design.

The values of $k_L a$ are practically independent of the gas velocity up to a critical value which depends on nozzle type, column diameter, and physical properties (P7, M11); then $k_L a$ increases with increasing gas velocity. Typical results are given in Fig. 24 for a shower nozzle. Moreover $k_G a$, k_G, and a all increase as the gas velocity increases independently of the type of nozzle, the column size, the liquid flow rate, and probably the physical properties, as shown in Fig. 25. The interfacial area and true liquid-side coefficient increase with the liquid flow rate L (Fig. 24), owing to increased surface area, higher drop velocity, and increased circulation or

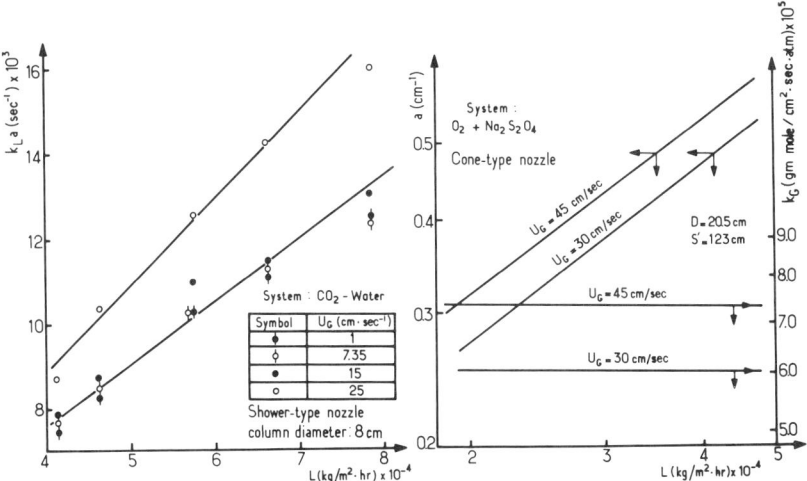

FIG. 24. Influence of gas and liquid flow rates on $k_L a$ and a in spray towers (M11).

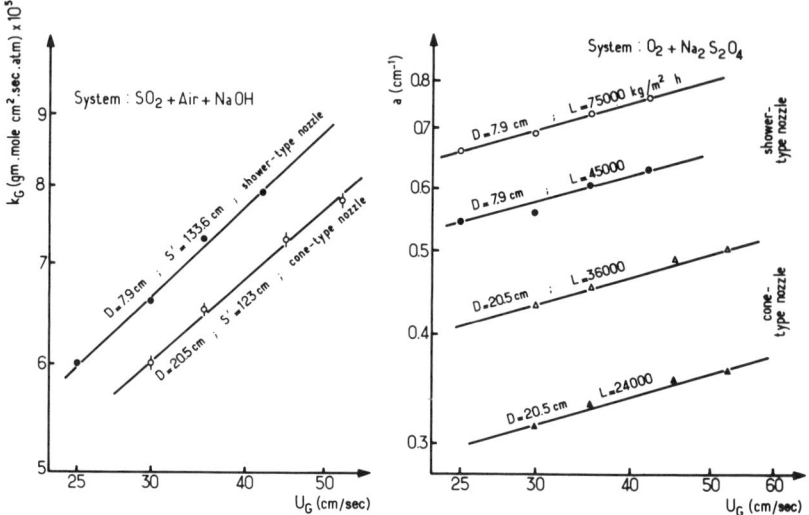

FIG. 25. Influence of gas velocity on k_G and a in spray towers (M11).

turbulence within the drops. The interfacial area varies exponentially with the liquid flow rate, the exponent increasing with decreases in the nozzle orifice diameter, that is, decreases in droplet size. In contrast, k_G is not influenced by the liquid rate.

The values of interfacial area and of overall mass-transfer coefficient increase with decreasing distance S' between the spray nozzle and gas inlet, whatever the nozzle type, column dimensions, and flow rates. Indeed the spray provides a large interfacial area in the vicinity of the nozzle, where there is intensive circulation. Then a decreases quickly away from the nozzle, as a result of both coalescence of droplets and collection of liquid on the column walls. $k_G a$ and a are approximately proportional to $(S')^{-0.4}$ (P7, H12, M11) for absorption and desorption processes, which shows that k_G is practically independent of the column height. Moreover Mehta and Sharma (M11) indicate that a is unaffected by ionic strength and viscosity but may decrease about 20% when solids are generated by the reaction of gas with the liquid. Thus the following correlations may be used for design (M11):

$$a = \alpha' L^{\beta'} u_G^{0.3}(S')^{-0.4} \qquad (122)$$

$$k_G = m u_G^{0.5} \qquad (122a)$$

where a is in cm²/cm³, k_G is in gm moles/cm² sec atm, L is in kg/m² hr, u_G is in cm/sec, and S' is in cm. Constants α', β', and m are listed in Table XXIII for nozzle type, orifice diameter, and other variables investigated by Mehta and Sharma. For other specific cases, the books of Ramm (R1) and Kohl and Riesenfeld (K9) will be of help. Scale-up often utilizes a

TABLE XXIII

Values of α', β', and m for the Spray Tower Mass-Transfer Equation (M11)

D(cm)	S'(m)	Type of nozzle	Orifice diameter (mm)	$\alpha'(\times 10^4)$	β'	$m(\times 10^5)$
8	1.3	Shower	69 holes of 1.2 mm	246	0.38	1.02
21	1.2	Solid cone	5.5	2.12	0.81	0.95
21	1.2	Solid cone	4.4	0.51	0.93	2.2
39	2.8	Solid cone	5.5	8.97	0.62	2.2
39	2.8	Solid cone	4.4	4.9	0.70	2.2
39	2.8	Solid cone	8.4	42.8	0.47	2.2

general correlation in which the number of transfer units $k_G aH/G$ is proportional to $(P_L/P_G^{0.1})$, where P_L and P_G are the power introduced in the liquid and in the gas, H is the height, and G is the superficial molar gas flow rate.

Values of the true liquid-side coefficient k_L are comparable to those in packed columns.

F. Mechanically Agitated Bubble Reactors

Mechanically agitated bubble contractors are very effective with viscous liquids or slurries and at very low gas flow rates and large liquid volumes. They are also noted for the ease with which the intensity of agitation can be varied and the heat can be removed. Their principal disadvantage is that both the liquid and the gas phase are almost completely back-mixed. Among the many types of agitators employed, the most common are paddles with straight or inclined blades, and turbines. Gas may be introduced through a perforated tube or through a porous or perforated plate.

A considerable amount of information is available in the literature on these contractors, and comprehensive reviews have been published by Sideman et al. (S28), Valentin (V1), Nagata (N1), and L'Homme (L12). The most systematic experimental work has been carried out by Westerterp et al. (W4), Reith (R7), Mehta and Sharma (M12), and Miller (M16), and a synthesis of the results presented in these studies will be reported now and compared with the sparse and more recent data for a sparged contracting system of dispersion.

1. Aqueous Liquid

At low stirrer speeds the gas holdup and specific interfacial area are essentially dependent on the gas flow rate, while at high stirrer speeds these fundamental parameters depend more upon the stirrer speed. Thus it is only at high speeds of agitation that the stirrer is fully effective in dispersing the gas in small bubbles into the liquid. There is a transition region where both the gas flow rate and the stirrer speed are important. The minimum speed of agitation n_0, sometimes called the critical speed, beyond which stirring is fully effective and agitator speed is the dominant factor, corresponds to a tip velocity of about 2.25 m/sec and is given by

$$n_0 d_A / (\sigma_L g / \rho_L)^{0.25} = A + B(D/d_A) \qquad (123)$$

where n_0 is in sec^{-1}, agitator diameter d_A is in m, reactor diameter D is in m, liquid density ρ_L is in kg/m^3, surface tension σ_L is in kg/sec^2, and acceleration of gravity g is in m/sec^2. A and B are functions of the type of agitator. For various vessel diameters ($14 < D < 90$ cm) and 12 geometrically similar turbine impellers of different sizes ($0.2 < d_A/D < 0.7$), Westerterp et al. (W4) have reported $A = 1.22$ and $B = 1.25$. This means that the linear tip velocity $\pi n_0 d_A$ must be at least 8–30 times higher than the rising velocity of the gas bubble for a good dispersion, depending on the ratio d_A/D. For two- and four-bladed agitators Westerterp et al. (W4) also reported $A = 2.25$ and $B = 0.68$. The existence of n_0 is confirmed by recent studies on aqueous liquids (M12, M16) and organic liquids (G1, S36). Moreover it appears that values of $H/D = 1$, d_A/D in the range 0.4–0.5, and h_A/D in the range 0.33–0.5, with the gas superficial velocity u_G limited to less than 5 cm/sec, are likely to be most desirable for gas–liquid contacting in a mechanically agitated bubble reactor with a single impeller.

Above the minimum speed of agitation, the effective interfacial area a''' and liquid mass-transfer coefficient $k_L a'''$ (based on unit clear liquid volume) increase linearly with the speed of the agitator n and are affected less by the gas velocity u_G. Below the minimum speed, mass transfer depends more on gas flow rate and less on agitator speed. The experimental data of Sharma et al. (M12, J9) have been regrouped by Laurent and Charpentier (L3) into Figs. 26 and 27. With the present state of knowledge, these diagrams give a''' in aqueous liquids when the reactor has a standard configuration and the agitator does not suck in gas from the surface.

The true liquid-side mass-transfer coefficient does not change significantly with agitator speed, depending only on the physicochemical properties of the gas–liquid system. For bubbles smaller than 2.5 mm in diameter (typical industrially with impeller tip speeds of 2.25 m/sec or more), the correlation of Calderbank and Moo-Young (C3) is valid for k_L (in cm/sec) in pure liquids or solutions of nonionic solutes:

$$k_L = 0.31 (\rho_L D_A / \mu_L)^{2/3} [(\rho_L - \rho_G) g \mu_L / \rho_L^2]^{1/3} \qquad (124)$$

This relationship suggests that k_L is independent of the power input. Such

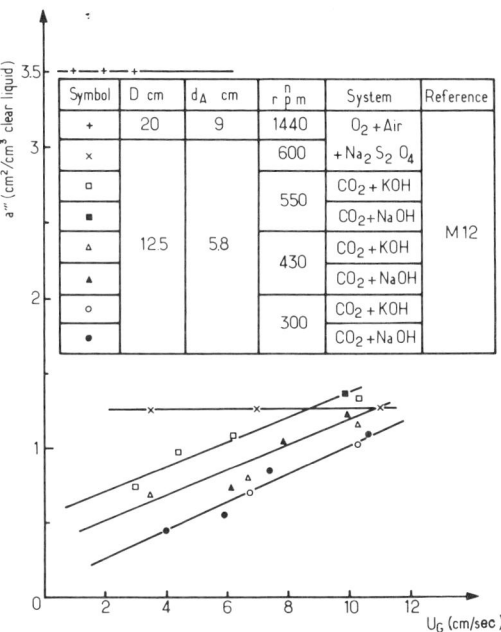

FIG. 26. Influence of gas velocity on specific interfacial area at low agitator speed in a mechanically agitated reactor (M12).

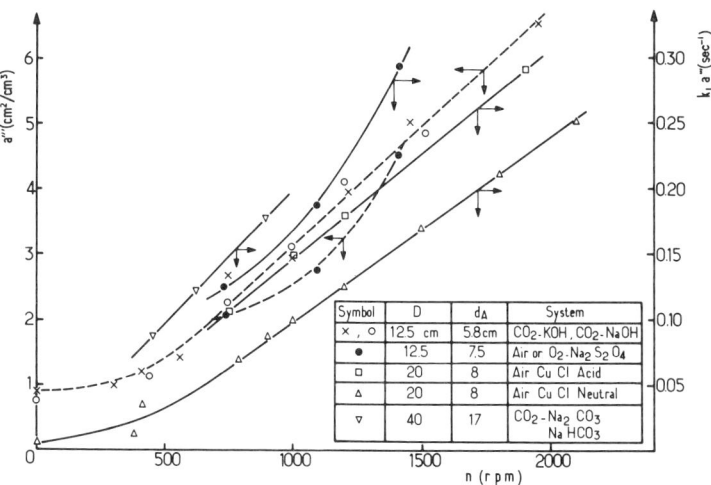

FIG. 27. Influence of agitator speed on superficial area and liquid transfer coefficient per unit volume of clear liquid in a mechanically agitated reactor (M12, J9).

is not the case in electrolyte solutions. In particular, here k_L decreases with increasing power input per unit clear liquid volume P_m/V_L, corresponding to reduced bubble diameter d_b, as shown by Robinson and Wilke (R14). Below a d_b value of 2.5 mm, k_L seems roughly proportional to d_b. Such behavior is consistent with results for bubbles of the same size range dispersed in viscous Newtonian liquids and electrolyte-free aqueous solutions (C3).

The effective interfacial area a''' is increased by increases in ionic strength, ion valence number, or viscosity, by the presence of a solid or immiscible liquid, and by a decrease in liquid surface tension. Thus it is nearly impossible to predict a priori the interfacial area. However, scale-up is practicable from experiments carried out with the actual gas–liquid system in a small agitated contactor (D = 10–20 cm). The experimental work of Sharma et al. (M12, S23) shows that a scale-up basis of equal $nd_A/\sqrt{D}$ or $(n - n_0)d_A/\sqrt{D}$ (when d_A/D = 0.4–0.5) can be used with a fair degree of confidence (respectively, 10 and 16% average deviations) for agitated vessels with diameters up to 60 cm.

At higher vessel diameters, to ensure the same specific interfacial area, scale-up should be based upon constant power input per unit volume of liquid, geometrically similar vessels, and the same superficial gas velocity. For example, Fig. 28 shows specific interfacial areas measured by Reith (R7, R8) in three geometrically similar mixing tanks, with D = 19, 45, and 120 cm, at u_G = 4.7 cm/sec. The use of $n^3 d_A^2$ as an independent parameter, which is functionally related to dimensions of power input per unit mass and has the same dimensions, yields a reasonable correlation. At the left of Fig. 28, at lower agitator speeds and increasing tank diameter, the curves deviate from the general solid line toward lower values of a'''. Consistent with the findings of Westerterp et al. (W4) and Reith (R7), stirring is effective only when the stirrer's linear tip velocity $\pi n d_A$ exceeds 2.25 m/sec for d_A/D in the range 0.4–0.5 (solid line in Fig. 28).

2. Organic Liquids

Results of Figs. 27 and 28 are confined to aqueous and/or electrolyte solutions. Many industrially important gas–liquid reactions involve polar or nonpolar organic media. Mass-transfer rates in agitated bubble reactors using viscous and organic solvents are substantially different from those using aqueous media, so data for the latter cannot be extended directly.

Recent work with nonaqueous nonelectrolyte liquids, in reactors with diameters of 10–20 cm, has been published by Ganguli (G1) and Sridharan (S36). Ganguli measured specific interfacial area and liquid-side mass-transfer coefficients for the hydrogenation of edible oil in a sparged mechanically agitated hydrogenator using a Ziegler–Natta catalyst, with sol-

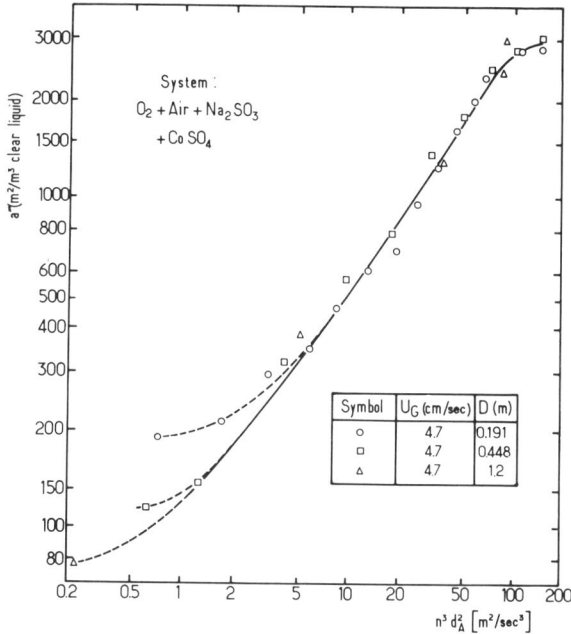

FIG. 28. Scale-up in mechanically agitated reactors: a''' as a function of $n^3 d_A^2$ (R8).

ids either present or absent. In the same type of contactor, Sridharan measured the absorption of CO_2, with and without chemical reaction, into solvents such as xylene and toluene containing cyclohexylamine, benzyl alcohol containing diethanolamine and benzylamine, diethylene glycol containing mono- and diethanolamine, and cyclohexanol containing cyclohexylamine; also, into viscous aqueous solutions such as those of diethylene glycol or polyethylene glycol, containing mono- or diethanolamine.

Organic solvents behave qualitatively the same as aqueous solvents. A critical speed n_0 exists, above which a''' and $k_L a'''$ depend only on n, and below which a''' and $k_L a'''$ depend on both n and u_G; $n_0 \simeq 1000$ rpm for viscous solvents with $\mu_L > 10$ cP, in vessels of small diameter. However, values of the mass-transfer parameters are quite different from those in aqueous liquids (Figs. 27 and 28). As seen in Fig. 29, values of a''' in organic solvents (toluene, xylene) with nearly the same viscosity as water ($\mu_L = 0.6$–1.4 cP) are high compared with those in aqueous systems under similar conditions. The high values of $k_L a'''$ are explained by the low surface tension ($\sigma_L \simeq 27$ dyn/cm) which leads to a decrease in bubble coalescence and thus to smaller bubbles and to higher values of a'''. As expected, k_L increases with increasing diffusivity of solute gas and with decreasing viscosity of the medium.

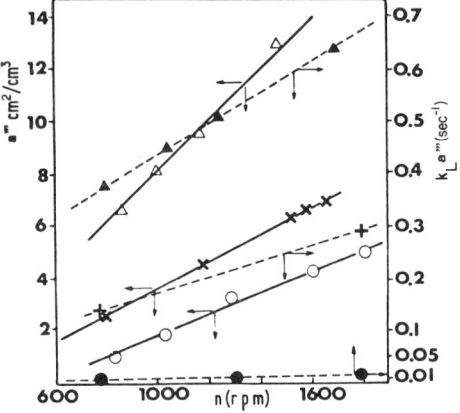

Symbol	System	μ_L (cP)	σ_L (dyn/cm)	D_{CO_2} 10^5 cm²/sec	D (cm)	d_A (cm)	Ref.
△	CO_2–cyclohexylamine in xylene	0.6	27	4.35	13	5	S36
▲	CO_2–xylene	0.6	27	4.35	13	5	S36
×	CO_2–NaOH	1.4	72	1.8	13	5.8	M12
+	CO_2–Na_2CO_3–$NaHCO_3$	1.4	72	1.5	20	8	M12
○	CO_2–cyclohexylamine in cyclohexanol	37	33	0.16	13	5	S36
●	CO_2–cyclohexanol	41	33	0.16	13	5	S36

FIG. 29. Mass-transfer data for organic and viscous solvents in mechanically agitated reactors.

The values of liquid-side mass-transfer coefficients fall drastically as the liquid viscosity increases, because of low values of both k_L and a''' (G1); k_L and d_b not vary significantly either with u_G or with n. However, k_L and a''' are decreased by the presence of solids, which serve simultaneously to decrease the interface mobility and increase the effective viscosity, especially at low agitator speeds. Table XXIV gives some representative data. It is interesting to note that, even if the gas dispersion characteristics d_b and a''' for aqueous solutions in agitated tanks are not systematically different from those of nonaqueous and viscous nonelectrolytic liquids, k_L and $k_L a'''$ will still depend on the physicochemical properties.

It appears that the scale-up may again be done from experiments carried out with the actual gas–liquid system in a small, geometrically similar, agitated contactor ($D = 10$–20 cm) on the basis of equal total power input per unit volume of liquid in vessel and equal superficial gas flow. More-

TABLE XXIV
Typical Mass Transfer Data for Organic and Aqueous Liquids in Agitated Reactor with Six-Blade Turbines

System	n (rpm)	T (°C)	u_G (cm/sec)	d_b (mm)	D (cm)	d_A (cm)	D_A (cm²/sec) × 10⁵	ρ_L (gm/cm³)	μ_L (cP)	σ_L (dyn/cm)	a''' (cm⁻¹)	k_L (cm/sec) × 10²	$k_L a'''$ sec⁻¹	Ref.
H₂–edible oil	900	70	3	1.1	21	8.5	5.8	0.9	12	30	7.5	1.6	0.12	Ganguli (G1)
CO₂–diethylene glycol	900	30	3	—	20	8	0.25	1.1	22	48	3.3	0.6	0.02	Sridharan (S36)
O₂–Na₂SO₃	900	30	4.7	1	19	7.6	2.1	1.1	1.1	69	7.4	4.2	0.31	Reith (R7)

over, the work of Perez and Sandall (P4) suggests that a correlation of $k_L a$ developed for Newtonian liquids holds also for non-Newtonian liquids when the liquid viscosity term is replaced by an effective viscosity taking into account a flow behavior index, a consistency index, and stirrer speed.

3. Power Input and Gas Holdup

When experiments on small agitated contactors are impractical, it is necessary to know the power dissipated per unit volume of liquid:

$$P_e/V_L = P_e/[(1 - \alpha)V_R]$$

where V_R is the total volume of gas and liquid, for insertion in an equation for a specific interfacial area. Such a relation has been given by Calderbank (C1), for pure liquids without suction of gas from the surface, at high gas-sparging rates:

$$(1 - \alpha)a''' = 1.44(P_e/V_L)^{0.4}\rho_L^{0.2}\sigma_L^{-0.6}[u_G/(u_t + u_G)]^{0.5} \quad (125)$$

Here u_t is the terminal velocity of bubble rise in m/sec. The effective power input term P_e combines both sparged gas (P_s) and mechanical energy (P_m) contributions in W (M16):

$$P_e = P_m + C_1 P_s \quad (126)$$

$$P_s = \rho_G Q_G (RT/M) \ln(P_0/P) \quad (126a)$$

For six-bladed turbines or impellers centered both vertically and horizontally, by introducing a blade width term into the results of Rushton *et al.* (R20), Al Mojil *et al.* (A8) obtained the relations

$$P_m = YRP_{m0} \quad (126b)$$

with

$$\begin{aligned} P_{m0} &= 20n^3 d_A^4 w_A \rho/g_c, \quad R = 0.36 + 0.64 \exp(-2.0X) \\ X &= [(nd_A^2\rho/\mu)(d_A/d_T)(h/w_A)]^{0.25}(n^2 d_A^2/g_c d_T) \\ Y &= \exp[-90Z(M^{-130}Z^{1-M})^{1/(1+50Z)}] \end{aligned} \quad (126c)$$

$$M = 22.4[(n^2 d_A/g)(w_A/h)]^{0.5}(d_A/d_T)^2, \quad Z = u_T(hd_T/d_A g)^{1/2}(h/w_A)^{2/3}$$

where R, X, Y, and M are all nondimensional functions; $n^2 d_A/g$ is the Froude number (dimensionless), w_A is the impeller width, d_A is the impeller diameter, h is the tank height, d_T is the tank diameter, ρ_G is the gas density in kg/m³, Q_G is the gas flow rate in m³/sec, M is the gas molecular weight, T is the absolute temperature in K, R is the gas constant in Nm/kg mole K, P_{m0} is the unaerated power input by mechanical agitation in W, p_0 is the pressure at the sparger in N/m², and P is the absolute pressure.

An alternate correlation for P_m has been reported by Michel and Miller (M14); in SI units,

$$P_m = C(P_{m0}^2 n d_A^3 / Q_G^{0.56})^{0.45} \qquad (126d)$$

Coefficient C varies between 0.7 and 1.2, depending chiefly on the impeller geometry. Once C has been determined in a reactor volume of about 10 liters, Eq. (126d) can be used with $\pm 30\%$ accuracy for scale-up to a geometrically similar volume of several cubic meters.

Miller (M16) gives values of the correction term C_1 in Eq. (126) for spargers of vessel sizes in the range of 2.5–250 liters obtained by comparing mean bubble diameters for gas sparging with and without mechanical agitation. This means that agitation imposed in a sparged system augments the energy content of the liquid phase, reduces the bubble size, and increases the specific interfacial area. Thus the separate power inputs of gas sparging and mechanical agitation combine into an effective overall value for predicting the interfacial area. The term $C_1 P_s$ is especially useful in the prediction of specific interfacial area at low agitator speeds ($n < 250$ rpm), where gas sparging is in fact the stronger effect. Miller (M16) also proposed a further correction term to account for total gas input inclusive of both sparging and surface entrainment.

As explained in Sections IV,F,2 and 3, a reliable design can be made from experiments carried out in a small-scale apparatus, with scale-up based on total power input ($P_T = P_m + P_s$). The scale-up rule is that the same specific interfacial area requires the same total power input per unit volume of liquid. Strictly speaking, for a given u_G and geometric similarity, scale-up must be based on P_T and not on $n^3 d_A^2$ or only P_m. For example, the specific interfacial areas for three gas–liquid systems are compared with $n^3 d_A^2$ and P_T / V_L in Fig. 30. It is seen that the scatter of the experimental data is smaller for low values of n when the total power input is used. Practically, mechanically agitated reactors are operated at high values of n, where agitation is considered more effective and where P_m is much larger than P_s; this explains the good results with scale-up based on $n^3 d_A^2$ (which varies with P_m / V_L instead of P_T / V_L).

Several relationships exist for predicting gas holdup α in mechanically agitated gas–liquid reactors (C1, W4, V3, M16), but all seem inadequate in terms of the scatter of the data points relating interfacial area and bubble diameter. Gas holdup α varies linearly with the speed of agitation above the minimum speed to a maximum value. Then it remains independent of n and may even decrease when u_G increases; the stirrer becomes ineffective because it rotates in a region containing mainly trapped gas. Below the minimum speed, α is more dependent on u_G but may increase linearly with both u_G and n. A satisfactory fit might result when holdup is plotted against P_T / V_L, but this has not been tried.

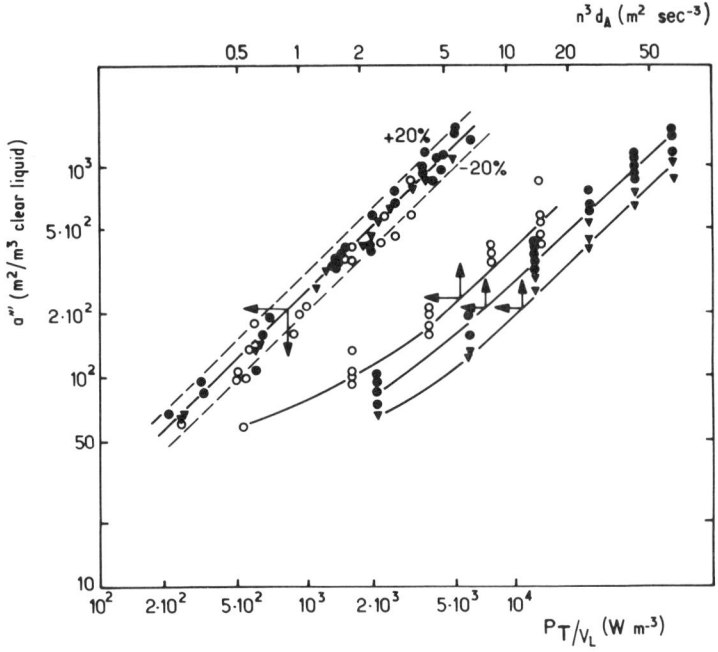

Symbol	System	μ_L (cP)	σ_L (dynes/cm)	ρ_L (gm/cm³)	D (cm)	d_A (cm)	u_G (cm/sec)	n (sec⁻¹)	Ref.
●	O_2–Na_2SO_3	1.5	61	1.08	19	6.4	0.64–5.50	8.33–25	L23
▼	O_2–CuCl(HCl)	1.25	44	1.15	19	6.4	0.64–4.70	8.33–25	L23
○	O_2–Na_2SO_3	1.3	61	1.09	29	12	0.28–1.62	3.33–10	L12

FIG. 30. Scale-up in mechanically agitated reactors: a''' as a function of P_T/V_L or $n^3 d_A^2$.

4. Some Comments on Gas-Side Mass-Transfer Coefficient and Particle–Liquid Mass Transfer

Almost no quantitative information is available on the gas-side mass-transfer coefficient $k_G a'''$ for mechanically agitated vessels. This reflects the uncertainty in the residence-time distribution of the gas phase during chemical mass transfer (M12). The gas-phase mass-transfer coefficient is usually quite high, leading to relatively negligible gas-phase resistance. The qualitative information available shows that, when u_G is increased, $k_G a'''$ increases to some extent, reaches a maximum, and then decreases. The rate of aeration at the maximum $k_G a'''$ seems to be larger at higher agitator speeds. There is no accurate method for predicting $k_G a'''$ from u_G and P_T/V_L. Specific data for the absorption of air or oxygen into Na_2SO_3

have been given by Cooper *et al.* (C16) and Ramm (R1), for absorption of CO_2 from air into aqueous NaOH or Na_2CO_3 by Nagata (N1), and for absorption of SO_2 from air into aqueous NaOH by Mehta and Sharma (M12). These data concern the overall mass-transfer coefficient and may be used for geometrically similar apparatus.

Also, data on particle–liquid mass transfer from suspended solids in gas–liquid mechanically agitated vessels are practically nonexistent (R18). However, many studies have been published on mass-transfer experiments *in the absence of gas*, which give an idea of the magnitude of k_s. Recent reviews by Nienow (N9) and Blasinski and Pyc (B17, B18) indicate two fundamentally different approaches to the prediction of k_s: the Kolmogoroff theory, which implies equal k_s at equal power input per unit volume (B17) and the terminal velocity–slip velocity theory which relates k_s to the value that would apply if the solid particle moved at its terminal velocity (H2). As explained by Nienow (N9), the resulting values of k_s are approximately the same. Use may be made of the graphical correlation given by Brian *et al.* (B29).

G. Jet Reactors

During the last few years, interest has increased in reactors in which high mass transfer is realized by means of liquid jet or injection devices providing gas or liquid entrainment: jet reactors (N3, B30), hydrocyclones (B11), venturis (S16, J2), and other high-velocity gas–liquid contactors.

1. *Submerged-Jet and Plunging-Jet Reactors*

The plunging liquid jet reactor utilizes the flow geometry of a coherent liquid jet plunging through a reactive gaseous atmosphere into a bath of the same liquid (Fig. 31). As explained by Burgess *et al.* (M1, B30, B31), entrainment and subsequent reaction of the gas in the subsurface gas–liquid region is the phenomenon of interest. The area surrounding the jet plunge point in the receiving bath resembles a venturi formed by free liquid surfaces. The jet and bath surface interact at the "venturi throat" where cavitation and closure occur in the same manner as in a jet ejector pump with liquid drive.

The physical entrainment rate of gas varies over orders of magnitude as a function of jet stability, which acts as the surface shape generator. The controllable parameters of the system are nozzle geometry, nozzle velocity, nozzle height, jet length, and jet velocity. McCarthy *et al.* (M1) have analyzed gas entrainment in such contactors in terms of the surface roughness of the jet and have proposed an entrainment ratio:

$$q_G/q_L = (d_E/d_N)^2 - 1$$

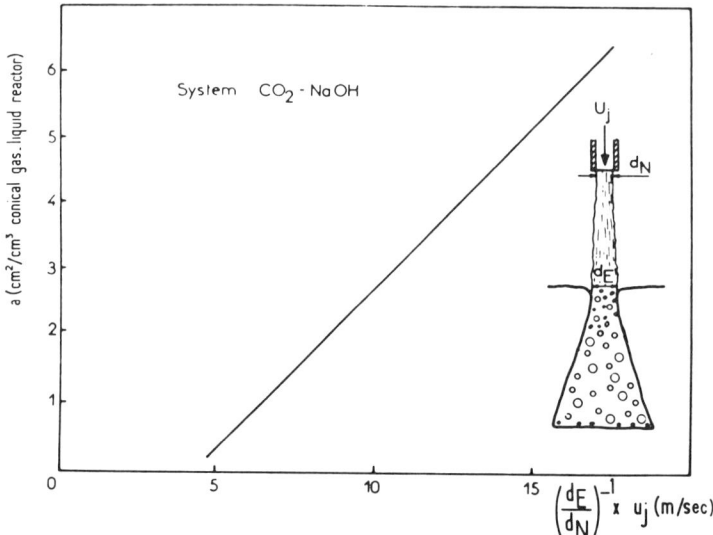

FIG. 31. Interfacial area related to jet conditions in the submerged-jet reactor (B30).

where q_G and q_L are the volumetric flow rates, d_E is the diameter of the gas envelope totally enclosing the jet, and d_N is the diameter of the nozzle (Fig. 31). This equation fits three different types of nozzles with diameters in the range 2.5–12.5 mm, jet lengths in the range 1–25 cm (5- to 80-cm nozzle diameters), and jet velocities up to 30 m/sec. The entrainment ratios were varied between 0 and 3. Complementary data are found in the extensive work by Van de Sande and Smith (V2).

In the bath, the gas–liquid mixture is assumed to occupy a conical volume in which the interfacial area depends on nozzle design, free jet length, and jet velocity. Burgess et al. (B30) showed that the interfacial area per unit conical volume was a function of the jet surface roughness d_E/d_N and jet superficial velocity u_j (Fig. 31); it was visually observed that the size distribution of bubbles in the bath was a function of these two factors. The highest values of the specific interfacial area so obtained (up to 7 cm^{-1}) are of the same order of magnitude as in the sonic gas velocity bubble column. However, the conical gas–liquid volume has boundaries defined by the extremities of the submerged jet originating at the jet plunge point and the line at which total absorption of the gas occurs, which may be difficult to characterize other than visually. So the specific interfacial area based on that submerged column may not be very accurate.

Burgess and Molloy (B31) have extended this technique to a system where the jet extends over the entire bath and the reactor volume is defined by the solid walls of the container; that is, total absorption of the entrained gas does not occur in the vessel. In such a case the plunging jet reactor is analogous to a gas-sparged, stirred-tank contactor with the plunging jet acting as both reactor agitator and gas-bubble generator. Thus specific interfacial areas are in the range 0.2–1.1 cm^{-1}, smaller than those of the submerged jet contactor because of low values of gas holdup ($\alpha < 0.06$).

2. Ejector Reactor

The ejector reactor conceptually is a bubble reactor into which a liquid jet entraining gas is injected (N2) (Fig. 32). The ejector provides high liquid velocities (over 20 m/sec) which entrain gas by suction through a mixer device placed inside the reactor where there is intense gas–liquid contact. Also, gas and liquid are circulated inside the reactor by a pumping effect in such equipment, and the ejector and mixer devices act jointly as the reactor agitator and gas bubble generator. The liquid leaves the equipment by overflow in the presence of the gas phase, and the two phases are then separated. Nagel *et al.* (N2, N5) have performed complete and detailed studies on this type of reactor, measuring the interfacial area by the chemical method. Simultaneous measurement of phase flow rates, entrainment, and pressure drop has enabled these authors to compare the values of interfacial area with those of stirred tanks (R7) in terms of same specific input power relative to the volume of the reactor (Fig. 32). Values obtained with the ejector are always higher for similar operating condi-

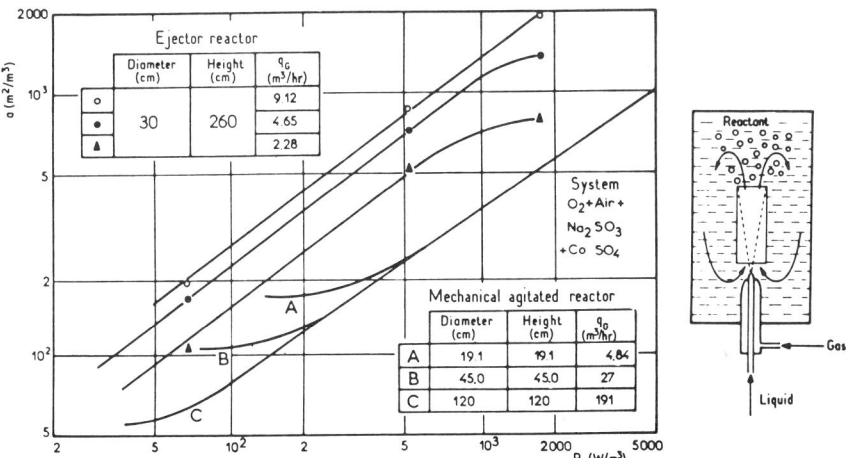

FIG. 32. Interfacial area in ejector reactor: comparison with mechanically agitated reactor (N2, N5).

tions. Recently Zehner (Z1) has proposed mixer devices able to increase still more the interfacial area values for this type of contactor.

3. *Venturi Scrubber*

In venturi equipment, the liquid is injected through a nozzle into a high-velocity gas stream. The liquid is then atomized by the formation and subsequent shattering of attenuated, twisted filaments and thin cuplike films which provide a high degree of turbulence and large interfacial areas for heat and mass transfer. In the breakup, nearly spherical droplets are formed which also provide a large surface area per unit volume of liquid, but then the degree of turbulence is decreased. Venturi scrubbers are often used for simultaneous removal of gaseous and particulate pollutants. Their advantages are high volumetric flow rate; simplicity, compactness, and absence of moving parts, all leading to a low first cost of equipment; and the ability to handle slurry absorbents. Disadvantages are the short gas–liquid contact time, high pressure drop on the gas side, the need for a phase separator after the scrubber, and the limitation to applications with large volumetric gas/liquid ratios (at small ratios, efficient atomization of the liquid does not occur.)

Two principal modes of operation commonly employed are liquid injection into the throat of the venturi and liquid introduction so as to wet the entire convergent section. The second mode of operation gives a lower pressure drop under similar operating conditions (S15). Excellent reviews of the literature pertaining to pressure drop, atomization, and efficiency of venturis are given by Uchida (U1) and Wen and Fan (W10). Most of the experimental data are given in terms of gas absorption efficiency, and only a few studies lead to mass-transfer correlations. The overall pressure drop across the venturi scrubber is an important practical factor that must be considered in reactor design, since the mass-transfer rates are thus directly related to the power consumed. The pressure drop can be estimated by considering friction along the wall, which depends largely upon the geometry of the scrubber, and acceleration of the injected liquid, which is insensitive to the geometry and is often predictable theoretically. Most of the available pressure-drop correlations have been obtained on relatively small-scale, specific equipment (V12, B15, N7). Only Calvert (C6) and Boll (B19) have developed equations applicable to venturi scrubbers in general. Calvert's equation is

$$\Delta P_{LG} = 10^{-8} \times u_G^2 r \qquad (127)$$

where ΔP_{GL} is the pressure drop across the venturi in cm H_2O, u_G is the gas velocity relative to the duct in cm/sec, and r is the liquid/gas ratio in liters/m³. Though this equation neglects frictional loss at the wall, it predicts the pressure drop with fair accuracy. If, for reactor design, it is necessary to estimate the pressure profile within the venturi, the more detailed equations of Boll (B19) and Uchida (U1) can be used, which take into consideration the different terms of the momentum balance.

Another important parameter is the mean surface-to-volume droplet diameter d_s, which affects the velocity and the absorption behavior of the liquid phase. Among the available correlations the most widely used is that of Nukiyama and Tanasawa (N13),

$$d_s = 585 \times 10^{-3} \frac{\sigma_L^{1/2}}{v_t(\rho_L^{1/2})} + 597 \times 10^{-3} \left(\frac{\mu_L}{(\sigma_L \rho_L)^{1/2}}\right)^{0.45} \left(\frac{q_L}{q_G} \times 10^3\right)^{1.5} \quad (127a)$$

where d_s is, in mm, q_L and q_G are the liquid and gas volumetric flow rates in cm³/sec, μ_L is the liquid viscosity in P, ρ_L is the liquid density in gm/cm³, σ_L is the surface tension in dyn/cm, and v_t is the relative velocity of gas and liquid at the throat in cm/sec. This correlation is suitable in the range of high relative velocities (100 m/sec to sonic), a large gas/liquid mass ratio (from 1 to 50), and viscosities up to 50 cP.

As already mentioned, many studies have been reported on mass transfer in venturi scrubbers in terms of efficiencies for specific equipment and specific problems (absorption of SO_2, NO_2, NH_3; desorption of CO_2, O_2, and so on). So only qualitative relations of a, $k_L a$, and $k_G a$ to the flow parameters can be proposed.

The gas-side mass-transfer coefficients $k_G a$ and k_G increase with liquid feed rate or with gas velocity at each given position in the venturi scrubber and decrease at constant liquid rate and gas velocity with increasing distance from the point of liquid injection (J7, V11). The values of $k_L a$ generally increase with increasing liquid flow rate or gas velocity (often referred to as the velocity at the throat). However, $k_L a$ will sometimes exhibit a maximum when the gas velocity increases; the explanation is that, at higher gas velocities, an increase in turbulence in the throat of the venturi results in the formation of droplets smaller than the thin filaments first formed at lower gas velocities. Internal circulation is reduced in these smaller droplets, and there is also a reduction in the size of the zone of intense turbulence. These two phenomena lead to a maximum for the values of k_L, as found experimentally by Kuznetsov and Oratovskii (K15) and Virkar and Sharma (V11). The values of the effective interfacial area a increase with both gas and liquid flow rates.

Virkar and Sharma are the only investigators to have reported systematic measurements of a, $k_L a$, and $k_G a$ by the chemical method. Their work utilized laboratory-scale vertical venturis, operating with liquid injection either at the throat or ahead of the convergent section. For example, for wetted-approach operation, in the range of gas velocities from 50 to 90 m/sec and liquid flow rates from 7 to 30 cm^3/sec, the following equations were proposed and are given here for illustration:

$$a = 2(\Delta P/\Delta h) + 0.65 \tag{128}$$

and

$$k_G a \times 10^{+10} = 0.25 \times 10^{-2} \Delta P \sqrt{r} + 1.56 \tag{128a}$$

where ΔP and Δh are the pressure drops across the venturi and across the convergent section in N/m^2, r is the liquid/gas ratio in liters/m^3, a is in cm^2/cm^3, and $k_G a$ is in kg moles/m N sec). The reactor volume is taken to be the volume of the divergent cone plus the volume of the spray zone in the separator. The venturi had a circular throat of 16 mm diameter and 9 mm length, and the angles of the convergent and divergent sections were 35° and 8°, respectively. The entrance and exit sections were each 5 cm in diameter. These equations are specific to one type of equipment of laboratory scale and are of doubtful validity for other types and sizes. With the present state of knowledge, the mathematical model recently proposed by Wen and Fan (W10) and Uchida (U1) will be of great help in describing the hydrodynamics and heat and mass transfer in industrial-scale venturis.

It is important to note that large extents of mass and heat transfer take place near the liquid nozzle. Usually the operation of the venturi scrubber is almost isothermal, except for a few centimeters from the point of liquid injection. The effect of any solid phase in the liquid may be considered negligible; the liquid residence time is so short that very little dissolution can take place. An interesting feature is that decreasing the surface tension results in an increase in a because of a decrease in the mean droplet diameter.

H. Conclusions Regarding Scale-Up Problems

In this section, data for the mass-transfer parameters most often measured by the chemical method, that is, the integral values of a, $k_L a$, and $k_G a$ (and their dependence on fluid and equipment parameters), have been presented. In practice, their mode of use depends upon which of the following problems confronts the design engineer:

1. *The equipment has already been in operation,* but the problem is to change the gas–liquid system in the existing reactor for economic reasons. Although the procedure is time-consuming, it seems reasonable to predict the performance from experiments carried out in a small-scale laboratory apparatus with the same k_L, k_G, and a/β as the existing industrial equipment (see Section V). In most practical cases, these laboratory experiments simultaneously will provide the needed knowledge of the reaction kinetics.

2. *The type of gas–liquid reactor has been chosen* (packed column, plate column, mechanically agitated tank, and so on), and the problem is to size the equipment, using published mass-transfer data. In such case, the diagrams and correlations given above for contactors with diameters smaller than 0.4 m can be used with a fair degree of confidence to scale up tubular packed, spray, and plate columns to 2 or 3 m in diameter. For mechanically agitated or bubble reactors, small-scale experiments are recommended with the given gas–liquid system in the laboratory apparatus ($D = 5$–20 cm) similar in shape, agitation, and contact time to the chosen reactor type; and then, to scale up the system, the same specific interfacial area is ensured by a constant total power input per unit liquid volume; that is, $P_T/a'''V_L$ and u_G are constant, in a mechanically agitated reactor, and $\rho_L g\, u_G/a'''\beta$ is constant, in a tubular bubble reactor without agitation.

3. *A suitable type of reactor has to be chosen prior to sizing.* This is an economic problem, with competition between the value of the interfacial area and the energy expense required to create it. In such cases, Fig. 33 in Nagel *et al.* (N4) provides a comparison between the extents of interfacial area provided by the major types of equipment and their energy costs. Once the choice is made, the remaining part of the design can be carried out as in case 2.

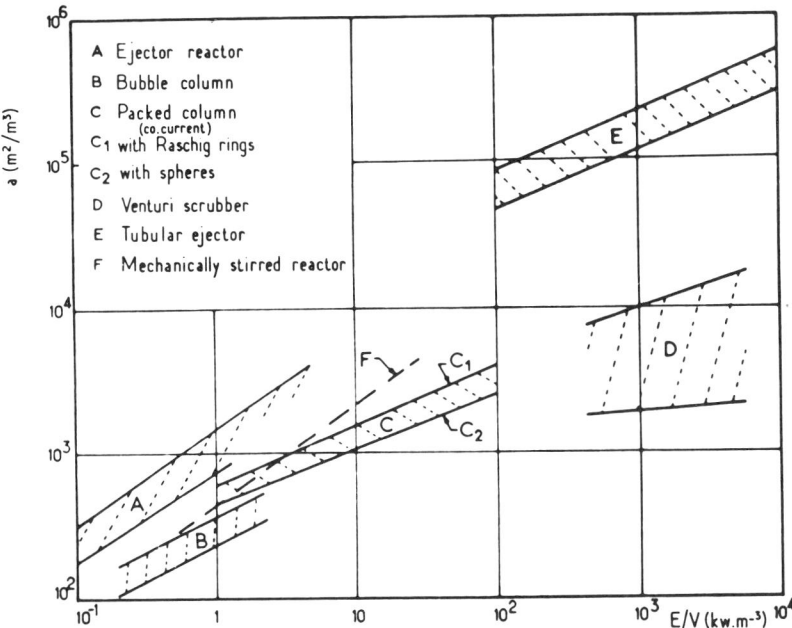

FIG. 33. Interfacial area in several types of reactors (N4).

Axial Dispersion. In designing absorption columns, the common practice is to assume plug flow, ignoring axial dispersion. As explained by Dunn et al. (D17), the phenomenon of axial mixing arises from the fact that, for various reasons, packets of fluids of one or both phases do not all move through the reactor at a constant and uniform velocity. This tends to reduce the concentration driving force for mass transfer from that which would exist for piston flow, and this can lead to unsafe designs; that is, to achieve a given separation more transfer units are required, or for a given column under conditions of axial mixing the overall mass transfer coefficients based on a difference in the solute concentration measured in the inlet and outlet streams are smaller than the actual mass transfer coefficients.

The diffusion model is generally used to take the effect of axial dispersion in mass-transfer equipment into account. When the dispersion is characterized in each phase, analytical solutions of the set of equations of this model, or tabulated numerical results or approximate equations, are proposed by Mecklenburgh and Hartland (M9a) and Miyauchi and Vermeulen (M15a). Note that in many practical cases it is the dispersion in the phase where there is a large concentration change that is important. However, Miyauchi and Vermeulen (M15a) have provided a calculation procedure with axial dispersion and concentration change in both phases that will be useful for the design of a gas–liquid reactor with backmixing in using the values of the mass-transfer coefficients presented in this section.

V. Simulation of Absorbers in Laboratory-Scale Apparatus Using Chemical Reaction

In this section, we are concerned with laboratory apparatus in which absorption experiments are carried out to simulate the behavior of industrial absorbers. Among the different steps in designing such equipment, the determination of solubility and diffusivity of one or several solutes in a reacting solution with unknown kinetics can be a challenging problem. These difficulties have justified making relatively simple laboratory models with a well-defined interfacial area and carrying out experiments to obtain data directly applicable to design. The aim is thus to predict the effect of chemical reaction in an industrial absorber from tests in a laboratory model with the same gas–liquid reactants, or to predict the reactor length for a specified task, using data from the laboratory model, even though the means of agitating the liquid in the two types of equipment is quite different.

A. Criteria for Simulating an Industrial Absorber

Consider a small but representative volume element $\Omega\, dh$ of an industrial tubular absorber in which a gas–liquid reaction occurs:

A (dissolved solute) + zB (liquid reactant) → products

The material balance equations for this case, when the gas contains only one soluble component and the liquid only one reactant, are

$$\varphi(k_L, k_G, C_A^*, C_{A0}, C_{B0}) a\Omega\, dh = -\frac{\Omega u_G}{V_M P} dp_h$$

$$= u_L \Omega\, dC_{A0} + r(c_{A0}, C_{B0})\beta\Omega\, dh \quad (129)$$

$$-u_L \Omega\, dC_{B0} = zr(C_{A0}, C_{B0})\beta\Omega\, dh \quad (130)$$

where P is the pressure in the absorber, u_L and u_G are the superficial liquid and gas velocities, β is the liquid holdup, p_h is the bulk partial pressure of solute gas at length h, V_M is the molar volume of the gas, $r(C_{A0}, C_{B0})$ is the reaction rate, C_{A0} and C_{B0} are the bulk concentration of reactants, z is the moles of B reacting with each mole of gas, and $\varphi(k_L, k_G, C_A^*, C_{A0}, C_{B0})$ is the rate of absorption per unit interfacial area in the volume element.

Rearranging these two equations leads to double numerical integration over the length h of the absorber:

$$\int_{in}^{out} \frac{a\, dh}{u_L} = \int_{in}^{out} \frac{dC_{A0}}{\varphi(k_L, k_G, C_A^*, C_{A0}, C_{B0})} - \frac{1}{z}\int_{in}^{out} \frac{dC_{B0}}{\varphi(k_L, k_G, C_A^*, C_{A0}, C_{B0})}$$

$$= -\frac{1}{V_M P}\int_{in}^{out} \frac{u_G dp_h}{u_L \varphi(k_L, k_G, C_A^*, C_{A0}, C_{B0})} \quad (131)$$

$$\int_{in}^{out} \frac{\beta\, dh}{u_L} = \frac{1}{z}\int_{out}^{in} \frac{dC_{B0}}{r(C_{A0}, C_{B0})} \quad (132)$$

Consider now a laboratory absorber in which the liquid and gas are agitated in a way that gives rise to mass-transfer coefficients k_G and k_L of the same magnitude as in the industrial absorber. Also assume that the ratio u_G/u_L, the concentrations C_A^*, C_{A0}, and C_{B0}, the temperature, and the pressure in the laboratory absorber are the same as those of the volume element located at length h of the industrial absorber. Then the specific rate of absorption $\varphi = \varphi(k_L, k_G, C_A^*, C_{A0}, C_{B0})$ is the same in both apparatuses, and the right side of Eq. (131) is also the same, for the same concentration or partial pressure limits (that is, entrance and exit bulk concentrations).

It follows that the ratio ah/u_L is identical for both absorbers, hence

$$\frac{h}{h_m} = \frac{u_L/a}{(u_L/a)_m} \qquad (133)$$

where the subscript m refers to the laboratory model. Similar considerations applied to Eq. (132) lead to

$$\beta h/u_L = (\beta_m h_m)/(u_L)_m$$

which means that the space time is the same in the volume element of the industrial absorber and in the laboratory model. Combining Eqs. (133) and (134) gives

$$a/\beta = a_m/\beta_m = A_m/\beta_m V_m = A_m/V_L \qquad (135)$$

where A_m is the geometric interfacial area of the model, and V_L is the liquid volume of the model having total volume V_m. Equation (135) shows that, per unit volume of absorber or reactor, the ratio of interfacial area to liquid holdup is the same in the laboratory model and the industrial absorber.

Therefore the three criteria for simulation are identical values of k_L, k_G, and a/β in the industrial and the laboratory absorbers. "Simulation" means that, if the bulk compositions of gas and liquid in the laboratory absorber are the same as in a volume element of the industrial absorber, the absorption rate per unit interfacial area φ in this element will be the same as in the laboratory model φ_m, whatever the means of agitating the gas and the liquid in the two absorbers. Thus $\varphi_m = \varphi$ can be determined experimentally as a function of C_{A0}, C_{B0}, or p, and Eq. (130) can be integrated numerically step by step between the limit compositions at the entrance and the exit of the industrial absorber to find the length h of the absorber.

The third criterion (constant a/β) is required whenever the reaction between dissolved gas and a reactant in solution is slow. Indeed reactions will proceed in the bulk liquid, and the rate of absorption in the industrial or laboratory absorber will depend upon the volume of bulk liquid available per unit area of interface.

However, until now, published studies on this topic (D5, L4) have been confined to cases in which all the reactions that determine the rate of absorption occur essentially in the liquid diffusion film and the concentration of dissolved gas in the bulk of the liquid is zero.

B. PRACTICAL SELECTION OF A SIMULATIVE LABORATORY MODEL

1. *Absorption and Reaction with a Pure Gas*

The first criterion for simulation involves identical values of k_L in both apparatuses. Indeed the average rate of absorption (per unit interfacial areas) of a (pure) gas solute into a liquid with which it reacts is determined partly by physicochemical factors (solubility, diffusivity, kinetics) and partly by hydrodynamic factors (gas and liquid flow rates and shape and size of the absorber). It is known that simple theoretical models of absorption with chemical reaction lead to conclusions about the effect of the chemical reaction that are numerically close, whether the film model or a surface-renewal model is used as the basis of calculation (Section II,B). This means that, for a fixed value of k_L, the different theories give approximately the same value for the average rate of absorption. It follows that, to design a laboratory model of an industrial pure-gas absorber, the essential feature to reproduce is the true liquid-side mass-transfer coefficient k_L (that is, the same mean contact time).

Several laboratory models for simulating large-scale absorbers are shown in Fig. 8. The ranges of contact time for different industrial and laboratory absorbers are given in Table XXV. This table should aid in the selection of laboratory equipment to give a contact time (or k_L) that simulates the industrial absorber. For instance, a wetted wall or stirred vessel can simulate a packed column.

TABLE XXV

Absorber Simulation Based on Equality of k_L

Laboratory model	Contact time (sec)	Industrial absorber	Contact time (sec)	Laboratory model for simulation
Cylindrical wetted wall	0.1–2	Plate column	0.019–0.04	Laminar jet, rotating drum
Spherical wetted wall	0.1–1	Packed column	0.03–1	Cylindrical wetted wall, string of disks or of spheres, stirred vessel
Conical wetted wall	0.2–1	Venturi	0.004	Laminar jet
Laminar jet	0.001–0.1	Spray column	0.03–0.1	Laminar jet, rotating drum
String of disks or of spheres	0.1–2	Well-stirred tank	0.01	Laminar jet, rotating drum
Rotating drum	0.0002–0.1	Bubble column	0.01	Laminar jet, rotating drum
Stirred vessel	0.06–10	—	—	—

2. Absorption and Reaction with a Dilute-Gas Solute

When the absorption involves a *dilute* solute in the gas phase, there is resistance in the gas phase as well. Then simulation also requires identical values of the true gas-side mass-transfer k_G. In the laboratory model, the values of k_G may be adjusted to that of the industrial equipment either by varying the gas flow rate (D2) or by stirring the gas phase (D2, D6, B16, H8).

Experimental values of k_G in laboratory models are listed in Table XXVI. An example of absorption with equal values of k_L and k_G in both apparatuses will be given in Section V,C.

3. Absorption and Complex Reaction with a Dilute-Gas Solute

In the two previous cases, reaction between the dissolved gas and the liquid reactant was assumed fast enough to occur only within the liquid film. But when the reaction and mass-transfer resistances are of the same magnitude, that is, when reaction occurs both within the film and within the bulk liquid (or only in the bulk liquid), the third criterion for simulation is necessary, that is, equality of the ratios of interfacial area to liquid holdup. This criterion may be satisfied by the laboratory double-mixed contactor suggested by Levenspiel and Godfrey (L11) shown on the right side of Fig. 8—a steady-flow contactor with uniform gas and liquid compositions maintained by stirring each phase independently, which provides independent control of the interfacial area and the surface/volume ratio of the phases by using replaceable interface plates having holes which give from 2 to 30% open area.

A string of spheres with holes full of liquid could be considered an

TABLE XXVI

VALUES OF k_G IN LABORATORY MODELS (L2)

Laboratory model	k_G (10^5 moles/cm² sec atm)	Varying parameter	Reference
String of disks or spheres	7–25	Relative velocity of the fluids	S38
Laminar jet	10–40	Stirrer speed in the gas phase	B16
Stirred vessel	1–15	Stirrer speed in the gas phase	H8
Cylindrical wetted wall	1–9	Gas velocity and stirrer speed in the gas phase	K13

alternative to this type of laboratory model (L2), which has been proposed for simultaneous absorption of several gases into a liquid containing several reactants.

C. EXAMPLE: STIRRED CELL SIMULATING A PACKED COLUMN

The use of a stirred cell to simulate a packed column is widespread (D5, K7, D7, C11, L5). Let us consider its use in predicting absorption rates at different levels of a column, when absorption occurs with a fast second-order irreversible chemical reaction and with mass-transfer resistance in both phases.

Variations in k_L and k_G with the respective phase stirring rates are determined first. Determination of k_L as a function of liquid-side stirrer speed N_L is carried out by physical absorption, absorption with slow chemical reaction, or absorption with an instantaneous chemical reaction of a pure gas for a well-known geometric interfacial area (generally the cross-sectional area of the stirred cell minus that of the blade-type stirrer). To determine k_G as a function of gas-side stirrer speed N_G, absorption with instantaneous chemical reaction of a dilute solute is carried out. These values of k_L and k_G are specific to the geometry of the laboratory apparatus and to the solute gas. For other gases, k_G under the same conditions can be estimated by assuming it to vary as the square root of the diffusivity D_G of the solute gas. In the same way, k_L for other liquids under the same conditions may be assumed to vary as the square root of the diffusivity D_L of the dissolved gas.

Similarly, when values of a, $k_L a$, and $k_G a$ are known for the packed column, the variations in k_L and k_G can be determined as functions of the respective flow rates u_L and u_G, as discussed in Section IV,A. Various stirring speeds for each phase in the laboratory model can then be related to the phase flow rates (in the packed column) which have, respectively, identical k_L and k_G, as outlined in Fig. 34. Thus pairs of values of u_L and u_G are simulated by each couple N_L and N_G.

The next step is measurement, or theoretical calculation when possible, of the average rates of absorption per unit interfacial area φ_m of the chemical system in the laboratory model where k_L and k_G are adjusted to be the same as in the packed column. These measurements are carried out for different liquid and gas compositions representative of different levels in the column and are reported as plots of φ_m versus p for different reactant concentration contours. Knowledge of these absorption rates is essential for predictive calculation of the column length h, as the consecutive values of φ_m from the stirred cell must be used to integrate Eq. (131) between the inlet and outlet conditions:

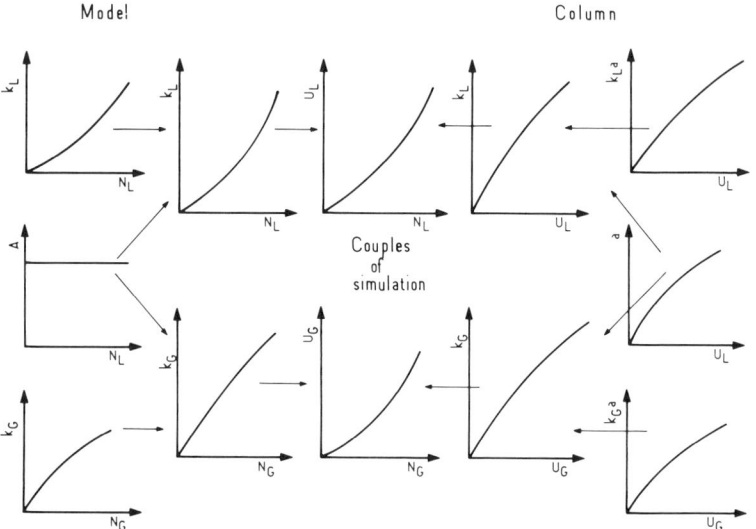

FIG. 34. Schematic procedure for establishing model column couples for simulation between (U_L, N_L) and (U_G, N_G).

$$h = \frac{u_G}{aV_m P} \int_{\text{out}}^{\text{in}} \frac{dp_h}{\varphi_m(k_L, k_G, C_A^*, C_{B0})} = \frac{u_L}{za} \int_{\text{out}}^{\text{in}} \frac{dC_{B0}}{\varphi_m(k_L, k_G, C_A^*, C_{B0})} \quad (136)$$

Laurent (L2) considers the use of a 10-cm-i.d. double-stirred cell to simulate a 30-cm-i.d. column packed with 20-mm glass Raschig rings to a height of 1.92 m. The values of k_L and k_G of the model leading to the simulating couple N_L, u_L, and N_G, u_G in Fig. 34 were determined by absorption of pure CO_2 into different liquids (water, Na_2SO_3, NaOH) and absorption of dilute SO_2 into NaOH. Then CO_2 from air was absorbed into NaOH and Na_2CO_3 solutions, both in the stirred cell and in the packed column, so as to compare the predictions from the model with the packed-column results for different values of u_L, u_G, and gas and reactant concentrations. The experimental values of φ_m measured in the stirred cell for different bulk compositions of gas and liquid, given in Fig. 35, were used to integrate Eq. (136) for given feed and product conditions. The resulting values of h are the predicted height in the table accompanying Fig. 35. The predicted and actual heights differed by less than 20% in all cases, indicating that this method is quite sound. Danckwerts and Alper (D7) have obtained even better prediction (within 10%) using different packing heights in a 10-cm-i.d. column where the fluid flow rates were kept constant.

D. Applications

It appears that laboratory equipment of the types shown in Fig. 8 can be used to predict the effects of chemical reaction on absorption rates in industrial absorbers with a degree of accuracy representing an improvement over present design procedures. This promising simulation method does not require detailed knowledge of solubilities, diffusivities, or reaction kinetics. Usually the method requires knowledge of k_L and k_G, which can be easily determined experimentally in a laboratory model.

This technique, suggested by Peaceman (P3a) for absorption and desorption of chlorine in water in a packed column or a short wetted-wall column, and fully developed by Danckwerts and Gillham (D5), has been successively used by Jhaveri and Sharma (J6) for absorption of O_2 into solutions of sodium dithionite, by Yano et al. (Y2) for liquid-phase oxida-

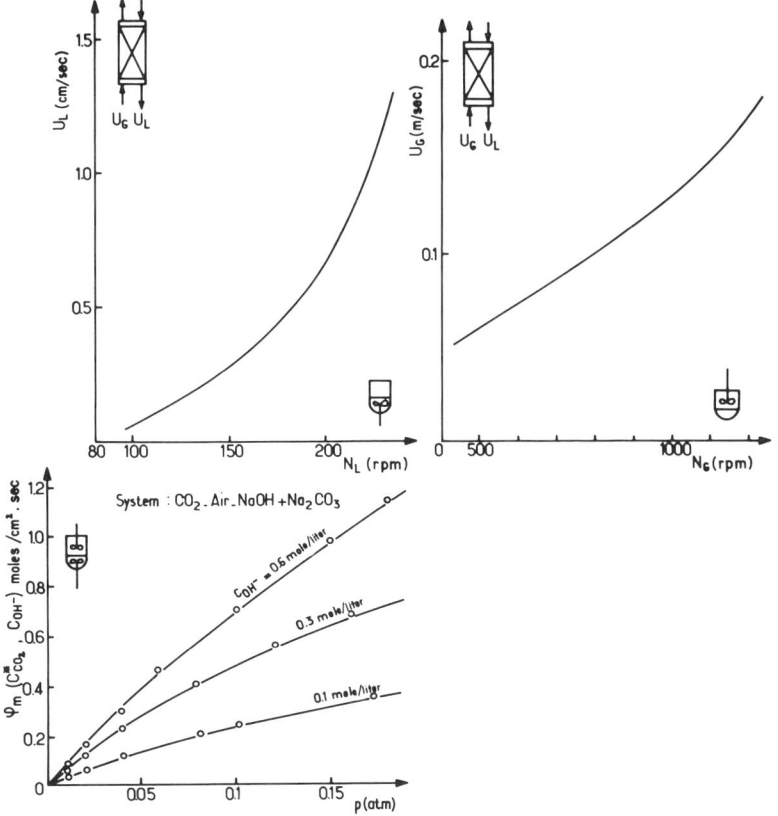

FIG. 35. Simulation of a packed column by a stirred cell (L2). (See tabulation on p. 122.)

TABULATION FOR FIGURE 35

RESULTS OF ABSORPTION OF CO_2–AIR INTO AQUEOUS SOLUTIONS OF $NaOH$ + Na_2CO_3 EXPERIMENTS IN A PACKED COLUMN[a]

u_G (cm/sec)	$p_{CO_2} \times 10^2$ (atm) Inlet	$p_{CO_2} \times 10^2$ (atm) Outlet	u_L (cm/sec)	$C_{BO} = C_{OH^-}$ (moles/liter) Inlet	$C_{BO} = C_{OH^-}$ (moles/liter) Outlet	$\int_{IN}^{OUT} \frac{dc_{BO}}{\varphi_m}$ (sec/cm)	H (m)	Predicted height (m)	Difference (%)	Error in material balance (%)
7.5	6.2	0.4	0.5	0.60	0.52	654		1.72	10.4	9.9
7.5	3.2	0.01	0.5	0.60	0.55	578		1.52	20.8	25
6.9	2.9	0.05	0.23	0.59	0.52	1120		1.84	4.2	4
6.9	5.5	0.6	0.23	0.59	0.47	1102		1.81	5.7	3.5
6.9	7.3	0.75	0.23	0.59	0.42	1144		1.88	2.1	2.4
7.6	2.5	0.4	0.5	0.30	0.27	673	1.92	1.77	7.8	11.9
7.5	2.8	0.45	0.23	0.30	0.24	1096		1.80	6.2	6.7
10.9	2.0	0.5	0.23	0.30	0.24	1145		1.88	2.1	0
7.5	2.6	0.3	0.5	0.28	0.25	692		1.82	5.2	4
7.4	4.2	0.4	0.5	0.28	0.23	722		1.90	1.0	5.9

[a] $D = 30$ cm, $d = 20$ mm, $T = 25°C$, $P = 1$ atm; a, k_L, and k_G are experimentally determined.

tion of propylene with PdCl$_2$, and by Gehlawat and Sharma (G5) and Shaffer *et al.* (S14) for absorption of butenes into H$_2$SO$_4$ or into trifluoroacetic acid. In all cases reported, the chemical absorption rates measured for a pure gas in a stirred vessel as a model have made it possible to predict the rate in a packed column to within 20%.

For the absorption of a dilute gas solute, this technique has been used for absorption of CO$_2$ into NaOH and Na$_2$CO$_3$ solutions (D7, L2). Thus the path has been prepared for simulation to be used, whenever desired, for slow, complex, or simultaneous reactions whenever these are met in practice. For such systems, more elaborate laboratory equipment is required that makes it possible to vary the ratios of interfacial area to liquid holdup (L11).

VI. Nomenclature

In particular equations, the units that apply to a variable may be different from the units shown here, and if so are indicated in each specific case. When correlations in the literature are dimensional, the units are sometimes English and sometimes metric.

A, B, C   Components
- A_m Geometric interfacial area of a laboratory model, cm^2
- a Effective interfacial area per unit packed volume, cm^{-1}
- a' Effective interfacial area per unit volume of froth, or dispersion, cm^{-1}
- a'' Effective interfacial area per unit area of plate, cm^2/cm^2
- a''' Effective interfacial area per unit volume of clear liquid, cm^{-1}
- a_c Total dry area of packing per unit of packed volume, cm^{-1}
- a_d Effective interfacial area based on dispersion volume in packed bubble columns, cm^{-1}
- a_g Specific surface area of packing, cm^{-1}
- C_A^* Saturated concentration of gas A, moles/liter
- C_{A0}, C_{B0} Bulk concentration of dissolved gas and of reactant, moles/liter
- D Diameter of reactor, cm
- D_A Molecular diffusivity in very dilute solution, cm^2/sec
- D_A^m Molecular diffusivity in very dilute solution of mixed solvents, cm^2/sec
- $(D_A)_{conc}$ Molecular diffusivity in concentrated solutions, cm^2/sec
- d Diameter of jet or tube, cm
- d_A Diameter of agitator, cm
- d_b Bubble diameter, cm
- d_{SM} Sauter mean diameter, cm
- d_T Tank diameter, cm
- E Enhancement factor
- E_i Enhancement factor for instantaneous reaction
- F Percentage free area of the plate
- f_A Fugacity of component A, atm
- G Gas mass superficial flow rate, kg/m^2 sec
- g Gravitational acceleration, cm/sec^2
- g_c Gravitational conversion fac-

	tor, force/(mass)(acceleration)
H, h	Length of reactor or depth of clear liquid
H	Henry constant, atm cm^3/gm mole
h_G, h_+, h_-	Contributions of gas, positive ion, and negative ion to salting coefficient h, liters/gm mole
h_A	Height of agitator from base of reactor
I	Ionic strength, gm moles/liter
k_B	Boltzmann's constant
k_c	Catalyst rate constant, liters/gm mole sec
K_H	Henry constant, atm
k, k_1, k_2	Rate constant; for first and second order, sec^{-1} and cm^3/gm mole sec
k_{mn}	Rate constant for reaction mth order in A, nth order in B, (cm^3/gm mole)$^{m+n-1}$/sec
k_G, k_L	True gas liquid-side mass-transfer coefficient, cm/sec
k_S	Liquid-to-solid mass transfer coefficient, cm/sec
L	Liquid mass superficial flow rate, gm/cm^2 sec
l	Length of the generatrice of a cone, cm
M	Molecular weight
m	Molality, moles/1000 gm solvent
N_L, n	Rate of revolution of liquid stirrer, rpm
N_G	Rate of revolution of gas stirrer, rpm
n_o	Minimum speed of agitator
Os	Ostwald coefficient
p	Partial pressure, atm
P	Total pressure, atm
P_m	Mechanical agitation power, W
P_s	Vapor pressure, atm, or power expense to sparge the gas, W
Q	Amount of gas absorbed by unit area during time of contact θ, gm moles/cm^2
Q_G	Gas flow rate, cm^3/sec
Q_L	Liquid flow rate, cm^3/sec
r	Rate of reaction, gm moles/cm^3 sec
R	Rate of transformation in the bulk of liquid, gm moles/cm^3 sec
S	Submergence, cm
s	Fractional rate of surface renewal, sec^{-1}
T	Absolute temperature, K
T_c	Absolute critical temperature, K
u_L, u_G	Liquid and gas superficial velocity, cm/sec
$V°$	Molecular liquid volume, cm^3/gm mole
$\bar{V}_B$	Partial molecular volume of solvent in solution, cm^3/gm mole
V_L	Volume of liquid held in a laboratory model or in a reactor, m^3
V_m	Molecular volume of pure liquid component at its normal boiling temperature, cm^3/gm mole
V_R	Aerated volume or reactor volume, m^3
X_A, X_B, X_C	Mole fractions of A, B, and C
X_A^*	Mole fraction solubility of gas in solution
x	Associated factor for the solvent
Z	Height of clear liquid, cm
Z_a	Height of aerated liquid, cm
Z_D	Concentration diffusion parameter
z	Number of moles of B reacting with 1 mole of A
z_+, z_-	Absolute values of cation and anion valences
α	Gas holdup
β	Liquid holdup
γ_A	Activity coefficient of the solute
$\gamma_\pm$	Mean molar activity coefficient
ΔE	Latent heat of vaporization, cal/gm or cal/gm mole
ΔH	Heat of solution, cal/gm or cal/gm mole
ΔP	Pressure loss: ΔP_{LG} for two-

	phase flow, ΔP_L for single-phase flow of liquid, ΔP_G for single-phase flow of gas, N/m² or mH₂O/m
δ	Solubility parameter, (cal/cm³)^(1/2)
δ_L, δ_G	Thickness of diffusion film in liquid and in gas, cm
δ_r	Distance from interface to reaction plane, cm
ϵ	Porosity (void fraction)
θ	Time of exposure of liquid to gas, sec
$\lambda_+^\circ, \lambda_-^\circ$	Electrolytic conductance at infinite dilution, mho cm²/gm equivalent
μ_{AB}, μ_{ABC}	Viscosity of the solution, cP
μ_B	Viscosity of the solvent, cP
μ_{WT}	Viscosity of water at T, cP
σ	Surface tension, dyn/cm
σ_c	Critical surface tension of liquid for a particular packing material, dyn/cm
Φ	Rate of absorption per unit volume of reactor, gm moles/cm³ sec
Φ'	Volumetric rate of absorption, cm³/sec
φ	Average rate of absorption per unit interfacial area, gm moles/cm² sec
φ'	Volumetric average rate of absorption, cm³/cm² sec
Ω	Cross-sectional area of the column, cm²

ACKNOWLEDGMENT

The author wishes to thank Dr. A. Laurent for the helpful advice and fruitful discussions, and Dr. B. I. Morsi for reading and correcting the proofs. Thanks are due also to Miss Claude Poulain for her wholehearted support and patience during preparation of this article.

References

A1. Abergel, A., Thesis, University of Nancy, France, 1974.
A2. Akgerman, A., and Gainer, J. L., *Ind. Eng. Chem. Fundam.* **11**, 373 (1972).
A3. Akita, K., and Yoshida, F., *Ind. Eng. Chem. Process. Des. Dev.* **1**, 76 (1973).
A4. Alper, E., *Trans. Inst. Chem. Eng.* **51**, 159 (1973).
A5. Ashley, M. J., and Haselden, G. G., *Trans. Inst. Chem. Eng.* **50**, T 119 (1972).
A6. Astarita, G., "Mass Transfer with Chemical Reaction." Elsevier, Amsterdam, 1967.
A7. Astarita, G., Marruci, G., and Coletti, L., *Chim. Ind.* **9**, 1021 (1964).
A8. Al Mojil, S. I., Clark, M. W., and Vermeulen, T., personal communication.
B1. Bakos, M., and Charpentier, J. C., *Chem. Eng. Sci.* **25**, 1822 (1970).
B2. Baldi, G., and Sicardi, S., *Chem. Eng. Sci.* **30**, 617 (1975).
B3. Baldi, G., and Sicardi, S., *Chem. Eng. Sci.* **30**, 769 (1975).
B4. Banerjee, S., Rhodes, E., and Scott, D. S., *Can. J. Chem. Eng.* **48**, 542 (1970).
B5. Barona, N., and Prengle, H. W., *Hydrocarbon Process.* **52**, 63 March (1973).
B6. Barona, N., and Prengle, H. W., *Hydrocarbon Process.* **52**, 73 Dec. (1973).
B7. Barrett, P. V. L., Gas absorption on a sieve plate. Ph.D. Thesis, Cambridge University, 1966.
B8. Barron, C. H., and O'Hern, H. A., *Chem. Eng. Sci.* **21**, 397 (1966).
B9. Bartholomai, G. B., Gardner, R. G., and Hamilton, W., *Br. Chem. Eng. Process. Technol.* **1**, 48 (1972).
B10. Battino, R., and Clever, H. L., *Chem. Rev.* **60**, 395 (1966).
B11. Beenackers, A. A., and Van Swaay, W. P. M., *Proc. Eur. Chem. Eng. Symp., Heidelberg*, 1976.
B12. Beimesch, W. E., and Kessler, D. P., *AIChE J.* **17**, 1160 (1971).
B13. Bennet, A., and Goodvidge, F., *Trans. Inst. Chem. Eng.* **48**, T 241 (1970).
B14. Bernard, J. D. T., and Sargent, R. W. H., *Trans. Inst. Chem. Eng.* **44**, T 314 (1966).

B15. Beskin, L. Z., Strel'tsov, V. V., and Demshin, V. Ya., *Int. Chem. Eng.* **1**, 88 (1969).
B16. Bjerle, I., Bengtsson, S., and Färnkvist, K., *Chem. Eng. Sci.* **27**, 1853 (1972).
B17. Blasinski, H., and Pyc, K. W., *Inz. Chem.* **3**, 647 (1973).
B18. Blasinski, H., and Pyc, K. W., *Int. Chem. Eng.* **15**, 409 (1975).
B19. Boll, R. H., *Ind. Eng. Chem. Fundam.* **12**, 40 (1973).
B20. Bourne, J. R., Von Stockar, U., and Goggan, G. C., *Ind. Eng. Chem. Process. Des. Dev.* **13**, 115, 124 (1974).
B21. Bossier, J. A., III, Farritor, R. E., Hughmark, J. A., and Kao, J. T. F., *AIChE J.* **19**, 1065 (1973).
B22. Boxhes, W., and Hofmann, H., *Chem. Ing. Technk.* **44**, 882 (1972).
B23. Boxkes, W., Doctoral Thesis, University of Erlangen-Nuremberg, W. Germany, 1973.
B24. Bretsnajder, "Prediction of Transport and Other Physical Properties of Fluid." Pergamon, New York, 1971.
B25. Brian, P. L. T., *AIChE J.* **10**, 5 (1964).
B26. Brian, P. L. T., Hurley, J. F., and Hasseltine, E. H., *AIChE J.* **7**, 226 (1961).
B27. Brian, P. L. T., Baddour, R. F., and Matiatos, D. C., *AIChE J.* **10**, 727 (1964).
B28. Brian, P. L. T., Vivian, J. E., and Matiatos, D. C., *AIChE J.* **13**, 28 (1967).
B29. Brian, P. L. T., Hales, H. B., and Sherwood, T. K., *AIChE J.* **15**, 727 (1969).
B30. Burgess, J. M., Molloy, N. A., and McCarthy, M. J., *Chem. Eng. Sci.* **27**, 442 (1972).
B31. Burgess, J. M., and Molloy, N. A., *Chem. Eng. Sci.* **28**, 183 (1973).
B32. Burgess, J. M., and Calderbank, P. H., *Chem. Eng. Sci.* **30**, 743 (1975).
B33. Burgess, J. M., and Calderbank, P. H., *Chem. Eng. Sci.* **30**, 1107 (1975).
B34. Byrne, J. E., Battino, R., and Danforth, W. F., *J. Chem. Thermodyn.* **6**, 245 (1974).
C1. Calderbank, P. H., *Trans. Inst. Chem. Eng.* **37**, 443 (1958).
C2. Calderbank, P. H., *in* "Mixing, Theory and Practise" (W. Uhl and J. B. Gray, eds.), p. 2. Academic Press, New York, 1967.
C3. Calderbank, P. H., and Moo-Young, M. B., *Chem. Eng. Sci.* **16**, 39 (1961).
C4. Calderbank, P. H., and Rennie, J., *Trans. Inst. Chem. Eng.* **40**, 3 (1962).
C5. Calderbank, P. H., Evans, F., and Rennie, J., *Proc. Int. Symp. Distill. (Int. Chem. Eng.)*, p. 51 (1960).
C6. Calvert, S., *AIChE J.* **16**, 392 (1970).
C7. Cameron, J. F., *Int. Conf. Radio Isotop. Sci. Res.* **1**, 426 (1957).
C8. Carleton, A. J., Flain, R. J., Rennie, J., and Valentin, F. H. H., *Chem. Eng. Sci.* **22**, 1839 (1967).
C9. Chappelow, C. C., and Prausnitz, J. M., *AIChE J.* **20**, 1097 (1974).
C10. Charpentier, J. C., *Chem. Eng. J.* **11**, 161 (1976).
C11. Charpentier, J. C., and Laurent, A., *AIChE J.* **20**, 1029 (1974).
C12. Charpentier, J. C., and Favier, M., *AIChE J.* **21**, 1213 (1975).
C13. Charpentier, J. C., Prost, C., and Le Goff, P., *Chem. Eng. Sci.* **24**, 1777 (1969).
C14. Chen, B. H., and Vallabh, R., *Ind. Eng. Chem. Process. Des. Dev.* **9**, 121 (1970).
C15. Clever, H. L., and Battino, R., *in* "Solutions and Solubilities" (M. Dack, ed.), p. 379. Wiley (Interscience), New York, 1975.
C16. Cooper, C. M., Fernstrom, G. A., and Miller, S. A., *Ind. Eng. Chem.* **36**, 504 (1944).
C17. Coulson, J. M., and Richardson, J. F., "Chemical Engineering," Vol. 3. Pergamon, New York, 1971.
C18. Cukor, P. M., and Prausnitz, J. M., *J. Phys. Chem.* **76**, 598 (1972).
D1. Danckwerts, P. V., *Ind. Eng. Chem.* **43**, 1460 (1951).
D2. Danckwerts, P. V., "Gas-Liquid Reactions." McGraw-Hill, New York, 1970.
D3. Danckwerts, P. V., and Kennedy, A. M., *Chem. Eng. Sci.* **8**, 201 (1958).
D4. Danckwerts, P. V., and Sharma, M. M., *Chem. Eng. (London)* 244 (1966).

D5. Danckwerts, P. V., and Gillham, A. J., *Trans. Inst. Chem. Eng.* **44**, T 42 (1966).
D6. Danckwerts, P. V., and Rizvi, S. F., *Trans. Inst. Chem. Eng.* **49**, T 24 (1971).
D7. Danckwerts, P. V., and Alper, E., *Trans. Inst. Chem. Eng.* **53**, 34 (1975).
D8. Davies, G. A., Porter, A. B., and Graine, K., *Can. J. Chem. Eng.* **45**, 372 (1967).
D9. Deckwer, W. D., Burckhart, R., and Zoll, G., *Chem. Eng. Sci.* **29**, 2177 (1974).
D10. De Coursey, W. J., *Chem. Eng. Sci.* **29**, 1867 (1974).
D11. De Kee, D., and Laudie, H., *Hydrocarbon Process.*, p. 224, Sept. (1974).
D12. De Waal, K. J. A., and Van Nameren, A. C., *AIChEI. Chem. Eng. Symp. Ser. 6, London* p. 60 (1965).
D13. De Waal, K. J. A., and Okeson, J. C., *Chem. Eng. Sci.* **21**,559 (1966).
D14. De Waal, K. J. A., and Beek, W. J., *Chem. Eng. Sci.* **22**, 585 (1967).
D15. Dim, A., Gardner, G. R., Ponter, A. B., and Wood, T., *J. Chem. Eng. Jpn.* **4**, 92 (1971).
D16. Dodds, W. S., Stutzman, L. F., Solami, B. J., and Carter, R. J., *AIChE J.* **6**, 197 (1960).
E1. Eckert, J. S., *Chem. Eng. Prog.* **66**, (3), 39 (1970).
E2. Eisenklam, P., and Ford, L. H., *Proc. Symp. Interact. Fluids Particles, Inst. Chem. Eng.* p. 333 (1962).
E3. Eissa, S. H., and Schügerl, K., *Chem. Eng. Sci.* **30**, 1251 (1975).
E4. Ertl, H., Ghai, R. K., and Dullien, F. A., *AIChE J.* **20**, 1 (1974).
F1. Fair, J. R., *Chem. Eng.* **74**, 47 (1967).
F2. Fair, J. R., Lambright, A. J., Andersen, J. W., *Ind. Eng. Chem. Process. Des. Dev.* **1**, 34 (1962).
F3. Field, L. R., Wilhem, E., and Battino, R., *J. Chem. Thermodyn.* **6**, 237 (1974).
F4. Fleury, D., and Hayduck, W., *Can. J. Chem. Eng.* **53**, 195 (1975).
G1. Ganguli, K. L., Ph.D. Thesis, Delft, Holland, 1975.
G2. Gardner, R. G., and McLean, A. Y., *AIChE I. Chem. Eng. Symp. Ser. 32* p. 239 (1969).
G3. Geddes, R. L., *Trans. Am. Inst. Chem. Eng.* **42**, 79 (1946).
G4. Gestrich, W., and Krauss, W., *Int. Chem. Eng.* **16**, 10 (1976).
G5. Gehlawat, J. K., and Sharma, M. M., *Chem. Eng. Sci.* **23**, 1173 (1968).
G6. Ghai, R. K., Ertl, H., and Dullien, F. A., *AIChE J.* **19**, 881 (1973).
G7. Gianetto, A., Baldi, G., and Specchia, V., *Ing. Chim. Ital.* **6**, 125 (1970).
G8. Gianetto, A., Specchia, V., and Baldi, G., *AIChE J.* **19**, 916 (1973).
G9. Gildenblat, I. A., *Teor. Osnory Chim. Technol.* **47**, 325 (1969).
G10. Glasstone, S., "Thermodynamics for Chemists." Van Nostrand-Reinhold, Princeton, New Jersey, 1947.
G11. Golding, J. A., and Mah, C. C., *Can. J. Chem. Eng.* **33**, 414 (1975).
G12. Gordon, A. R., *J. Chem. Phys.* **5**, 522 (1937).
G13. Goto, S., and Smith, J. M., *AIChE J.* **21**, 706 (1975).
G14. Goto, S., Levec, J., and Smith, J. M., *Ind. Eng. Chem. Process. Des. Dev.* **14**, 421 (1975).
G15. Greenhalgh, S. H., McManamey, W. J., and Porter, K. E., *Chem. Eng. Sci.* **30**, 155 (1975).
G16. Gregory, G. A., and Scott, D. S., *in* "Co-Current Gas-Liquid Flow" (E. Rhodes and D. S. Scott, eds.), p. 633. Plenum, New York, 1969.
G17. Gregory, G. A., and Scott, D. S., *Chem. Eng. J.* **2**, 287 (1971).
G18. Gurumurthy, C. V., and Govindarao, W. M. H., *Ind. Eng. Chem. Fundam.* **13**, 9 (1974).
H1. Hamilton, W., and Bartholomai, G. B., *Br. Chem. Eng. Process. Technol.* **16**, 1133 (1971).
H2. Harriott, P., *AIChE J.* **8**, 93 (1962).

H3. Harris, I. J., and Roper, G. H., *Can. J. Chem. Eng.* **41**, 158 (1963).
H4. Hayduk, W., and Laudie, H., *AIChE J.* **19**, 1233 (1973).
H5. Hayduk, W., and Laudie, H., *AIChE J.* **20**, 611 (1974).
H6. Higbie, R., *Trans. Am. Inst. Chem. Eng.* **35**, 365 (1935).
H7. Hikita, H., and Asai, S., *Int. Chem. Eng.* **4**, 332 (1964).
H8. Hikita, H., Asai, S., Ishikawa, H., and Saito, Y., *Chem. Eng. Sci.* **30**, 607 (1975).
H9. Hildebrand, J. H., Prausnitz, J. M., and Scott, R. L., "Regular and Related Solutions." Van Nostrand-Reinhold, Princeton, Jersey, 1970.
H10. Hildebrand, J. H., and Lamoreaux, R. H., *Ind. Eng. Chem. Fundam.* **13**, 110 (1974).
H11. Hirose, T., Toda, M., and Sato, Y., *J. Chem. Eng. Jap.* **7**, 187 (1974).
H12. Hixson, A. W., and Scott, C. E., *Ind. Eng. Chem.* **39**, 808 (1947).
H13. Hughmark, G. A., *Ind. Eng. Chem. Process. Des. Devp.* **6**, 218 (1967).
H14. Hwa, C. S., and Beckmann, R. B., *AIChE J.* **6**, 359 (1960).
J1. Jackson, G. S., and Marchello, J. M., *J. Chem. Eng. Jap.* **2**, 263 (1970).
J2. Jackson, M. L., *AIChE J.* **10**, 836 (1964).
J3. Jepsen, J. C., *AIChE J.* **16**, 705 (1970).
J4. Jhaveri, A. S., and Sharma, M. M., *Chem. Eng. Sci.* **22**, 1 (1967).
J5. Jhaveri, A. S., and Sharma, M. M., *Chem. Eng. Sci.* **23**, 1 (1968).
J6. Jhaveri, A. S., and Sharma, M. M., *Chem. Eng. Sci.* **24**, 189 (1969).
J7. Johnstone, H. F., Feild, R. B., and Tassler, M. C., *Ind. Eng. Chem.* **46**, 1601 (1954).
J8. Johnson, P. A., and Babb, A. L., *Chem. Rev.* **56**, 387 (1956).
J9. Juvekar, V. A., and Sharma, M. M., *Chem. Eng. Sci.* **28**, 976 (1973).
J10. Joosten, G. E. H., and Danckwerts, P. V., *Chem. Eng. Sci.* **28**, 453 (1973).
K1. Kamal, M. R., and Canjar, L. N., *Chem. Eng. Prog.* **62**, 82 (1966).
K2. Kastanek, F., Rylek, M., Kratochvil, J., and Hartman, M., *Congr. CHISA Session H3, Absorpt. Distill., Prague* (1975).
K3. Kasturi, G., and Stepanek, J. B., *Chem. Eng. Sci.* **29**, 713 (1974).
K4. Kasturi, G., and Stepanek, J. B., *Chem. Eng. Sci.* **29**, 1849 (1974).
K5. Kawecki, W., Reith, T., Van Heuven, J. W., and Beek, W. J., *Chem. Eng. Sci.* **22**, 1519 (1967).
K6. Kishinenevskii, M. K., Kormenko, T. S., and Popa, T. M., *Theor. Found. Chem. Eng.* **4**, 641 (1971).
K7. Koch, R., Kuciel, E., and Kuzniar, J., "Stoffaustausch in Absorptionskolonnen" Beiträge zur Verfahrenstechnik. VEB Deutscher Verlag für Grundstoffindustrie, Leipzig, 1969.
K8. Kohl, A. L., *AIChE J.* **2**, 264 (1956).
K9. Kohl, A. L., and Riesenfeld, F. C., "Gas Purification," 3rd. Ed. Gulf Publ., Houston, 1979.
K10. Kolev, N., *Verfahrenstechnik* **7**, 3, 71 (1973).
K11. Kritchevsky, I., and Ilinskaya, A., *Acta Physicochim. USSR* **20**, 327 (1945).
K12. Kulic, E., and Rhodes, E., *Can. J. Chem. Eng.* **52**, 114 (1974).
K13. Kulov, N. N., and Malyusov, V. A., *Dokl. Akad. Nauk. SSSR* **173**, 4, 876 (1967).
K14. Kuong, J. F., *Chem. Eng.* **68**, 258 (1961).
K15. Kuznetsov, M. D., and Oratovskii, V. J., *Int. Chem. Eng.* **2**, 185 (1962).
L1. Larkins, R. P., White, R. R., and Jeffrey, D. W., *AIChE J.* **7**, 231 (1961).
L2. Laurent, A., Doctoral Thesis, University of Nancy, France, 1975.
L3. Laurent, A., and Charpentier, J. C., *Chem. Eng. J.* **9**, 85 (1974).
L4. Laurent, A., Charpentier, J. C., and Prost, C., *J. Chim. Phys.* **71**, 613 (1974).
L5. Laurent, A., Prost, C., and Charpentier, J. C., *Congr. CHISA Section H, Paper H.3, Prague* Aug., (1975).
L6. Laurent, A., Prost, C., and Charpentier, J. C., *J. Chim. Phys.* **72**, 236 (1975).

L7. Leffler, J., and Cullinan, H. T., *Ind. Eng. Chem. Fundam.* **9**, 84 (1970).
L8. Leffler, J., and Cullinan, H. T., *Ind. Eng. Chem. Fundam.* **9**, 88 (1970).
L9. Lemay, Y., Pineault, G., and Ruether, J. A., *Indian Eng. Chem. Process. Des. Dev.* **14**, 280 (1975).
L10. Levenspiel, O., "Chemical Reaction Engineering," 2nd Ed. Wiley, New York, 1972.
L11. Levenspiel, O., and Godfrey, J. H., *Chem. Eng. Sci.* **29**, 1723 (1974).
L12. L'Homme, G. A., Collection des Publications, Faculté des Sciences Appliquées, Université de l'Etat, Liège, **51**, 91 (1975).
L13. Linek, V., *Chem. Eng. Sci.* **21**, 777 (1966).
L14. Linek, V., *Chem. Eng. Sci.* **26**, 491 (1971).
L15. Linek, V., *Chem. Eng. Sci.* **27**, 627 (1972).
L16. Linek, V., and Mayrhoferova, J., *Chem. Eng. Sci.* **24**, 481 (1969).
L17. Linek, V., and Mayrhoferova, J., *Chem. Eng. Sci.* **25**, 787 (1970).
L18. Linek, V., and Turdik, J., *Biotech. Bioeng.* **13**, 353 (1971).
L19. Linek, V., Sobotka, M., and Prokop, A., *Chem. Eng. Sci.* **29**, 637 (1974).
L20. Linek, V., Stay, V., Machon, V., and Krivsky, Z., *Chem. Eng. Sci.* **29**, 1955 (1974).
L21. Linke, W. F., and Seidell, A., "Solubilities: Inorganic and Metal-Organic Compounds," 4th Ed. Van Nostrand-Reinhold, Princeton, New Jersey, Vol. I, 1958; Vol. II, 1965.
L22. Lockhart, R. W., and Martinelli, R. C., *Chem. Eng. Prog.* **45**, 39 (1949).
L23. Loiseau, B., Doctoral Thesis, University of Nancy, France, 1976.
L24. Lopez Cardoso, R., *Chem. Eng. Sci.* **17**, 783 (1962).
L25. Lusis, M. A., and Ratcliff, G. A., *Can. J. Chem. Eng.*, **46**, 385 (1968).
M1. McCarthy, M. J., Henderson, J. B., Molloy, N. A., *Proc. Chemeca 1970 Conf. Australia*, Section 2, pp. 86–100. Butterworths, London, 1970.
M2. McLachlan, C. N. S., and Danckwerts, P. V., *Trans. Inst. Chem. Eng.* **50**, 300 (1972).
M3. McMillan, W. P., Ph.D. Thesis, University of London, 1962.
M4. McNeil, K. M., *Can. J. Chem. Eng.* **48**, 252 (1970).
M5. Maffiolo, G., Vidal, J., and Renon, H., *Ind. Eng. Chem. Fundam.* **11**, 100 (1972).
M6. Maloney, D. P., and Prausnitz, J. M., *AIChE J.* **22**, 74 (1976).
M7. Marucci, G., and Nicomedo, L., *Chem. Eng. Sci.* **32**, 1257 (1967).
M8. Mashelkar, R. A., *Br. Chem. Eng.* **10**, 1297 (1970).
M9. Mashelkar, R. A., and Sharma, M. M., *Trans. Inst. Chem. Eng.* **48**, T 162 (1970).
M9a. Mecklenburgh, J. C., and Hartland, S., "The Theory of Backmixing," Wiley (Interscience), New York, 1975.
M10. Mehta, V. D., and Sharma, M. M., *Chem. Eng. Sci.* **21**, 361 (1966).
M11. Mehta, K. C., and Sharma, M. M., *Br. Chem. Eng.* **15**, 1440, 1556 (1970).
M12. Mehta, V. D., and Sharma, M. M., *Chem. Eng. Sci.* **26**, 461 (1971).
M13. Meisner, J., *Indian Eng. Chem. Fundam.* **11**, 83 (1972).
M14. Michael, B. J., and Miller, S. A., *AIChE J.* **8**, 262 (1962).
M15. Midoux, N., Favier, M., and Charpentier, J. C., *J. Chem. Eng. Jap.* **9**, 350 (1976).
M15a. Miyauchi, T., and Vermeulen, T., *Ind. Eng. Chem. Fundam.* **2**, 113 (1963).
M16. Miller, D. N., *AIChE J.* **20**, 445 (1974).
M17. Mohunta, D. M., Vaidyanathan, A. S., and Laddha, G. S., *Indian Chem. Eng.* **11**, 39 (1969).
N1. Nagata, S., "Mixing." Wiley, New York, 1975.
N2. Nagel, O., Kurten, H., and Sinn, R., *Chem. Ing. Technk.* **42**, 474 (1970).
N3. Nagel, O., Kurten, H., and Sinn, R., *Chem. Ing. Technk.* **44**, 367 (1972).
N4. Nagel, O., Kurten, H., and Sinn, R., *Chem. Ing. Technk.* **44**, 899 (1972).
N5. Nagel, O., Kurten, H., Hegner, B., and Sinn, R., *Chem. Ing. Technk.* **42**, 921 (1970).
N6. Narayanan, S., Subramanian, L. P., and Guha, D. K., *Indian Chem. Eng.* **10**, 75 (1968).

N7. Nedoborov, Yu.P., and Semenov, P. A., *Int. Chem. Eng.* **13**, 430 (1973).
N7a. Nicklin, D. J., Wilkes, J. O., and Davidson, J. F., *Trans. Inst. Chem. Eng.* **40**, 61 (1962).
N8. Nienow, A. W., *Br. Chem. Eng.* **10**, 827 (1965).
N9. Nienow, A. W., *Chem. Eng. J.* **9**, 153 (1975).
N10. Nijsing, R. A. T. O., and Kramers, H., *Chem. Eng. Sci.* **8**, 81 (1958).
N11. Nijsing, R. A. T. O., Hendriksz, R. H., and Kramers, H., *Chem. Eng. Sci.* **10**, 38 (1959).
N12. Nonhebel, G., "Gas Purification for Air Pollution Control." Butterworths, London, 1972.
N13. Nukiyama, S., and Tanasawa, Y., *Trans. Soc. Mech. Eng.* **5**, 63 (1939).
N14. Null, H. R., "Phase Equilibrium in Process Design." Wiley (Interscience), New York, 1970.
N15. Nyvlt, V., and Kastanek, F., *Coll. Czech. Chem. Commun.* **40**, 1853 (1975).
O1. O'Connell, J. P., and Prausnitz, J. M., *Ind. Eng. Chem. Fundam.* **3**, 347 (1964).
O2. Onda, K., Takeuchi, H., and Okumoto, Y., *J. Chem. Eng. Jpn.* **1**, 56 (1968).
O3. Onda, K., Takeuchi, H., and Maeda, Y., *Chem. Eng. Sci.* **27**, 449 (1972).
O4. Onda, K., Sada, E., Kobayashi, T., and Fujine, M., *Chem. Eng. Sci.* **25**, 753 (1970).
O5. Onda, K., Sada, E., Kobayashi, T., and Fujine, M., *Chem. Eng. Sci.* **25**, 761 (1970).
O6. Onda, K., Sada, E., Kobayashi, T., and Fujine, M., *Chem. Eng. Sci.* **25**, 1023 (1970).
O7. Onda, K., Sada, E., Kobayashi, T., and Fujine, M., *Chem. Eng. Sci.* **27**, 247 (1972).
O8. Onda, K., Sada, E., Kobayashi, T., Kito, S., and Ito, K., *J. Chem. Eng. Jpn.* **3**, 18 (1970).
O9. Onda, K., Sada, E., Kobayashi, T., Kito, S., and Ito, K., *J. Chem. Eng. Jpn.* **3**, 137 (1970).
O10. Orentlicher, M., and Prausnitz, J. M., *Chem. Eng. Sci.* **19**, 775 (1964).
O11. Othmer, D. F., and Thakar, M. S., *Ind. Eng. Chem.* **45**, 589 (1953).
P1. Pangarkar, V. G., and Sharma, M. M., *Chem. Eng. Sci.* **29**, 561 (1974).
P2. Paradis, G., and L'Homme, G. A., Memoire de fin d'études, Université de Liège, 1972.
P3. Pasiuk-Bronikowska, W., *Chem. Eng. Sci.* **24**, 1139 (1969).
P3a. Peaceman, D. W., Sc.D. Disertation, M.I.T., Cambridge, Massachusetts, 1951.
P4. Perez, J. F., and Sandall, O. C., *AIChE J.* **20**, 770 (1974).
P5. Perry, J. M., "Chemical Engineers' Handbook," 5th Ed. McGraw-Hill, New York, 1973.
P6. Pexidr, V., and Charpentier, J. C., *Coll. Czech. Chem. Commun.* **40**, 3130 (1975).
P7. Pigford, R. L., and Pyle, C., *Ind. Eng. Chem.* **43**, 1649 (1951).
P8. Pohorecki, R., *Chem. Eng. Sci.* **23**, 1447 (1968).
P9. Porter, K. E., *Trans. Inst. Chem. Eng.* **44**, T 25 (1966).
P10. Porter, K. E., *Chem. Eng. Sci.* **30**, 155 (1975).
P11. Porter, K. E., King, M. E., and Varshney, K. C., *Trans. Inst. Chem. Eng.* **44**, T 274 (1966).
P12. Porter, K. E., Davies, B. T., and Wong, P. F. Y., *Trans. Inst. Chem. Eng.* **45**, T 265 (1967).
P13. Prasher, B. D., *AIChE J.* **21**, 407 (1975).
P14. Prengle, H. W., and Barona, N., *Hydrocarbon Process.* **49**, 106 (1970).
P15. Preston, G. T., Funk, E. W., and Prausnitz, J. M., *Phys. Chem. Liquids* **2**, 193 (1971).
P16. Puranik, S. S., and Vogelpohl, A., *Chem. Eng. Sci.* **29**, 501 (1974).
R1. Ramm, V. M., "Absorption of Gases." Israel Program for Scientific Translations, Jerusalem, 1968.
R2. Ratcliff, G. A., and Holdcroft, J. G., *Trans. Inst. Chem. Eng.* **41**, 315 (1963).
R3. Ratcliff, G. A., Richards, G. M., and Danckwerts, P. V., *Chem. Eng. Sci.* **19**, 325 (1964).

R4. Reddy, K. A., and Doraiswamy, L. K., *Ind. Eng. Chem. Fundam.* **6**, 77 (1967).
R5. Reid, R. C., and Sherwood, T. K., "The Properties of Gases and Liquids," 2nd Ed. McGraw-Hill, New York, 1966.
R6. Reiss, L. P., *Ind. Eng. Chem. Process. Des. Dev.* **6**, 486 (1967).
R7. Reith, T., Ph.D. Thesis, Delft University, 1968.
R8. Reith, T., *Br. Chem. Eng.* **15**, 1559 (1970).
R9. Reith, T., and Beek, W. J., *Chem. Eng. Sci.* **28**, 1331 (1973).
R10. Reith, T., Renken, S., and Israel, B. A., *Chem. Eng. Sci.* **23**, 619 (1968).
R11. Roberts, D., and Danckwerts, P. V., *Chem. Eng. Sci.* **17**, 961 (1962).
R12. Robinson, R. A., and Stokes, R. H., "Electrolyte Solutions," 2nd. Ed. Academic Press, New York, 1959.
R13. Robinson, R. L., Jr., Ph.D. Thesis, Oklahoma State University, Stillwater, 1964.
R14. Robinson, C. W., and Wilke, C. R., *AIChE J.* **20**, 285 (1974).
R15. Rodionov, A. I., and Radikovski, V. M., *J. Appl. Chem. USSR* **40**, 2751 (1967).
R16. Rodionov, A. I., and Vinter, A. A., *Int. Chem. Eng.* **3**, 460 (1967).
R17. Rodionov, A. I., and Sorokin, *Most. Inst. Khim. Mashinost.* **4**, 2753 (1970).
R18. Ruether, J. A., and Pari, P. S., *Can. J. Chem. Eng.* **51**, 345 (1973).
R19. Ross, S. L. (with R. L. Curl), Ph.D. Dissertation, University of Michigan, 1971.
R20. Rushton, J. H., Costich, E. W., and Everett, H. J., *Chem. Eng. Prog.* **46**, 395, 467 (1951).
S1. Saada, Y., *Chim. Ind. Gen. Chim.* **102**, 1283 (1969); **105**, 20 (1972).
S2. Sahay, B. N., and Sharma, M. M., *Chem. Eng. Sci.* **28**, 41 (1973).
S3. Sandall, O. C., *Can. J. Chem. Eng.* **53**, 702 (1975).
S4. Sankholkar, D. S., and Sharma, M. M., *Chem. Eng. Sci.* **30**, 729 (1975).
S5. Sato, Y., Hirose, T., and Ida, T., *Kagaku Kogaku* **38**, 534 (1974).
S6. Sato, Y., Hirose, T., Takahashi, F., and Toda, M., *Pacif. Chem. Eng. Conf., Session 8*, Paper 8.3 (1972).
S7. Sato, Y., Hirose, T., Takahashi, F., and Toda, M., *J. Chem. Eng. Jpn.* **6**, 147 (1973).
S8. Sato, Y., Hirose, T., Takahashi, F., Toda, M., and Hashiguchi, Y., *J. Chem. Eng. Jpn.* **6**, 315 (1973).
S9. Satterfield, C. N., *AIChE J.* **21**, 209 (1975).
S10. Sawicki, J. E., and Barron, C. H., *Chem. Eng. J.* **5**, 153 (1973).
S11. Scheibel, E. G., *Ind. Eng. Chem.* **46**, 2007 (1954).
S12. Scott, D. S., and Hayduck, W., *Can. J. Chem. Eng.* **44**, 130 (1966).
S13. Seidell, A., and Linke, W. F., "Solubilities of Inorganic and Organic Compounds," Suppl. to 3rd Ed. Van Nostrand-Reinhold, Princeton, New Jersey, 1952.
S14. Shaffer, D. L., Jones, J. H., and Daubert, T. E., *Ind. Eng. Chem. Process. Des. Dev.* **13**, 14 (1974).
S15. Shah, I. S., *Chem. Eng. Prog.* **67**, 51 (1971).
S16. Shah, A. K., and Sharma, M. M., *Can. J. Chem. Eng.* **53**, 572 (1975).
S17. Sharma, M. M., Ph.D. Thesis, Cambridge University, 1964.
S18. Sharma, M. M., *Trans. Faraday Soc.* **61**, 681 (1965).
S19. Sharma, M. M., and Danckwerts, P. V., *Chem. Eng. Sci.* **18**, 729 (1963).
S20. Sharma, M. M., and Gupta, R. K., *Trans. Inst. Chem. Eng.* **45**, T 69 (1967).
S21. Sharma, M. M., and Mashelkar, R. A., *AIChE I. Chem. Eng. Symp. Ser. 28*, p. 10 (1968).
S22. Sharma, M. M., and Danckwerts, P. V., *Br. Chem. Eng.* **15**, 522 (1970).
S23. Sharma, M. M., Mashelkar, R. A., and Mehta, V. D., *Br. Chem. Eng.* **1**, 14, 70 (1969).
S24. Shende, B. W., and Sharma, M. M., *Chem. Eng. Sci.* **29**, 1763 (1974).
S25. Sherwood, T. K., and Wei, J. L., *AIChE J.* **1**, 522 (1955).
S26. Shrier, A. L., *Chem. Eng. Sci.* **22**, 1391 (1967).

S27. Shulman, H. L., Ullrich, C. F., and Wells, N., *AIChE J.* **1**, 247 (1955).
S28. Sideman, S., Hortacsu, O., and Fulton, J. W., *Ind. Eng. Chem.* **58**, 32 (1966).
S29. Simons, J., and Ponter, A. B., *Can. J. Chem. Eng.* **53**, 541 (1975).
S30. Sitaraman, R., Ibrahim, S. H., and Kuloor, N. R., *J. Chem. Eng. Data* **8**, 198 (1963).
S31. Skelland, A. H. P., "Diffusional Mass Transfer." Wiley (Interscience), New York, 1974.
S32. Snider, J. W., and Perona, J. J., *AIChE J.* **20**, 1172 (1974).
S33. Solomakha, G. P., Azizov, A. G., and Planorski, *Mosk. Inst. Khim. Mashinost.* **4**, 315 (1970).
S34. Specchia, V., Sicardi, S., and Gianetto, A., *AIChE J.* **20**, 646 (1974).
S35. Sridharan, K., and Sharma, M. M., *Congr. CHISA Section G, Distill. Absorpt.*, Prague, Aug., 1975.
S36. Sridharan, K., Ph.D. Thesis, Bombay, India, 1975.
S37. Srivastava, R. D., MacMillan, A. F., and Jarris, I. J., *Can. J. Chem. Eng.* **46**, 181 (1968).
S38. Stephens, E. J., and Morris, G. A., *Chem. Eng. Prog.* **47**, 232 (1957).
S39. Stiel, L., and Harnish, D. F., *AIChE J.* **22**, 117 (1976).
S40. Sylvester, N. D., and Pitayagulsarn, P., *Ind. Eng. Chem. Process. Des. Dev.* **14**, 473 (1975).
T1. Takeuchi, H., Fujine, M., Sato, T., and Onda, K., *J. Chem. Eng. Jap.* **8**, 252 (1975).
T2. Tamir, A., and Wisniak, J., *Chem. Eng. Sci.* **30**, 335 (1975).
T3. Tamir, A., Danckwerts, P. V., and Virkar, P. D., *Chem. Eng. Sci.* **30**, 1243 (1975).
T4. Tiepel, E. W., and Gubbins, K. E., *Ind. Eng. Chem. Fundam.* **12**, 18 (1973).
T5. Tiepel, E. W., and Gubbins, K. E., *Ind. Eng. Chem. Fundam.* **14**, 143 (1975).
T6. Todtenhauft, E. K., *Chem. Ing. Technk.* **43**, 336 (1971).
T7. Tokunaga, J., *Jap. J. Chem. Eng.* **8**, 7 (1975).
T8. Tokunaga, J., *Jap. J. Chem. Eng.* **8**, 326 (1975).
T9. Towell, G. D., Strand, C. P., and Ackerman, G. H., *AIChE I. Chem. Eng. Symp. Ser.* **10**, p. 97 (1965).
T10. Tripathi, G., Shakla, K. N., and Pandey, R. N., *Can. J. Chem. Eng.* **52**, 691 (1974).
T11. Turpin, J. L., and Huntington, R. L., *AIChE J.* **13**, 1196 (1967).
T12. Tyn, M. T., *Chem. Eng. J.* **6**, 27 (1971).
U1. Uchida, S., Ph.D. Thesis, West Virginia University, Morgantown, 1973.
U2. Ufford, R. C., and Perona, J. J., *AIChE J.* **19**, 1223 (1973).
V1. Valentin, F. H. H., "Absorption in Gas-Liquid Dispersions." Spon, London, 1967.
V2. Van de Sande, E., and Smith, J. M., *Chem. Eng. Sci.* **31**, 219 (1976).
V3. Van Dierendonck, L. L., Fortuin, J. K. H., and Venderbos, D., *Proc. Eur. Symp. Chem. Eng., 4th* p. 205 (1971).
V4. Van Krevelen, D. W., and Hoftijzer, P. J., *Int. Congr. Chim. Ind., 21st, Brussels* Special No., p. 168 (1948).
V5. Verma, S. L., and Delancey, G. B., *AIChE J.* **21**, 96 (1975).
V6. Vermeulen, T., Williams, G. M., and Langlois, G. E., *Chem. Eng. Prog.* **51**, 85F (1955); Langlois, G. E., and Vermeulen, T., *Rev. Sci. Instrum.* **25**, 360 (1954).
V7. Vidwans, R. D., and Sharma, M. M., *Chem. Eng. Sci.* **22**, 673 (1967).
V8. Vinges, A., *Ind. Eng. Chem. Fundam.* **5**, 189 (1966).
V9. Vinograd, J. R., and McBain, J. W., *J. Am. Chem. Soc.* **63**, 2008 (1941).
V10. Vinolur, Y. G., and Dil'man, V. V., *Khim. Prom.* **7**, 619 (1959).
V11. Virkar, P. D., and Sharma, M. M., *Can. J. Chem. Eng.* **53**, 512 (1975).
V12. Volgin, B. P., Efimova, T. F., and Gofman, M. S., *Int. Chem. Eng.* **1**, 113 (1968).
V13. Voyer, R. D., and Miller, A. I., *Can. J. Chem. Eng.* **46**, 335 (1968).

W1. Wales, C. E., *AIChE J.* **12**, 1166 (1966).
W2. Wen, C. Y., O'Brien, W. S., and Fan, L. T., *J. Chem. Eng. Data* **8**, 42 (1963).
W3. Wesselingh, J. A., and Van't Hoog, A. C., *Trans. Inst. Chem. Eng.* **T69**, 48 (1970).
W4. Westerterp, K. R., Van Dievendonck, L. L., and De Kraa, J. A., *Chem. Eng. Sci.* **18**, 157 (1963).
W5. Whitman, W. G., *Chem. Metall. Eng.* **29**, 147 (1923).
W6. Wilhem, E., and Battino, R., *Chem. Rev.* **73**, 1 (1973).
W7. Wilhem, E., and Battino, R., *J. Chem. Thermodyn.* **5**, 117 (1973).
W8. Wilke, C. R., and Chang, P., *AIChE J.* **1**, 264 (1955).
W9. Wise, D. L., and Houghton, G., *Chem. Eng. Sci.* **21**, 999 (1966).
W10. Wen, C. Y., and Fan, L. T., "Models for Flow Systems and Chemical Reactors." Dekker, New York, 1975.
Y1. Yagi, S., and Inoue, H., *Chem. Eng. Sci.* **17**, 411 (1962).
Y2. Yano, T., Suetaka, T., Umehara, T., and Yamashita, T., *Int. Chem. Eng.* **13**, 371 (1973).
Y3. Yoshida, F., and Akita, K., *AIChE J.* **11**, 9 (1965).
Y4. Yoshida, F., Yamane, T., and Miyamoto, Y., *Ind. Eng. Chem. Process. Des. Dev.* **9**, 571 (1970).
Z1. Zehner, P., *Chem. Ing. Technk.* **47**, 209 (1975).

Note Added in Proof

The present chapter on the topics of gas–liquid separation was written in 1975. Since then many results and data have been published in the sense proposed by the author. This complementary information will be obtained in the following papers:

Laurent, A., Fonteix, C., and Charpentier, J. C., Simulation of a pilot scale, liquid motivated, venturi jet scrubber by a laboratory scale model. *AIChE J.* **26**, 282 (1980).
Charpentier, J. C., Gas–liquid reactors. Chemical Reaction Engineering Reviews (D. Luss and V. W. Weekman, eds.), Amer. Chem. Soc. Symp. Ser. 72, 223–261 (1978).
Schugerl, K., Lucke, J., and Oels, U., Bubble column bioreactors. *Advan. Biochem. Eng.* **7**, 1 (1977).
Shah, Y. T., "Gas–Liquid–Solid Reactor Design." McGraw-Hill, New York (1979).
Sovova, H., A correlation of diffusivities of gases in liquids. *Collect. Czech. Chem. Commun.* **41**, 3715 (1976).
Stichlmair, J., and Mersmann, A., Dimensioning plate columns for absorption and rectification. *Int. Chem. Eng.* **18**, 223 (1978).
Van't Riet, K., Review of measuring methods and results in nonviscous gas–liquid mass transfer in stirred vessels. *Ind. Eng. Chem. Process Des. Dev.* **18**, 357 (1979).
Yorizane, M., and Miyano, Y., A generalized correlation for Henry's constants in nonpolar binary systems. *AIChE J.* **24**, 181 (1978).
Zlokarnik, M., Sorption characteristics for gas–liquid contacting in mixing vessels. *Advan. Biochem. Eng.* **8**, 133 (1978).

THE INDIAN CHEMICAL INDUSTRY—ITS DEVELOPMENT AND NEEDS

Dee H. Barker* and C. R. Mitra

Birla Institute of Technology and Science
Pilani, Rajasthan, India

I.	Summary	136
II.	Development to 1800	137
III.	Development from 1800 to 1947	142
IV.	Structure of the Chemical Industry	147
V.	Development since Independence	154
	A. Heavy Chemicals	157
	B. Organic Chemicals	166
	C. Soaps and Detergents	168
	D. Pesticides	170
	E. Plastics	170
	F. Iron and Steel	172
	G. Nonferrous Metals	175
	H. Pharmaceutical Industry	180
VI.	Awards for Progress	184
VII.	Professional Societies and Education	187
VIII.	Needs of the Chemical Industry	189
	References	195

In India the development of chemistry, chemical technology, and chemical engineering has taken place over a vast span of time. The industry has developed from the realm of secret practice and magic to the use of the latest technology and the ability to export this technology and its associated equipment to lands where development has not proceeded as rapidly or as far.

The progress in chemical engineering stems not so much from the ex-

---

* Present address: Chemical Engineering Department, Brigham Young University, Provo, Utah 84601.

tension of theory into new fields, or from the discovery of new ideas, but from adaptation to the conditions, materials, and methods available in a developing country. Innovation consists of changing capital-intensive practice to labor-intensive practice, of finding a path through the network of regulations, of training workers, engineers, and individuals who have had little association with technology in everyday life, and of adaptation.

Few people outside India are aware of the progress being made in that country and in the rapid advance of technology, education, and attainment of self-sufficiency. It is this progress and attainment that will be described in this report.

I. Summary

A study of archeological findings and records indicates that chemistry has been applied for the betterment of humanity in India for over 4000 yr. However, this use of chemistry has been based on a technological rather than a scientific standpoint. The earliest users concerned themselves with metals for use in war and as money, ceramics for many and varied uses, and mixtures of herbs for the betterment of health.

Applied chemistry will be considered here in its widest possible connotation. The discussion will include not only chemistry but also mineralogy, metallurgy, and pharmacology. Chemical developments from ancient times to the present are considered, each of which has contributed to the overall use of chemistry and chemical technology.

India, because of its location, climate, and people, has always been a rich source of raw material. The British, like all the previous conquerors, exploited these resources by taking raw materials to their own country for processing rather than by developing industry within India itself. As a result of the conquerors' entering and leaving India, the culture absorbed the traits and characteristics of the conquering people, so that technology and science, too, were assimilated. The industrial revolution in England made it more profitable to use the low-cost labor and mechanization of England rather than the bountiful low-cost labor of India; and as a consequence the British used tariffs and other governmental actions to suppress development of the industrial capacity of India.

The real history of the chemical industry in India began with the independence of the country in 1947. Initially growth was slow, but later it became very rapid as the effects of advanced planning began to be felt. An understanding of this development requires a knowledge of the structure of industry within India. The interrelationships among research, development, and production, the control exercised by the government, and the financial control exercised both by the government and by private industry all have a critical bearing on the ability of the industry to progress.

Up to the present, the chemical industry has tended to concentrate in several areas within India rather than being generally distributed. Plant location is complicated by the lack of a proper water supply, even on the plains of the Ganges. The amount of water in the Ganges is great, but its location is not always fixed because the river wanders. For this reason, elaborate means have to be provided to ensure a water supply for a chemical plant. Thus, even though social planning dictates that plant location benefit the overall development of the country, a geographical concentration of industry has occurred.

The needs of the chemical industry are discussed in regard to the educational system, research and development, changes in the structure of the industry, and so on. Among the outstanding needs are a change in the educational setup, the development of more self-sufficiency in the construction of chemical plants, and the development of chemical processes. As part of a great effort in this direction, study groups have published documents (S4, S5, S6) summarizing the findings in regard to the research needs of the country.

The bulk of research and development work is done by government laboratories. Changes need to be made to encourage private companies to carry out research. This will require a change in licensing and tax policies. The appropriate independent technology is being developed and will result in phenomenal growth. The per capita consumption of chemical-related products today is relatively small within India. For this reason growth of the chemical industry over the coming years will be rapid, much greater than throughout the world as a whole.

II. Development to 1800

India is an ancient country with a long and glorious history. Chemistry, in its many ramifications, was one of the sciences used extensively by the ancient inhabitants. Well-documented and comprehensive treatments of the use of chemical technology in India are available (B1, B2, B3, C2, W1), on which the material presented here is heavily based. A great deal can be inferred as to the technical capability of the early inhabitants from the archeological sites being excavated throughout India. Neither archeological records nor more recent written records actually describe the technology, so that inferences have to be made as to its usage in everyday life.

For the purpose of study, the Indian historical scene is divided into several different ages or periods. The first of these is the pre-Vedic age which includes all time prior to 1500 BC, an age of which our knowledge depends almost exclusively on archeological findings. The pre-Vedic era was followed by the Vedic, from 1500 to 600 BC. During this period Aryans entered from the north and eventually settled over most of the northern part of India, bringing with them their culture, technology, and science where they mixed with the prevailing culture.

The Vedic period was followed by the classic period extending from 600 BC to AD 1200. During this period, the country was continually overrun by outsiders whose cultures were assimilated into the Indian pattern. The classic period was followed by the medieval, running from approximately 1200 AD to the end of the eighteenth century by which time the British began to have a major influence in India. The nineteenth century saw many changes take place in India, as in the rest of the world, in regard to the development of sciences and their use for the aid of humanity. The early twentieth century is classified as the preindependence period, and the time since 1947 as the postindependence period.

The rise and fall of Indian industry in general and the chemical industry in particular have been caused by many competing factors. India's location and climate have made it a veritable storehouse of raw materials—both natural and agricultural—valuable for the development of humanity. The people were peace-loving and not warlike. For this reason, envious neighbors moved in and conquered and exploited the resources and the people. These influences brought new ideas, new sciences, and a new technology which were assimilated and improved upon by the local inhabitants. Because India lived so long under foreign rule, there was little freedom to choose the direction of expansion, growth, and development. Under these conditions, the technological progress made in ancient India is remarkable.

Humans have used chemical technology since the beginning of recorded history. The first evidences of the use of chemistry as a practical and purposeful art are found in the use of clay which was fashioned into useful articles and hardened by fire. The development and improvement of the potter, from the standpoint of both artistic skill and use, led the chemical technologist to develop processes involving prolonged heating, fusion, evaporation, and often the treatment of minerals (B3).

The preservation and maintenance of health has also been a subject of concern to humanity since early times. Archeological evidence from about 4000 BC shows that in city planning the inhabitants were well aware of the problems of health and sanitation and therefore were probably also aware of the treatment necessary to cure certain diseases. Written records from later times show they were well informed as to the operations used in the manufacture of various chemicals, drugs, pharmaceuticals, and dyes. These included emulsification, filtration, decantation, sedimentation, concentration, precipitation, dehydration, evaporation, and chemical combination. These were carried out on a small scale, not on the huge manufacturing scale of modern chemical industries. Other early chemical technologies were in the areas of brewing, dyeing, metalworking, glassmaking, the use of pigments, and the preparation of building materials such as burnt limestone and cement.

In the earliest times, pottery was made without benefit of the potter's wheel and was baked without adequate kilns. The low firing temperatures made it impossible to develop high strength or elaborate designs. The pottery was somewhat porous and was evidently the subject of continued experimentation. During the Harappan period which lasted from about 2300 to 1750 BC in northern India, or just preceding this era, the potter's wheel was developed and most vessels were made by this method. The firing procedure had been developed to take place under either oxidizing or reducing conditions, and the many colors show an understanding of the use of chemical change in developing the required colors. Many furnace designs have been found indicating that wood and charcoal were used as fuels.

Some of the earliest examples of glazed pottery appeared in India during the Harappan period. (They did not appear in other countries, such as Mesopotamia, until about 1500 yr later.) The use of terracotta was also well known in early India. With the coming of the Indo-Aryans and the Vedic age, pottery became refined in shape, size, color, and strength, indicating an increasing awareness of the quality control needed to develop utensils for use.

From the evidence available, the glass industry was well established in the period from 1000 to 750 BC. By the post-Vedic, classic age the development of pottery had reached a stage that included glasses and ceramics. Glass was made in many different colors and molded for many different uses. (Glass may have been introduced into India from Egypt, and some Indian glass objects have been determined to be of Roman origin.) The earliest uses of glass were for ornamentation; this was later followed by use as vessels. Some evidence has been found of glass factories; blocks of glass up to 120 lb have been discovered. It was also during this period that northern black polished ware made its appearance. Its production indicates a degree of technology that allowed firing under both oxidizing and reducing conditions so as to obtain the desired colors.

The ability to control temperature in a furnace or oven as exemplified by the appearance of high-fired ceramics and glasswork was also evidenced in the flourishing trade and production of metals that began in ancient times. The earliest metal used was copper. There is also evidence of the use of bronze, lead, silver, and gold, although some of these metal objects may have been manufactured outside India and brought in by traders and conquerors. Most of the earlier finds are copper or bronze implements, tools, and weapons of warfare. The refining of copper was so carried out that almost pure copper could be obtained, as analysis of some copper ingots shows. Melting was carried out near the mines in relatively small furnaces. There are many places in India today where slag deposits can be seen, but no remains of a furnace. In the early nineteenth century copper smelting furnaces were described which were evidently much like the earlier furnaces. These consisted of a small vertical cylinder about 15 in. in diameter and 3 ft tall, operated with two leather bellows which supplied the blast air.

Copper and its alloys continued to be important commercial products throughout all of India's earlier history. Today the ancient skills are being reactivated, and the production of copper is becoming an important industry. During the classic age metal workers were sufficiently skillful and technically educated to construct a copper image of Buddha 80 ft high and temple bells 100 ft high. They were well versed in casting methods, including the lost wax casting process. By the Mogul period, techniques had developed to such an extent that a brass cannon over 14 ft long and weighing 53 tons could be cast.

Iron was apparently unknown in the earliest of the Indian cultures. However, by the Vedic age, it had been introduced or developed in India, and there is evidence of slag piles and furnaces. The iron objects found encompass the entire range of tools, weapons, and utensils and extend throughout India. Following 800 BC, Indian iron and steel objects were recognized and praised for their quality throughout the Western world. When Alexander reached India, one of the gifts to him was 100 talents of steel, a prized commodity. The Indians were also adept in using steel in cutlery and for armor.

The production of iron and steel in India progressed rapidly, and during the classic age some of the largest castings in the world were made. An outstanding example is an iron pillar located near New Delhi, which is over 24 ft high and weighs over 6 tons. By chance or by design this pillar was made of nonrusting iron. It is a marvel to behold, standing after centuries with no sign of rust and covered uniformly with a blackish coating except near the base where it has become bright and shiny after being touched by the hands of countless visitors. Iron was used for construction purposes, as shown in the temple at Konarak in Orissa. Over 239 iron beams were used in its construction, some as long as 35 ft and 1 ft sq, which in most cases have rusted. Another pillar located at Dhar was probably over 43 ft long. In most cases the remains of blast furnaces show them to have been very small, and it is difficult to understand how the large pieces were cast.

Evidence indicates that the ancient people of India from as early as 4000 BC were well acquainted with the production of building materials such as lime and plaster. By using crude equipment they fashioned the materials they needed, some of which have lasted until the present day. Present-day methods used in India probably resemble those used in ancient times. In the production of bricks, it is still not uncommon to find that sun-dried bricks are heaped or stacked in piles alternating with coal and arranged in the form of a beehive oven for firing.

The production of pigments in ancient times was also well known and is evident in the remains of paintings in various ruins throughout India. In later times the development and exportation of dyes and pigments was an important segment of the industry, particularly in regard to the native indigo.

Another example of the development of Indian chemical technology lies in the production of pharmaceuticals, cosmetics, and perfumes. Early chemical endeavors in India, as in the rest of the world, revolved around the purposes of seeking an elixir for life and changing base metals into gold. However, many developments in medicine were also made. Many of the preparations in India depended on mercury, and it is small wonder that these compounds did not kill more people than they cured. The preparation of pharmaceutical compounds primarily depended on the collection of plants and herbs. These were used either individually or in elaborate compounds. Many of them are still used today; investigation of the plants used in medicine is being made, with the result that many new drugs and preparations are being found.

The use of pottery led to the development of a large amount of special-purpose laboratory equipment for the preparation of cosmetics, medicines, metals, and so on. Drawings show the existence of equipment for evaporation, sublimation, prolonged heating, steeping, and distillation, all made from pottery vessels of various sizes and shapes. Sometimes the equipment was fired from below and sometimes from above. In all cases a high degree of skill in chemical and laboratory preparation was indicated.

By the eleventh century AD, the art of making paper was well developed in India. Indian artisans were highly skilled, and their paper was in great demand throughout the world. Although much of the paper was made of rags, a great deal of effort went into using many types of plants, some of which are still grown in India but are not now used in paper manufacture. Perhaps with the present shortage of paper some of these techniques should be revived.

A later development, occurring in the beginning of the fifteenth century AD, was the introduction of gunpowder and pyrotechnics, probably from China. Indians became very skilled in their use and compounded many different kinds of gunpowder for various applications such as in fireworks, rockets, and cannons. They developed many formulas; in the literature, 98 different formulas for 20 different types of fireworks are given. They were adept at using metal dust and other chemicals to give color and light to fireworks.

The ancient history of Indian chemical technology mostly concerns the production of small amounts of material in many different places. However, in the case of iron, as shown above, the ability to cast larger objects was not exceeded elsewhere in the world. Exports in early times included salt, sugar, and some metal products. Under the British during the nineteenth century, both a development of and a decline in industrial chemistry took place. Only after the departure of the British did the chemical industry really begin to flourish in India.

III. Development from 1800 to 1947

Although the British occupied parts of India during the eighteenth century, it was not until the nineteenth century that their influence spread to such an extent that it had a marked effect on the industrialization of the country. As will be shown, several attempts were made in specific industries to establish chemical processing, but without success. Except for one or two products, the chemical industry did not flourish until World War I which created a necessity for developing certain chemical industries in India. The period of British rule in India was much like that of other ages during which the people of the country were exploited. (In fairness to the British, however, it is recognized that British rule was a major factor in welding India into a united nation.) India was considered to be primarily a source of raw materials, agriculturally produced (jute, raw cotton, raw silk, indigo, and raw drugs such as opium), and in some cases minerals. In the eighteenth century the Indian people became highly skilled in their ability to spin thread and weave textiles, and India exported a considerable quantity of finished goods, particularly cotton and silk textiles. The manufacture of textiles was "the most widespread industry of the country. The tropical climate made this article [cotton] suitable as a garment of the vast majority of the people. On the Coast of Coromandel and in the Province of Bengal, when at some distance from a high road or principal town, it was difficult to find a village in which every man, woman or child was not employed in making a piece of cloth." (C2)

As a measure of the decline in industry, in 1880 approximately 60% of the people were engaged in agriculture, while in 1921 73% were thus engaged. In the eighteenth century, the weaving industry in England had not advanced to such stages that it required protection. With the advent of steam power and mechanization, the cheap labor available in India was offset by the combined cheap labor and mechanical power in England. For this reason, the British instituted a program that strangled the developing Indian industries. This was done primarily by imposing high tariffs on the importation of Indian-made goods into Britain. It was only in fields such as indigo, salt, and sugar that any chemical industrial development occurred during this period. Basu (B1) gives the following reasons for the decline in all Indian industries: "Forcing of British free trade in India; imposing heavy duties on Indian manufacturers in England; export of raw products from India; exacting factory acts; transit and customs duties; special privileges to Britishers in India; compelling Indian artisans to divulge their trade secrets." As another example, steel and iron were exported at the beginning of the British period, while later only ore was exported.

The iron and steel industry did not again become an important manufacturing process until after India gained independence. Much earlier the direct process for the production of iron cementation had been developed in India (W1), which was utilized in Europe only several centuries later. Damascus steel was derived from India.

The blast furnaces used in India were small, 3–4 ft in height and 2–3 ft in external diameter, with the blast furnished by goat skin bellows. Two to three hours' operation yielded about 50% iron (B3). The charge consisted primarily of iron ore and charcoal, sometimes with flux and sometimes without. The use of charcoal proved in the long run to be uneconomical, especially in view of the rapid advances in steel production in England. The largest native blast furnaces were capable of producing about 1500 lb of iron per week. At one location (in 1852) there were 70 furnaces in operation, which were capable of producing about 1700 tons of iron per year. The native steel and iron industry was virtually wiped out soon after this time.

In about 1830 attempts to introduce steel and iron manufacture based on European practice were at first unsuccessful, the main problem being a short supply of high-quality iron and of coking coal. The first successful steel mill was established in 1889 by a company that is still in operation. In 1902 the Tata Iron and Steel Company (TISCO) came into being and produced the first steel in India in 1913. In order to encourage growth of the Indian iron and steel industry, protective tariffs were imposed in 1923, which aided in development of the industry. By 1930 four companies were producing steel, with pig iron production reaching about 1.6 million tons in 1939 and about 500,000 tons being exported to Japan. Some early investigations indicated that shipping iron ore to England for processing would be uneconomical because of the low iron content (W1). However, this was later shown to be a mistake in sampling rather than in the quality of the ore. Like the British earlier, the Japanese soon found it better to import the ore and export iron and steel; hence today Japan is a major importer of Indian iron ore.

Sugarcane is native to India, and the process of extracting sugar from the cane originated early in Indian history. Early methods consisted of crushing the cane in stone or wooden mortars and then processing the juice by boiling. Gradually the crushing process was developed to include the use of stone or iron rollers to express the juice. Boiling was still carried out in open pans, the product (gur) needing very little subsequent refining.

The sugar industry flourished in India up until about the time of Napoleon (M1). At that time the development of sugar beets in Europe interfered with the growth of cane sugar manufacturing in India. Sugarcane plants were exported from India to the West Indies, where the climate and the labor supply made it possible to supply sugar at a cheaper rate to England than from India. In addition, the British Empire imposed tax duties that favored the West Indian merchants and utilized slave labor in the West Indies. In 1903, however, attempts were made in India to start a sugar factory, and in 1920 the government formed a sugar committee. By 1929, it was recognized that a tariff reduction by England would be required to help the Indian industry grow.

Sugar was available in India during this period in approximately seven different grades from cane juice through refined sugar. The processes employed including crushing, filtering, boiling, clarification, and subsequent refining steps. The finest white sugar was made by boiling, clarifying, removing the scum, and draining molasses from the crystals. Further clarification was carried out in a long, conical basket lined with fine cloth and suspended so as to drain freely. Water was dripped over the top through wet waterweeds. The many methods of refining sugar depended on local materials, and each village or area had its own experts.

When steel became available wooden crusher rollers were replaced by steel ones. Production went up, but the industry was hampered by the inefficiency of juice extraction. This was improved in later years by the use of larger central mills which could extract a larger fraction of the juice, but the industry remained in a depressed state until after independence. Today India is one of the largest sugar-producing nations in the world, although the bulk of the product is used at home. Since a large part of the sugar is still produced by crude methods by local farmers, the production per acre is very low.

As an industry in India, glassmaking is new even though the art is of ancient origin. About 1892, five modern glass factories were established, all of which were unsuccessful. Between 1906 and 1913, 16 other factories were attempted unsuccessfully; they all utilized European or Japanese techniques and did not take into account local conditions. The main use of glass was for bangles and other decorative objects. By 1933 successful glass factories were established, and there were about 20 glass factories in operation in various parts of India. The primary problems were inadequate financing, lack of skilled labor, and competition from abroad. Only after India became independent did glass manufacture become important. Raw materials for glass were plentiful, and all that was needed was to develop the indigenous manufacturing capability.

The use of cement and lime is of ancient origin. The earliest archeological sites show a knowledge of their use in construction and structural decoration. Even today lime is produced in small conical ovens charged with limestone and charcoal or coke. The first production of portland cement was started in Madras about 1904, based on imported technology. Large-scale production began after World War I. By 1930 approximately 5.5 million tons of cement were being produced annually, almost entirely for use inside India.

Salt has been of economic importance in India and is produced in large quantities. About two-thirds of the supply comes from solar evaporation and the remainder from natural salt deposits in dry lake beds. Solar production is carried out in a series of open ponds, using hand labor and making very little effort to purify or recrystallize the salt. In addition, in some desert regions, evaporation of subsoil brine is carried out. The development of the salt industry in India was curbed by the tax on its production; in fact, the first confrontation between Gandhi and the British government involved the salt tax.

An early chemical industry, of little importance today, was the production of saltpeter or KNO_3. The highly populated Bihar area of India has climatic conditions that favor the growth of nitrifying bacteria in manure. In addition, the population used wood and cow dung for fuel and scattered the ashes over the fields. Following a monsoon, the scattered materal effloresced, and the top $\frac{1}{2}$ in. of nitrous earth was scraped up and placed in long, shallow pits where it was leached with water. The resulting crude KNO_3 was then transferred either to an iron evaporating pan about 5 ft in diameter or to a solar pond. The rough material contained about 60% KNO_3 with common salt as the major impurity. The purification process (conducted in about 400 places in 1905) consisted of redissolving, boiling, removal of scum and the first-formed crystallized material, and recrystallization in vats containing bamboo latticework. The resulting white crystals were placed in bags, water was trickled through to wash out the final impurities, and the crystals were then spread out and dried. In 1905, about 20,000 tons of saltpeter were produced.

Up until the 1940s the manufacture of heavy chemicals in India was insignificant. For example, the production of H_2SO_4 was only 18,000 tons as contrasted with 7 million tons in the United States. Until the establishment of the first contact acid plant in 1948, all production involved the chamber process. The installed annual capacity in 1948 was 175,000 tons, distributed among 49 different factories. Because India does not have a source of free sulfur, the production of H_2SO_4 is one of the critical chemical industries. Development of the large deposits of iron pyrite in the northern part of the country would alleviate this problem.

Drugs and pharmaceuticals were produced mainly at the cottage or village level until well after independence. The chief production of drugs was in the agricultural field, and they were given primary processing and then shipped to other countries for further processing. Quinine, opium, and belladonna were among the raw materials exported. Oils were produced in great quantity and involved crushing and expressing at the village level.

Until the advent of modern dye-making techniques developed in Germany, the production of indigo was an important industry. Accounts of indigo production date from as early as 1607. The process at that time consisted of steeping the leaves in water, putting the liquid in great pots where it was beaten to allow oxidation to take place, and then settling the resultant material. This process was repeated over a period of days, after which the remaining solid was taken out, formed into balls, and laid on the sand to dry. In the nineteenth century, efforts were expended to improve the plant and to increase the yield and quality of dyestuff extracted. Indigo was grown widely over most of northern India, usually with two crops a year. Water supply was considered in factory location. Water was carried through the plant by gravity. The steeping tanks were about 20 ft sq and 4–5 ft deep, and there were 15–20 in each factory. The vats were loaded with plants, filled with water, and allowed to steep for 12–14 hr. After this, the water was run into beating baths where the water was beaten with primitive hand sticks for about 2 hr; in some cases mechanical means of beating were arranged, or NH_3 and air were blown through the mixture. After beating, the mixture was allowed to settle, and the supernatant liquid was drummed off. The residual material was mixed with more water, boiled, filtered through heavy canvas sheets, formed into thick cakes in a press, allowed to dry, and then packed for shipment. In 1894, about 2.4 million lb were exported by India.

IV. Structure of the Chemical Industry

The organizational structure of the chemical industry in India has been an inherent element in its progress and development and has played a significant role in establishing India's self-reliance. To quote (S4):

> The existence of a protected economy, rising out of a controlled market because of the matching of life and capacity with domestic demands, is one of the main reasons for the lack of R&D inputs by industry in India. Another has been the ease with which foreign technology could be imported in the period 1950 to 1970. In India nearly 94 percent of research funding is by government. The government, therefore, has a vital stake in assuring that these funds are spent in a manner which will accelerate economic development and promote human welfare.

In 1948, the government of India announced a policy of planned development and regulation of industries. The objective was to arrive at a "mixed economy" which overall would be beneficial to the nation. Since India is dedicated to a form of social control through democratic procedures, the term "mixed economy" refers to ownership of industrial units supported partly by public funds and partly by private funds. Modifications have been made from time to time in the policy and in the actual administrative structure.

Under this industrial policy, Indian industries were classified into three groups (M1):

> The first category included arms and ammunitions, atomic energy, river valley projects and railways. These were to be directly under the management of the State. The second category included coal, iron and steel, aircraft, telephones, telegraphs, ship building and mineral oils which were also the responsibility of the State. The private undertakings in these industries were, however, to continue for at least 10 years. The third included the remaining industries which were to be developed by private enterprise.

Lists of the specific industries included under these three categories have been issued from time to time by the government. Although the first two categories were to be exclusively state-owned, considerable private ownership is found in these categories at the present time. For industry as a whole 30% is public and 60% is private, the remaining being in the joint sector. The chemical industry shows the same pattern.

Overall industrial policy is under the control of the central government and is the direct responsibility of a cabinet-level minister who works through a director general and through the agencies shown in Fig. 1. A central advisory council, with broad representation, advises the government on all matters concerning the development and regulation of industries. Development councils for individual industries were also set up to aid in the development policy.

The development councils for specific areas comprising the chemical

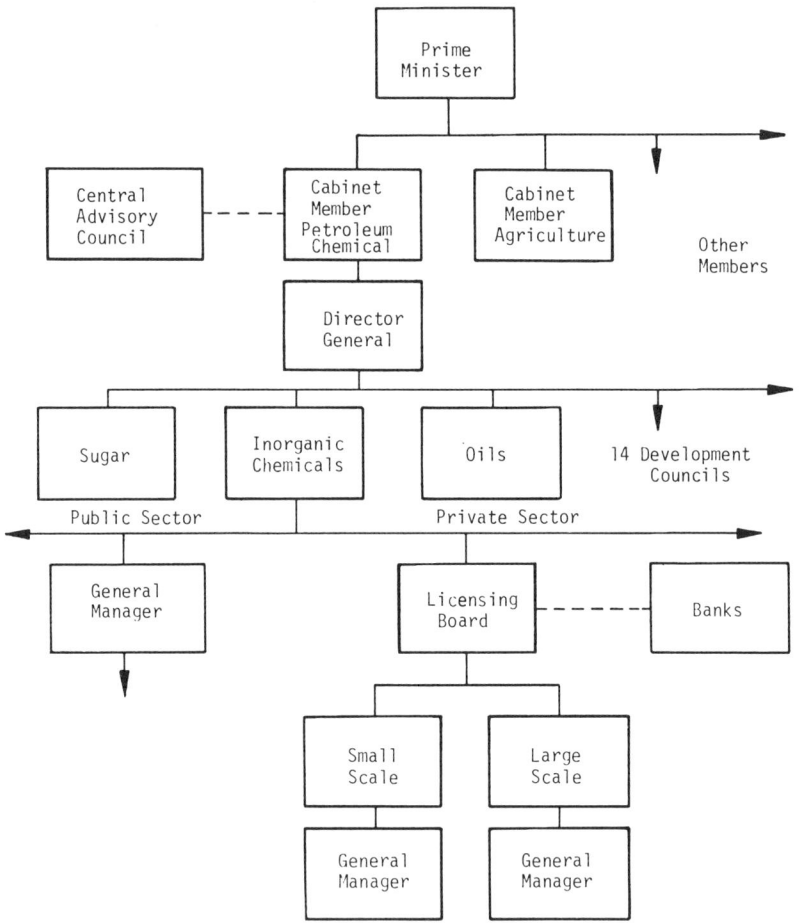

FIG. 1. Control of industrial development.

industry are inorganic chemicals; sugar, drugs and pharmaceuticals; nonferrous metals; oils, soaps, and paints; cosmetics; food processing; organic chemicals; paper, pulp, and industries; leather and leather goods; and cloth and ceramics (see Fig. 1).

Three types of ownership of Indian industry are the public sector, owned completely by the government; the private sector, supported by private funds and private investors; and a joint group owned both privately and by the government. Certain areas are classified as core industries in which greater emphasis is placed on achieving public ownership.

There is a further classification based on the size of industrial units. These are so-called large business houses on the one hand and small industries on the other. A large number of industrial processes have been rsserved primarily and specifically for small-scale users. The scale is somewhat dependent on the amount of investment necessary for establishment of the industry, with greater than 200 million rupees ($2.5 million) being classified as large-scale.

All industrial production, both large-scale and small-scale, is controlled strictly by the issuance of licenses devised to control the type of production, to prevent monopolies, and to ensure that production meets the nation's social needs. Until 1973 the licensing procedure was very complicated and took an inordinate amount of time, up to 3–4 yr, between application and approval. The steps which were, and are, required to obtain a license and financing are shown in Fig. 2. In 1973, licensing was streamlined and put under a single agency. Since then the backlog of applications has been greatly reduced; approximately 75% of the applications received have been processed. The time lag is still great (about 2 yr), as there is a bureaucratic tendency to shift the responsibility for decision to some other part of the structure.

The licensing procedure has not been entirely successful. Many apply for a license and then do not use it, thus blocking the intended economic benefit. In addition, there is widespread evidence that materials that

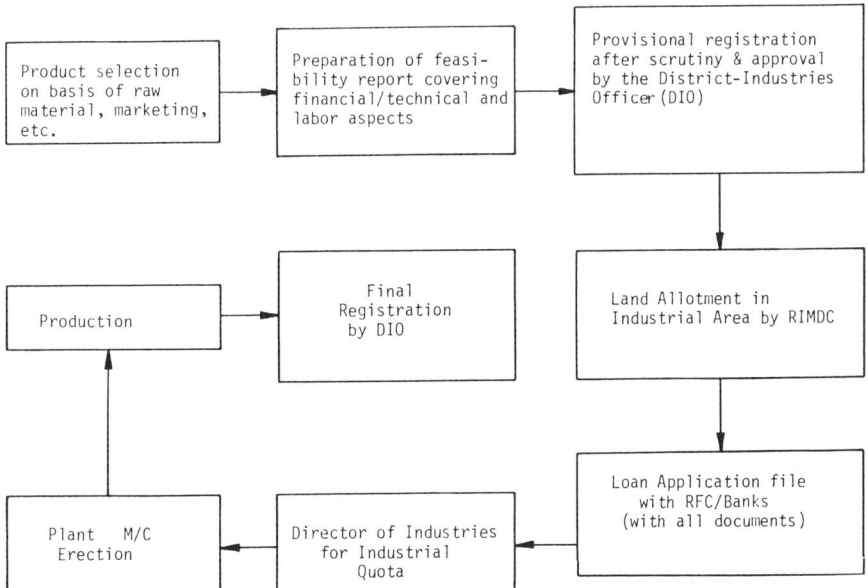

FIG. 2. Licensing and financing flow chart (small-scale industry): RFC, Rajasthan Financial Corp.; RIMDC, Rajasthan Industrial Mineral Development Corp.; DIO, District Industrial Office.

should have gone into construction of a licensed capacity have been diverted to the black market. That is, materials that should have been used in building and operating a plant have been sold on the black market rather than used in the manner intended. This points out that there is no adequate check on the use of licenses, a matter that is beginning to receive attention in government circles.

In 1958, the National Productivity Council was established, and five regional productivity directors were named. The production capacity and growth of the chemical industry are also controlled to a large extent by financing. Through the Indistrial Finance Corporation Act of 1957, the Industrial Credit and Finance Corporation was set up to help private investors, particularly small-scale investors, in various industries such as paper, chemical, pharmaceutical, sugar, metal ore, lime, cement, and glass manufacture. In 1970, the banks of India were nationalized with a view to making more money available to small investors.

There has never been a great incentive for Indian industrialists to expend funds for the research and development necessary to maintain a lead or to expand the chemical industries. All processes developed by any industry have to be licensed, and there is no protection for a manufacturer who develops a new method. These licensing policies have made it more advantageous to hire foreign technology and then to use the equipment as long as it can be used, without regard to replacement or improvement of the processes. These policies are to be reviewed during the next 5-yr plan.

The bulk of industrial research within India is controlled through government funds, and a separate organization has been developed for the pursuit of research and development. Figure 3 is an organization chart of scientific and technical research in India as it applies to the chemical industry, showing interrelationships among the various organizations doing research. Most research institutions such as the Council of Scientific and Industrial Research (CSIR) are autonomous bodies. The organization of the CSIR is shown in Fig. 4 together with the names of some of the research laboratories. Similar organizations exist in the various states; a schematic of one is shown for Andhra Pradesh in Fig. 5. In practice, there exists a wide separation between research and development and actual factory production. There is a great need for research and development to be more closely tied to industry and to provide assistance in solving the recurrent problems faced by production managers. An experimental scheme being carried out in India, called the practice school, will be described in a later section.

Most research is carried out either in educational research institutes or in national research laboratories, which are structured and operated much like graduate research facilities at American universities. The

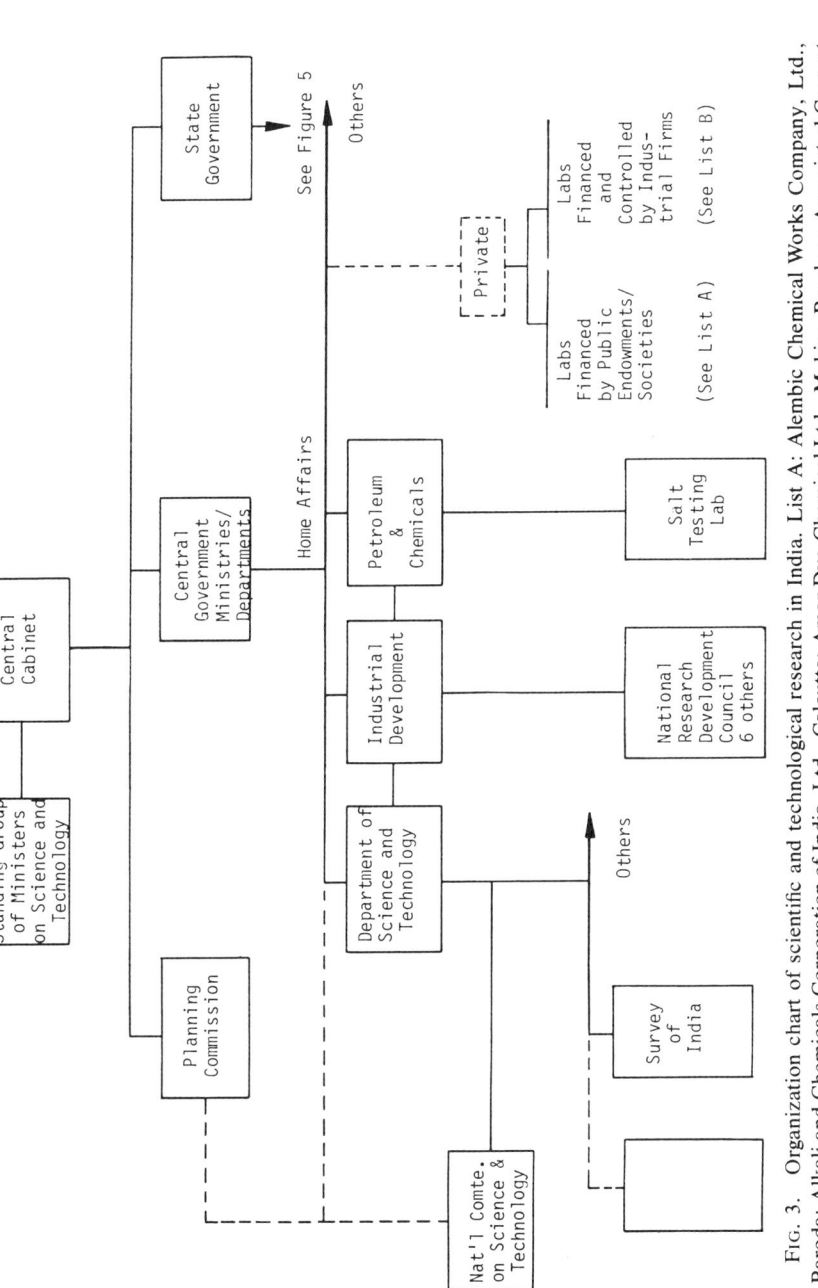

FIG. 3. Organization chart of scientific and technological research in India. List A: Alembic Chemical Works Company, Ltd., Baroda; Alkali and Chemicals Corporation of India, Ltd., Calcutta; Amar Dye-Chemical Ltd., Mahim, Bombay; Associated Cement Companies, Ltd., Thana, Bombay; ATIC Industries Ltd., Atul (Gujarat state); Bengal Chemical and Pharmaceuticals Works, Ltd., Calcutta; Bombay Chemical Pvt., Ltd., Bombay; Calcutta Chemicals Company, Ltd., Calcutta; plus 50 others. List B: Birla Research Institute for Applied Sciences, Nagda; Drugs Research Laboratory, Indian Drugs Research Association, Poona; Engineering and Mineral Industries Research Laboratory, Bangalore; Shri Ram Institute for Industrial Research, Delhi.

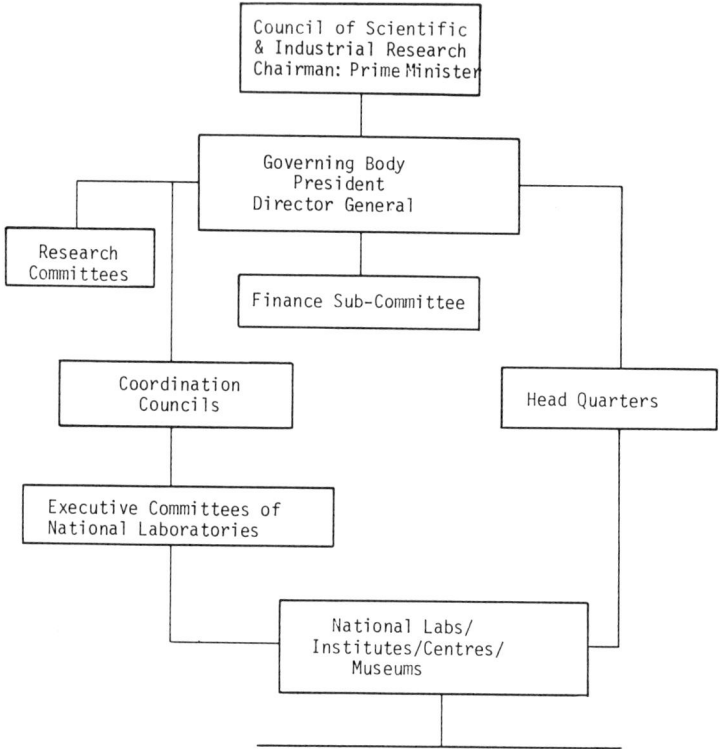

FIG. 4. Organization chart of the CSIR.

government-operated laboratories and their users have been somewhat isolated. The use of processes developed at these institutions is small compared with total production within the chemical industry, owing to charges for use of the process and burdensome licensing procedures. Recent steps have been taken to overcome these deficiencies through the

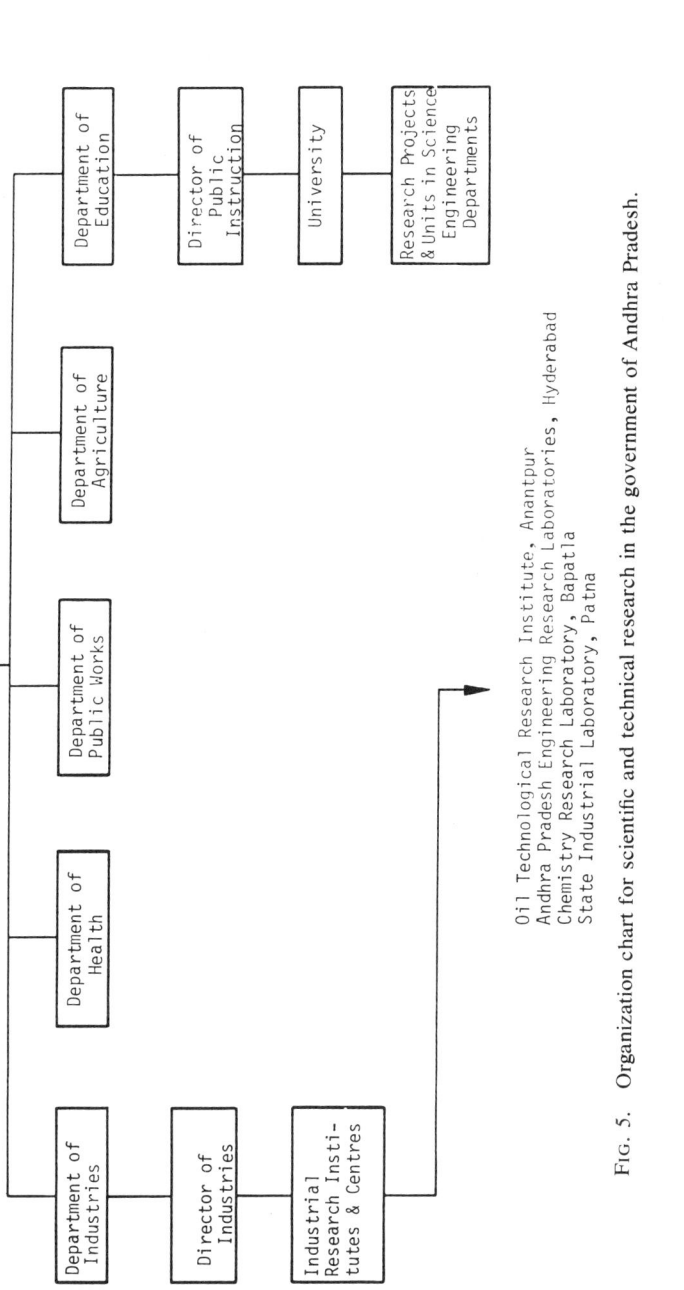

FIG. 5. Organization chart for scientific and technical research in the government of Andhra Pradesh.

formation of advisory boards. The National Committee on Science and Technology (NCST) made an extended study to determine the types of research that are necessary. Many questionnaires were circulated throughout the industry, and the results analyzed. Based on this study, NCST technological plans (S4, S5, S6) have been published, which outline the research needs of the industry. These have been supplemented by two volumes on the chemical industry (J3, J4). These documents are used for national planning and influence the allocation of research funds.

Among the major governmental research centers, particular mention should be made of the following. The National Chemical Laboratory (NCL) at Poona is charged with the development of methods for the manufacture of a large number of chemicals, materials, and devices. Similarly, the Indian Institute of Petroleum at Dehra Dun, the Central Fuel Research Institute of Dhanbad, the Central Salt and Marine Chemicals Research Institute at Bhatnagar, and the region research laboratories at Hyderabad, Jorhat, and Jamma provide support for development of the chemical industry. The emphasis is on research based on the immediate needs of the country.

The NCL has the best record for adoption of developed processes, with 47 out of 75 in production. In 1973, the value of goods produced by these techniques was $5.7 million. This represents a small but important contribution to the progress of the indigenous technology.

V. Development since Independence

Since 1947, the development of Indian industry, particularly the chemical industry, has been rapid and dramatic. Chemical construction firms in both the private and public sectors are now capable of building, on a turnkey basis, many important types of chemical manufacturing plants. Great emphasis is placed on the use of indigenous materials and of talent developed within India. For many commodities the rate of growth is more rapid than the rate of growth in developed countries. However, the primary deterrents are the difficulty of obtaining foreign help and the reluctance of both industry and government to look far enough ahead to identify and meet the real needs of the industry and the country in general.

One of the main problems in the development of the industry has been the lack of suitable materials and adequate instrumentation. For example, one manufacturer had considerable difficulty in manufacturing gear boxes needed for the preparation of rayon staple fiber because a suitable steel was not available. An alternate solution was to redesign the gearbox to use another type of steel or to import steel with the proper qualities.

Foreign exchange was in short supply, and obtaining import licenses was very difficult. Originally, the manufacturer thought he lacked the skill to redesign, but faced with the fact that he had to make a gearbox, a new design was successfully carried out. Considerable progress is now being made in developing the plastics industry and specialty metals so that in the future this type of material shortage will be less of a problem in the chemical industry.

The location of the various chemical industries in India is shown in Fig. 6. As shown by this map, chemical industries have been concentrated in a few particular areas, which does not correspond to the widespread distribution of industry desired by the government. Although planners have assumed it desirable to locate industry in a way that will help utilize surplus agricultural labor, other factors have been more important and have influenced the location of plants. Industry tends to expand in heavily populated urban areas, especially adjacent to port facilities and along established rail lines. Even the location of raw materials has played a small part in siting a manufacturing unit. Planners have tried to locate

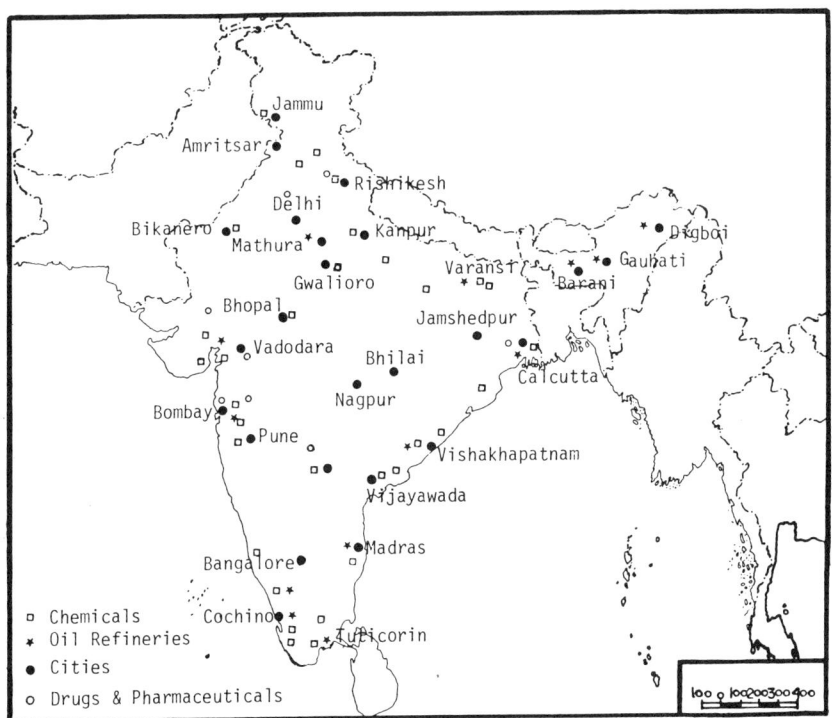

FIG. 6. Chemical industries in India.

plants in accordance with social needs and not from the standpoint of optimizing the economics involving water, materials, market, labor supply, and so on. For example, an attempt was made to locate a steel plant based only on a large population and the availability of a seaport (Vishakhapalanam). All raw materials would have had to come overland by rail. The location of this plant has still not been decided.

A vital factor in siting chemical industries, perhaps the determining factor, is the availability of suitable water for processing and cooling needs. There are several major waterways in India such as the Ganges, Indus (and its five main tributaries), Brahmaputra, Narmade, Mahanadi, Godavari, Kushna, and Kaveri rivers. Two of these, the Indus and the Brahmaputra, flow through Pakistan and Bangladesh. The control of these waterways and the building of dams and canals are being disputed by the various countries involved, with a consequent delay in effective utilization. Other river systems flow through two or more states. Just as in the western United States, the distribution and usage of these waters is highly contested. The result is limited usage of the water for power, irrigation, or industrial production. It is estimated that only 25% of the hydroelectric potential is utilized. Large canal systems are slowly being developed, especially in the northwest desert areas.

Most of India's water supply depends on the monsoons, which occur in the summer months between July and September and greatly affect industry as well as agriculture. The waterways are seasonal and erratic. The primary streams such as the Ganges flow over broad, flat plains with relatively little fall between the mountains and the outlet at the ocean. The flow of the Ganges in the central portion reaches a volume of 1 million sec ft at flood crest. At Varanasi flood crests of 36 ft have been recorded. These have been marked with a white painted line on the steeples of riverside temples. The Ganges enters the plains at Rishekesh, having fallen 12,800 ft. From this point, the 1200-mi path to the sea lies over a gently falling plain from an elevation of about 1000 ft. The average fall is about 0.8 ft/mi (9.5 cm/km). Since the Ganges sometimes is very high and meanders, it can greatly change its channel over the flat plain country. Because of the small fall, it is difficult to make storage facilities to store the water during the monsoon period, the time when water is really available. This has impeded development of the chemical process industries along the Ganges. Near the mouth of the river the flow is much more stable, and it is easier to ensure a supply of water for a chemical plant. In certain cases it has been possible to develop large underground infiltration galleries to ensure a water supply. The eventual development of flood control systems and dams should be a great aid in the development and location of chemical process industries.

Chemical development in India, as well as other industrial development, has been carried out in five plan periods, nominally of 5 yr each. The sixth 5-yr plan (D1) is currently under review, and development in the ensuing 5 yr will be guided by this plan when it is complete. During the plan period, the overall growth pattern is projected on the basis of broad social objectives. Control is maintained by the use of licensing and financing, both of which are in the public domain, as discussed above.

The periods covered by each plan are shown in the accompanying table. During the period 1966–1968 the war with Pakistan, general drought, and political instability dictated the use of annual plans. The sixth and next plan is in the final stages of development.

Plan	Years covered
I	1951–1955
II	1956–1960
III	1961–1965
Annual plans	1966–1968
IV	1969–1973
V	1974–1978
VI (in planning)	1978–1993

Various degrees of success have been obtained; generally the investment requirements are met, but production does not usually meet the target for a variety of reasons, including the weather.

The progress of several important parts of the chemical industry is outlined below under several general broad categories. This outline indicates the areas of rapid growth and documents the continuing development of Indian self-sufficiency.

Since 1947, many important chemical industries have been established. Examples of the beginning of the manufacture of various chemicals within India can be seen by consulting the chronology given in Table I. Many more new chemicals are being produced each year. In many cases, dependence on imports from other countries has been entirely removed.

A. HEAVY CHEMICALS

1. *Sulfuric Acid*

The production of H_2SO_4 shown in Table II has risen from about 107,000 tons in 1950 to about 1.5 million tons in 1977. This rate of growth exceeds the rate of growth in developed countries. The utilization factor for the

TABLE I

THIRTY YEARS OF CHEMICAL PROGRESS—INTRODUCTION
OF NEW PROCESSES AND PRODUCTS

Year	Process and products
1949	Benzene, H_2SO_4, HCl, $(NH_4)_2SO_4$
1950	Caustic soda, soda ash
1951	TiO_2, penicillin, rayon
1952	Carbon or C_2
1953	BHC, acetic acid
1954	NH_4Cl, sulfa drugs
1956	H_2O_2, DDT, antituberculosis drugs
1960	Urea, tetracyclines
1961	C.A.N.
1962	Hydrosulfite, streptomycin, nylon
1965	Polyester, polyethylene, polystyrene
1966	PVC
1967	Catalysts, synthetic rubber
1968	Phthalic anhydride, methane
1969	Acetamide, cellulose acetate
1970	Ingrain dyes, ethylene oxide
1973	Hydrazobenzene, benzine dihydrochloride, white phosphorus
1974	$NaNO_3$, $NaNO_2$
1975	Rayon from eucalyptus, terpene-based chemicals

installed capacity is one of the highest of any industry within India. Because H_2SO_4 is used in phosphate fertilizer, rayon staple fiber, organic dyestuffs, explosives, pickling of steel, petroleum refining, and the production of other acids, the increase in its production has brought about rapid growth in other industries.

The first contact H_2SO_4 plant went into production in 1948. Out of 49 plants in 1951, 30 utilized the chamber process. With the introduction of ceiling prices on all sales of H_2SO_4 in excess of 1 ton in January 1956, the chamber plants were gradually replaced by contact plants of larger capacity. By 1968–1969 all the units were contact plants, with a total capacity of 1.121 million tons/year. To reduce the dependence on imported sulfur or internal sulfur sources, the Fertilizer Corporation of India (FCI) has developed techniques in which the use of H_2SO_4 is greatly reduced. These processes include: (1) the use of electrolytic instead of wet processes for the manufacture of elemental phosphorus and H_3PO_4, (2) the manufacture of H_3PO_4 and phosphate fertilizers by using HCl, (3) the manufacture of phosphate fertilizer using HNO_3.

TABLE II

Sulfuric Acid Production

Year	Capacity (1000 metric tons)	Production (1000 metric tons)	Utilization (%)
1951	201	107	53.1
1952	192	96	50.1
1953	189	109	57.7
1954	193	151	76.5
1955	208	166	74.9
1956	245	165	67.4
1957	273	196	71.8
1958	290	227	78.1
1959	374	292	78.1
1960	476	354	74.4
1961	564	423	74.9
1962	702	470	66.9
1963	821	568	69.2
1964	1011	580	67.2
1965	1082	685	63.3
1966	1328	690	51.9
1967	1529	805	50.8
1968	1955	1008	51.5
1969	1921	1121	58.4
1970	1930	1051	54.4
1971	1963	1022	52.1
1972	1963	1028	52.4
1973	2112	1434	67.7
1977[a]	2640	1900	72.0

[a] Estimated.

Under the sixth 5-yr plan H_2SO_4 production will be expanded to 3.79 million tons/yr, an increase of 44%.

The contact H_2SO_4 plants are designed and their components built entirely in India. These plants have been exported to other countries. However, the technology has not developed to such a point that the V_2O_5 catalyst can be produced in India. Therefore, it must be imported. In addition, equipment for use in the production of oleum and furnaces for roasting sulfide ores must also be imported. These design capabilities await only production of the proper type of steel for India to be able completely to design, fabricate, construct, and operate H_2SO_4 plants within the country.

A major problem is the lack of elemental sulfur deposits, already mentioned. Until a few years ago, all H_2SO_4 plants were based on elemental sulfur imported chiefly from the United States. Several plants have been or are being built to obtain SO_2 from the indigenous sulfide ore deposits chiefly in the vicinity of the nonferrous metallurgical industries. A 400-ton/day plant based on pyrites from Amjhore in Bihar was established near the Sindri fertilizer factory (see Fig. 6). The Cominco-Binani zinc smelter at Alwaye, the Udaipur smelter of Hindustan Zinc, and the Indian Copper Corporation smelter at Ghatsila are other examples. The flash smelter of the Khetri copper plant in Rajasthan will have a H_2SO_4 plant to utilize the SO_2 from the flash smelter. When all these plants go into production by 1977, they will provide an annual H_2SO_4 production of about 250,000 tons.

There has been some effort to recover elemental sulfur from high-sulfur petroleum crudes. The Madras refinery has a capacity of 20,000 tons of sulfur per year. As a resource not yet fully utilized, Amjhore in Bihar has iron pyrite deposits estimated at 400 million tons. Development of these deposits could make India entirely independent of imported sulfur. The ore beds are in relatively narrow seams, so that the overall quality of the ore is somewhat low. The technology needs to be developed to utilize this low-grade ore, and the NCST (S4) lists this project as one of the primary research needs for H_2SO_4 production. Location of plants utilizing the acid should be considered for location near the source of the sulfur. Amjhore is an area where there is also iron ore, coal, and bauxite, providing the opportunity for constructing a well-integrated industrial complex.

2. *Fertilizer*

Fertilizer continues to be a critical need for the progress of India as a nation. The acreage under treatment with fertilizer has been increasing at about 1.4% per year. The application level in kilograms per hectare remains low, as does the per capita use. For the world, the application rate is 47.4 kg/ha and the per capita rate is 18.3 kg (F2, S1). The range is 13–749 kg/ha. The corresponding figures for India are 13.2 kg/ha and 4 kg per capita. This indicates the great need for fertilizer in India if it is to meet its food production targets. The production levels and percent of capacity since 1947 are shown in Table III. The rate of increase in production over the period since independence has varied from 12 to 13% per year against a target of about 23%. The percent utilization of installed capacity, indicated in Table III, is relatively poor even though the need for fertilizer has not been met. The underutilization of capacity in recent years is caused by a lack of adequate power, by labor problems, and by plant obsolescence. Before 1951 the only indigenous fertilizer produced was $(NH_4)_2SO_4$ made mainly by coal carbonization at an annual rate of about 46,000 tons.

In 1951, the Sindri fertilizer plant went into production with a capacity of

TABLE III

Fertilizer Production[a] (P and N)

Year	Capacity	Production	Utilization (%)
1961	548	347	63
1965	909	533	59
1966	1145	662	58
1969	2130	1392	65
1970	1855	1311	71
1971	3626	2237	62
1972	3650	2650	73
1973	4004	2622	65
1974	4333	2598	60
1977[b]	4400	2430	55

[a] 1000 tons total nutrients.
[b] Estimated.

1000 tons/day of $(NH_4)_2SO_4$ based on gypsum and synthetic NH_3. This plant is scheduled for extensive remodeling during the next plan period. The Sindri plant was followed by the Rourkela fertilizer unit in Orissa in 1961 and the Nangal complex in Punjab in 1963, which together introduced calcium NH_4NO_3 to the country. These units differ in their production of hydrogen for NH_3 synthesis. The Sindri unit uses the coke–steam reaction, the Nangal unit electrolysis, and the Rourkela plant coke oven by product gas.

Since 1947, the suitability of basic chemical ingredients for particular soils has engaged the attention of fertilizer technologists. This led to a stabilized fertilizer production strategy starting in 1956. It was thought desirable to use refinery gases rather than coke or electrolytic hydrogen. To reduce the price of available nitrogen, concentrated fertilizers such as urea were stressed. Larger quantities of balanced fertilizers such as $(NH_4)_2HPO_4$ also were considered important to agricultural development. The government then planned at least one large fertilizer factory in each state and recommended six sites: (1) Namrup in Assam for $(NH_4)_2SO_4$ based on natural gas, (2) Kotagudem in Andhra Pradesh for urea based on coal, (3) Itarsi in Madhya Pradesh for urea based on coal, (4) Mangalore in Mysore for $(NH_4)_2SO_4$ based on naphtha, (5) Haumangarh in Rajasthan for urea based on lignite and later on naphtha, and (6) Gorakhpur in Utter Pradesh for urea based on naphtha from the Barouni refinery.

Units 1 and 6 were commissioned in 1965–1966. In Rajasthan a urea plant, unit 5, has been put into operation by the private sector. In Andhra Pradesh, a private-sector unit has been built at Visakhapatnam. Other public and private projects have been started and are in different stages of completion. At present there are 70 fertilizer plants of all sizes in India operating at about 60% capacity. There are 18 new or enlarged plants under implementation with a total capacity of 1.88 million tons. This will bring the total installed capacity to 8.22 million tons.

There was a decline in the demand for fertilizer because of the great increase in cost caused by the higher cost of oil. This demand, however, is again increasing. There are three large plants (with a total capacity of 0.8 million tons) under construction, which will each use about 1 million tons of coal a year. Other coal-based units are planned to make use of the plentiful supply of coal.

A plant for NH_4Cl (along with Na_2CO_3) utilizing the modified Solvay process was developed by the private sector in Varanasi in Utter Pradesh. NH_4Cl was not effective for the soils present in this area. Recent tests have shown that NH_4Cl is an ideal fertilizer for the rice crop, and it is now being used by rice farmers in Bengal. Therefore, the production rate of the factory in Varanasi was doubled in 1976, and an additional plant was built in Maharastra the same year. At the present time four more plants are under construction. Although production is small, this represents a use of local materials, coke and salt, to help meet the needs of India.

It was planned that this large program for fertilizer production would reduce the need for imports substantially. Actually, in 1971–1972, the total production of indigenous nitrogen fertilizers was about 1 million tons and the demand had risen by 1 million tons, so that it was still necessary to import large quantities. In 1975–1976, it was necessary to import 1.54 million tons as against the production of 2.75 million tons (a total demand of 4.29 million tons). This deficit was brought about largely by the "green revolution" which resulted in improved strains of grain requiring a larger fertilizer input. The worldwide fertilizer demand has increased, which makes it more difficult for India to import it, resulting in an overall decrease in grain production.

A study committee has determined for the next few years, through 1979, the amounts of nitrogen and phosphate fertilizers that will be required by Indian farmers. Table IV shows the estimated production and the deficit in terms of fixed nitrogen and of P_2O_5. By 1979, it is expected that the total demand for nitrogen fertilizer will exceed 6 million tons and the demand for phosphate fertilizer 2 million tons. This will leave a deficit of about 3 million tons to be imported by the beginning of 1979. In recognition of this, a crash program was instituted to boost fertilizer production to 8 million tons by the end of 1979. The progressive design and construction technology available in India is now adequate to handle all the manufacture of fertilizer plants within India.

TABLE IV

ESTIMATED PRODUCTION AND DEFICIT OF FERTILIZER COMPONENTS

Year	Estimated production (1000 tons)		Estimated deficit (1000 tons)	
	Nitrogen	P_2O_5	Nitrogen	P_2O_5
1974–1975	2010	549	860	599
1975–1976	2465	752	980	625
1976–1977	2956	897	1074	755
1977–1978	3725	983	1685	1001

With the technology available in India, it is possible to undertake the construction of additional fertilizer factory projects in oil-rich developing countries in Asia and Africa. Plans are currently underway to manufacture fertilizer in these countries and then import it into India, rather than importing crude oil and then making fertilizer. (In India, only a small fraction of the imported crude is used for transportation, the bulk being used for fertilizer and petrochemicals.)

3. *Chlor-Alkali Industry*

Soda ash manufacture is another fast-growing heavy-chemical industry. The actual production for India and the percent utilization are shown in Table V. Again, the underutilization of capacity is caused primarily by labor problems and by the energy shortage. (Development of the coal reserves within India would be of great help in this connection.) Until 1951, only a plant at Dhrangadhra and another at Mithapur were in production, both using the Solvay process and having a combined annual output of 47,000 tons of soda ash. By 1956, their production had reached about 90,000 tons through improvement and modernization. A plant at Porbandar with a capacity of 200 tons/day was commissioned in 1969 and has since increased its capacity. Maharashtra in western India provides an ideal location in regard to availability of raw materials, that is, limestone and salt. However, locating all the plants in Maharashtra was not considered desirable because of the difficulty in distributing products to the eastern and southern regions.

After 1960, licenses were issued for soda ash plants in noncoastal regions. This introduced pollution problems from the release of by-product $CaCl_2$ into the river systems. The plant in Varanasi that went into produc-

TABLE V

SODA ASH PRODUCTION

Year	Capacity (1000 metric tons)	Production (1000 metric tons)	Utilization (%)
1951	55	48	87
1956	91	86	95
1961	91	176	193
1966	363	349	96
1961	435	421	97
1970	470	446	94
1971	470	477	101
1972	500	486	97
1973	508	470	93
1974	508	510	100
1977[a]	704	530	75

[a] Estimated.

tion in 1959 therefore adopted a modified Solvay process; as already mentioned, it recovers by-product NH_4Cl for use as fertilizer.

The soda ash industry has nearly achieved self-sufficiency and has potential for export. Sufficient know-how has been developed to design and fabricate plants within the country and to establish similar plants in other developing countries.

a. *Caustic Soda.* The annual production of caustic soda since about 1940 is shown in Table VI together with the percent utilization of the installed capacity. This represents a growth rate of about 17% per year. In general, utilization is excellent and not like that in many other industries.

Caustic soda, which has much wider application than soda ash in the chemical industry, was not produced in India until about 1940, presumably because it was little needed in industry until that time. It was used chiefly by the textile and soap industries, for which 25,000 tons were imported in 1938–1939. The emergence of three major products in later years increased the demand for caustic soda. These were viscous rayon, paper pulp, and alumina from Bayer process extraction of bauxite.

Small units employing the electrolytic process were established in 1941 at Calcutta and Methur. Later, several plants were established using the caustification of soda ash; total production inside India reached 15,000 tons in 1950–1951, but total imports at this time remained high at 63,000 tons. Mercury cells went into production in 1952, and the country achieved self-sufficiency in 1967. By 1976–1977, total production in-

TABLE VI

PRODUCTION OF CAUSTIC SODA

Year	Capacity (1000 metric tons)	Production (1000 metric tons)	Utilization (%)
1951	27	14	53.2
1952	34	17	49.0
1953	38	23	60.3
1954	41	29	70.4
1955	44	34	77.2
1956	56	39	70.0
1957	45	42	93.8
1958	66	57	86.7
1959	98	69	70.7
1960	17	78	81.2
1961	124	119	96.3
1962	124	126	101.4
1963	187	152	81.1
1964	203	184	90.5
1965	268	214	80.0
1966	323	230	71.3
1967	377	251	66.6
1968	366	318	86.9
1969	367	347	94.7
1970	367	373	101.7
1971	372	374	100.5
1972	428	397	92.8
1973	439	404	92.2
1977[a]	704	530	75.0

[a] Estimated.

creased to 530,000 tons. Nearly all states in India now have caustic soda–chlorine plants. Complete caustic soda plants using mercury cells are now designed and produced entirely within India. All parts are manufactured indigenously, and plants are being designed and constructed for other countries.

b. *Chlorine.* Chlorine is produced (see Table VII), mainly as a by-product of the electrolytic manufacture of caustic soda. At the present time, the liquid chlorine produced is not completely utilized, so that the utilization is lower than indicated. Auxiliary industries using chlorine as an input have not been fully developed, and much of the chlorine produced is transformed into $CaCl_2$ and released into the river systems, sometimes by way of large holding ponds which are discharged during the monsoons when the river flow is high.

TABLE VII

Production of Liquid Chlorine

Year	Capacity (1000 tons)	Production (1000 tons)	Utilization (%)
1951	13	5	40
1956	21	15	72
1961	33	33	101
1966	136	62	46
1969	222	129	59
1970	293	147	63
1971	231	162	76
1972	231	147	64
1973	267	122	46
1974	267	142	53

The problem of chlorine is being solved by the rapid growth of chlorine-based industries. The bulk of the chlorine is utilized for pesticides, disinfectants, and bleaching powder. The remainder is converted into HCl. A notable growth in the polyvinyl chloride (PVC) industry has increased the utilization of chlorine. As the chemical industry in the country grows during the next decade, it is expected that the demand will double and that production will increase accordingly, with little being discharged as waste.

B. Organic Chemicals

Like most chemical industries in India, organic chemical production was in its infancy at the time of independence. There were a number of plants engaged in coke manufacture, particularly for the steel industry. However, in only a few were the valuable coal tar by-products recovered. In the mid-1950s when the production of benzyl hexachloride (BHC) and dichlorodiphenyldichloroethane (DDT) was started, the demand for benzene went up steeply and the coke ovens used in the public sector steel plants were rearranged so as to recover by-product organic chemicals.

In the early 1950s, the country had a moderately developed fermentation industry with an annual production of alcohol of about 10 million gal (compared with 88 million gal in 1977). Nearly 50% of this was used for fuel. The production of glycerine at this time from soap manufacturing was nearly 5000 tons. A little methanol was produced by wood distillation at Bhadravati, Mysore, while CaC_2 was nonexistent. During that period, important chemicals such as phenol, phthalic anhydride, urea, acetic acid, and organic and various solvents required by the dyestuff, pharmaceutical, and plastics industries were imported, involving over $19 million of foreign exchange.

The need for a strong organic and fine chemical industry in India was recognized; therefore, several governmental committees were established to promote such an industry. Concurrently, several private bodies investigated the profitability of this industry and instituted several production projects. European manufacturing organizations were instrumental in collaborating with government and private-sector plants to begin the manufacture of organic chemicals. The times these started are shown in Table I. Thus, a beginning was made in the field of organic chemistry in the late 1950s and early 1960s. Plants based on coal tar products, CaC_2, alcohol, and naphtha were brought into production.

The annual growth rate of many of these industries has been between 15 and 20% per year. Table VIII compares the installed capacity for a few chemicals in 1963 with those for 1973 and estimated for 1978–1979 (S4). The production of organic chemicals doubled between 1973 and 1978. Even with this increase, the need remains to import many of these chemicals. The political priority of other sectors of the industry make more rapid growth impossible.

As a result of the recommendation of the Petrochemical Advisory Committee, several petrochemical complexes were planned, each with a naphtha throughput on the order of 200,000 tons/yr. The first of these complexes, set up by National Organic Chemicals in Bombay, went into production in 1967. This complex mainly produces polyethylene oxide (20,000 tons), butanol, butadiene, and benzene. Another complex at Koyali near Baroda (1974) produces aromatics and has a capacity of 24,000 tons of DMT, 21,000 tons of o-xylene, and 25,000 tons of mixed xylene. A

TABLE VIII

INSTALLED CAPACITY ESTIMATED REQUIREMENTS

Substance	1963 (tons)	1973 (tons)	1978–1979 (tons)
Acetone	1500	16,000	28,000
Acetic acid	6000	17,000	40,000
Plasticizers	900	20,000	—
Ethylene oxide	—	12,000	33,000
Ethylene glycol	—	10,000	25,000
Butanol	3500	12,000	12,000
Phenol	—	16,000	25,000
Formaldehyde	7000	33,000	—
Phthalic anhydride	—	10,000	60,000

large olefin project in this complex produces feedstocks required by the polymer industry. One other similar project is underway—at Bongaigaon (1979) in Assam.

C. Soaps and Detergents

Soap making is one of the country's oldest industries and originated as early as 1879. The industry grew rapidly during and after World War II. The total production of soap in 1948 was estimated to be 180,000 tons. Soap and detergent manufacturing has increased markedly since independence, as shown in Tables XI and X. Many of the units are small and of the village type, so that the total installed capacity and percent utilization are based on the larger units. This part of the chemical industry has an outstanding record of utilization.

TABLE IX

Production of Soap

Year	Capacity (1000 metric tons)	Production (1000 metric tons)	Utilization (%)
1951	192	83	43
1952	193	86	44
1953	193	82	42
1954	193	88	45
1955	240	99	41
1956	253	110	43
1957	253	111	44
1958	253	123	48
1959	253	130	51
1960	257	142	55
1961	257	149	57
1962	242	164	67
1963	232	164	70
1964	232	158	68
1965	232	169	72
1966	232	181	77
1967	232	176	75
1968	212	189	89
1969	212	224	105
1970	217	231	106
1971	217	259	119
1972	217	297	136
1973	218	235	107
1977[a]	233	290	124

[a] Estimated.

TABLE X

PRODUCTION OF SYNTHETIC DETERGENTS

Year	Capacity (1000 metric tons)	Production (1000 metric tons)	Utilization (%)
1961	7	8	100
1965	7	8	116
1966	15	8	57
1968	30	18	61
1969	30	22	75
1970	30	31	103
1971	47	53	113
1972	55	61	111
1973	55	60	108
1974	88	84	96
1977[a]	210	104	50

[a] Estimated.

With the ban on soap imports in 1957 and continued facilities for importing the raw materials, the industry has recorded phenomenal growth during recent years. However, the annual per capita availability of soap remains low, 1.5 kg as against about 10 kg in developed countries, and there is considerable room for development.

There is a worldwide tendency to replace conventional soaps (made from fats and oils derived from animals and plants) with synthetic detergents (syndets) from petroleum sources. In developed countries syndets have replaced soaps to the extent of 70–80%. In India, the situation is nearly the reverse; the total estimated production of soaps exceeds 700,000 tons, while that of syndets is only 100,000 tons. With the increasing demand for soap, the import of oils and fats tends to increase tremendously, affecting not only the soap industry but also the edible oils industry which is closely linked to it and also is subject to an ever-increasing demand (based on the improving standards of living and the increase in population—625 million in 1977).

The combined demand has greatly increased the price of edible oils in the world's markets. Imports from Western countries must now be paid for in hard foreign currency and have therefore been curtailed greatly. For this reason, it is considered desirable to switch from soaps to syndets as soon as possible and to undertake further large-scale production of syndets. This would make more edible oils available for use in cooking and in manufacturing hydrogenated oils. If soaps are replaced rapidly by syndets, another problem which is social in nature will develop. As mentioned above, much of the soap production is carried out in small-scale units, over 3000 in all, employing perhaps 200,000 people. If soap production were cut drastically, over 100,000 people would be jobless and the machinery presently used would become idle. Since syndet plants require heavy investments, small-scale producers cannot easily switch over to syndet production. In consideration of the socioeconomic problem, soap production should be phased out over a number of years but not stopped abruptly. Plans could also be made for small-scale producers to begin compounding and packaging soaps and detergents rather than only manufacturing them from edible oils.

D. Pesticides

It is estimated that about one-fourth of the crops in India are damaged by either rodent or insect infestation (I1, I2). This means that there is a great need for pesticides and rodent poisons. The 1977 production was about 39,000 metric tons at about 78% utilization. Increased production will help to utilize the excess chlorine production within India.

With emphasis on the need for pesticides in agriculture a BHC plant was commissioned in the private sector in 1952. A DDT plant in the public sector was established near Delhi in 1955. Production, which was only 462 tons in 1954–1955, increased to about 28,000 tons in 1968–1969. Other related pesticide products the country produces are organic phosphates, parathion, Al_3P, methyl bromide, and ethylene dibromide, along with about 20 other compounds.

In 1965–1966, the government spent $5 million to import both pesticide (45%) raw materials and finished products (55%). In 1975–1976, the use of pesticides was 47,000 metric tons. Thus, the policy of the government in the matter of importing pesticides has been liberal, corresponding to its classifying pesticides as "key" and "priority" materials.

The recent growth of petrochemical complexes has improved the availability of numerous basic raw materials and intermediates for pesticides. The possibility of reducing the use of scarce metals such as copper, mercury, and nickel in pesticides by using nonmetallic alternatives has been considered and appears to be feasible.

E. Plastics

In India, as in the rest of the world, the use of plastics has increased rapidly. It is possible to replace many metals and other scarce materials with plastics. The production of selected plastics is shown in Table XI for the years since independence, along with the percent of utilization of capacity. This table does not include the many existing fabrication plants.

TABLE XI

PLASTICS (MOLDING POWDER PLUS RESINS PLUS PHENOFORMALDEHYDE MOLDING POWDER)

Year	Capacity (1000 tons)	Production (1000 tons)	Utilization (%)
1951	0.457	0.208	45
1956	0.896	1.001	112
1961		16.1	
1966	42	43	102
1969	108	89	92
1970	112	102	91
1971	119	117	98
1972	132	117	89
1973	142	124	87
1974	157	114	73

The plastics industry was in its infancy in the years just following independence. A few small-scale industries existed, which mainly converted molding powders into finished products by compression or injection molding. The production of plastic goods has rapidly increased to a point where it furnishes considerable employment in small-scale industrial units. This will increase rapidly in the future as the capability for producing plastics increases.

The total investment in the plastics industry rose from $31 million in 1961 to $125 million in 1970 and, in the same period, production rose from 13,000 tons to over 100,000 tons. The manufacture of processing machinery has also increased rapidly. Bulk production is increasing, but the manufacture of sophisticated materials and quality products is still not being carried out. All major raw materials are now produced in the country with the establishment of petrochemical complexes, so that the industry will play a vital role in future economic development. In spite of the rapid growth, the annual per capita production in 1971 was only 0.2 kg as against 81 kg in West Germany, 59 kg in Japan, and 40 kg in the United States. Thus, there is an ample market in the country and room for growth. Acrylic plastics, Teflon, silicones, and so on, are yet to be developed and manufactured. Such specialty plastics are essential for the replacement of more critical materials and therefore need to be developed within India. The processes and know-how for the manufacture of these types of plastics should be imported from more highly developed countries, and efforts should then be made to adapt this information to local economic and labor conditions.

F. IRON AND STEEL

India's modern steel industry is about 75 yr old. The first fully successful iron and steel mill was established in 1907 by Jamshedji Tata. There were as many as 17 earlier unsuccessful attempts to establish Indian ironworks using Western-type technology. Some of the reasons for failure were as follows:

(1) In the Western technology adopted, charcoal was used by most plants, even though coke was widely available. The importance of coke was not perceived by either the British or the Indians. Had coke been used, production costs would have been much lower and Indian iron would have been more competitive.

(2) Most plants produced pig iron for export at a time when world prices were falling.

(3) Most projects were started with too little capital.

(4) Some plants were established in areas too inaccessible for profitable marketing.

(5) Central and provincial governments neglected to support or encourage these ventures.

(6) Those who attempted these ventures neither possessed the necessary managerial and technical skills nor did they employ persons who did.

Such mistakes were avoided by the Tata company. Coke, not charcoal, was used. The mill did not produce pig iron for export, but steel for internal consumption. The mill was sited at a location accessible to the important raw materials. The most important factor that contributed to the success of the Tata ventures was the employment of highly trained personnel; technical skills were imported, and the advice of foreign experts was sought. Finally, full cooperation of the government was obtained. The favorable environment created by the Indian government, including tariff protection and guaranteed purchase of much of TISCO's steel for government-owned railways, contributed greatly to the company's rapid growth.

TISCO grew rapidly. In 1917, it undertook expansions which resulted in a fourfold increase in output. By the advent of World War II, its production capacity had doubled again. By 1939, TISCO had become one of the largest steel mills in the British Empire and also one of the lowest-cost producers in the world. In 1939, TISCO produced three-fourths of the steel consumed in India.

About 1918, two more iron mills were established. The Indian Iron and Steel Company (IISCO), founded by British interests, erected a mill at Burnpur in West Bengal. Mysore Iron and Steel Works (MISW) was founded by the maharaja of Mysore, and its mill was erected at Bhadravati. At first, IISCO restricted production to pig iron, primarily for export to the United Kingdom and Japan. In 1936, it undertook an expansion and acquired the Bengal Iron Company which had previously failed. The management of IISCO formed the Steel Corporation of Bengal (SCOB) to undertake the construction of a steel mill adjacent to IISCO's blast furnace. SCOB went into production in 1939, and in January 1953 IISCO and SCOB were formally merged.

The MISW plant was originally designed to use charcoal produced by a wood distillation plant owned by the maharaja. About 1920, synthetic products were developed which rendered the maharaja's wood distillation plant obsolete and caused it to close. However, pig iron continued to be produced by MISW with charcoal from other sources. By the mid-1930s, MISW had added facilities for the production of steel, which have never been economical because of the use of charcoal. In the earlier years, the losses were absorbed by the Mysore state government, and recently subsidies have been given to this plant by the government. Even though it is uneconomical, it still affords a source of steel for which scarce foreign exchange does not have to be expended.

By World War II, steel imports were replaced with the output from SCOB. India's production of steel since 1951 is shown in Table XII.

India was self-sufficient in steel until about 1954. The inability of the industry to keep pace with the demand for steel after 1954 is attributed to failure to expand during the immediate postwar period.

In 1945, the government established the Iron and Steel Panel to look into India's steel industry and suggest future programs. The panel recommended, in addition to expansion of the existing mills, that one or two new mills with a combined capacity of 1 million tons of steel be erected. In 1947, the new government commissioned three non-Indian firms to investigate the feasibility of this proposal, and all three firms concluded that two mills, with an initial capacity of $\frac{1}{2}$ million tons each, should be established.

The industrial policy resolution of 1948 dictated that the new steel plants should be the responsibility of the government. Under the industrial act, private steel companies were allowed to continue operating and were given licenses to increase their production capacity.

During the first 5-yr plan, inaugurated in 1950, one new steel mill was to be erected by the government, IMSW was to triple its production, and the private-sector mills were to increase their combined capacities by nearly 50%. The most significant increase during this period resulted from expansion programs at IISCO. TISCO modernized its plant, which had deteriorated during World War II, so as to increase production. The MISW

TABLE XII

Production, Imports, and Exports of Steel Since 1951

Year	Production (1000 tons)	Imports (1000 tons)	Exports (1000 tons)
1951–1952	1091	140	4
1952–1953	1118	160	3
1953–1954	1040	188	4
1954–1955	1264	251	5
1955–1956	1280	834	2
1956–1957	1359	1256	1
1957–1958	1438	1286	1
1958–1959	1439	806	—
1959–1960	1795	793	—
1960–1961	2337	1238	2
1961–1962	2939	1002	3
1962–1963	3864	870	8
1963–1964	4347	888	32
1964–1965	4508	929	76
1965–1966	4604	734	140
1966–1967	4551	405	251
1967–1968	4078	452	525
1968–1969	4801	385	682
1969–1970	5078	367	4758
1971–1976	13,300	—	—

expansion programs were abandoned as being uneconomical. To replace this capacity, three new mills were started in the public sector.

In 1953, a steel mill was established at Rourkela with German collaboration. In 1955, a steel mill at Durgapur was started with British collaboration and, in 1956, one at Bhillai with Russian collaboration. Most of the foreign exchange requirements of these three mills were to be met by loans from the participating governments.

The second-plan targets for steel were ambitious, namely, to triple the actual production between the beginning and the end of the plan period. Both the public and private sectors were to participate, the role of the public sector being more predominant. TISCO and IISCO were also allowed to expand under the second plan. Although the output of steel in 1960–1961 was far short of the target, the country's industrial expansion programs were mostly successful.

Under the third 5-yr plan which started in 1961, the three public-sector steel mills were to be expanded and a fourth one to be started at Bokaro, while the private-sector mills were not to be expanded. This entire program was delayed until the first year of the fourth 5-yr plan. Originally it was thought that the United States government would participate in the

Bokaro project, but the U.S. Congress objected. In early 1964, the Soviet Union agreed to help the Bokaro project as well as the expansion of the one at Bhillai. Construction of the Bokaro plant started in 1966, and the production of pig iron and ingot steel began in 1973. This plant is still under construction and has expanded from a capacity of 1.7 million to 4 million tons/yr.

At the present time, there are six integrated steel plants and two specialty steel plants with a total installed capacity of 10.6 million tons of steel and pig iron and 137,000 tons of specialty steel. In 1976 the integrated production was 8.4 million tons or 79% of capacity. Alloy steel produced was 103,000 tons at 75% of capacity. India now stands thirteenth in steel production in the world.

Construction is expected to start in 1979 on a 6-million-ton plant at Salem in TAMIL Nadu (southern sector). This plant will be entirely of Indian design. Two other plants are under consideration, one at Visakhapatna in Andhra Pradesh and the other at Vijay Yagar in Karuataka. It is expected that the investment decisions will be made during the sixth plan and that construction will be started at the same time.

The domestic demand for finished steel and pig iron by 1973–1974 was assessed at 7.12 million and 2.0 million tons, respectively. The fourth plan took into account the need for increasing output to meet this demand and to ensure additional capacity for meeting future requirements during the fifth plan.

G. NONFERROUS METALS

1. *Aluminum*

Among metals, aluminum is second only to iron in industrial importance. It is so popular now that aluminum utensils for household purposes can be found in the average Indian home, having displaced iron and brass. Today aluminum is widely used in the aircraft, electrical, and building industries. In the electrical industries, it has almost completely replaced copper, not only as a bare conductor but also in insulated cables. The greatest single use of aluminum is in high-voltage electrical house wiring—about 50%. The production and percent utilization for aluminum are shown in Table XIII.

The consumption of aluminum in India rose from 12,000 tons in 1951 to 220,000 tons in 1977. The estimated demand in 1984 is to 400,000 metric tons. The demand is likely to increase dramatically, since the per capita use in India is 0.4 kg versus 22 kg in the United States and 2.9 kg elsewhere in the world.

The first production of aluminum in India was at the Indian Aluminum

TABLE XIII

Production of Aluminum

Year	Capacity (1000 metric tons)	Production (1000 metric tons)	Utilization (%)
1951	4	3	96
1956	7	6	87
1961	22	18	83
1966	73	65	89
1969	117	132	112
1970	147	161	109
1971	167	178	107
1972	177	179	101
1973	196	154	179
1974	196	128	66
1977[a]	275	180	65

[a] Estimated.

Company at Alupuram, Kerala, in 1943. The production that year was 1300 metric tons. At present, there are five producing plants in the country. Of these, four are in the private sector and one in the public sector. The largest plant of the Hindustan Aluminum Corporation, Ltd., at Renekoot in Bihar has an installed capacity of 95,000 metric tons. The newest plant in the public sector at Korda in Madhya Pradesh will have an installed capacity of 100,000 tons. The first unit was commissioned in 1975. A sixth plant with a capacity of 50,000 tons is in the planning stage and will be located at Ratnagiri in Maharashtra.

The aluminum plants have operated at less than 65% capacity since 1974, the primary problem being a lack of adequate power. Although the government recognizes the critical need for aluminum for use in power transmission, this industry was the first to undergo power cuts in favor of farming and other industries. There are also many labor problems which further reduce productivity.

Another serious problem in the aluminum industry is the pricing policy. By law, 50% of the production goes to the government at a fixed price, so-called levy metal. The current price is $903/metric ton versus a production cost of $1084/metric ton.

It was estimated that the existing bauxite resources would be exhausted by 1990. This led to the need for a further geological survey of India, and new deposits were located in various parts of the country; for example, in Ranchi (Bihar), Kutui Jabalpur, Bhopal, Mysore, Bombay, Salem, Jammu and Kashmir, Kerala, and Goa.

In addition to bauxite, other raw materials used in the manufacture of the metal are AlF_3, fluorspar, cryolite, caustic soda, coke, and coal. At present, most of the cryolite and AlF_3 have to be imported, since there is no adequate supply within India.

The present installed capacity for the production of cryolite is 4830 metric tons, with a production of 4000 metric tons. The process is based primarily on the recovery of fluorine compounds from phosphate fertilizer. About 1000 metric tons/yr are being recovered from pot cases and old pot linings. There are two plants producing AlF_3 at the rate of 2500 metric tons/yr versus a demand of 8000 metric tons/yr.

2. *Copper*

Copper has been known in India from prehistoric times. The technology of its production was also known and is evident from archeological discoveries. In modern times, with increased demands for the metal, there is an urgent need for increased production. The first major attempt to locate and work copper mines in modern times was made by the Indian Copper Corporation which was established in 1924 and started production in 1928.

Major supplies of copper are converted into alloys such as bronze and brass. The production of copper pipes and tubes, arsenical copper rods, and so on, was started only during World War II and, at the same time, metal recovery from commercial scrap was initiated. It was only in the third 5-yr plan that plans for adequate production of copper were formulated. The demand in 1971 was 85,000 metric tons, and the estimated demand in 1979 is 116,000 metric tons. Estimated production and scrap recovery are 45,000 and 16,000 metric tons. This leaves 57,000 metric tons for import.

Copper represents the largest tonnage of any metal imported. The electrical cable industry came into existence quite early and created a large demand for copper metal. It is vital for the production of much electrical generating machinery, even though in power transmissions it has been supplanted by aluminum. Until recently, the bulk of the imported copper supply came from Rhodesia. However, with economic sanctions against Rhodesia by the United Nations and by India, this importation was stopped. Therefore the principal recent supplier of copper has been the United States. The large amount of foreign exchange now required for the purchase of copper has increased the urgency for developing indigenous resources for the production of copper metal.

The chief deposits of copper ores are found in Singhbhumi (Bihar), and other sources have been located in Utter Pradesh, Rajasthan, West Bengal, Kashmir, Madkya Pradesh, Mysore, and Andhra Pradesh. The better known deposits are found near Jaipur-Alwar in Rajasthan, Kumyun and the Kangra Valley in the Himalayan region, Sindu-Bara-Buunda in Sikkim, and Mysore. In Rajasthan rich deposits have been located in the Khetri Dariboo area. The total reserves are estimated at 26 million tons with an average content of 1.8%.

A geological survey of India recently discovered three major deposits at Nallakonda, the lead–copper belt in Andhra Pradesh. The total reserves of all three are about 60.7 million tons. The interesting part of the investigation was that the samples indicated the presence of significant amounts of silver, cobalt, arsenic, and nickel, recovery of which was considered economically possible.

3. *Lead*

Resources for the production of lead are very small; India contributes only 0.2% of the estimated world output of lead. India's principal resources are the Zawar mines and Banjare mines in Rajasthan. Other sources are in Utter Pradesh, Andhra Pradesh, Bihar, Madhya Pradesh, Gujarat, and Jammu-Kashmir, but these are not commercially significant. Recently in South Arcot, Madras, and Mamandur near Madras, lead ores have been located. The geological survey of India has estimated that this deposit contains 300,000 tons of mixed metal ores. This ore deposit is under the control of the government of Madras, and a private company, Cominco Binani, Ltd., has been licensed to exploit it commercially. A 150,000-ton/yr plant has been proposed by Hindustan Copper, Ltd., to utilize the Nallakonda (Andhra Pradesh) deposit mentioned above. A further project, based on the refining of imported concentrated ores, has been sanctioned by the government at Vishakpataman.

The demand in 1974 was 47,000 metric tons of which only 2500 metric tons were produced. Thus, a large part of the lead must be imported. A single lead smelter exists at the present time, located at Junda in Bihar. The installed capacity is 3600 metric tons and is being increased to 6000 metric tons. Two additional plants are in the design stage—one at Vizag and one at Dariba.

Lead is very important for development of the various sectors of India's economy. For example, it is necessary for advanced development of the electrical industry, the paints and pigments industry, and other areas. The chief need of the industry is for the location and development of additional ore bodies.

4. *Zinc*

The main use of zinc is in the production of alloys, particularly brass. It is also used for galvanizing steel sheets and wires and for producing zinc oxide for the paint and pigment industries. The demand for galvanized sheets was estimated at about 700,000 tons at the end of the fourth plan.

The two main zinc smelters are located at Debari, Rajasthan, and at Alwaye, Kerala. The former is in the public sector, and the latter is in the private sector. The installed capacity is 38,000 metric tons/yr with a production of 25,000 metric tons. This leaves an import requirement of 75,000 metric tons. By 1979, the demand should be 150,000 metric tons. To help meet this demand, the Debari plant is being expanded to 45,000 metric tons from 18,000. Thus, the total installed capacity will be 65,000 metric tons. A new zinc smelter is under construction at Visha Khapatray, with a capacity of 30,000 metric tons. Completion should be in the early 1980s.

5. *Atomic Minerals*

Thorium is among the minerals that are useful in the production of atomic energy. Thorium deposits have been found in Kerala and in the Madras area. According to the survey made by the Atomic Energy Department, 30,000 tons of uranium are available in economic quantities in Saurastra and Kuch, and steps are being taken to develop these deposits.

Another atomic mineral available is beryl ore from which beryllium is produced. Reserves of this mineral are located mainly in Rajasthan, Bihar, Andhra Pradesh, and Mysore and are large enough to meet India's requirements for beryllium.

Cadmium is used in atomic reactors and in the electrical and steel industries. With the establishment of zinc smelters from which cadmium metal can be obtained, imports (80,000 tons/yr in 1970) have been reduced and are expected to be discontinued.

6. *Research and Development Needs of the Metal Industries*

The NCST (S6), which drafted the science and technology plan for India for the period 1974–1979, has observed that most of the plants set up are underutilized (S5). On the other hand, the demand for metals by various sectors of the economy has been growing rapidly, necessitating extensive imports. The slow growth rate in production has been attributed to poorer techniques of beneficiation and extraction of minerals, lack of indigenous maintenance capability, lack of materials that meet specifications, breakdowns, and lack of personnel who can solve production–breakdown problems. To alleviate these problems, the NCST in its science and technology plan (STP) (S4) has identified 243 research, development, and design projects with an allocation of about $112 million, to be completed within a period of 5 yr. These plans have been projected while keeping in view the gaps existing in exploration, mining, and production. The plans include proposals developed separately by the Department of Steel and the Department of Mines. The Department of Steel's proposal is to set up a $3.8 million electrolytic manganese plant based on the dioxide process developed by the National Metallurgical Laboratory at Jamshedpur. The proposals of the Department of Mines are development of gold mining, seismic and microseismic projects, and a materials testing laboratory. The NCST plan includes programs for the production of formed coke, magnesium special metals, and superalloys, as well as improved facilities for ore dressing. It has been suggested that the underlying research in the metallurgical sector be encouraged at universities and institutes of technology.

The metals considered by the NCST are aluminum, copper, lead, zinc, and magnesium. With a view toward the demand for these metals and their alloys, the plans proposed by the NCST panel include (1) facilities for hydrometallurgy and electrometallurgy, (2) production of magnesium, and (3) production of high-strength aluminum and magnesium alloys for the defense and space industries.

In 1975 and 1976 the state of emergency that existed in India slowed down or postponed many of the plans outlined above. The planned research and expenditures have been extended to the end of the sixth 5-yr plan.

In view of the unsatisfactory position of nonferrous metals in India, research is proposed on development of roasting techniques, treatment of residues for increased metal recovery, utilization of waste for conservation of metals, imported substitutes, treatment of lower-grade ores to supplement the indigenous supply, and preventing surface tarnishing of metal products.

High-purity metals and superalloys are required for the aeronautics, electronics, instruments, space, and defense industries; the raw materials are at present imported. Primarily, these special metals include nickel- and cobalt-based superalloys, high-strength iron-based alloys, titanium-based alloys, controlled-expansion alloys, and magnetic materials. Keeping in view the importance of these metals and alloys and the expertise available in India for making them, the NCST has identified two projects for their development: the setting up of a special metal and superalloys plant and the development of controlled-expansion alloys.

H. PHARMACEUTICAL INDUSTRY

The growth of the pharmaceutical industry in India, shown in Tables XIV and XV, has been much more rapid than elsewhere around the world

TABLE XIV

PHARMACEUTICAL SALES

Year	Millions of dollars
1948	12
1954	65
1966	84
1968	211
1975	651
1976	771

(I3, R1). Table XIV is based on sales, since the available data are in terms of dollars rather than weight or other units. Table XV compares 1977 and 1978 production figures for selected drugs. Utilization of capacity is good and ranges from 60 to 75%. The opportunity for further expansion is also very great. The percentage of pharmaceuticals in the chemical industry is above the world average (world average, 11.6%; Indian average, 31.1%).

The remarkable growth in India during the past 25 yr can be attributed to the strong base of the chemical industry and to favorable governmental policies. In the 1950s and early 1960s, foreign companies took advantage

TABLE XV

POSITION OF SOME IMPORTANT DRUGS

Compound	Production in 1972	Target for end of fifth plan (1978–1979)
Antibiotics		
Penicillin	230 MMU	780 MMU
Streptomycin sulfate	199 tons	825 tons
Tetracycline	71.4 tons	200 tons
Chloramphenicol	41.05 tons	390 tons
Sulfa drugs	1285 tons	2195 tons
Antituberculosis drugs	532.54 tons	1380.32 tons
Antimalarials		
Quinine	36.23	—
Chloroquin	23.82	150
Hormones		
Sex hormones, including corticosteroid	1747 kg	15.510
Antihistamines		
Diphenhydramine	2210 kg	16,000 kg

of the huge market and expanded rapidly. In addition, a large number of bulk drugs worth about $44 million were imported. With all this expansion, medicines have reached only about 20% of the population of the country, despite the fact that in the last 24 yr their production has increased 30-fold. The annual per capita consumption of drugs in India is on the order of 8 rupees ($1.00), compared to 235 rupees ($29.30) in Germany, 252 rupees ($31.50) in Japan, and 310 rupees ($38.70) in the United States.

At present, there are more than 2900 plants making drugs and pharmaceuticals. Of these, 116 are large-scale units which produce more than 80% of the total supply. Of these units, 82 produce basic pharmaceuticals and formulations, while others depend on the 82 for most of their raw materials. The industry needs to increase production, because it currently provides only $1\frac{1}{2}$ times the value of investment. In the advanced countries, this ratio is between 1 : 15 and 1 : 50.

The targets for the pharmaceutical industry by the end of the sixth plan period are ambitious. During this period, the industry aims at increasing production from the present $375 million to $750 million. The present output of $63 million of bulk drugs will be expanded three times, and exports from $12.5 million to $44 million. According to the Development Council of Drugs and Pharmaceuticals, the fourth-, fifth-, and sixth-plan targets for production are $456 million, $750 million, and $1500 million, respectively, at current prices. To achieve the targets of the fifth plan, an investment of $125 million between 1973 and 1976 was needed. An additional investment of $312 million between 1976 and 1980 will be needed to achieve the targets for the sixth plan.

To help control the price of drugs, the government of India instituted the Drug (price control) Order of 1970. The Bureau of Industrial Cost and Prices is continuing its work on the matter and has submitted further recommendations to the government. The position of the pharmaceutical industry is difficult, since it has to consider both fixed prices and growing costs. This will require strict control on spending and also steps to see that productivity increases.

The progress of small-scale units has been quite impressive. Of about 1900 units in this sector, 673 are located in Maharashtra, 232 in West Bengal, 192 in Tamil Nadu, 168 in Andhra Pradesh, 129 in Gujarat, and 108 in Madhya Pradesh. Some of these units prepare basic drugs and are suppliers of raw material to other countries. The export performance of the industry is quite good. From $4.5 million in 1967–1968, it reached $12.2 million in 1970–1971. It is estimated that by 1988–1989 the industry will be able to balance its foreign trade in drugs and formulations.

There is a need for intensive research and development in the drug and pharmaceutical industry, and more attention is now being given to these areas. Present annual research and development expenditures are about $10 million (2% of total turnover)—inadequate in view of the industry's annual turnover. In 1978–1979, the planned annual amount available for research and development was $37.5 million and at the end of the sixth plan should rise to $75 million per annum, representing about 5% of the present total turnover. A large number of units have started research and development activities and are trying to expand them. At present, research in the field of pharmaceutical and medicinal chemistry is carried out at a number of schools, as well as at a number of research institutes in both the public and private sectors.

There are five well-defined areas of research and development for the industry:

(1) Establishment of good quality control techniques and good manufacturing practices and of facilities for formulation and packaging development activity to undertake compatibility and stability tests as well as safety testing and bioavailability studies.

(2) Improvements in recovery procedures, cost reduction, process control, methods of assay, and so on.

(3) Substitution of native raw materials for imports in drug formulation and drug manufacture (recommended for top priority).

(4) Basic research involving the discovery of new products or new uses of existing projects.

(5) Applied research involving the translation of laboratory discoveries into profitable commercial processes.

The NCST in the science and technology plan for 1974–1979 (S6) has identified the following major problems for the industry.

(1) Imbalance with respect to the production of basic drugs and pharmaceuticals required for formulations. The current consumption of drugs is valued at $94 million, and the indigenous production is worth about $62.5 million.

(2) Inadequate supply of raw materials and intermediates for attaining the desired production levels.

(3) Inadequate field trials for ensuring extensive testing of the drug preparations proposed for formulation and marketing.

The NCST has also recognized that there is heavy domination of the drug industry by foreign companies and foreign-owned manufacturing units and that the next 5 yr will require substantial inputs in research and development. One of the major projects envisaged is a new Fermentation Technology and Enzyme Research Centre to be located at Hindustan

Antibiotics, Ltd., Poona. These areas of study have a great potential for industrial growth; antibiotics, of course, play an important role in the health program. Other proposed projects relate to newer routes for the synthesis of important drugs and pharmaceuticals, as well as the cultivation of aromatic plants for extracting perfumery aromatics and gums used in pharmaceutical preparations.

The pharmaceutical industry has been instrumental in generating extensive employment opportunities. In 1975, the industry gave direct employment to 2.04 million persons. Of these, 10% were technical and scientific, 4% executive, and 8% marketing personnel. Employment rose to about 3.06 million by 1976, an increase of 50% over the previous year. Employment figures continue to advance steadily with the increase in turnover.

In addition, the industry provides indirect employment to large numbers of people in the distribution trade and associated industries. In the distribution trade alone, more than 1.06 million persons are employed, and another 1.02 million work in industries producing containers, cartons, packaging cases, and so on. On an average, the industry provides indirect and indirect employment to about 5 million people.

The industry is taking big strides so far as the export of drugs and formulations is concerned. Fifteen years ago, the export figures were almost negligible, amounting to about $1.2 million per annum. The industry has recently ventured into the export market; in 1975 it exported products worth $30 million. Indian industry has created markets mainly in the developing countries and is vigorously striving to capture bigger and bigger shares of markets in the developing nations of Asia and Africa.

VI. Awards for Progress

Each year since 1964 the Indian Chemical Manufacturers' Association (ICMA) has presented awards to chemical industries in recognition of their progress and service to the chemical industry. Three awards are given, two of which are indicative of the progress being made in the chemical industry. The first of these is the Sri P. C. Ray Award which recognizes the chemical industries that have made the greatest progress during a particular year. This award is named after P. C. Ray who was an early leader in the chemical industry. The second is the ICMA award which recognizes and encourages industrial units showing outstanding examples of forward development of technology in the chemical industry in India. Tables XVI and XVII list some of these awards. The year, firm, and the reason for the award are detailed in these tables. As can be seen, they encompass a wide range of firms and technologies. These tables

TABLE XVI

Recipients of the P. C. Ray Award for Best Industrial Unit Developed in the Chemical Industry in India

Year	Name of firm	Basis of award
1964	Exeol Industries, Ltd., Bombay	Production of chemicals—H_3PO_4 B. P., oxalic acid, ethylene dichloride, and organo-mercurials for the first time in India
1965	Amar Dye Chemicals, Ltd., Bombay	Development and manufacture for the first time in India of hot and cold types of fiber-reactive dyestuffs
1967	M/s Sudarshan Chemical Industries, Ltd., Poona	Manufacture of inorganic and organic pigments; the company produced 26% of the country's production of organic pigments and 47% of its inorganic pigments and has started producing several *intermediates*
1968	M/s Alembic Chemical Works, Ltd., Baroda	Contribution to developments in the design and engineering of penicillin technology
1970	M/s Gharde Chemicals, Pvt., Ltd., Bombay	Outstanding achievement in developing its own process for making ingrain dyes
	M/s Hico Products, Pvt., Ltd., Bombay	Development of a varied chemical complex which supplies chemicals for drugs, pharmaceuticals, insecticides, paints, varnishes, and food; tamed ethylene oxide gas by developing a technique for its handling and harnessed the highly hazardous gas by manufacturing wetting agents, emulsifiers
1971	Dharamsi Morarji Chemicals, Ltd., Bombay	Research, developments, and import substitution; the 300-tons/day single-stream H_2SO_4 plant which started in 1964 was entirely designed, constructed, and directed by its own technical personnel; it utilizes process heat to generate high-pressure steam which is used to drive turbines; this steam is exhausted at low pressure, and its heating value is made use of in other processes
1973	M/s Gharda Chemicals, Pvt., Ltd., Bombay	Developed, with its own research and development efforts, a new process for the manufacture of hydrazobenzene and benzidine dihydrochloride; the unit has developed a novel catalyst for the reduction of nitro compounds such as nitrobenzene and nitrotoluene with methanol and caustic soda, the reduction reaction being highly exothermic; it successfully engineered the process
1973	M/s Cxel Industries, Ltd., Bombay	Setting up of a unique plant for the manufacture of white phosphorous at Bhavanagar involving the designing and building of a huge electric air furnace with monolithic carbon

(Continued)

TABLE XVI (*Continued*)

Year	Name of firm	Basis of award
		electrodes weighing more than 100 tons, with its own technical and engineering knowledge and with the least reliance on imported equipment and components; designing and erection of the Rs. 2.5 core plant was accomplished by engineers of the company; quite a few improvements in technique were introduced in construction of the furnace operating at working temp. of 1600°C
1974	M/s Deepak Nitrite, Ltd., Navsari, District Baroda	Successful completion of the first factory in India for commercial production of $NaNO_2$ and $NaNO_3$; although it was the first venture of the company based on its own efforts, it developed an efficient process with innovations in the basic technique obtained from the Fertilizer Corporation of India and set up its unit almost entirely with indigenous plant and equipment; it has succeeded also in ensuring the production of nitrite and nitrate in the favorable ratio of 1:1; this production has helped in shipping imports of $NaNO_2$
1975	M/s Harihar Polyfibres, Harihar, Karnstak	The production of rayon-grade pulp from hybrid eucalyptus; this development was probably the first successful effort with eucalyptus; the achievement is most impressive because initial research on the process as well as on the development of a new engineering design was done by the company in India and most of the equipment was fabricated within the country
1975	M/s Camphor and Allied, Ltd., Bombay	Development of an indigenous technology for the manufacture of terpene-based chemicals; the company has also developed the catalyst required for the production of capolyte CP Resin, and this catalyst is economical and is a perfect substitute for the imported varieties
1976	Hindustan Lever, Ltd., Bombay	Manufacture and distribution of soaps, synthetic detergents, toilet preparations, edible fats and other foods, animal nutrition products, chemicals, and so on
1977	Raymon Glues and Chemicals	Development of unique processes for converting low-value crushed bones and hides into higher-value products
1977	D. D. Shah and Company, Bombay	Adoption of a novel process for the manufacture of phenylacetamide and phenylacetic acid based on acetophenone; this plant is perhaps the only one of its kind in the world

TABLE XVII

INDIAN CHEMICAL MANUFACTURERS ASSOCIATION (BOMBAY) AWARD FOR THE MOST
OUTSTANDING EXAMPLE OF FORWARD DEVELOPMENT OF TECHNOLOGY IN
INDUSTRIAL CHEMISTRY IN INDIA

Year	Name of firm	Basis of award
1965	M/s Gwalior Rayon and Silk Manufacturing (Wig.) Company, Ltd.	Significant achievement in establishing commercial production of rayon-grade pulp for the first time, and from a nontraditional material, bamboo
1966	Tata Chemicals, Ltd.	
1967	M/s Synthetics and Chemicals, Ltd., Bareilly	First synthetic rubber complex company to establish local production facilities to meet its own basic requirements and to supply styrene and butadiene monomers needed to produce several grades of synthetic rubber
1975	Bhabba Atomic Research Centre, Isotope Group, Bombay	Development of radiation technology using a cobalt-60 source to sterilize prepackaged medical products
	M/s Fertilizer Corporation of India, Ltd.	Achievement in the implementation of various fertilizer projects using varied feedstocks and employing different sophisticated technologies
1977	D. C. M. Chemical Works, New Delhi	Production of dimensionally stable anodes for diaphragm cells for the electrolytic manufacture of caustic soda and chlorine

furnish a good indication of the progress being made and of the areas considered by the ICMA to be important to the chemical industry in India. Years missing in the tables are those in which no award was made.

VII. Professional Societies and Education

Part of the overall development and progress in the chemical industry is dependent on the education system and the activity of the professional societies associated with it.

The chemical engineering profession is represented by the Indian Institute of Chemical Engineers (IIChE). This institute was formed May 18, 1947, at Jadaupur University, near Calcutta (T2). In 1948, there were 101 members; in 1958, there were 384; and there was a total membership of over 2800 in 1978. This enrollment represents about 10% of the 18,000 chemical engineers in India. At present, the country is divided into 18 regional centers (sections). The institute moved into its own facilities on the Jadaupur University campus in 1973 and was recognized by the American Institute of Chemical Engineers, as well as other professional societies, in 1958. Publication of the society's journal, *Indian Chemical Engineers* was started in 1959.

The IIChE is a dynamic, growing organization. In 1959, an associate membership became available. To obtain this membership, a rigorous examination is given by the institute to chemical engineers who have not had the benefit of a formal education at a recognized university. This associate membership is recognized as equivalent to a degree by the government of India. The institute is a member of the Federation of Engineering Institutions representing chemical engineers. It promotes excellence among students by sponsoring an essay contest and aiding in the formation of student chapters. It is also active in setting standards for education, establishing continuing education programs, providing programs to promote the development of appropriate technology, and advising the government.

Another group that plays an important part in development of the chemical industry is the ICMA. The awards given by this group were discussed in the previous section.

The education problem in India is vast and complicated. In 1977 there were 625 million people in India, with an annual increase of 12–13 million/yr. The age distribution is heavily skewed toward the lower ages which constitute about 42% of the population (I1). Those of college age represent about 17% or 106 million persons. There are 118 universities with a total enrollment of about 3.2 million (R2). There are 32 chemical engineering departments with an annual capacity of 1400 students (in contrast to 8 departments with a 200-student capacity in 1948). Thus, only about 7000 students are chemical engineers—a very small fraction of the students in the colleges and universities.

At the postgraduate level, 25 institutions offer a M.E. degree, and 20 of these offer a Ph.D. Because of the competition with jobs and going to a larger university, postgraduate programs are limited, and the available students are not always the best. In addition, the wage structure does not favor a person with an M.E. degree.

In the past, the education pattern was 11 yr of grade school followed by a 5-yr program leading to a B.S. degree. In 1977, a uniform pattern of 12 yr in high school was initiated. The course of study in chemical engineering compares well with the subject matter offered in the institutions of the United States. In general, however, the curricula are rigid and must be passed year-wise rather than course-wise. That is, failure of a single subject requires repetition of the course for a given year. The number of lectures or contact hours is high, and laboratory time is limited. There is very little opportunity to interact with jobs in industry, summer jobs, or otherwise, before graduation.

A Chemical Engineering Development Center was set up in Madras in 1971 by the Ministry of Education. The center has developed and published a suggested first-degree course in chemical engineering.

The chemical engineer is supposed to spend 6–8 weeks in "practical training." This consists of an extended plant visit during which the student is often considered a hindrance by plant management (B4). Plans are under way to modify this procedure so as to involve faculty members in working with students at plant locations. A more expanded version of this program, called practice school, is also under study and development. An explanation of this plan in relationship to one institution is presented later. Suggestions have also been made to plan more elaborate plant trips and to establish regional pilot plants.

Another area under development is continuing education or collaborative education, also described later. Great effort is being expended to make the education fit the needs of the developing Indian chemical industry.

VIII. Needs of the Chemical Industry

The individual needs and research directions for specific industries have been discussed above. However, there are a number of needs common to all industries that, if supplied, will aid materially in their future development: education, more rational research and development schemes, initiation of more mission-oriented research, and certain organizational changes.

A foremost need is to change the educational pattern of the country; not only undergraduate education but also education and research projects leading to M.S. and Ph.D. degrees. Some of these have been indicated above. The problems of education in India are not new; in 1933, C. L. Dhawan wrote (M1):

> Indian universities are year-in and year-out turning out thousands upon thousands of graduates and undergraduates whose market value, these days, has gone down to practically nothing. These unfortunate young men are swelling the ranks of the unemployed, thus bringing about misery not only to themselves and their near ones, but also to the whole country. It is, indeed, a painful fact that almost every Indian university these days shelter more such youths for whom finding work in this world has become a problem. The loss to the nation on account of such widespread compulsory idleness can be better imagined than described.

In general, the teachers at universities and colleges and the workers in government research laboratories have not been exposed to industrial practices. In most cases, their information is based entirely upon textbook learning and on some small amount of laboratory work carried out under the usual university laboratory conditions. In addition, much of the research taking place in both the national laboratories and the universities is of a highly theoretical nature not strongly related to actual needs. This is borne out by the relative poor rate of acceptance by industry of the projects completed by the national laboratories. Part of industry's problem also lies in the high cost of doing its own research and development, relative to the potential financial return. The large business houses are mostly managerial or holding companies which have too little perception of the possible economic improvements of their processes. There are, however, some outstanding examples of modest-scale industrial laboratories which have been and are being established in India for research on specific problems (Fig. 3).

An outstanding need, therefore, for the development of self-reliance in the chemical industry is a better interchange between the scientific laboratory and the professional personnel of the industries. This lack of dialogue between industry and research personnel is magnified by the lack of awareness on both sides of the need for it.

Considerable study has gone into this problem both at the national and local levels. For instance, the Education Commission report of 1966 (R2) strongly emphasized the need for this type of interchange and also for an increase in vocational training. A later report (V1) has recommended the establishment of continuing education programs to be carried out at manufacturing plants. A successful and innovative program is being carried out at the Birla Institute of Technology and Science (BITS) located at Pilani on the eastern edge of the Rajasthani Desert. One of the few private engineering and science schools in India, established in 1964, the institute has continually recognized the need for coordination between industry and education. The Massachusetts Institute of Technology chemical engineering practice school, numerous cooperative programs elsewhere in the United States, and the sandwich programs in England were all studied to determine how they might be adapted to Indian conditions. As a result, the practice school concept was adopted and implemented in 1972.

The practice school programs now encompass all areas of engineering as well as other branches of study. BITS's chief sponsor is one of the largest industrialists in India, who has many chemical and industrial plants under his control. Therefore, the first practice school programs were established in Birla industries. Now, they are also in operation in other private industries and in the public sector. In all, there are 20 stations which service about 200 students every 6 months. These include banks, newspapers, design companies, manufacturing companies, and national laboratories. Most of these locations offer accommodations or pay up to 500 rupees/month ($65.60). Over 95% of BITS students participate in the practice school program.

New stations are located and organized by negotiation between the faculty and the personnel at the industrial location. Prior to the establishment of a station, selected faculty members spend 6 months at the site in order to identify and list the possible problems. An interdisciplinary team of teachers and students is then sent to the site for about 6 months to work on these problems to find their solutions.

In addition to the practical problems the students solve or attempt to solve, many extensions of the problems are brought back to the classroom and to the institute research laboratories for subsequent discussion. A large number of faculty members have become acquainted with industrial practices, and this has had a marked effect on the quality of teaching and on the students' enthusiasm for their courses. Industrial staffs have reported a positive contribution to their work by the students and faculty. Employers visiting campuses find that students who have completed the practice school program are much better prepared than students who have not. Other educational institutions are now beginning similar programs, and it is hoped that the overall procedure will have a measurably beneficial effect on the industrial capability of India.

The structural concept of the practice school program, shown in Fig. 7, indicates that each educational branch has from one to three components. At BITS, all students completing the third year enter the first phase of the practice school program, a 2-month interdisciplinary training period at a laboratory or factory, which is designed to familiarize them with industrial practices. Therefore, in subsequent treatment, all practice schools are considered equivalent.

After the fourth year, each student spends 6 months undertaking the full program (95% of all science and engineering students) at an industrial location and is expected to work on several problems. At the location, students are placed in groups and, in turn, each person assumes a managerial position within the group. The groups hold internal seminars, as well as both formal and informal discussions with plant personnel. Engineering students, after the fifth year of study, participate in a third practice school project which involves designing industrial equipment at a design firm. Other chemical engineering departments are also developing a similar concept but are not as far advanced.

A collaborative education program is also under development at BITS under which a M.E. degree will be given at selected practice school locations. This will involve both teachers from the BITS faculty and qualified personnel at the site. Degree work will be possible for plant personnel as well as students.

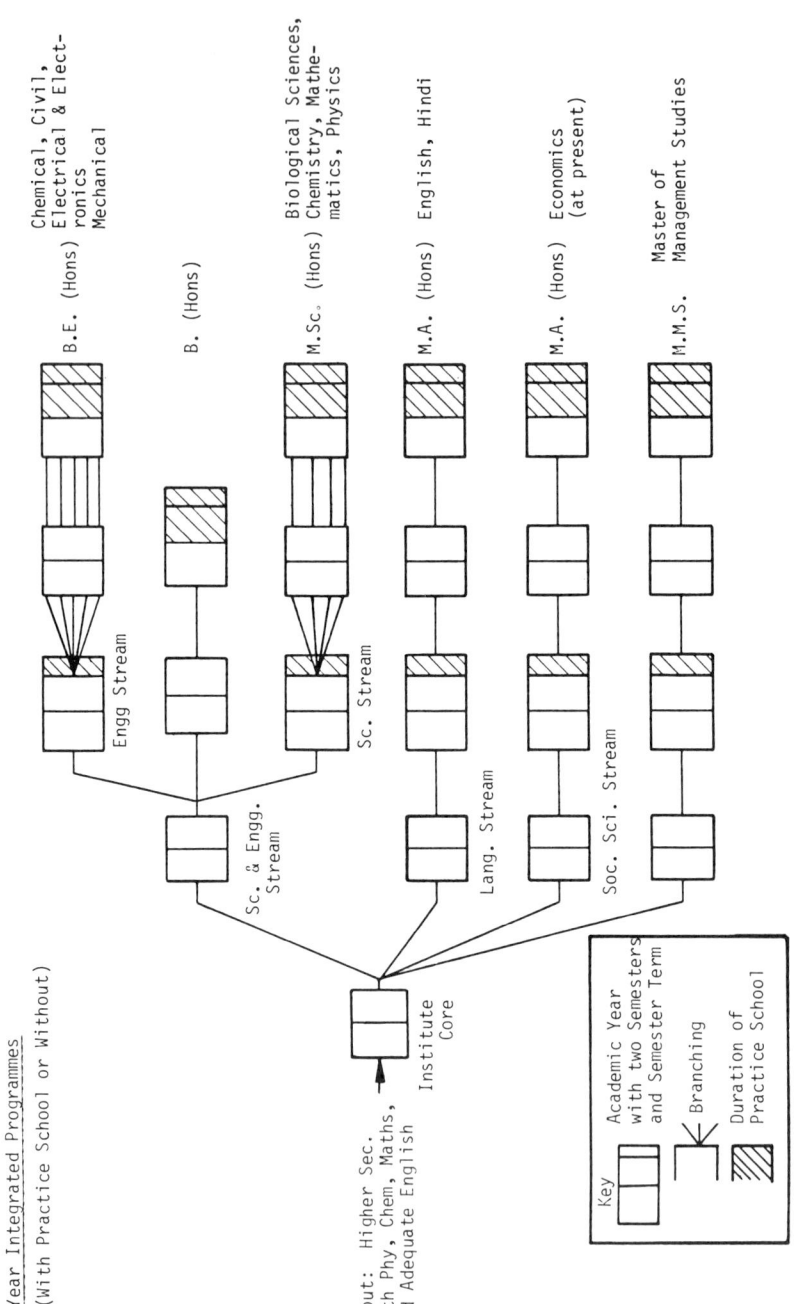

FIG. 7. Academic program at BITS.

Another way in which the relationship between industry and educational institutions can be improved is to have qualified personnel from industry spend a period of time at the various educational institutions. Some institutions have started a system of hiring adjunct professors for this purpose. Outstanding individuals in the fields of business management, systems analysis, history of science, materials science, and so on, are currently employed in this type of activity. The number of such persons is being increased yearly. In this system, the expert spends a minimum of 1 week each month on the campus teaching courses and helping the faculty to develop courses.

On a broader note, considerable effort and change in direction are needed in India's overall research and development activities. Since almost 94% of the research is funded by the government, there are only a few examples of industrial research in specific areas. In the past, a great deal of the research done in government laboratories was not connected with the immediate needs of the industry. In addition, the licensing policies make the processes developed by the national laboratories somewhat uneconomical and, thus, unused by Indian manufacturers. In the relationship between government laboratories and industry it would be beneficial to have some sort of interchange of personnel. One the one hand, the laboratories could give technical assistance to the industries by sending research workers to the plants for a year or so. This could be particularly effective in the public sector where the matter of proprietary information should not arise. On the other hand, the production person could benefit a research laboratory by going to the location and working on a specific problem related to his or her own industry and expertise. Recently considerable effort has been made in regard to industrial research in the report on the science and technology plan (S3) for the chemical industry and in the NCST forward plan (S4).

One area of needed organizational change involves the dissemination throughout industry of information developed both in industry and industrial laboratories. At the present time, publication of such material is fragmentary and slow. Consolidated publication, or abstracting of some type spread over a wide audience, should be considered.

Organizational changes in the chemical industry should be considered with the possibility of increasing competition, removing the sheltered production the industry has, and developing increased self-reliance. These organizational changes should include a broader look at the licensing policy and the actual effects it has upon the economy. Changes should then be made that will speed licensing so as to increase the production of critical items. Both licensing and research should be guided by the primary needs of the Indian economy: food, clothing, shelter, and health.

Often in the past, this was not done either in allocating industrial licenses or in setting research policy. Lack of follow-up to check how national needs are actually met by research and development efforts has been one of the detrimental factors in industrial development.

A major item needing serious consideration in India is the scale of operation. In general, the larger a chemical plant, the more economical its capital and its operation costs and the cheaper the resulting output. In general, a decrease in production costs is gained through a decrease in labor costs, as well as an increase in volume, and a decrease in costs can only come about when the plant is operating at almost full capacity. In India, because of problems involving materials supply, transport of finished goods, labor, a lack of developed skills, and so on, chemical plants often operate at less than 60% of capacity, and the necessary conditions are not met in regard to the size of the plant. Serious consideration should be given to the availability of supplies and the ability to operate at full capacity. It is certainly not economical to operate one or two large plants at 40% capacity when perhaps several smaller plants by judicious location and scheduling could be operated at 70 or 80% of capacity. With the prevailing shortage of transportation (railroads, trucks, and so on), power, and fuel and with the current labor situation, it is doubtful that large-scale plants will be able to operate at full capacity for the next 10–15 yr (a substantial fraction of the life of a normal chemical plant). Consideration should, therefore, be given to building smaller plants closer to the source of supply or to markets. This would utilize the effect of the scale factor while taking into account local conditions and transportation problems.

Some means also needs to be found to encourage the proper maintenance, updating, and improvement of existing plant facilities. This perhaps could be done in the form of tax relief.

The problem of substituting the natural resources of India for imports is of paramount importance in development of the chemical industry. The outstanding examples of this are the importation of oil to feed the fertilizer and petrochemical industries and the importation of sulfur for the manufacture of H_2SO_4. Extensive deposits of iron pyrites containing enough sulfur to supply India's needs for 400 yr are located in Bihar. Although considerable effort has been expanded in developing this resource, India still imports 60% of its sulfur as elemental sulfur from the sulfur market of the world. Proper development of the pyrite as a source of sulfur and H_2SO_4 would reduce this dependence and would allow the further expansion of industries requiring H_2SO_4 as a feedstock.

Because of the current prices of petroleum in world markets, with no sign of a reduction in the near future, the development of hydrocarbon resources in India in the form of coal should be of primary importance and research funds should be expended in this direction. India's coal resources are enormous, and the use of coal as a chemical feedstock would considerably reduce dependence on foreign oil. At present, about 70% of India's oil imports are used in the production of fertilizer. By suitable design, it would be possible to use the volatile material from the coal, forming a chemical feedstock, and still have coke left over to furnish energy for industry, for cooking in homes, and so on. In addition, Indian coal contains about 30% ash which results in a loss in heating value and an increase in shipping cost to the place of use. Research in the beneficiation of coal, using some of the processes being developed in other countries, might reduce the critical transportation problem as well as increase the value of coal as a fuel.

Another critical area that could occupy the chemical industry is the production of methane by biological conversion processes. Large amounts of organic materials are burned as fuel, including cattle manure, bagasse from the sugarcane industry, and many other materials. Proper utilization of these waste materials would help to ease the fuel shortage and would contribute significantly to the fertilizer capacity of India.

The problem of meeting the food, clothing, shelter, and health needs of the increasing population of India is critical. For example, an increase of 2 or 3 million tons of food grains is required to cover the population increase per year. The food crop in 1978 is sufficient to supply food for approximately 650 million people if there is no loss. Since about 25% of the grain is lost to insects and vermin, the present production of food is marginal. The development of insecticides and preservatives is a critical need for the chemical industry. Governmental policies and expenditure of funds must reflect this need. The question to be asked regarding all expenditures is whether they will lead to increased food or clothing production. Development of consumer goods industries and the like will not solve these problems.

The chemical industry is experiencing, and will experience, rapid growth in the Indian economy. From a consideration of the per capita consumption, the rapid growth in population and the developing aspirations of the Indian people, the growth and development of the Indian chemical industry offer exciting possibilities and will continue to be a challenge during the 1980s.

References

B1. Basu, D., Mager, D., and Chatterjee, R., "Ruin of Indian Trade and Industries, Calcutta." 1939.
B2. Bernel, J. D., "Science in History." Penguin, New York, 1969.
B3. Bose, D. M., Sen, S. N., and Subbarayappa, D. B., "A Concise History of Science in India." Indian National Science Academy, New Delhi, 1971.
B4. Barker, D. H., *J. Indian Inst. Chem. Eng.* May (1969).

B5. Bazaz, M. C., "India Pharmaceutical Guide." Pamposh Publ., New Delhi, 1978.
C1. Chaudhuri, M. R., "Indian Industries Development and Location," (4th rev. Ed.). Indian Economic Geographic Studies, 1970.
C2. Chaudhuri, R., "The Evaluation of Indian Industries." Univ. of Calcutta, 1939.
D1. "Draft, Five-Year Plan 1973–83." Planning Commission (Draft), Gov. of India, 1978.
F1. Sharma, L., "The Times of India, Directory and Year Book." Times of India Press, Bombay, 1978.
F2. Sahar, C., and Ankulkarui, "Fertilizer Statistics." Fertilizer Assoc. of India, New Delhi, 1972.
F3. "Fertilizer Statistics, 1973–74." Fertilizer Assoc. of India, New Delhi, 1979.
F4. "Fourth Five-Year Plan 1969 to 1974." Planning Commission, Gov. of India, Planning Commission.
I1. "India 1977–78." Publ. Div., Gov. of India, 1978.
I2. "India '76—A Reference Annual." Publ. Div., Gov. of India, 1976.
I3. "Indian Pharmaceutical Guide, 1978." Pamposh Publ., New Delhi, 1979.
J1. Jathar, G. B., and Jather, T. G., "Indian Economics." Oxford Univ. Press, London and New York, 1957.
J2. Johnson, W. A., "The Steel Industry of India." Harvard Univ. Press, Cambridge, Massachusetts, 1967.
K1. "Kothari's Economic and Industrial Guide of India" (31st Ed.). Kathari, Madras, 1976.
M1. Malik, K. B., and Dhawan, C. L., "Main Industries for Indians." The Youngmen's Own Institute, 1933.
M2. Mukherjee, R., and Dey, H. L., "Economic Problems of Modern India," Vol. II. Macmillan, New York, 1941.
M3. Mukherjee, S. K., *Indian Chem. Eng.* **20**(1), (Jan.–March 1978).
R1. "Report of the Committee on Drugs and Pharmaceutical Industry." Ministry of Petroleum and Chemical, Gov. of India, April, 1975.
R2. "Report of the Education Commission." Dept. of Education, Gov. of India, 1966.
R3. Rajan, T. P. S., *Chem. Process. Eng.* **1** (6), 78 (Dec. 1967).
R4. Rahman, Bhargava, R. N., Qureshi, M. A., and Pruthi, S, "Science and Technology in India." Indraprasthan Press News, Delhi, India, 1973.
S1. "Statistical Abstract India 1975." Central Statistical Organization, Dep. of Statistics, Ministry of Planning, Gov. of India, 1976.
S2. "Statistics for Iron and Steel Industry in India." Hindustan Steel, Ranchi, India, 1970.
S3. "Sanctioned Capacities in Engineering Industries." Natl. Council of Appl. Econ. Res., New Delhi, 1971.
S4. "Science and Technology Plan," Vol. 1. Natl. Committee on Sci. and Technol., Gov. of India, 1974.
S5. "Science Technology Plan," Vol. II. Natl. Committee on Sci. and Technol., Gov. of India, 1974.
S6. "Science and Technology Plan." Natl. Committee on Sci. and Technol., Gov. of India, 1974.
T1. "The Imperial Gazetter of India," The Indian Empire, Vol. 3, Economic. Oxford Clarendon Press, 1908.
T2. "Twenty-five Years of Chemical Engineering in India," Editorial, *Indian Chem. Eng.* **XV**, (1), (Jan.–March 1973).
T3. Tilak, B. D., "Report on Science and Technology Plan for the Chemical Industry, A General Overview," Vol. I. Gov. of India, 1973.

T4. Tilak, D. B., "Report on Science and Technology for the Chemical Industry," Status Report on Chemical Industry, Vol. II. Gov. of India, 1973.
T5. "Twenty Years of Indian Chemical Industry, 1949–1969," *Chem. Age India* **20** (11), (Nov. 1969).
V1. Venkateswarlu, D., *Indian Chem. Eng.* **XV,** (1), (Jan.–March, 1973).
W1. Watt, Sir G., "The Commercial Products of India." Today and Tomorrow's Printers and Publishers, 1908; reprinted Ed., 1966.

THE ANALYSIS OF INTERPHASE REACTIONS AND MASS TRANSFER IN LIQUID–LIQUID DISPERSIONS

Lawrence L. Tavlarides and Michael Stamatoudis*

Department of Chemical Engineering
Illinois Institute of Technology
Chicago, Illinois

I.	Introduction	200
II.	Flow Field in Agitated Dispersions	200
	A. Turbulent Dispersions	201
	B. Laminar/Transitional Flow Dispersions	205
III.	Behavior of Liquid–Liquid Dispersions	207
	A. Maximum and Minimum Drop Size in Dispersions	207
	B. Phenomenological Models for Drop Breakage Rates	209
	C. Phenomenological Models for Drop Coalescence Rates	215
IV.	Measurements and Analysis of the Properties of the Dispersion	221
	A. Interfacial Surface Area Measurements	221
	B. Drop Size Distribution Measurements	223
	C. Coalescence Frequency Measurements	228
V.	Mathematical Models for Mass Transfer with Reaction in Liquid–Liquid Dispersions	233
	A. Effective Interfacial Surface Area Models	234
	B. Drop Size and Residence Time Distribution Models	236
	C. Dispersed Phase Interaction Models	237
	D. Type I Interaction Models: Population Balance Techniques	238
	E. Type II Interaction Models: Monte Carlo Simulation Models	253
	F. Type III Interaction Models: Use of Macromixing and Micromixing Concepts	259
VI.	Conclusions	262
	Nomenclature	263
	References	266

* Present address: Department of Chemical Engineering, University of Thessaloniki, Thessaloniki, Greece.

I. Introduction

Many operations in chemical engineering require the contact of two liquid phases between which mass and heat transfer with reaction occurs. Examples are hydrometallurgical solvent extraction, nitrations and halogenations of hydrocarbons, hydrodesulfurization of crude stocks, emulsion polymerizations, hydrocarbon fermentations for single-cell proteins, glycerolysis of fats, and phase-transfer catalytic reactions. A most common method of bringing about the contact of the two phases is to disperse droplets of one within the other by mechanical agitation.

The rate of interphase mass transfer is affected by the physical and chemical characteristics of the system and the mechanical features of the equipment. The former include viscosities and densities of the phases, interfacial surface properties, diffusion coefficients, and chemical reaction coefficients. The latter include, for example, the type and diameter of the impeller, vessel geometry, the flow rate of each phase, and the rotational speed of the impeller.

The nature of the design and analysis problem depends on the phase in which the reaction occurs, whether multiple reactions are involved, the relative magnitudes of the rates of mass transfer, of reaction, and of the macromixing and micromixing processes of the dispersed and continuous phases. When slow reactions in either phase control the rate of transfer, micromixing of the dispersed or continuous phase is not important. The total interfacial surface area is important and can be related to the power expended. For reactions of intermediate velocity and of a rate near that of the mass transfer process or for fast reactions in the dispersed phase, micromixing and macromixing of the dispersed phase can significantly affect the extent of conversion and selectivity. Accordingly, dispersion phenomena such as coalescence and breakage of droplets, and drop size distribution are discussed below. Models for predicting conversion and selectivity are also discussed, but consideration is limited to those most suitable when the macromixing and micromixing of the dispersed phase is important.

II. Flow Field in Agitated Dispersions

A rational theoretical treatment of the dynamics of drop breakup and drop coalescence in a turbulent agitated dispersion requires a fundamental knowledge of the behavior of the flow field. Accordingly, in this section, an introduction to turbulence phenomena and isotropic turbulent behavior is presented with recent pertinent findings included. Studies on the actual hydrodynamics in an agitated vessel are discussed, with some attention given to the modification of the turbulent properties due to the presence of a second dispersed phase.

A. TURBULENT DISPERSIONS

1. *Concepts of Turbulence*

A thorough treatment of turbulence is given in the work of Batchelor, Brodkey, and Hinze (B5, B15, H9). Most of the work on modeling dispersed phase dynamics such as coalescence and breakage is based on models developed from concepts of isotropic turbulence. It seems appropriate to review these concepts here.

Turbulence is considered to be a somewhat random flow of eddies superimposed on the overall average flow. If $\mathbf{U} = \mathbf{i}U_x + \mathbf{j}U_y + \mathbf{k}U_z$ is the instantaneous point velocity in space, expressed in terms of its scalar coordinates U_x, U_y, U_z along the unit vectors $\mathbf{i}$, $\mathbf{j}$, $\mathbf{k}$, it may be written as

$$\mathbf{U} = \bar{\mathbf{U}} + \mathbf{u}$$

where $\bar{\mathbf{U}} = \mathbf{i}\bar{U}_x + \mathbf{j}\bar{U}_y + \mathbf{k}\bar{U}_z$ is the mean velocity at the point and $\mathbf{u} = \mathbf{i}u_x + \mathbf{j}u_y + \mathbf{k}u_z$ is the instantaneous velocity relative to the mean. The three components of the velocity fluctuation vector $\mathbf{u}$ are not necessarily equal.

The equation of motion for the turbulent flow of an incompressible fluid is obtained from the Navier–Stokes equations by replacing the instantaneous values of each point quantity by the sum of the average and its fluctuating component, and time averaging. This results in the Reynolds equations for incompressible turbulent motion in which there are more unknowns than available equations. Therefore additional relations are needed to solve the equations.

As turbulence fluctuations are random in nature, the problem is approached from statistical theory, into which are introduced various simplifying assumptions that will permit solution for some variables of interest. The intuitive model of turbulence, the "energy cascade of turbulence" model, assumes that fluid eddies range in size from the smallest to the largest, which is determined by the bounds of the vessel. Ideally the boundaries only influence the large eddies and transfer energy to or from them. The larger eddies transfer their energy to smaller eddies, until energy is transferred to the smallest. These eddies lose their energy by viscous dissipation. The smaller eddies become statistically independent from the larger ones. The turbulence generated throughout the vessel is assumed isotropic. This requires that the fluctuations in velocity be independent of the direction of reference, or that all root-mean-square values be equal:

$$u'_x = u'_y = u'_z \tag{1}$$

In isotropic turbulence there are no shearing stresses and no gradients of the mean velocity.

Taylor (T7) suggested that a statistical correlation could be applied to fluctuating velocity terms in turbulence. Regardless of what is denoted as the size of an eddy, a high degree of correlation exists between the velocities at two points in space if the separation between the points is small when compared with the eddy diameter. Only one scalar function is necessary to specify the velocity correlation in isotropic turbulence. Either Eulerian or Lagrangian correlation functions can be used. These correlations can be used to determine the velocity terms of the Reynolds stresses. The mathematics is considerably simplified if Fourier transformations are made of the equations. The transformed correlations have the form of an energy spectrum, and can provide insight into the distribution of the energy of turbulence over the frequency of velocity fluctuations. In this respect the energy spectrum function $E(k)$ is equal to the Fourier transform of the velocity correlation tensor from Eulerian space to wave number space. The wave number k is often considered to be the reciprocal of an eddy size.

For isotropic turbulence the following equation of the energy spectrum can be obtained by taking the Fourier transform of the Karman–Howarth equation (H9):

$$\partial E(k)/\partial t = T(k) - 2\nu k^2 E(k) \tag{2}$$

where $T(k)$ is associated with the transfer of energy between wave numbers or eddy sizes. Integration of $T(k)$ over all wave numbers gives zero. By integration of Eq. (2) over all wave numbers, the time rate of change of the kinetic energy of turbulent motion is obtained:

$$-\tfrac{1}{2}(\overline{\partial u_i^2}/\partial t) = -\tfrac{3}{2}(\partial u'^2/\partial t) = 2\nu \int_0^\infty k^2 E(k)\,dk = \epsilon \tag{3}$$

where ϵ is the rate of dissipation of turbulent energy. Equation (3) indicates that the change in kinetic energy is equal to the dissipation of energy into heat. The energy dissipation increases as the wave number increases or as the eddy size decreases. Thus, according to "the energy cascade of turbulence model," there is a continuous flux of energy through the wave number range and a continuous dissipation. A wave number k_d is associated with the size of the eddies that provide the main contribution to the total dissipation. A wave number k_e is associated with the size of energy-containing eddies.

For high levels of turbulence the range of energy containing eddies and the range of maximum dissipation eddies are sufficiently far apart. For this case Kolmogoroff (K10) postulated that a range of high wave numbers exists where the turbulence is statistically in equilibrium and uniquely determined by the energy dissipated per unit mass ϵ and the kinematic viscosity ν. This range is called the universal equilibrium range. This universal equilibrium range is subdivided into two subranges: the inertial subrange where the energy spectrum is independent of ν and solely dependent on ϵ, and the viscous dissipation range where the energy spectrum is dependent on both ϵ and ν.

The length and velocity parameters characteristic of this range were derived from dimensional reasoning by Kolmogoroff as

$$\eta = (\nu^3/\epsilon)^{1/4} \qquad (4)$$

$$u = (\nu\epsilon)^{1/4} \qquad (5)$$

The wave number k_d where viscous forces become very strong is of the same order as $1/\eta$. The wave number k_e of the energy-containing eddies is of the same order as $1/l_e$ if l_e is the linear scale of the energy-containing eddies.

It should be noted that the normal definition of Reynolds number is inadequate since it refers to some characteristic velocity and length of the mean flow. Alternatively, the microscale Reynolds number $(N_{Re})_\lambda$ and the macroscale Reynolds number $(N_{Re})_{l_e}$ are used:

$$(N_{Re})_\lambda = u'\lambda_g/\nu, \qquad (N_{Re})_{l_e} = u'l_e/\nu \qquad (6)$$

where u' is the turbulence intensity and λ_g is the dissipation scale (a measure of the size of eddies responsible for dissipation). For purposes of design the tank Reynolds number $(N_{Re})_T = N^*D^2/\nu$ is used. It has been shown (H9) that for the existence of the equilibrium range

$$(N_{Re})_\lambda^{3/2} \gg 1, \qquad (N_{Re})_{l_e}^{3/4} \gg 1 \qquad (7)$$

Kolmogoroff (K10) developed the following expression for the energy spectrum function $E(k)$ applicable to the inertial subrange:

$$E(k) = c_1 \epsilon^{2/3} k^{-5/3} \qquad (8)$$

Similarly, Heisenberg (H6, H7) obtained for the viscous subrange

$$E(k) = c_2 \epsilon^2 \nu^{-4} k^{-7} \qquad (9)$$

where c_1 and c_2 are constants.

The mean square of the relative velocity in an isotropic turbulent flow between two points separated by a distance r is given (B7, K10) by

$$\overline{u^2(r)} = c_1' \epsilon^{2/3} r^{2/3} \qquad (r \gg \eta) \qquad (10)$$

$$\overline{u^2(r)} = c_2' \epsilon r^2 / \nu \qquad (r \ll \eta) \qquad (11)$$

where c_1' and c_2' are universal constants.

These relations will be used later to derive expressions for the breakage and coalescence rates.

2. *Hydrodynamics in a Stirred Vessel*

The flow in an agitated vessel at present cannot be treated analytically due to the presence of turbulent flow and the complexity of the governing equations of motion. Reviews of the entire area of vessel hydrodynamics have appeared in published books (N3, U1). Most of the experimental work done with agitated vessels is directed toward measuring and modeling the velocity profiles of the impeller discharge stream (A2, A4, A5, C17, D8, F5, G13, H14, K7, M18, M23, N1, N5, R7, S1, S2). In addition, work has been done on measuring turbulent characteristics of agitated vessels (C17, F5, G13, G14, I2, I3, K7, K11, M18, M23, N1, N8, O1, O2, R7, S8, S9, S13, V9). At low vessel Reynolds number $(N_{Re})_T$ the flow is laminar. As $(N_{Re})_T$ increases, turbulence spreads from the impeller to all parts of the vessel. The energy spectrum measurements by various investigators (K7, K11, M23, R7) on the impeller stream show a short range of about one decade of $k^{-5/3}$ dependence for $(N_{Re})_T > 10,000$. Employing air as the fluid, Gunkel (G13) found a $k^{-5/3}$ dependence of about one decade in the impeller stream and k^{-7} dependence at higher frequencies. Very few experiments (G14, R7) have been done outside the impeller, but seem to indicate an energy spectrum behavior similar to the one in the impeller stream. It should be noted that an energy spectrum dependence of $k^{-5/3}$ does not always mean that the flow is locally isotropic (B15, R7). Recent studies (V9, V10) of turbine agitators show a strong periodic pseudoturbulence, due to the impeller, which may give incorrect values for the measured turbulence characteristics.

The energy dissipation throughout the vessel is not uniform (C17, G13, N3). Cutter (C17) found that about 70% of the power was dissipated in the impeller stream and only 30% in the rest of the vessel. On the other hand, Gunkel and Weber (G13) found that most of the energy dissipation takes place outside the impeller and impeller stream. Experiments (B8, R19) show that for high tank Reynolds number, $(N_{Re})_T > 10,000$, the average energy input of the impeller per unit mass of liquid, ϵ, is independent of the properties of the liquid and a function only of the vessel and impeller geometries. In this case

$$\bar{\epsilon} = K'N^{*3}D_I^2 \tag{12}$$

where K' is a constant dependent only on the vessel and impeller geometries, N^* is the rotational speed of the impeller, and D_I is the impeller diameter. The total power input into an open agitated vessel with flat turbine impeller has been correlated by Rushton *et al.* (R19) and Bates *et al.* (B8), and for a closed vessel by Nienow and Miles (N7).

3. *Effect of Dispersed Phase on Turbulent Flow*

In a turbulent flow, the entire spectrum of the continuous phase eddies is imparted to the droplets or particles of the dispersed phase present. Theory and experiments (K15, K16, L8, S14) indicate that small drops or particles follow the behavior of the fluid eddies very closely. Drops or particles larger than the integral scale of turbulence follow the mean fluid flow. Experiments (K16) with solid particles (0.013–0.20 cm in diameter) in an agitated vessel show that the particle velocity fluctuations are given by the Maxwell distribution. In addition, the micromotion of the particles was determined mainly by eddies of size comparable to particle diameters.

The presence of the dispersed phase, for small holdup fraction, should not have a significant effect on the turbulent characteristics of the dispersion. Investigators (C11, C12, R15, R16) found that for high holdup fraction their developed models for drop size distribution fitted experimental data better by taking into account a "damping" effect on turbulence by the dispersed phase. Doulah (D11) developed a theory for the increase of drop size due to this "damping" of turbulence effect. Experiments (L1) on two-phase jet flows show that the "damping" of turbulence can be approximated by

$$u'/u_0' = (1 + 0.2\phi)/(1 + \phi) \tag{13}$$

where u' is the root-mean-square turbulent fluctuating velocity in a dispersion of particles, u_0' is the corresponding value in the fluid with no particles, and ϕ is the dispersed-phase holdup fraction.

B. Laminar/Transitional Flow Dispersions

Agitated dispersions at low impeller speeds or high continuous phase viscosities are in a state of laminar or transition flow. At low impeller Reynolds number, $(N_{Re})_T < 15$, the flow is laminar around the impeller and stagnant away from it. At higher Reynolds number, $(N_{Re})_T > 15$, discharge flow develops. This region is still mostly laminar and extends up to $(N_{Re})_T \approx 200$ for turbine impellers. Beyond this point, the transition region begins with the flow becoming turbulent around the impeller, while further away the flow is still laminar. The transition region for disk-type turbine impeller extends up to $(N_{Re})_T \approx 10,000$.

In an agitated vessel there is a spectrum of shear rates and of local power dissipation rates (N3, O2). The average power supplied to an agitated vessel with a disk-type turbine impeller over all regions of flow has been correlated (B8, B9, R19). Thus, the average power input in an agitated vessel operating in the laminar region is given by

$$P \propto (\mu N^{*2} D_I^3)/g_c \qquad (14)$$

where μ is the viscosity, N^* is the impeller rotational speed, and D_I is the impeller diameter.

In the transition-range flow, Fig. 1 shows that the average power input in the vessel over short ranges is given by

$$P \propto (\rho N^* D_I^2/\mu)^m \rho N^{*3} D_I^5 \qquad (15)$$

where ρ is the density and m is the slope of the plot of log of power number versus log of $(N_{Re})_T$. With increasing $(N_{Re})_T$, m continuously increases from -1 [$(N_{Re})_T \approx 15$] to 0.16 [$(N_{Re})_T \approx 300$] and then decreases, approaching 0 [$(N_{Re})_T \approx 11,000$].

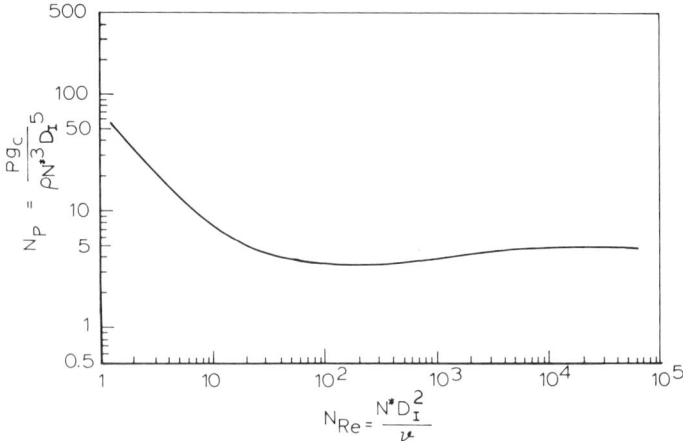

FIG. 1. Power number–Reynolds number correlation in Newtonian fluids, for disk-type 6-bladed impellers [after Metzner et al. (M12)].

III. Behavior of Liquid–Liquid Dispersions

Knowledge of the processes of droplet breakage and coalescence is the basis for developing rational fundamental models for describing the properties of liquid–liquid dispersions, such as drop size distributions, average drop size and interfacial surface area, and the rate of mixing of the dispersed phase. In this section the dynamics of drop breakage and drop coalescence in various flow fields is discussed. Earlier models based on Kolmogoroff's theories are shown to predict expected average minimum and maximum stable drop size in a mechanically equilibrated turbulent dispersion. An outline of the more recent significant phenomenological models which attempt to describe the rates of drop breakage and drop coalescence follows.

A. Maximum and Minimum Drop Size in Turbulent Dispersions

Agitation of two immiscible liquids results in a dispersion of one phase into the other in the form of drops. The drops are subject to shear stresses and to turbulent velocity and pressure variations along their surfaces. These processes cause the drops to deform, and if the deformation exceeds a certain minimum value, the drops break into smaller parts. At the same time, drops are also colliding with each other, and if they remain together for a long enough time so that the separating continuous-phase liquid film drains down to a certain minimum thickness, film rupture occurs and the drops coalesce. Breakages and coalescences take place simultaneously, and the entire dispersion reaches a dynamic equilibrium after a certain time. The drops in the dispersion are not uniform. Investigators usually assume that there is a maximum drop diameter, $a_{\max}$, above which no stable drop can exist and a minimum drop diameter, $a_{\min}$, below which no stable drop can exist.

1. *Maximum Stable Drop in an Isotropic Turbulent Field*

First Kolmogoroff (K9) and then Hinze (H10) obtained a relation for $a_{\max}$ for drops in locally isotropic turbulent fields. Their basic assumption was that in order for a drop to become unstable and break, the kinetic energy of the drop oscillations must be sufficient to provide the gain in surface energy necessary for breakup. The kinetic energy of the drop oscillations is assumed proportional to $\rho_c \overline{u^2(a)} a^3$, where $\overline{u^2(a)}$ is the mean-square of the relative velocity fluctuations between two diametrically opposite points on the surface of the drop. The minimum gain in surface energy can be assumed proportional to σa^2. The Weber number, N_{We}, which is defined as the ratio of the kinetic energy to the surface energy, has a critical value above which the drop becomes unstable. This critical Weber number depends on the system itself. For a local isotropic turbulent flow and a drop diameter $a \gg \eta$, $\overline{u^2(a)}$ is given by Eq. (10). Thus,

$$(N_{We})_{crit} = (c_1 \rho_c \overline{u^2(a)} a^3)/\sigma a^2 = (c_2 \rho_c \epsilon^{2/3} a^{5/3})/\sigma = \text{constant} \quad (16)$$

and

$$a_{max} = c_3 (\sigma/\rho_c)^{3/5} \epsilon^{-2/5} \quad (a \gg \eta) \quad (17)$$

Shinnar (S22) derived the following relation for a_{max} in a turbulent agitated vessel by using Eq. (12):

$$a_{max} = c_4 (\sigma/\rho_c)^{3/5} N^{*-6/5} D_I^{-4/5} \quad (18)$$

In contrast to this, when a drop is smaller than η the dominant forces acting on the drop are the viscous shear forces. Taylor (T4) derived the corresponding equation for breakup due to viscous shear stresses only for which

$$(N_{We})_{crit} = \mu_c (\partial u/\partial r)(a/\sigma) = f(\mu_d/\mu_c)$$

Shinnar (S22) then, using the relation for locally isotropic flow $(\partial u/\partial r)^2 = 2\epsilon/15\nu$, derived the expression

$$a_{max} = c_5 (\sigma \nu_c^{1/2}/\mu_c) \epsilon^{-1/2} f(\mu_d/\mu_c)$$
$$= c_6 \sigma \nu_c^{1/2} \mu_c^{-1} N^{*-3/2} D_I^{-1} f(\mu_d/\mu_c) \quad (a \ll \eta) \quad (19)$$

2. Minimum Stable Drop in an Isotropic Turbulent Field

The local turbulent velocity fluctuations of the dispersion causes the drops to collide with each other. If following a collision, two drops stay together for sufficient time in order for the continuous-phase liquid film separating them to drain out, then a coalescence occurs. Shinnar (S22) assumed that there is an adhesive force, which tends to hold two colliding drops together. Assuming that this force is a function of the droplet diameter, he proposed that there is a minimum drop diameter below which the eddies will not be able to separate the two colliding drops and therefore will not be able to prevent their coalescence. The kinetic energy of the droplets is proportional to $\rho \overline{u^2(a)} a^3$. Shinnar (S22) calculated the adhesion energy as

$$E_a = A(h_0) a \quad (20)$$

where $A(h_0)$ is a constant. The ratio of the kinetic energy to the adhesion energy determines if two drops will coalesce or separate again. The critical ratio is given by

$$(\rho_c \overline{u^2(a)} a^3)/E_a = (\rho_c \overline{u^2(a)} a^2)/A(h_0) = \text{constant} \tag{21}$$

Using this equation and Eqs. (10) and (12), Shinnar (S22) obtained

$$a_{\min} = c_7 \rho_c^{-3/8} \epsilon^{-1/4} A(h_0)^{3/8} = c_8 \rho_c^{-3/8} N^{*-3/4} D_I^{-1/2} \qquad (a \gg \eta) \tag{22}$$

In contrast to this, when $a \ll \eta$ the force for preventing coalescence is the viscous shear force. The critical ratio which determines if two drops will coalesce is given by Sprow (S29) as

$$(\mu_c \nabla u a^2)/F = \text{constant} \tag{23}$$

where ∇u is the local velocity gradient and F is the adhesion force. Substituting $\nabla u = c_8 \epsilon^{1/2} \nu_c^{-1/2}$ (S30) and Eq. (12) into Eq. (23), Sprow (S29) obtained

$$\begin{aligned} a_{\min} &= c_9 F^{1/2} \mu_c^{-1/2} \nu^{1/4} \epsilon^{-1/4} \\ &= c_{10} F^{1/2} \mu_c^{-1/2} \nu^{1/4} N^{*-3/4} D_I^{-1/2} \qquad (a \ll \eta) \end{aligned} \tag{24}$$

B. Phenomenological Models for Drop Breakage Rates

Equations (18) and (19) predict an $a_{\max}$ based on the drop breakage phenomena. Investigations concerning drop breakage have primarily been focused on the maximum drop size $a_{\max}$ above which a drop will break (H10, K9, P6, S22, S26, S30, S36) or on the deformation of a drop in simple shear fields (K2, T4, T6, T9). Although these are helpful in understanding the breakage phenomena, they do not give us any information about the rate or frequency of drop breakage.

The basic expressions describing the breakage rates are

$$\begin{aligned} r(a') \, da &= g(a') N A(a') \, da' \\ r(a, a') \, da \, da' &= g(a') \nu(a') \beta(a', a) N A(a') \, da \, da' \end{aligned} \tag{25}$$

where $r(a') \, da$ is the number of drops of diameter between a' and $a' + da'$ breaking per unit volume per unit time; $r(a', a) \, da \, da'$ is the number of daughter drops of diameter between a and $a + da$ produced per unit volume per unit time by breakage of drops of diameter between a' and $a' + da'$; N is the total number of drops of all sizes present per unit dispersion volume; $A(a') \, da'$ is the fraction of drops with diameters between a' and $a' + da'$; $g(a')$ is the fraction of drops of diameter a' breaking per unit time; $\nu(a')$ is the number of daughter particles produced by breaking a drop of diameter a'; and $\beta(a', a) \, da$ is the fraction of drops of diameter between a and $a + da$ produced by breaking drops of diameter a'. The suggestions of various investigators for the evaluation of the several factors in Eq. (25) will be discussed in this section.

1. Breakage Frequency $g(a')$

Valentas and Amundson (V1) arbitrarily assumed $g(a')$ to be proportional to the droplet surface area:

$$g(a') = K_1 a'^2 \qquad (26)$$

where K_1 is a constant.

Shiloh *et al.* (S21) concluded that this expression fitted their experimental data best.

Curl and co-workers (R15, R16, V11) assumed an analogy between drop breakage and molecular decomposition. They assumed a drop forms an "activated complex" due to imparted kinetic energy, which can be represented schematically:

$$\text{Normal drops} \rightleftarrows \text{"Activated" or unstable drops} \xrightarrow{K''} \text{Breakage products}$$

where the normal drops are in equilibrium with the unstable drops and K'' is the rate constant for the decomposition or breakup process. Thus, the breakage rate of drops of diameter a' is given by

$$g(a')NA(a')\,da' = K''K^*NA(a')\,da' \qquad (27)$$

where K^* is the equilibrium constant for the normal-activated drop exchange. Thus, $g(a') = K''K^*$. From similarity to the molecular decomposition theory

$$K^* = \exp(-\text{ activated energy/kinetic energy}) \qquad (28)$$

The kinetic energy responsible for breakup comes from eddies smaller than the drop diameter, since larger eddies would presumably only carry the drop along with them without breaking it. The kinetic turbulent energy is proportional to $(\pi/6)a'^3\rho E$, where for drops of size within the inertial subrange of an isotropic turbulent flow:

$$E = \int_{1/a'}^{\infty} E(k)\,dk \propto \int_{1/a'}^{\infty} \epsilon^{2/3} k^{-5/3}\,dk \propto \epsilon^{2/3} a'^{2/3} \qquad (29)$$

Thus, the kinetic turbulent energy of drop of size a' is given by

$$E_T \propto \rho_c \epsilon^{2/3} a'^{11/3} \qquad (30)$$

The assumption is made that

$$\text{activated energy} \propto \sigma a'^2 \qquad (31)$$

where σ is the interfacial tension. They assumed also that the rate of decomposition of the activated drop is a function of the rate of shear of the drop. Or,

$K'' \propto$ energy input per unit time/kinetic turbulence energy

$$\propto \rho_c(\pi/6)a'^3\epsilon/\rho_c(\pi/6)\epsilon^{2/3}a'^{11/3} \propto \epsilon^{1/3}a'^{-2/3} \qquad (32)$$

Ross (R15) and Ross and Curl (R16) combined Eqs. (27)–(32) and obtained

$$g(a') = K_2\epsilon^{1/3}a'^{-2/3}\exp[-K_3\sigma/(\rho_c\epsilon^{2/3}a'^{5/3})] \qquad (33)$$

where K_2 and K_3 are constants.

For turbulent agitated vessels using Eq. (12), the previous equation is transformed to

$$g(a') = K_4 N^* D_1^{2/3} a'^{-2/3}\exp[-K_5\sigma/(\rho_c N^{*2} D_1^{4/3} a'^{5/3})] \qquad (34)$$

where K_4 and K_5 are constants.

Coulaloglou and Tavlarides (C12), in contrast to the above development, derived an expression for $g(a')$ based solely on the hydrodynamics of the dispersion. They defined $g(a')$ as

$g(a') = (1/t_b)$(fraction of drops of size a' breaking during time t_b)

$$= (1/t_b)\Delta[NA(a')\,da']/(NA(a')\,da) \qquad (35)$$

where t_b is the breakage time. They assumed that the fraction of drops of diameter a' breaking is proportional to the fraction of drops of size a' which have a total kinetic energy greater than a minimum value necessary to overcome the surface energy holding the drop intact. This minimum surface energy was taken as

$$E_c \propto \sigma a'^2 \qquad (36)$$

Assuming that the distribution of the total kinetic energy of the drops is proportional to the distribution of the kinetic energies of the turbulent eddies and using the two-dimensional normal distribution for the velocity fluctuations of the eddies, they derived a relation for the fraction of drops with kinetic energies exceeding E_c:

$$\int_{E_c}^{\infty} P(E)\,dE = \exp(-E_c/E_T) \qquad (37)$$

Substituting Eqs. (30) and (36), the fraction of drops of diameter a' breaking is

$$\Delta[NA(a')\,da']/(NA(a')\,da') = \exp[-K_6\sigma/(\rho_d\epsilon^{2/3}a'^{5/3})] \qquad (38)$$

Furthermore, the assumption is made that the motion of the centers of mass of daughter droplets to be formed (binary breakage) is similar to the relative motion of two lumps of fluid in a turbulent flow field as described by Batchelor (B6). Thus, for the inertial subrange eddies

$$A'B'^2(t) \propto (AB \cdot \epsilon)^{2/3} t^2 \quad (39)$$

where $A'B'(t)$ is the separation distance of lumps of fluid at time t, AB is the initial separation distance, and K_7 is a constant. Assuming that $AB \propto a$, and that at breakup time t_b the separation distance will be $A''B'' = K_9 a$, they arrived at

$$t_b \propto a'^{2/3} \epsilon^{-1/3} \quad (40)$$

Combining Eqs. (35)–(40),

$$g(a') = K_7 a'^{-2/3} \epsilon^{1/3} \exp[-K_8 \sigma/(\rho_d \epsilon^{2/3} a'^{5/3})] \quad (41)$$

For an agitated vessel, with uniform energy dissipation this becomes

$$g(a') = K_9 a'^{-2/3} N^* D_I^{2/3} \exp[-K_{10} \sigma/(\rho_d D_I^{4/3} N^{*2} a'^{5/3})] \quad (42)$$

At this time, it is instructive to note that Eqs. (34) and (42) predict that the large drops will continuously break down to a size $a_{\max}$, where the breakage function $g(a')$ approaches some arbitrary low value close to zero. At this point, the exponential term dominates the behavior of the breakage rate expression and a relation for $a_{\max}$ is obtained similar to the one given by Eq. (18), which was derived on the basis of maximum stable drop theory.

Delichatsios (D4) and Delichatsios and Probstein (D5) also derived a corresponding expression for $g(a')$. Following Levich (L7), they proposed that a breakup will occur whenever the instantaneous turbulent velocity differences across a drop diameter exceed u_b, the velocity necessary to break the drop. From similarity arguments,

$$u_b/a \simeq (\sigma/\rho a^3)^{1/2} \quad (a \gg \eta) \quad (43)$$

Assuming that the local statistical properties of the velocity field are known, the breakup rate would be equal to the number of crossings per unit time of the velocity u_b (with positive acceleration) by the statistical field of relative velocity differences. Using Rice–Kac's formula (R11) and assuming unskewed probability distribution function, they obtained

$$\begin{aligned} g(a') = W_b &= \int_0^\infty P(u_b, \dot{u}) \dot{u}\, d\dot{u} + \int_0^\infty P(u_b, \dot{u})\, d\dot{u} \\ &= 2P(u_b) \int_0^\infty P(\dot{u}/u_b) \dot{u}\, d\dot{u} \end{aligned} \quad (44)$$

where W_b is the number of crossings of velocity u_b per unit time, $P(u, \dot{u})$ is the joint probability density of the local velocity u and acceleration $\dot{u}$, between two points a distance a' apart, $P(u)$ is the probability distribution function for u, and $P(\dot{u}/u_b)$ is the conditional probability of the local acceleration, when the corresponding velocity difference is u_b. The last integral represents the mean positive acceleration difference between two points a distance a' apart, when the corresponding velocity difference is u_b. From Rotta (R17),

$$(\overline{\dot{u}^2})^{1/2} \propto \overline{u^2}/a' \qquad (45)$$

Experiments (G6, V4) show that the probability density distribution in the inertial subrange can be written as Gaussian with variance $\overline{u^2}$ and cutoff velocity u_c. Thus,

$$g(a') = (K_{11}(\overline{u^2})^{1/2}/a')[\exp(-u_b^2/2\overline{u^2}) - \exp(-u_c^2/2\overline{u^2})] \qquad (46)$$

where K_{11} is a constant; u_c, u^2, and u_b are given (R17, T5, V4), respectively, by

$$u_c = 3(\overline{u^2})^{1/2}, \qquad \overline{u^2} = 1.88(\epsilon a')^{2/3}, \qquad u_b = (\sigma/\rho a')^{1/2}$$

All previous models for $g(a')$ were derived on the basis of existence of an isotropic turbulent field. On the other hand, Stamatoudis (S32) has developed models for breakage frequencies in agitated vessels operating in the laminar or transition range of flow. It was assumed that the breakage of drops in a laminar or transition range of flow is mainly a result of the shear stresses on the drop. The model of Curl and co-workers (R15, R16, V11) given by Eqs. (27), (28), and (31) was used with the exception of the kinetic energy, which is replaced by the shear energy given by

$$E_S \propto \mu_c G^2 a'^3 t_{br} \qquad (47)$$

where t_{br} is the time during which a drop is subjected to the shear rate G. This t_{br} is assumed proportional to the circulation time t_c, the time necessary for the liquid dispersion to circulate around the vessel once. It has been found that for a wide range of operating conditions (H14)

$$t_{br} \propto t_c \propto (1/N^*)D_T^2/D_I^2 \qquad (48)$$

where N^* is the impeller speed, D_I is the impeller diameter, and D_T is the vessel diameter. The shear rate in laminar flows is given by

$$G \propto (P/\mu_c)^{1/2} \qquad (49)$$

where G is the shear rate in the dispersion, P is the power input to the vessel by the impeller, and μ_c is the continuous-phase viscosity. Substituting Eq. (14), the following expression is obtained for the laminar region:

$$G \propto (P/\mu_c)^{1/2} \propto N^* D_I^{3/2} \qquad (50)$$

In the transition range of flow, a mixture of turbulent and laminar flow exists. The assumption is made in this range that the total power into the vessel supplied by the impeller is distributed in the form of shear power with an effective shear rate G_{eff}. By making an analogy with the laminar range and using Eq. (15), the following expression is obtained for the transition flow:

$$G_{\text{eff}} \propto (P/\mu_c)^{1/2} \propto (\rho_c^{m+1} N^{*m+3}/\mu_c^{m+1})^{1/2} D_I^{(2m+5)/2} \qquad (51)$$

where m increases with N_{Re} from -1 [$(N_{\text{Re}})_T \approx 15$] to a maximum 0.16 [$(N_{\text{Re}})_T \approx 1300$] and then decreases to 0 [$(N_{\text{Re}})_T \approx 11{,}000$]. This effective shear rate G_{eff} should not be confused with the average G calculated from velocity profiles, which is proportional to N^* for flows in the laminar (M12, M13), turbulent (O1), and transition range by extrapolation.

Combining these equations, the following expressions for the breakage frequencies of drops of size a' are obtained:

$$g(a') = K_{12} \exp(- K_{13}\sigma/\mu_c a' N^* D_I D_T^2) \qquad (52)$$

for the laminar range and

$$g(a') = K_{14} \exp(- K_{15}\sigma/\bar{\mu}^m \rho_c^{m+1} N^{*m+2} a' D_I^{2m+3} D_T^2) \qquad (53)$$

for the transition range, where K_{15} is constant over short ranges of the transition range. It should be noted that the breakage rate increases with increasing a' and N^* and decreases with increasing σ, as expected.

Using arguments similar to those in the case of turbulent flows, a maximum drop size $a_{\max}$ can be obtained from Eqs. (52) and (53):

$$a_{\max} \propto \mu_c^{-1} \sigma N^{*-1} D_I^{-1} D_T^2 \qquad (54)$$

for the laminar range and

$$a_{\max} \propto \mu_c^m \rho_c^{-(m+1)} \sigma N^{*-(m+2)} D_I^{-(2m+3)} D_T^{-2} \qquad (55)$$

for the transition range.

2. Daughter Droplets Formed

Experiments (B4, F4, K2, R18, S36, T9) show that breakage of a drop results in various numbers and sizes of daughter drops. To simplify computation, most investigators assume a fixed number of daughter drops.

Although Valentas et al. (V1) found that their model is very sensitive to the numerical value of $\nu(a')$, Ross (R15) concluded that nothing can be gained by considering $\nu(a') > 2$. Various other investigators (C10, C11, C12, D5, S21) use the value $\nu(a') = 2$ in their models.

3. Daughter Droplets Probability Density Function

Experiments show that the daughter drop distribution is not uniform. Valentas and co-workers (V1, V2, V3) Coulaloglou and Tavlarides (C12) have assumed a normal distribution of the daughter drop volumes

$$\beta(a', a) = [\sigma'(2\pi)^{1/2}]^{-1} \exp[-(v - \bar{v})^2/2\sigma'^2] \quad (56)$$

where v' is the volume of the drop breaking, $\bar{v}$ is the mean volume of the daughter drops, and σ'^2 is the variance chosen in such a way that the density lies entirely within the volume range 0 to v'.

Ross (R15) and Ross and Curl (R16) assumed the $\beta(a', a)$ term to be given by the beta function, whereas Shiloh *et al.* (S21) and Delichatsios and Probstein (D5) assumed that a drop breakage results in two equal daughter drops.

C. Phenomenological Models for Drop Coalescence Rates

Equations (22) and (24) predict an $a_{\min}$ based on the drop coalescence phenomena. They do not give any information about the rate of coalescence. The coalescence of two drops is accomplished through the draining and rupture of the film of the continuous phase. The factors affecting the coalescence process have been summarized by Jeffreys and Davies (J2). Most researchers have concentrated on studying the film drainage between a drop and a flat interphase (A3, B20, B21, B22, C5, C6, C9, E2, G7, H2, H3, H4, H5, H11, H12, J3, L3, L5, L9, M8, N6, P7, V15, W5). Reviews of these are given elsewhere (B16, G1, J2, L10, T1, V14, W5). Models for film drainage between pairs of drops immersed in an incompressible stagnant liquid have been proposed (L6, M1, M5, M6, M8, M24, P7, R20, R21, S10). Reviews of these also appear elsewhere (J2, V14, V16). In this section, phenomenological models describing the rate of coalescence between two drops in various flows will be discussed.

The basic expressions describing drop coalescence rate are

$$\begin{aligned} F(a, a') \, da \, da' &= \lambda(a, a')z(a, a') \, da \, da' \\ z(a, a') \, da \, da' &= h(a, a')NA(a)NA(a') \, da \, da' \end{aligned} \quad (57)$$

where $F(a, a') \, da \, da'$ is the number of coalescences per unit volume of dispersion per unit time, $\lambda(a, a')$ is the collision efficiency of a collision between droplets of sizes a and a', $z(a, a') \, da \, da'$ is the number of binary collisions between drops of sizes a and a' per unit volume of dispersion per unit time, $h(a, a')$ is the collision frequency between drops of sizes a and a' for a binary collision process based on number concentrations, and $NA(a)\,da$ is the number of drops of size a to $a + da$ per unit volume of dispersion.

A discussion will be given on the most important developments by various investigators for the terms $z(a, a')$ and $\lambda(a, a')$.

1. *Collision Rate*

Coalescence between drops or particles may occur only after their collision with other drops or upon collision. Collision rates have been developed for different systems and conditions.

Under uniform shear flow, Smoluchowski (O6, S27) derived

$$z(a, a')da\,da' = \tfrac{4}{3}(\tfrac{1}{2}a + \tfrac{1}{2}a')^3 GNA(a)NA(a')\,da\,da' \tag{58}$$

where G is the velocity gradient in a direction perpendicular to the direction of particle travel. Camp and Stein (C3) applied the previous equation for turbulent flows by relating the turbulent velocity gradient by $\bar{G} = (\epsilon/\nu)^{1/2}$, obtaining

$$z(a, a')da\,da' = \tfrac{4}{3}(\tfrac{1}{2}a + \tfrac{1}{2}a')^3(\epsilon/\nu)^{1/2}NA(a)NA(a')\,da\,da' \tag{59}$$

A more rigorous approach to collision rates in turbulent flows necessitates knowledge of drop velocities in such fields. As discussed in Section III, drops or particles smaller than the microscale of turbulence are completely contained within the eddies whereas larger drops or particles are being acted upon by the turbulent eddies.

The collisions of drops or particles smaller than the microscale of turbulence are caused by two independent and essentially different mechanisms. These mechanisms differ depending upon whether the droplet has a density the same as or different from that of the surrounding fluid. Drops or particles that have essentially the same density as the surrounding fluid follow the motion of fluid completely. Thus, droplet velocity fluctuations can be described by the fluid velocity fluctuations. Under these conditions Saffman and Turner (S3) obtained the following expression by employing the gradient mechanism:

$$z(a, a')\,da\,da' = (8\pi/15)^{1/2}(\tfrac{1}{2}a + \tfrac{1}{2}a')^3(\epsilon/\nu)^{1/2}NA(a)NA(a')\,da\,da' \tag{60}$$

Levich (L7) treated the collision of drops and particles smaller than η by assuming diffusion of drops or particles towards a "sink" drop or particle, and found a similar form to Eq. (59), but with an arbitrary constant coefficient.

Howarth (H15) assumed a mechanism analogous to Brownian motion. But his derivation for collision rates is wrong because he introduced a variable particle diffusion coefficient in the Smoluchowski equation (O6, S27) which had been derived on the basis of a constant coefficient.

The second collision mechanism comes about only if there is a significant difference between the densities of the fluid and the drops or particles. Because of this significant difference, the drops or particles are not completely entrained by turbulent eddies. Drops or particles with different diameters move with different velocities, which results in collisions between them. Researchers (E1, L7, P3, S3) have accounted for this "acceleration" collision mechanism in their derivation of collision expressions for drops in air. It should be noted that for liquid–liquid dispersions (small density differences) this "acceleration" mechanism is insignificant.

In contrast to the above, drops or particles are not completely entrained by the eddies when they are larger than η. Eddies impact on them at all directions, causing them to move in a random fashion. This led researchers to assume that the collisions are analogous to the collisions of molecules in the gas kinetic theory. Thus, Rietema (R12) assumed that

$$z(a, a)\, da\, da = (\sqrt{2}/2)\pi a^2 \bar{u}(r) N^2 A^2(a)\, da\, da \tag{61}$$

where $\bar{u}(r)$ is the average turbulent velocity fluctuation at the average distance between the drops. Kuboi et al. (K17) also used a similar equation from the kinetic theory of gases:

$$z(a, a)\, da\, da = 2(\pi/3)^{1/2} a^2 (\overline{u^2(a)})^{1/2} N^2 A^2(a)\, da\, da \tag{62}$$

where $\overline{u^2(a)}$ is the mean square velocity fluctuation for eddies of size a, which was found by them (K16) to be $\overline{u^2(a)} = 2.0(\epsilon a)^{2/3}$. Thus

$$z(a, a)\, da\, da = (8\pi/3)^{1/2} a^2 (\epsilon a)^{1/3} N^2 A^2(a)\, da\, da \tag{63}$$

This equation was found to describe collisions of droplets in turbulent pipe flow very well (K17). Coulaloglou and Tavlarides (C13) assumed that collision frequencies between drops of nonequal sizes are given by the following equation by analogy to kinetic theory of gases:

$$\begin{aligned}&z(a, a')\, da\, da' \\ &= (\pi/2)(a^2 + a'^2)[\overline{u^2(a)} + \overline{u^2(a')}]^{1/2} NA(a) NA(a')\, da\, da'\end{aligned} \tag{64}$$

where the mean fluctuating velocities are given by Eq. (10). The kinetic theory of gases also was employed by Delichatsios and Probstein (D6) to obtain

$$z(a, a')\, da\, da' = (\pi/8)(a + a')^2 \{u^2[\tfrac{1}{2}(a + a')]\}^{1/2} NA(a) NA(a')\, da\, da' \tag{65}$$

Recently Abrahamson (A1) also derived collision rate expressions for particles in a turbulent gas phase by making an analogy with the gas molecular collisions.

Gillespie (G8) has given a method for calculating the influence of nonuniform drop size distribution on the collision rates.

It should be noted that gravity can play an important role in collision rates for systems with large differences in density between the dispersed and continuous phases. Researchers (A1, E1, S3) have also developed collision rate models taking into account gravity effects. In turbulent liquid–liquid dispersions with small fluid density differences, gravity is not an important factor in the collision rate model.

Brownian motion also contributes insignificantly to the collision rates due to large drop or particle sizes in actual industrial operations.

2. Collision Efficiency

There are many factors that determine whether a collision results in a coalescence. The processes by which two drops coalesce are those of film thinning and final rupture of the intervening film. These processes are determined by factors such as surfactants, mass transfer, surface tension gradients, physical properties, Van der Waals forces, and double-layer forces. In a turbulent flow field the situation is more involved. The droplets must first collide and remain in contact for a sufficient time for the coalescence to take place. A realistic coalescence efficiency will account for these factors.

The coalescence efficiency $\lambda(a, a')$ is defined as the fraction of collisions between drops of diameter a and a' that result in coalescences.

Howarth (H15) developed an expression for collision efficiencies by assuming an analogy to bimolecular gas reactions. He assumed that a critical relative velocity W^* exists along the lines of centers of two colliding drops which must be exceeded for a collision to result in a coalescence. By assuming that the three-dimensional Maxwell's equation describes the drop turbulent velocity fluctuations, he obtained the coalescence efficiency as the fraction of drops which have kinetic energy exceeding W^*. Thus,

$$\lambda(a, a') = \exp(-3W^{*2}/4\overline{u^2}) \tag{66}$$

where $\overline{u^2}$ is the mean-square turbulent velocity fluctuation.

A more fundamental approach to the coalescence efficiency modeling necessitates taking into account all the important steps which lead to drop coalescence. In order for a collision to result in a coalescence, two drops must remain together long enough so that the intervening liquid film drains to its critical thickness, whereupon film rupture occurs. The coalescence time τ is defined as the time required for coalescence after a collision between two drops occurs. The contact time t is defined as the time that two drops would stay together under the effect of pressure or velocity fluctuations, until either a coalescence or a reseparation occurs. It is obvious that in order for a coalescence to occur, the contact time t must exceed the coalescence time τ. Ross (R15) and Ross and Curl (R16) have derived an expression for coalescence efficiency by assuming that the coalescence time τ and the contact time t are random variables and that the coalescence time is normally distributed. This expression was simplified by assuming that the coalescence time is not distributed although the contact time remains a random variable. Thus they obtained

$$\lambda(a, a') = \exp(-\bar{\tau}/\bar{t}) \tag{67}$$

where $\bar{\tau}$ and $\bar{t}$ are averages. The coalescence time $\bar{\tau}$ was assumed proportional to the time that it takes for the intervening liquid film to go from thickness h_0 to h_c, where coalescence takes place under a force compressing the two drops proportional to the mean-square velocity at either end of the two colliding drops. Thus, using the model for the approach of two deformable drops, they obtained

$$\bar{\tau} \propto (\mu F/\sigma^2) f(h_0, h_c)[aa'/(a + a')]^2 \tag{68}$$

where F is the force compressing the two colliding drops together and assumed to be given by

$$F \propto \rho_d \overline{u^2(a + a')}[aa'/(a + a')]^2 \tag{69}$$

and $f(h_0, h_c)$ is given by

$$f(h_0, h_c) = 1/h_c^2 - 1/h_0^2 \tag{70}$$

The contact time $\bar{t}$ was assumed to be given by

$$\bar{t} \propto (a + a')/[\overline{u^2(a + a')}]^{1/2} \tag{71}$$

Combining Eqs. (67)–(69) and (71), Ross and Curl obtained the coalescence efficiency for two deformable drops

$$\lambda(a, a') = \exp[-K_1 f(h_0, h_c)\mu_c \rho_d \epsilon/\sigma^2 \cdot (a a'/a + a')^4] \tag{72}$$

Coulaloglou and Tavlarides (C13) assumed that the contact time will be proportional to the characteristic period of velocity fluctuation of an eddy of size $a + a'$, given by Levich (L7) as

$$\bar{t} \propto (a + a')^{2/3}/\epsilon^{1/3} \quad (a + a' > \eta) \tag{73}$$

Combining Eqs. (67)–(69) and (73), they obtained the same expression as Eq. (72) for two deformable drops.

Coulaloglou (C11), using the model for the approach of two rigid drops, assumed that

$$\bar{\tau} \propto (\mu_c/F)[aa'/(a + a')]^2 \tag{74}$$

Thus for the case of two rigid drops

$$\lambda(a, a') \propto \exp[-K_2\nu_c/\epsilon^{1/3}(a + a')^{4/3}] \quad (a + a' \gg \eta) \quad (75)$$

and

$$\lambda(a, a') \propto \exp[-K_3\nu_c^{3/2}/\epsilon^{1/2}(a + a')^2] \quad (a + a' \ll \eta) \quad (76)$$

At this point, it is instructive to note that Eqs. (72), (75), and (76) predict that the small drops will continuously coalesce up to a minimum size $a_{\min}$, where the coalescence rate approaches some arbitrary value close to zero. In this case, the exponential term dominates the behavior of the coalescence efficiency function. Thus,

$$a_{\min} \propto D_I^{-1/2} N^{*-3/4} \quad (77)$$

which is similar to Eqs. (22) and (24) derived on the basis of coalescence prevention theory.

Valentas et al. (V1) also used a coalescence efficiency of exponential form, but gave no rigorous arguments to support their model.

The coalescence efficiency in a turbulent pipe flow was correlated by Kuboi et al. (K17) with the modified kinetic energy of collision $U_c = a^3 v_c^2$ as

$$\lambda(a, a) = 0.09 U_c^{0.5} \quad (78)$$

for $0.15 < U_c < 1.0 \text{ cm}^5/\text{sec}^2$, $0.1 < a < 0.3$ cm, and $1 < u_c^* < 30$ cm/sec. Here u_c^* is the relative velocity of approach along the lines of centers just before the instant of collision.

All previous models for drop coalescences assume the existence of a turbulent field. Stamatoudis (S32), on the other hand, derived models for drop coalescences for the case of laminar and transition flow regions. The same models given by Eqs. (67) and (68) were used with the following modifications for F and $\bar{t}$:

$$F \propto \mu_c G[aa'/(a + a')]^2 \quad (79)$$

$$\bar{t} \propto t_c/G \quad (80)$$

Here t_c is given by Eq. (48) and G' is given by either Eq. (50) or (51), depending on whether the range of flow is laminar or transitional. Thus, he obtained the following expression for the coalescence efficiency between rigid drops in dispersions under laminar and transition range flows

$$\lambda(a, a') = \exp(-c_3 N^* D_I^2/D_T^2) \quad (81)$$

and the following expressions for the case of deformable drops

$$\lambda(a, a') = \exp\left[-\frac{c_4 \mu_c^2 N^{*3} D_I^5}{\sigma^2 D_T^2}\left(\frac{aa'}{a+a'}\right)^4\right] \quad (82)$$

in the laminar range and

$$\lambda(a, a') = \exp\left[-\frac{c_5 \mu_c^{1-m} \rho_c^{m+1} N^{*m+4} D_I^{2m+7}}{\sigma^2 D_T^2} \left(\frac{a\,a'}{a+a'}\right)^4\right] \quad (83)$$

in the transition range. Using arguments similar to those in the case of turbulent flows, a minimum drop size $a_{\min}$ can be obtained from Eqs. (82) and (83). Thus, for deformable drops,

$$a_{\min} \propto \mu_c^{-0.5} N^{*-0.75} D_I^{-1.25} \sigma^{0.5} D_T^{0.5} \quad (84)$$

in the laminar range and

$$a_{\min} \propto \mu_c^{-(m-1)/4} \rho_c^{-(m+1)/4} N^{*-(m+4)/4} D_I^{-(2m+7)/4} \sigma^{0.5} D_T^{0.5} \quad (85)$$

in the transition range.

IV. Measurements and Analysis of the Properties of the Dispersion

Knowledge of interfacial areas, drop size distributions, and dispersed phase coalescence rates is essential for accurate description and prediction of mass transfer and chemical reaction rates in liquid–liquid dispersions. In this section, a review of the experimental methods and techniques developed for describing and measuring interfacial area, drop size distributions, and coalescence rates will be given; in addition, summaries of important results and correlations are presented.

A. Interfacial Surface Area Measurements

The measurement of the interfacial area in a liquid–liquid dispersion has been the focus of interest of many researchers, due to its importance in describing and predicting mass transfer and reaction rates involving species transferring from one phase to the other. The direct method to obtain the interfacial area is through knowledge of the dispersion drop size distribution. The measurement of drop size distributions will be discussed in the next section. In this section, two important indirect methods for obtaining interfacial area measurements will be discussed: the chemical method and the light transmittance method. Also an outline will be given of how interfacial areas are obtained from drop size distributions.

1. *Indirect Methods*

a. *Chemical.* The use of chemical means to measure the interfacial area has been used extensively for gas–liquid dispersions (L11, M10, R9, R10, S19, S31, W4). Chemical methods (S19) for determining the interfacial area of liquid–liquid dispersions involve a reactant A of relatively unchanging dispersed-phase concentration diffusing to the continuous phase, where a first- or pseudo-first-order reaction occurs. Models assume the reaction is fast enough to keep the concentration of A in the bulk of continuous phase equal to zero, but not fast enough for any appreciable amount of A to react in the diffusion film at the surface of the continuous phase. Under these conditions Dankwerts (D1) has shown that for a first- or pseudo-first-order reaction

$$Ra^* = a^* C_A^* (D_A k' + K_L^2)^{1/2} \tag{86}$$

where Ra^* is the rate of extraction per unit area, a^* is the effective interfacial area per unit volume of dispersion, C_A^* is the solubility of A in the continuous phase, D_A is the diffusion coefficient of A in the continuous phase, k' is the reaction rate constant, and K_L is the continuous phase mass-transfer coefficient. The effective interfacial area a^* can be obtained from Eq. (86) by substituting the total extraction rate and the physicochemical properties of the system.

The chemical method for the measurement of interfacial area in liquid–liquid dispersions was first suggested by Nanda and Sharma (S19). They calculated the effective interfacial area a^* by sparingly extracting soluble esters of formic acid such as butyl formate, amyl formate, etc., into aqueous solutions of sodium hydroxide. This method has been employed by a number of workers, using esters of formic acid, chloroacetic acid, and oxalic acid, which are sparingly soluble in water (D9, D10, F1, F2, F3, O4, P8, S15, S20). Sankholkar and Sharma (S5) employed the extraction of diisobutylene into aqueous sulphuric acid. Sankholkar and Sharma (S6, S7) have also found that the extraction of isoamylene into aqueous solutions of sulphuric acid, and desorption of the same from the loaded acid solutions into inert hydrocarbons such as n-heptane and toluene, can be used for determining the effective interfacial area. Recently, Laddha and Sharma (L2) employed the extraction of pinenes into aqueous sulphuric acid.

The disadvantage of this method is the effect of the mass transfer on the physicochemical properties of the dispersion. It has been observed that mass transfer can affect the interfacial tension and thus the interfacial area.

b. *Light Transmittance.* Another indirect method for measuring the interfacial area of dispersions is the light transmittance method. This method has the advantage over the chemical method of not having any influence on the system being measured.

The light transmittance method is based on the fact that when a parallel beam of light is passed through a dispersion only a fraction of it is transmitted because of scattering. The amount of light transmittance is a function of total interfacial area. Calderbank (C2) developed an expression for light transmittance through liquid–liquid dispersions assuming uniform drops

$$I/I_0 = \exp[-\tfrac{1}{4}(a^*l)/4] \qquad (87)$$

where I is the amount of light transmitted, I_0 is the amount of incident light, l is the length of the optical path, and a^* is the interfacial area per unit volume. McLaughlin and Rushton (M9) and Curl (C15) have shown that this expression is independent of drop size distribution. The light transmittance method has been used widely by various authors for estimating the interfacial area (B1, B17, B18, C1, L4, M20, R13, R14, S11, T13, V5, V13, W3). Interfacial area has also been measured by a modified electroresistivity probe (H13).

2. *Direct Methods*

Direct methods for obtaining the interfacial area are based on experimental measurement of the drop size distribution. The interfacial area per unit volume is given by

$$a^* = 6\phi/a_{32} \qquad (88)$$

where ϕ is the holdup fraction, and a_{32} is the Sauter mean diameter defined by

$$a_{32} = \Sigma na^3/\Sigma na^2 \qquad (89)$$

B. Drop Size Distribution Measurements

The drop size distribution is an important characteristic of a liquid–liquid dispersion. The drop size distribution not only affects the physical behavior of the dispersion, but also determines mass transfer and chemical reaction rates in the system.

Experimental methods for obtaining drop size distributions have been reviewed by some authors in the past (G12, S16). In this section, various methods for measuring the drop size distribution will be outlined, experimental drop size distributions and appropriate models will be discussed, and the most important expressions developed for predicting a_{32} will be given.

1. *Experimental Methods*

   a. *In Situ Measurements.* This technique is more direct and gives measurements of the dispersion in the actual dynamic conditions without changing the nature of the dispersion as a result of sampling. When photo probes are used, they can be designed to minimize the distortion of hydrodynamic conditions within the vessel. The most important methods are still photography or high-speed cinephotography. Still photographs are taken using a flash light source (1–100 μsec duration). The short-duration flash is necessary to freeze the drop motion. Various investigators (B14, C8, C10, K6, K8, L12, R13, S11, S21, S26, T13, V13) have taken photographs through a window of the vessel. This technique cannot be used for high holdup fraction except at distances very close to the vessel wall due to the interference of drops with the optical path. For distances further in the vessel, photographs can be taken through a probe (B18, B19, C2, M19). Some researchers (L4, S35, W1) have taken photomicrographs through a vessel window. Giles et al. (G9) inserted a photomicrographic probe assembly into the vessel. Coulaloglou and Tavlarides (C13) also used a photomicrographic probe assembly (shown in Fig. 2) which enabled them to measure the drop size distributions in various depths in the vessel. A special extension cup results in a reduction of the drop interference on the optical path of the microscope assembly making the technique suitable for measurements of drops in dispersions of higher holdup fractions. Another interesting technique was developed by Park and Blair (P4) using a fiber optic probe (shown in Fig. 3). Through this probe they measured drop size distributions by using high-speed cinephotography.

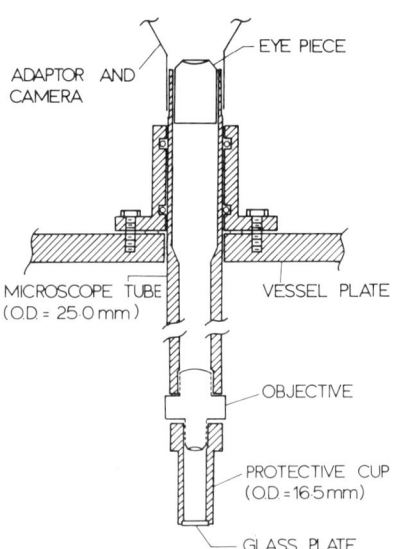

FIG. 2. Photomicrographic probe assembly [after Coulaloglou (C11)].

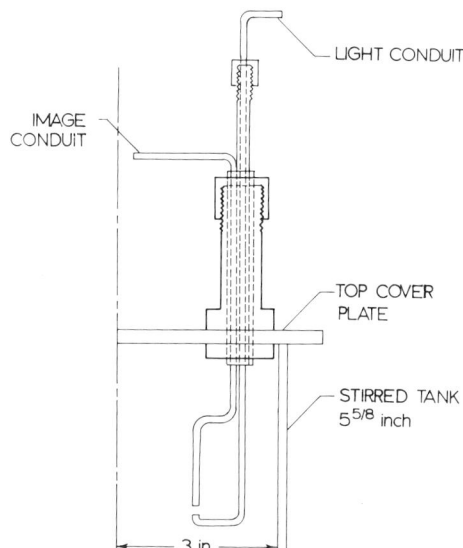

FIG. 3. Fiber optic probe assembly [after Park and Blair (P4)].

b. *Sample Withdrawal Measurements.* In contrast to the previous methods in which measurements can be made in the actual dynamic condition of the experiments, other techniques have been developed whereby a sample of the dispersion is withdrawn for measurements. Usually surfactants are added to the sample to prevent drop coalescence. Drop size distributions are obtained from the sample by microscopic examination, photography, or photomicrography (K5, M20, N2, S22, T14, V6, V7, V16).

Other researchers (M19, V6, V7) encapsulated the drops of the sample by interfacial polymerization and then obtained the drop distribution by photomicrography.

Curl and co-workers (R15, R16, V2) forced the stabilized dispersion through a capillary, where a specially designed photometer assembly (shown in Fig. 4) measured both the drop size distribution and the dye concentration of the drops. Sedimentation techniques have also been used for measuring the drop size distribution (B13, G11, K1, S4).

The sample-withdrawal methods are relatively easy to perform, but they have the disadvantage that they disturb the dispersion hydrodynamics. The drops of the withdrawn sample are in a completely different hydrodynamic state than is the actual dispersion. Also the addition of surfactants and other chemicals to stabilize the droplets affect the physicochemical properties of the dispersion.

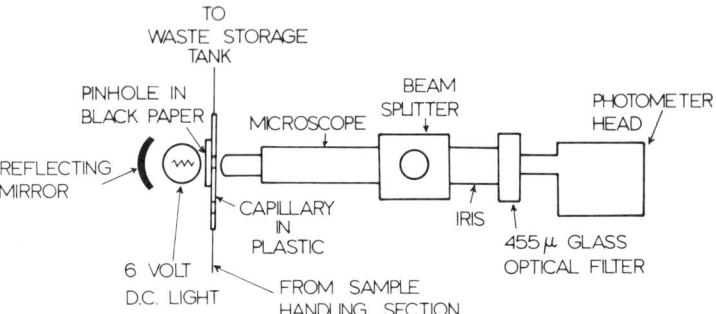

FIG. 4. Microscope assembly for drop size and dye concentration measurements [after Verhoff (V11)].

c. *Other Methods.* Methods that do not fall under either of the previous two categories include the encapsulation of the drops of the whole vessel without first removing a sample (M7, V6, V7). The limitations and uncertainties mentioned above apply to these techniques also.

2. *Experimental Results*

a. *Distributions.* Researchers have used various functions to fit their experimental distributions. The most common one appears to be the lognormal (G9, K6, N2, T15, Y1) given by

$$A(a) = [\log s(2\pi)^{1/2}]^{-1} \exp[-(\log a - \log \bar{a})^2/2(\log s)^2] \quad (90)$$

where $\bar{a}$ is the arithmetic mean drop diameter and s is the standard deviation. Other workers (B19, C8) have found a volume normal distribution given by

$$F_V(a) = [s(2\pi)^{1/2}]^{-1} \exp[-(a - \bar{a})^2/2s^2] \quad (91)$$

Figure 5 shows experimental results described by a normal distribution function. The Schwarz–Bezemer distribution given below was found to describe size distribution very well (S12, S29)

$$\ln V\% = \ln 100 + a_c/a_{\max} - a_c/a \quad (92)$$

Here $V\%$ is the cumulative volume percent of particles below diameter a, a_c is a characteristic diameter related to the maximum of the distribution function, and $a_{\max}$ is the largest drop diameter in the dispersion. Figure 6 illustrates experimental results described by a Schwarz–Bezemer distribution. Gal-Or and Hoelscher (G2) gave the following relation for drop size distribution:

$$f(a, \alpha) = 4(\alpha^2/\pi)^{1/2} a^2 \exp(-\alpha a^2) \quad (93)$$

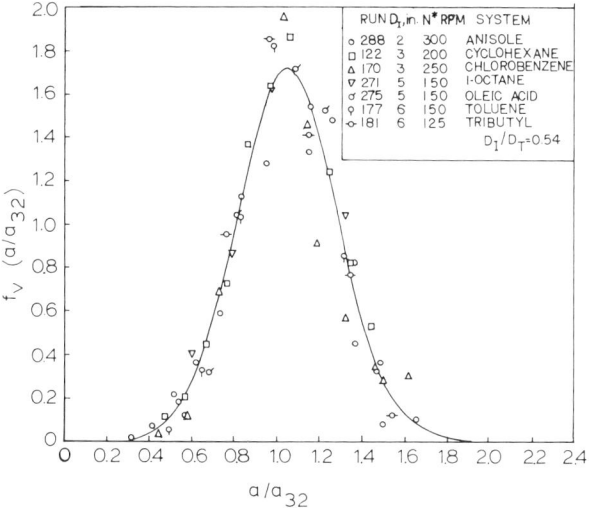

FIG. 5. Comparison of volume normal distribution function with data for $\phi = 0.001 - 0.005$ [after Chen and Middleman (C8)].

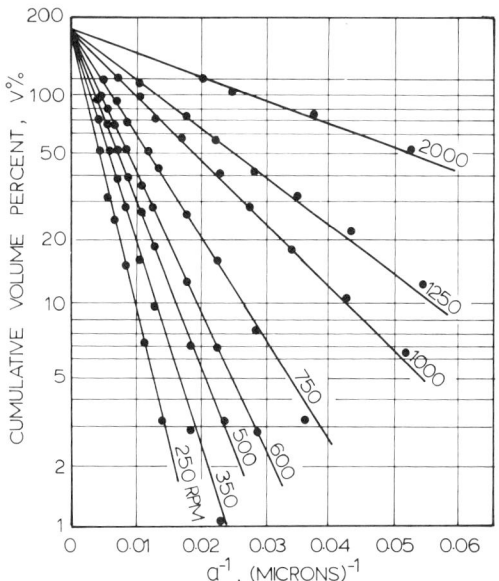

FIG. 6. Comparison of Schwartz–Bezemer distribution with data for $\phi = 0.005$ [after Schwartz and Bezemer (S12)].

in which

$$\alpha = (16\pi^{1/2}N/3\phi)^{2/3} > 0$$

The volume distribution is given by an exponential curve by Bajpai *et al.* (B2). Van Heuven and Hoevenaar (V6) found that the area distribution was described by

$$f_A = (KA/b(K - 1)!)(Ka/b)^{K-1} \exp(-Ka/b) \tag{94}$$

where $b = a_{32}$, $K = (1 - a_{21}/a_{32})^{-1}$, and A is the specific surface area of dispersed phase. Other investigators (G11, L12, S34) developed different distributions.

It should be pointed out that various investigators have reported a local variation of the drop sizes (C14, M19, P4, S11, S29) throughout the vessel. Thus when modeling mass transfer in dispersions, either a global-average drop size distribution or models with spatial dependence of drop size distribution may be needed for accurate analysis of such rate processes.

b. *a_{32} Correlations.* The Sauter mean diameter in an agitated vessel has been correlated by many investigators. The Sauter mean diameter depends on the physicochemical properties of the system, the turbulence intensity, and the holdup fraction. Table I gives the most important reported correlations and the ranges of operating variables and physical parameters over which they are valid. The most frequently reported correlation is of the form

$$a_{32}/D_1 = k_1(1 + k_2\phi)(N_{We})^{-k_3} \tag{95}$$

where k_3 is usually 0.6. Other investigators have used different correlations to fit their data (B3, B14, G9, S21, T8, Y3). It should be noted that the exponent -0.6 of the Weber number (i.e., $a_{32} \propto N^{*-6/5}$) indicates that the drop size is determined primarily by the breakage process as given in Eq. (18). Finally, holdup effects on a_{32} become important only at high holdup fractions.

C. Coalescence Frequency Measurements

Coalescence frequencies can have a pronounced effect on the rate of mass transfer or chemical reaction in a liquid–liquid dispersion. Various investigators have attempted to model and measure coalescence frequencies in agitated vessels. A review of the experimental techniques is given by Rietema (R12) and Shah *et al.* (S16).

The general expression for coalescence frequency is given by

$$\omega_v = \phi^{-1} \int_0^\infty \int_0^{v/2} vF(v - v', v') \, dv' \, dv \tag{96}$$

where ω_v is the volume fraction of dispersed phase coalescing per unit time, and $F(v - v', v')$ is the coalescence rate expression previously discussed and expressed now in terms of drop volumes. Coalescence frequency can also be represented in terms of number of coalescences. In this section, the experimental procedure and measurements for the coalescence rates will be discussed.

1. *Indirect Methods*

All coalescence frequency work reported in the literature with the exception of that of two groups of investigators has used indirect means of obtaining an average coalescence frequency for the entire mixing vessel. The indirect methods can be divided into two categories: chemical and physical.

a. *Chemical.* In this method, a reactant is initially dissolved either in the continuous or the dispersed phase. Another reactant soluble only in the dispersed phase is added at the beginning of the experiment. The coalescence frequency is obtained by monitoring the reaction rate and in conjunction with an appropriate model. Madden and Damerell (M2) obtained coalescence frequencies by monitoring reaction rates through analysis of withdrawn samples. Kramers and co-workers (as reported in R12), Hillestad (H8), and Shiloh *et al.* (S21) monitored the reaction rates by light transmittance techniques.

Chemical reaction methods have been used by other investigators (H1, M4, O5, V16) for coalescence frequency studies in a continuous-flow system. A reaction other than one of first-order is taking place in the dispersed phase and its extent of conversion is related to the coalescence frequency.

The chemical method for obtaining coalescence frequencies has the disadvantage of the influence of the mass transfer that takes place on the physicochemical properties of the dispersion. Further, the intrinsic chemical kinetics must be known.

b. *Physical.* The most important of the physical methods for obtaining coalescence frequencies is based on the measurement of dye dispersion in the dispersed phase. If a small amount of dye soluble only in the dispersed phase is added to a batch-agitated dispersion, the light transmittance in the dispersion versus time is related to coalescence frequency for uniform drops by (M14)

$$\frac{\ln[(I(t)/I(0)]}{\ln[(I(\infty)/I(0)]} = 1 - \frac{1}{2}\left(\frac{\beta \bar{a}}{\bar{c}}\right)\mu_2(0)e^{-\omega t/2} \tag{97}$$

where $I(0)$, $I(t)$, and $I(\infty)$ are the light transmittance values for time 0, t and

TAB

SAUTER MEAN DIAMETER CORRELATI(

Investigators	Correlation	ρ_d(g/cm³)	ρ_c(g/cm³)	μ_d(cp)	μ_c(cp)	σ (dy cm
Vermeulen (V12)	$\frac{a_{32}}{D_1} = Kf_\phi(N_{We})^{-0.6}$	0.693–1.595	0.693–1.595	0.52–184	1.81–65.4	3.1–
Roger et al. (R13)	$\frac{a_{32}}{D_1} = K(N_{We})^{-0.36}\left(\frac{D_1}{D_T}\right)^{-\lambda}$	0.761–1.101	1.0	0.578–3.91	1.0	2.1–
Calderbank (C1)	$\frac{a_{32}}{D_1} = 0.06(1 + 3.75\phi)$ $(N_{We})^{-0.6}$	—	—	—	—	35–
	$\frac{a_{32}}{D_1} = 0.06(1 + 9\phi)(N_{We})^{-0.6}$	—	—	—	—	35–
Shinnar (S22)	$\frac{a_{32}}{D_1} = K(N_{We})^{-0.6}$ (breakage control)					
	$\frac{a_{32}}{D_1} = K(\sigma D_1)^{-3/8}(N_{We})^{-3/8}$ (coalescence control)	—	—	22.5	0.4	—
Chen and Middleman (C8)	$\frac{a_{32}}{D_1} = 0.053(N_{We})^{-0.6}$	0.703–1.101	0.997–1.001	0.52–25.8	0.890–1.270	4.75–
Sprow (S30)	$\frac{a_{32}}{D_1} = 0.0524(N_{We})^{-0.6}$	0.692	1.005	0.51	0.99	41.
Brown and Pitt (B18)	$\frac{a_{32}}{D_1} = 0.051(1 + 3.14\phi)$ $(N_{We})^{-0.6}$	0.783–0.838	0.972–0.998	0.59–3.30	1.0–1.28	1.9–
Van Heuven and Beek (V7)	$\frac{a_{32}}{D_1} = 0.047(1 + 2.5\phi)$ $(N_{We})^{-0.6}$	—	0.998	—	—	8.5–
Mlynek and Resnick (M19)	$\frac{a_{32}}{D_1} = 0.058(1 + 5.4\phi)$ $(N_{We})^{-0.6}$	1.055	1.0	—	1.0	41
Weinstein and Treybal (W3)	$a_{32} = 10^{(-2.316+0.672\bar{\phi})}$ $\nu_c^{0.0722}\epsilon^{-0.194}$ $(\sigma g_c/\rho_c)^{0.196}$	0.831–0.997	0.831–0.997	0.722–7.43	0.722–7.43	3.76–3
	$a_{32} = 10^{(-2.066+0.732\bar{\phi})}$ $\nu_c^{0.047}\epsilon^{-0.204}$ $(\sigma g_c/\rho_c)^{0.274}$	0.831–0.997	0.831–0.997	0.722–7.43	0.722–7.43	3.76–3
Brown and Pitt (B17)	$a_{32} = K\left(\frac{\sigma}{\bar{\rho}\bar{\epsilon}t_c}\right)^{0.6}$	0.783–0.838	0.972–0.998	0.59–3.30	1.0–1.28	1.9–5
Coulaloglou and Tavlarides (C13)	$a_{32} = 0.081(1 + 4.47\phi)$ $(N_{We})^{-0.6}$	0.972	1.0	1.3	1.0	43

" B, batch; C, continuous.

∞, respectively, β is the extinction coefficient of dye, ā and c̄ are the average diameter and dye concentration, respectively. The dye dispersion–light transmittance technique in conjunction with the previously mentioned model have been used by various investigators for studying coalescence frequencies in batch systems (K12, K14, M14). The dye dispersion–light transmittance technique was also used by Vanderveen (V5) for batch systems with the exception of the use of a different model based on an analogy with a first-order chemical reaction.

)R STIRRED LIQUID–LIQUID CONTACTORS

pe[a]	D_I(cm)	D_T(cm)	ϕ	N^*(rps)	τ^*(sec)	Type of impeller	Measurement techniques and comments
B	—	25.4, 50.8	0.10–0.40	1.80–6.67	—	4-bladed paddles	In situ (light transmittance)
B	5.1–30.0	15.5, 45.7	0.5	1–20	—	6-bladed turbine	In situ (photography, light transmittance)
B	5.8–25.4	17.8, 38.1	0–0.2	—	—	4-bladed paddle	In situ (light transmittance)
3	5.8–25.4	17.8, 38.1	0–0.2	—	—	6-bladed turbine	In situ (light transmittance)
							No experiments
	12.7	29.0	0.05	2.6–10.5	—	Paddle turbine	Sample withdrawal
	5.1–15.2	10.0–45.7	0.001–0.005	1.33–16.7	—	6-bladed turbine	In situ (photography)
	3.2–10.0	22.2, 30.5	0–0.015	4.2–33.4	—	6-bladed turbine, modified turbine	Sample withdrawal (Coulter Counter)
	10	30	0.05–0.3	4.2–7.5	—	6-bladed turbine	In situ (photography)
	3.75–40.0	12.5–120	0.04–0.35	—	—	6-bladed turbine	Encapsulation, sample withdrawal
	10	29	0.025–0.34	2.3–8.3	—	6-bladed turbine	In situ (photography) sample withdrawal (drop encapsulation)
	7.62–12.7	24.5, 37.2	0.079–0.593	2.5–10.33	—	6-bladed turbine unbaffled vessel	In situ (light transmittance)
	7.62–12.7	24.5, 37.2	0.079–0.593	2.5–10.33	9.9–251	6-bladed turbine unbaffled vessel	In situ (light transmittance)
	10, 5	30	0.05	2.1–7.5	—	6-bladed turbines	In situ (light transmittance)
	10.0	24.5	0.025–0.15	3.2–5.2	600	6-bladed turbines	In situ (photomicrography)

These methods for obtaining coalescence frequencies in batch systems have been extended to continuous-flow systems (C13, K13). Komasawa (K13) derived the appropriate model for flow systems assuming uniform drops present and two dispersed-phase feed streams entering the vessel, one with dye and the other without. The result is

$$\frac{\ln(\bar{I}_a/\bar{I}_0)}{\ln(\bar{I}_b/\bar{I}_0)} = \frac{2 + \omega\theta}{2(1 - \phi_f)c_1\beta\bar{a}} \ln\left(1 + \frac{2(1 - \phi_f)c_1\beta\bar{a}}{2 + \omega\theta}\right) \quad (98)$$

where $\bar{I}_0$, $\bar{I}_a$, $\bar{I}_b$ are for the average steady-state light transmittance in the vessel prior to the addition of dye, for the case of dispersed-phase streams fed separately, and for the case of premixed dispersed-phase feed streams, respectively, θ is the residence time, ϕ_f is the fraction of dispersed-phase feed streams containing dye, and c_1 is the dye concentration in the dye-containing feed stream.

Curl and co-workers (R15, R16, V11) have measured coalescence frequencies in a continuous flow vessel by introducing two dispersed-phase feed streams containing different concentrations of dye. Samples removed from the vessel are analyzed by a specially designed photometer that measures the bivariate drop volume–dye concentration distribution in the dispersion (see Fig. 4). The appropriate equation for obtaining ω in this case is

$$c_r = [1 + \tfrac{1}{2}(\omega\theta)]^{-1} \tag{99}$$

where c_r is the ratio of variance of concentration in the vessel to the variance of concentration in the feed streams, and θ is the residence time in the vessel. The method of two dispersed-phase feed streams has also been used by Villermaux and Devillon (V16), where the tracer was salt. The bivariate drop volume–salt concentration distribution was obtained from withdrawn samples which were analyzed with a model developed by them.

Howarth (H16) developed another method for obtaining coalescence frequencies without introducing a dye or reactants which may modify the coalescence behavior. This method is based on the fact that the rate of establishment of a new steady-state mean drop size following a reduction in agitation intensity is related to the coalescence frequency in the vessel. The coalescence frequency in this case is given by

$$\omega = \frac{(da_{32}/dt)_{t=0}}{(2 - 2^{2/3})(a_{32})_{t=0}} \tag{100}$$

The course of the unsteady state was followed by light transmittance techniques. Similar methods based on sudden reduction in agitation intensity have been used by other investigators (B12, D12, M19), where the change of a_{32} versus time was followed by either light transmittance (B12, D12) or photographic means (M19). This method presents the problem that the step change of agitation intensity may in reality not be instantaneous. In addition, the change of drop diameter with time after a step change of agitation occurs is not only a result of increased coalescence rate but is also a function of the decreased breakage rate.

Treybal and co-workers (M20, S11) calculated "minimum" coalescence frequencies in continuous-flow vessels by taking into account the increase of drop diameter around the circulation path in the agitated vessel. In addition, coalescence frequencies were measured in an unbaffled vessel (M20).

Groothuis and Zuiderweg (G10) measured coalescence frequencies for continuous-flow dispersed-phase systems by introducing two streams of dispersed phase feed with different densities such that if a drop of one stream coalesces with a drop of the other stream, the new drop will be heavier than the continuous phase. Coalescence frequencies were then estimated by measuring the change in dispersed-phase fraction heavier than water as it passes through the vessel.

This method does not take into account the coalescences that occur between two light drops and between two heavy drops.

2. Direct Methods

There is little work on direct determination of coalescence frequencies in a turbulent dispersion. Kuboi *et al.* (K17) and Park and Blair (P4) have used high-speed cinephotography to directly observe and measure the drop coalescence frequencies in pipe flow and in an agitated vessel, respectively. The difficulty with this method is that only a few coalescences are observed at great expense of cine film after tedious examination. Thus there is doubt regarding the statistical representation of coalescence in the dispersion from a small sample.

3. Experimental Results and Observations

Table II summarizes coalescence frequency correlations found experimentally by various investigators. In general, with the exception of three works (D12, H8, M19), all investigators found that the coalescence frequency increases with N^*. In addition, ω increases with ϕ. Mass transfer out of the dispersed phase seems to increase ω. It is observed also that aqueous dispersions in organics have much higher ω than organic dispersions in aqueous continuous phase.

V. Mathematical Models for Mass Transfer with Reaction in Liquid–Liquid Dispersions

A variety of models are employed to predict extent of reaction and selectivity for complex reactions which occur in liquid–liquid dispersions in agitated vessels. The nature of the models depends on which phases the reaction occurs in, the relative magnitude of the time scales of the mass transfer and reaction processes compared to the mixing processes, com-

TABLE

COALESCENCE FREQUENCY CORRELATIONS IN

Investigators	Correlation	Systems	σ (dynes/cm)	Type
Vanderveen (V5)	$\omega = KN^*(N_{We})^{0.6}(N_{Re})_T^{-0.2}$	Organic in aqueous	—	B
Madden and Damerell (M2)	$\omega = KN^{*2.4}\phi^{0.5}$	Aqueous in organic		B
Miller (M14)	$\omega = KN^{*1.5-3.3}\phi^{0.5-1.1}$	Organic in aqueous		
Howarth (H15)	$\omega = 2.35 \times 10^{-5}\phi^{0.5}N^{*2.2}\exp(-2 \times 10^3/N^{*2})$			
Hillestad (H8)	$\omega = 2.88 \times 10^{-8}N^{*0.73}D_I^{0.81}D_T^{-0.2}\phi^{1.0}$ $\mu_c^{-0.98}\mu_d^{0.024}\rho_c^{1.23}\sigma^{1.74}$ For $(N_{We}) < 0.343 \times 10^{0.075(D_T 10.5)^2}$ $(N_{Re})_T^{0.75}$ $\omega = 4 \times 10^{-25}N^{*-3.5}D_I^{-7.7}D_T^{5.05}\phi^{1.58}$ $\mu_c^{-2.6}\mu_d^{-1.3}\sigma^{21.39}$ For $(N_{We}) > 0.343 \times 10^{0.075(D_T 10.5)^2}$ $(N_{Re})_T^{0.75}$	Aqueous in organic $\mu_d = 0.95 - 16$ cp $\mu_c = 0.652 - 13.2$ cp	26.3–37.6	B
Howarth (H16)	$\omega = KN^{*1.9-2.25}\phi^{0.6}$	Organic in aqueous	—	B
Doulah and Thornton (D12)	$\omega = K\phi N^*\left(\dfrac{D_I}{a_{32}}\right)^{2/3}\exp(-K_2\rho_d(\bar{\epsilon})^{2/3}a_{32})$	Organic in aqueous	—	B
Komasawa et al. (K12)	$\omega = 2 \times 10^{-10}N^*\phi^{0.7}(N_{Re})_T^{2.0}(N_{We})^{-0.25}$	Organic in aqueous	3.8–46.0	B
Komasawa et al. (K13)	$\omega = KN^{*2.0}\phi^{0.8}$	Organic in aqueous	—	C
Mlynek and Resnick (M19)	$\omega = KN^{*-1.8}\phi^{0.8}a_{32}^{-0.9}$	Organic in aqueous	41	B
Shiloh et al. (S21)	$\omega = K\phi^{1.0}$	Aqueous in organic	—	B
Blanch and Fietcher (B12)	$\omega = KN^{*2.0}$	Organic in aqueous	—	
Coulaloglou and Tavlarides (C13)	$\omega = K\phi^{0.5}N^{*2.85}$	Organic in aqueous	43	C

[a] B, batch; C, continuous.

plexity of reactions, thermal effects, and scale-up considerations. The models which will be considered in detail in the remaining sections are those which account for the dispersed-phase mixing effects. To place these models in context with all approaches, a brief discussion of models which do not account for dispersed-phase interactions is given.

A. Effective Interfacial Surface Area Models

These models represent the simplest approach to model mass transfer in liquid–liquid dispersions in that an effective interfacial surface area be-

STIRRED LIQUID–LIQUID CONTACTORS

D_I(cm)	D_T(cm)	ϕ	N^*(rps)	τ^*(sec)	Type of Impeller	Exptl. Techniques and Comments
2.7	25.3	0.10–0.40	2.57–6.45	—	4-bladed paddles	Addition of dye in conjunction with light transmittance techniques
7.6	14	0.00137–0.0109	2.5–5	—	6-bladed turbine	Chemical reaction method employed—analysis of reaction by sample withdrawal
5.1–33	10.2–66	0.1–0.4	—	—	turbine, propeller	Addition of dye in conjunction with light transmittance techniques Madden and Damerell (M2), data correlated
4.7–21.9	14.1–43.8	0.05–0.15	1.5–11.3	—	6-bladed turbine	Chemical reaction method employed—analysis of reaction by light transmittance techniques
7.6	15.2	0.1–0.25	1.84–6.0	—	6-bladed turbine	Step change of rotational speed in conjunction with light transmittance techniques
7	33.0	0.05–0.40	4.17–4.84	—	6-bladed turbine	Step change of rotational speed in conjunction with light transmittance techniques
1	11.0	0.05–0.40	5.67–25.0	—	Turbine	Addition of dye in conjunction with light transmittance techniques
1	11.0	0.07–0.16	10.0–20.4	48–72	6-bladed turbine	Addition of dye in conjunction with light transmittance techniques
	29	0.025–0.34	2.3–6.4	—	6-bladed turbine	Step change of rotational speed in conjunction with photographic techniques
	20	0.003–0.04	5.34–8.34	—	6-bladed turbine	Chemical reaction method employed—analysis of reaction by light transmittance techniques
—	—	—	—	—	—	Step change of rotational speed in conjunction with light transmittance techniques
	24.5	0.05–0.15	3.17–5.17	600	6-bladed turbine	Addition of dye in conjunction with light transmittance techniques

tween the two liquid phases is estimated based on power input or obtained experimentally by methods discussed in Section IV,A. Mass transfer with reaction is then described using models such as the two-film, penetration, film-penetration, and surface renewal theories (A6, D2, N4, R10, S25, T12).

Various investigators have recently employed this approach (C4, C14, K18, M3, M11, S20). For example, Mashkar and Sharma (M3) and Merchuk and Farina (M11) studied simultaneous diffusion with various chemical reactions occurring in both phases. This approach dismisses completely the fact that the dispersed phase consists of a distribution of droplet sizes which may coalesce or redisperse. Each drop may have different concentrations of the reacting species. No account is taken of the effects of the operating variables on these dispersion properties. Nevertheless, this approach has found great use in design of contacting equipment. Scale-up of equipment from laboratory data using this approach is very uncertain. Significant errors in prediction of conversion and selectivity can arise due to variations in drop size distribution and droplet mixing. These factors are not accounted for in these models.

B. Drop Size and Residence Time Distribution Models

A second level of model development includes the use of drop size distributions and residence time distributions of the droplets in the vessel. Notable contributions to this approach are those by Gal-Or and co-workers (G2, G3, G4, G5, P1, P2, R10, T2, T3, W2) and similar approaches by others (C7, D3, M21). These models assume that the number of drops in the vessel is constant and may have a distribution of sizes. The continuous phase is divided into an equal number of fluid elements assigned to each droplet. Expressions are developed for mass transfer with reaction between the drop and fluid element. The average moles transferred per unit volume of dispersion is obtained by integration of the flux over the distribution of droplet sizes for an average residence time of the drops in the reactor. The residence time distribution of droplets can also be accounted for by taking the "integral transformation" of the continuity equation to the dispersed phase with respect to a kernel. The kernel comprises the residence time and droplet size distributions. Solutions of the resulting transformed equations give the average mass transfer rate for the dispersion directly. These workers were also able to consider the more general problem of multicomponent mass transfer with general reversible reaction and coupled heat and mass transfer in dispersions (P1, T2, T3). Various drop size distribution functions were proposed which attempted to account for the operation parameters on drop distributions. These include that proposed by Baynes (B11) and adapted by Gal-Or and Hoelscher (G2), and that proposed by Mugele and Evans (M22) and adapted by Olney (O3) for drops. The exponential distribution was taken as the residence time distribution for the case of perfect mixing of the dispersed phase. Interactions among adjacent particles was accounted for through a "crowding" effect. As the holdup increases, for a given average droplet size, the total number of droplets increases and thus the volume of the continuous phase assigned to a given particle decreases. The concentration profiles in the continuous phase amongst particles thus change and accordingly mass transfer is affected.

This modeling approach accounts for the variation of droplet sizes on mass transfer, an interaction effect amongst droplets on mass transfer through the "crowding effect," the residence time distribution of droplets in the reactor assuming the droplets are segregated entities, and the fact that the total mass-transfer process is a cumulative effect of single droplet phenomena. However, there are limitations to these models. It is significant that coalescence and breakage of droplets are not accounted for. These micromixing effects on conversion and selectivity cannot be incorporated only with the residence time of the dispersed phase in the reactor. In fact, only a small number of droplets entering a vessel do not undergo breakage and coalescence. A more realistic temporal parameter might be the "age" for which a droplet retains its identity upon entering the vessel in the feed, or upon formation due to breakage of a parent droplet, or upon coalescence of two other droplets. These models also do not account for the possibility of a feed stream with a size distribution other than that within the reactor. The feed drops are assumed to break instantaneously into the distribution within the reactor.

It becomes evident that an important class of problems cannot be rationally handled by these models.

C. Dispersed Phase Interaction Models

There are a variety of reaction systems where dispersed phase mixing has significant effects on conversion and selectivity. These include:

(a) Reactions in the drop phase with time scale of the order of coalescence and breakage time scale.

(b) Bimolecular reactions of intermediate rate in the dispersed phase with unmixed feed streams.

(c) Reaction in the continuous phase with diffusing species from the drop phase where the reaction and diffusion time scales are of the same order as the coalescence and breakage time scale.

(d) Bimolecular interfacial reactions where each reactant is soluble only in one phase (e.g., metal chelation reactions in hydrometallurgical processes and single-cell protein fermentation).

(e) Complex reactions in the dispersed phase of intermediate rate.

(f) Instantaneous bimolecular reactions in the dispersed phase with unmixed feeds.

Coalescence and redispersion models applied to these reaction systems include population balance equations, Monte Carlo simulation techniques, and a combination of macromixing and micromixing concepts with Monte Carlo simulations. Most of the last two types of models were developed to account for the micromixing effects on conversion and selectivity for homogeneous reactions but can be applied to dispersions.

The problem of applying these models to predict scale-up criteria depends upon obtaining parameters for the droplet mixing which relate to the operating conditions. For example, in a turbulently agitated vessel the energy dissipation can vary greatly with position in the vessel. Also, circulation times vary as vessel size increases although power input per unit volume remains constant. Therefore, the spatial homogeneity of the dispersion as well as the local interaction rates of the droplets may vary significantly. If the above models do not account for these factors, they can only be used indirectly to determine the variation in conversion or selectivity one can expect for a given level of droplet mixing. The mixing encountered must be determined from experiments on the particular reactor in question for a given set of operating conditions and dispersion properties. On the other hand, if the dispersed-phase mixing parameters can be related to the operation conditions, physical properties, and variations in vessel geometry, one may directly predict the effects of scale-up on conversion. (Albeit that we may not be able to obtain the same performance in the large-scale vessel as the laboratory vessel due to the significant energy cost to maintain the same level of turbulence and circulation times).

A number of fine reviews have appeared recently which address in part the problems mentioned and the models employed. Rietema (R12) discusses segregation in liquid–liquid dispersion and its effect on chemical reactions. Resnick and Gal-Or (R10) considered mass transfer and reactions in gas–liquid dispersions. Shah *et al.* (S16) reviewed droplet mixing phenomena as they applied to growth processes in two liquid-phase fermentations. Patterson (P5) presents a review of simulating turbulent field mixers and reactors in which homogeneous reactions are occurring. In Sections VI, D–F the use of these models to predict conversion and selectivity for reactions which occur in dispersions is discussed.

D. Type I Interaction Models: Population Balance Techniques

The population balance approach is employed for the description of droplet dynamics in various flow fields. A significant advantage of the method is that a vehicle is provided to include the details of the breakage and coalescence processes in terms of the physical parameters and conditions of operation. A predictive multidimensional particle distribution theory is at hand which, in the case of well-defined droplet processes, can be employed for *a priori* prediction of the form and the magnitude of the particle size distribution. The physical parameters which affect the form of the drop size distribution are contained within the integro-differential equation. The theory can thus be a predictive tool and permit the analysis of transient behavior of dispersed-phase systems.

Conceptually, the framework of the theory permits description of interphase heat and mass transfer with reaction occurring in either or both phases. In theory one can use this approach to study the affects of partial mixing of the dispersed phase on extent of reaction for non-first-order reactions which occur in the droplets. Analyses can be made for mass-transfer-controlled reactions and selectivity for complex reactions. Difficulties in the solution of the resulting integro-differential equations have restricted applications at present to partial solutions. For example, the effects of partial droplet mixing on extent of reaction were studied for uniform drops. Mass transfer from nonuniform drops for various reactor geometries was studied for dispersions with drop breakage only or drop coalescence only.

The sections which follow outline the general multidimensional distribution theory. Applications of the theory are discussed to describe droplet size distributions and mixing frequencies in chemically equilibrated systems, the effects of droplet mixing on the extent of reaction, the analysis of mass transfer with and without chemical reaction, hydrocarbon fermentation, and emulsion polymerization.

1. *Conservation Equations in Particle Phase Space*

Hulburt and Katz (H17) developed a framework for the analysis of particulate systems with the population balance equation for a multivariate particle number density. This number density is defined over phase space which is characterized by a vector of the least number of independent coordinates attached to a particle distribution that allow complete description of the properties of the distribution. Phase space is composed of three external particle coordinates x_e and m internal particle coordinates x_i. The former (x_{e1}, x_{e2}, x_{e3}) refer to the spatial distribution of particles. The latter coordinate properties $(x_{i1}, x_{i2}, \ldots, x_{im})$ give a quantitative description of the state of an individual particle, such as its mass, concentration, temperature, age, etc. In the case of a homogeneous dispersion such as in a well-mixed vessel the external coordinates are unnecessary whereas for a nonideal stirred vessel or tubular configuration they may be needed. Thus $n(\mathbf{x}; t)d\mathbf{x}$ represents the number of particles per unit volume of dispersion at time t in the incremental range $\mathbf{x}, \mathbf{x} + d\mathbf{x}$, where $\mathbf{x}$ represents both coordinate sets. The number density continuity equation in particle phase space is shown to be (H18, R6)

$$[\partial n(\mathbf{x}; t)/\partial t] + \nabla \cdot [\dot{\mathbf{x}} n(\mathbf{x}; t)] = h^+(\mathbf{x}; t) - h^-(\mathbf{x}; t) \qquad (101)$$

where $\dot{\mathbf{x}}$ is the coordinate velocity in phase space. For example, if x_α represents the average concentration of solute in a drop, then $\dot{x}_\alpha$ represents the rate of change of the average concentration in a drop of state (x; t). The terms $h^+(\mathbf{x}; t)\,d\mathbf{x}$ and $h^-(\mathbf{x}; t)\,d\mathbf{x}$ are the number of droplets which are produced and destroyed per unit volume of dispersion at time t in the incremental range x to x + $d\mathbf{x}$ in phase space. Equation (101) along with mass and energy balances, which constrain the values of the internal coordinates, and flow conditions, which describe influx or efflux of dispersion from the region of interest, completely determine the formation and dynamics of multidimensional particle distributions.

Other forms of the number density equation or population balance equation are useful. For a spatially homogeneous dispersion, such as in a well-mixed vessel, with flow of dispersion and no density changes:

$$\partial n(\mathbf{x}_i; t)/\partial t + \nabla \cdot [\dot{\mathbf{x}}_i n(\mathbf{x}_i; t)] = h^+(\mathbf{x}_i; t) - h^-(\mathbf{x}_i; t) - \sum_k Q_k n_k / V \qquad (102)$$

where V is the volume of the dispersion, Q_k is the volumetric flow rate of stream k (taken positive for flow out of V), and n_k is the number density of stream k.

Often for engineering purposes the complete characterization of the dispersion by $n(\mathbf{x}; t)$ is not necessary and knowledge of average properties such as average size, surface area, or mass concentration is adequate for design purposes. It then may be expedient to reformulate the population balance equation in terms of the moments of the distribution.

To illustrate the method, assume a spatially distributed dispersion which can be described adequately with one internal coordinate. The moments are characterized by

$$\mu_\alpha(\mathbf{x}_e, t) = \int x_\beta^\alpha n(\mathbf{x}_e, x_\beta, t)\,dx_\beta \qquad (103)$$

where integration extends over all values of x_β. For the case when x_β is the droplet diameter a, $\mu_0(\mathbf{x}_e, t)$ denotes the total number of particles per unit volume of dispersion at $\mathbf{x}_e$ and time t. Also, μ_3/μ_2 is the Sauter mean diameter. The instantaneous local dispersed-phase holdup for spherical particles is given by

$$\phi(\mathbf{x}_e, t) = \tfrac{1}{6}\pi \int a^3 n(\mathbf{x}_e, a, t)\,da = \tfrac{1}{6}\pi\mu_3 \qquad (104)$$

The mean value in x_β of any function P^* at $\mathbf{x}_e, x_\beta$, and t can be denoted by

$$\langle P^* \rangle = \int P^*(\mathbf{x}_e, x_b, t) n(\mathbf{x}_e, x_\beta, t)\,dx_\beta \qquad (105)$$

The moment transformation of the population balance equation for one internal coordinate can be obtained by multiplying Eq. (101) by x_β^α and integrating over all values of x_β:

$$\partial\mu_\alpha/\partial t + \nabla_e \cdot v_e\mu_\alpha - \alpha\langle x_\beta^{\alpha-1} X_\beta(\mathbf{x}_e, x_\beta, t)\rangle$$
$$= \int x_\beta^\alpha h^+(\mathbf{x}_e, x_\beta, t)dx_\beta - \int x_\beta^\alpha h^-(\mathbf{x}_e, x_\beta, t)dx_\beta \qquad (106)$$
$$\langle \alpha = 0, 1, 2, \ldots \rangle$$

where $v_e = d\mathbf{x}_e/dt$ and $X_\beta = dx_\beta/dt$. Equations (106) have the same dimensionality as the transport equations and can be solved simultaneously with the equations of conservation of mass, momentum, and energy to yield a complete description of dispersed-phase processes. Average properties of the distribution can be obtained directly from the time–space dependent moment. The number of moment equations needed to obtain a satisfactory solution is about three or four.

2. Analysis of Drop Breakage and Coalescence in Dispersed Phase Systems

The population balance equation is employed to describe the temporal and steady-state behavior of the droplet size distribution for physically equilibrated liquid–liquid dispersions undergoing breakage and/or coalescence. These analyses also permit evaluation of the various proposed coalescence and breakage functions described in Sections III,B and C. When the dispersion is spatially homogeneous it becomes convenient to describe particle interaction on a total number basis as opposed to number concentration. To be consistent with the notation employed by previous investigators, the number concentration is replaced as $n(\mathbf{x}_i; t) d\mathbf{x}_i = NA(\mathbf{x}_i; t)d\mathbf{x}_i$, where N is the total number of particles per unit volume of the dispersion, and $A(\mathbf{x}_i; t) d\mathbf{x}_i$ is the fraction of drops in increment $\mathbf{x}_i$ to $\mathbf{x}_i + d\mathbf{x}_i$. For spatially homogeneous dispersions such as in a well-mixed vessel, continuous flow of dispersions, no density changes, and isothermal conditions Eq. (102) becomes

$$dNA(v)/dt = \int_v^{v_{max}} \beta(v', v)\nu(v)g(v')NA(v', t) \, dv' - g(v)NA(v, t)$$
$$+ \int_0^{v/2} \lambda(v - v', v')h(v - v', v')NA(v - v', t)NA(v', t) \, dv'$$
$$- NA(v, t)\int_0^{v_{max}-v} \lambda(v, v')h(v, v')NA(v', t) \, dv'$$
$$+ [Q_0 N_0 A_0(v, t)]/V - [QNA(v, t)]/V \qquad (107)$$

with drop volume v as the only internal coordinate x_i in particle phase space. The first two terms on the right-hand side represent the rate of formation and loss of drops of volume v per unit volume of dispersion due to breakage. The second two terms represent the rate of formation and loss of droplets of volume v due to coalescence. The last two terms account for the flow of droplets into and out of the vessel. Here N_0 is the number feed rate of drops per volume of dispersion flowing into the vessel; Q_0 and Q are volumetric flow rates of dispersion into and out of the vessel, respectively. The dispersion is subjected to the constraint of constant total volume. The nature of the breakage and coalescence terms $\beta(v', v)$, $\nu(v)$, $g(v')$, $\lambda(v, v')$, and $h(v, v')$ is discussed in Section III.

Valentas et al. (V1-3) employed Eq. (107) in the form of a total number over droplet mass balance to describe droplet breakage and coalescence in two-phase liquid dispersions for a spatially homogeneous well-mixed vessel. This work is significant in that a detailed phenomenological description of these interaction processes was described for the first time. The authors propose models for the coalescence and breakage kernels, some of which are discussed in Section III. The resulting integro-differential equations were solved by a numerical integration procedure. The steady-state case for breakage only results in a Volterra integral equation. A thorough analysis was made of the effects of various forms of the breakage kernels on the steady size distribution. For the case of both coalescence and breakage the analysis included both transient and steady flow cases. Although the authors present an inclusive model, the collision rates were not related to the fluid dynamics and physical properties. The breakage models were related in part to the physics through the maximum stable drop concepts. No comparisons were made with experimental data to test the models.

The importance of developing coalescence and breakage kernels based on physical grounds became evident for the goal of predicting liquid–liquid dispersion properties. Subsequent workers devoted efforts in this direction.

Verhoff and Curl (V11) developed a series of integro-differential equations similar to Eq. (107) based on several different space balances in which the breakage and coalescence functions were combined into a single mixing kernel. The kernel was employed in a balance using the two-dimensional area of drop size and dye concentration. Singularity problems were encountered in the numerical inversion of the integral equation when applied to measured distributions. Resulting kernels obtained appear unreasonable and it is doubtful they have physical significance. The experimental techniques developed and the demonstrated limitations of this modeling procedure are among the contributions of this work. An important contribution was made by Ross and Curl (R15, R16). They adopted Eq. (107) in the total number over droplet volume balance for their studies of a continuous-flow vessel taken to be perfectly macroscopically mixed. The authors developed models for $g(v)$ and $\lambda(v, v')$ (presented in Section III). They assumed $\beta(v', v)$ is given by the beta function, $\nu(v)$ is two, and a coalescence frequency $h(v, v')$ based on dimensional arguments was employed. An analysis of the behavior of the drop size distribution predicted from the model is presented. Figure 7 shows the effects of power input on the steady-state size distribution for the flow system with breakage only and unimodel feed distribution. The important effect of "flow-stabilized" dispersions was delineated. The drop size distributions and mixing frequencies calculated from solutions of Eqs. (107) and (96) were compared to experimental results to obtain values of the parameters of the models. The experimental techniques for measurement of drop size distributions and mixing frequencies are discussed in Section IV. Six parameters were needed since an additional two were necessary to account for the effects of dispersed-phase holdup on the rate processes. Favorable agreement was obtained between experimental and theoretical results and the results were compatable with those of other workers. A strong dependence of mixing frequency on power input was observed over these experimental conditions. It should be noted that the results were correlated over a low-power input range due to experimental limitations.

Coulaloglou and Tavlarides (C11, C12) presented a similar analysis for turbulent dispersions. The total number over droplet volume balance form of Eq. (107) was employed in their analysis. The models for $g(v)$, $\lambda(v, v')$,

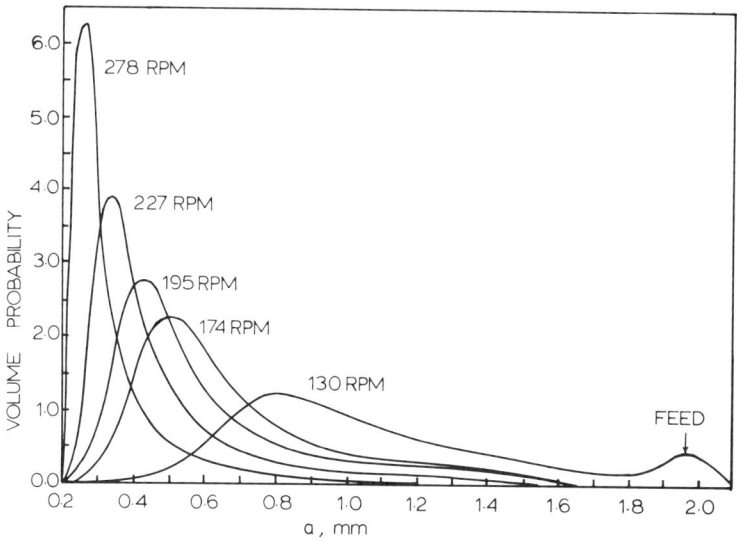

FIG. 7. Theoretical drop size distributions for breakage only for continuous-flow operation with uniform feed and $\phi = 0.05$ [after Ross (R15)].

and $h(v, v')$ were developed from phenomenological concepts based on turbulence theory and are presented in Section III. Comparison of experimentally determined drop size distributions and mixing frequencies permitted evaluation of the model parameters. Figure 8 shows such a comparison. Drop sizes were measured by microphotography and mixing frequencies by dye–light transmission techniques. Figures 9 and 10 compare their results for mixing frequency and average drop sizes as a function of experimental conditions with those of Ross and Curl (R15) and other investigators. Caution is needed in such comparisons due to the variety of operating conditions, range of physical parameters, and effects of trace amounts of surfactants. However, both groups of researchers are in general agreement with each other and the literature. A strong dependence of ω_v on power input is observed. In addition, Fig. 10 shows excellent agreement in prediction of average drop size and the effects of "flow stabilization" of dispersion drop size. Here the data of Mlynek and Resnick (M19) show smaller drop sizes for similar conditions during batch operation. These techniques show promise in describing interaction processes from fundamental bases.

Recently Stamatoudis (S32) investigated the effects of continuous phase viscosity on the dynamics of liquid–liquid dispersions in agitated vessels.

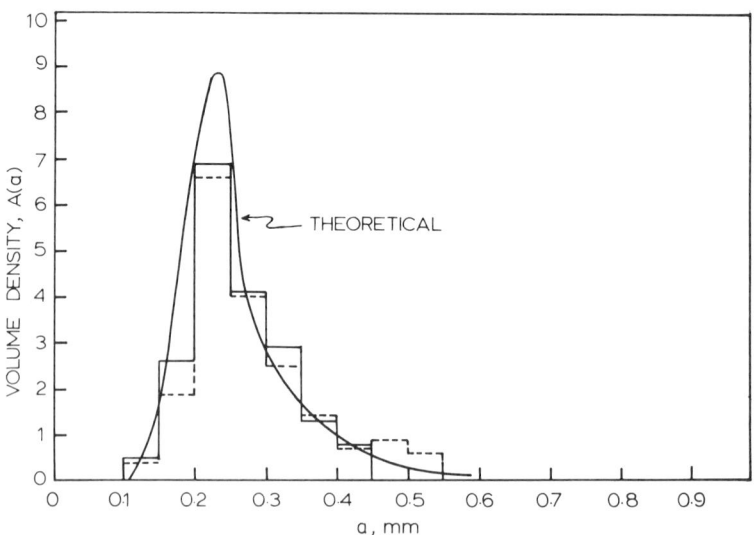

FIG. 8. Theoretical size distribution and experimental histograms: impeller region (solid line), circulation region (dashed line), $N^* = 310$ rpm, $\phi = 0.10$, $a_{32} = 0.255$ mm (experiment) versus 0.250 mm (calculation), $\omega_v = 0.980$ min^{-1} (experiment) versus 1.026 min^{-1} (calculation) [from Coulaloglou and Tavlarides (C12)].

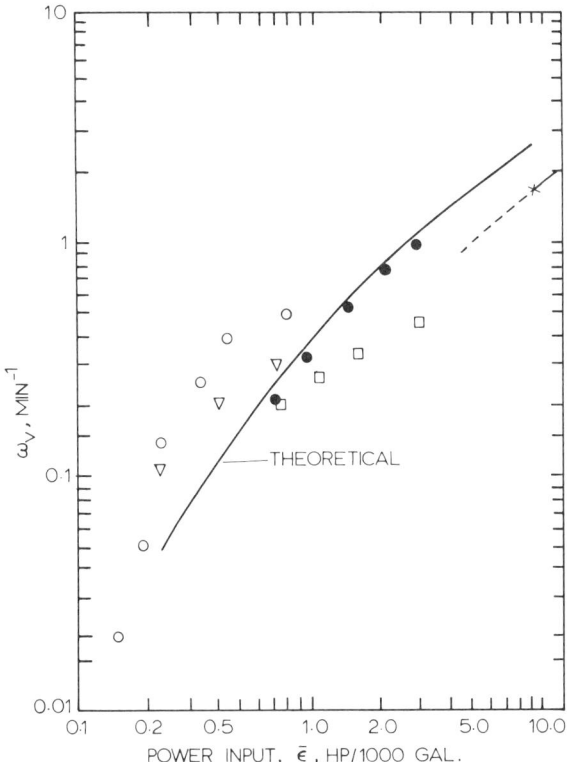

FIG. 9. Comparison of coalescence frequencies with theory: (—, ●) Coulaloglou and Tavlarides (C13) (theory, experiment), $\phi = 0.10$; (▽) Groothuis and Zuiderweg (G10), $\phi = 0.08$; (○) Ross (R15), $\phi = 0.10$; (□) Howarth (H16), $\phi = 0.10$; and (×) Miller et al. (M14) [after Coulaloglou and Tavlarides (C13)].

This work is the first to show experimentally the behavior of such systems and to propose phenomenological models to describe the coalescence and breakage processes. The system was modeled in the framework of the drop number over volume population balance equation. Depending on the viscosities of both phases, exceedingly long times are necessary to reach a steady state. This result may be partially attributed to the coalescence phenomena. Figure 11 indicates that at least 15 hours are needed to reach a stabilized drop size for a batch system subjected to a step change in power input. Other interesting results observed were that a maximum or minimum in drop size occurs, depending on power input, as outer fluid viscosity is increased. Also, as shown in Fig. 12, substantially different size distributions occur depending on viscosity at uniform power input.

Ramkrishna (R1) presented the number over droplet volume form of

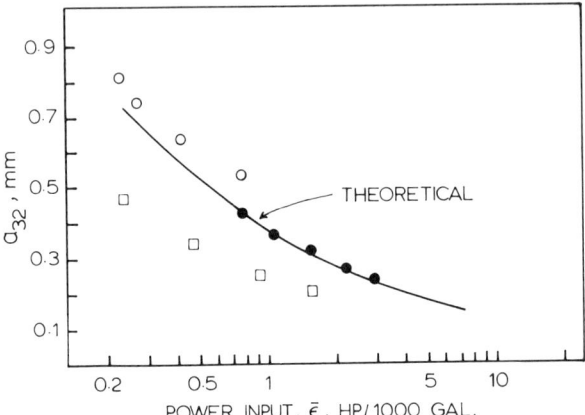

FIG. 10. Comparison of Sauter mean diameter versus power input with theory: (—, ●) Coulaloglou and Tavlarides (C12) (theory, experiment) $\phi = 0.10$, continuous system; (○) Ross (R15), $\phi = 0.10$, continuous system; and (□) Mlynek and Resnick (M19), $\phi = 0.10$, batch system [after Coulaloglou and Tavlarides (C12)].

drop population balance equation in terms of a cumulative distribution function to analyze drop breakage data of Madden and McCoy (M7). Under the model assumptions, evaluation of coefficients is possible for a power law model of the breakage frequency function. It should be noted that the data of Madden and McCoy (M7) are suspect due to the technique employed for drop measurements. Bajpai et al. (B2) developed a coalescence redispersion model for drop size distributions in an agitated vessel.

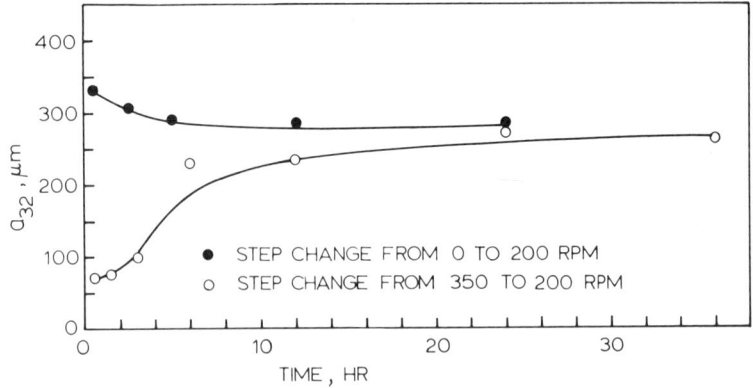

FIG. 11. Transient behavior of mineral oil dispersed in aqueous glycerol following step changes in impeller speeds; $\mu_c = 27.3$ cp, $\mu_d = 26.4$ cp, $\sigma = 33.0$ dynes/cm, $\phi = 0.05$, $\omega = 0.21$ hr^{-1} [after Stamatoudis (S32)].

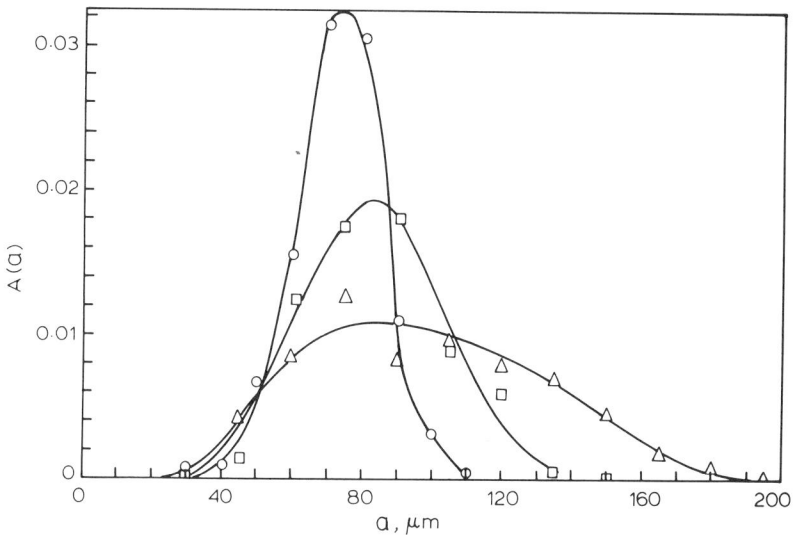

FIG. 12. Drop size distributions of dispersions of mineral oil in aqueous glycerol at various continuous-phase viscosities; $N^* = 300$ rpm, $\phi = 0.05$, $\sigma = 31.0$–35.7 dynes/cm, $\mu_d = 26.4$–26.7 cp, ($\triangle$) $\mu_c = 3.5$ cp, ($\square$) $\mu_c = 27.3$ cp, and ($\bigcirc$) $\mu_c = 53.8$ cp [after Stamatoudis (S32)].

The model assumes that drop coalescence is followed by immediate redispersion into two drops sized according to a uniform distribution. By assuming that the coalescence frequency is independent of drop size, the solution of the resulting form of Eq. (107) is exponential for the equilibrium drop volume distribution. Comparison of the distribution to experimental data is favorable. The analysis is useful in that a functional form for the distribution is obtained. The attendant simplifications necessary for solution, however, do not permit more rational forms of the interaction frequency of droplet pairs in order to account for the physical processes which lead to droplet coalescence and breakage as discussed in Section III. A similar work was presented by Inone et al. (I1).

Delichatsios and Probstein (D4–7) have analyzed the processes of drop breakup and coagulation/coalescence in isotropic turbulent dispersions. Models were developed for breakup and coalescence rates based on turbulence theory as discussed in Section III and were formulated in terms of Eq. (107). They applied these results in an attempt to show that the increase of drop sizes with holdup fraction in agitated dispersions cannot be attributed entirely to turbulence dampening caused by the dispersed phase. These conclusions are determined after an approximate analysis of the population balance equation, assuming the size distribution is approximately Gaussian.

Although it is recognized that the flow field in a turbulently agitated vessel is far from homogeneous, the aforementioned researchers assumed that spatial homogeneity of dispersion properties was adequate for modeling purposes. Depending on the liquid–liquid system under investigation, this assumption is reasonable if the circulation time is short relative to the coalescence time scale. Various investigators reported local variation of interfacial areas and drop sizes (M19, P4, S11, S29, V12, W3). Sprow (S29) has shown that for strongly coalescing systems, such as methyl isobutyl ketone (MIBK)–saltwater, the drop sizes are strongly dependent on the sampling position, being smaller near the impeller tip and larger at the bottom and top of the mixing tank. Results indicate that drop breakup predominates near the impeller, whereas coalescence can be controlling in other locations. These observations are reasonable when one recalls the nonhomogeneous energy dissipation throughout a turbulently agitated vessel (see Section II) and the importance of ϵ on the droplet interaction processes (see Section III). The work of Park and Blair (P4) is similar. Here droplets were observed in various vessel regions via a fiber optic probe coupled to high-speed cinephotography. The authors measured coalescence rates at various positions in the vessel directly for the MIBK–water system and concurred with the results of Sprow (S29). The variation in drop size distribution with position was modeled with the spatial dependent form of the population balance equation (107), where a one-dimensional circulation path model was employed.

It is evident from these discussions that population balance equations are important in the description of dispersed-phase systems. However, they are still of limited use because of difficulties in obtaining solutions. In addition to the numerical approaches, solution of the scalar problem has been via the generation of moment equations directly from the population balance equation (H2, H17, R6, S23, S24). This approach has limitations. Ramkrishna and co-workers (H2, R2, R6) presented solutions of the population balance equation using the method of weighted residuals. Trial functions used were problem-specific polynomials generated by the Gram–Schmidt orthogonalization process. Their approach shows promise for future applications.

It is encouraging that substantial progress has been made in analyzing the hydrodynamics of droplet interactions in dispersions from fundamental considerations. Effects of flow field, viscosity, holdup fraction, and interfacial surface tension are somewhat delineated. With appropriate models of coalescence and breakage functions coupled with the drop population balance equations, '*a priori*' prediction of dynamics and steady behavior of liquid–liquid dispersions should be possible. Presently, one universal model is not available. The droplet interaction processes (and thus the models) change with the particular hydrodynamic environment and physical properties. Effects of other important factors on dispersion properties such as interphase mass transfer, temperature, and surfactants are not adequately understood.

3. *Analysis of Mass Transfer or Reaction in Dispersions with the Population Balance Equation*

It was alluded to previously that droplet size distribution and droplet mixing can have significant effects on the extent of extraction and reaction. Consider first completely segregated drops with mass transfer out of the drops. If the drops were completely uniform they would be depleted of the extractant simultaneously. However, for nonuniform distribution the large drops would take a longer time to deplete than small drops due to their lower ratio of surface area to volume (assume that the mass-transfer mechanism is approximately the same for all sizes). Thus total extraction time is increased. The nature of the size distribution is important in determining total extraction time. Obviously, coalescence and redispersion reduce this time. If chemical reactions are occurring in the drops, then it is well known for non-first-order reaction that the droplet mixing processes affect the extent of reaction. Greater extent is realized for less than first order with increase in mixing. The reverse holds for greater than first order. In addition, selectivity is affected by mixing for complex reactions. An excellent summary of efforts to analyze dispersed-phase mixing up until 1964 is given by Rietema (R12). Recent attempts to account in part for dispersed-phase mixing and dispersed-phase size distributions using the population balance equation techniques are discussed below.

A simplified homogeneous dispersed-phase mixing model was proposed by Curl (C16). Uniform drops are assumed, coalescence occurs at random and redispersion occurs immediately to yield equal-size drops of the same concentration, and the dispersion is assumed to be homogeneous. Irreversible reaction of general order s was assumed to occur in the drops. The population balance equations of total number over species concentration in the drop were derived for the discrete and continuous cases for a continuous-flow well-mixed vessel. The population balance equation could be obtained from Eq. (102) by taking the internal coordinate to be drop concentration and writing the population balance equation in terms of number to yield

$$(\omega_r)^{-1}[\partial n(c)/\partial t]$$
$$= n_0(c) - n(c) + I^* \left[4 \int_0^c n(c+\alpha)n(c-\alpha)\,d\alpha - n(c) \right]$$
$$+ K_r\{\partial[csn(c)]/\partial c\} \qquad (108)$$

Where K_r is the reaction modulus, $k'/(\omega_r c_0)$; ω_r is the residence frequency; I^* is the dispersed-phase mixing modulus ω_i/ω_r; ω_i is the dispersed-phase mixing frequency; and $n(c)\,dc$ is the fraction of drops of concentration in the range c to $c + dc$.

Figure 13 shows the effect of dispersed-phase mixing on the conversion for a zero-order reaction in drops (which approximates diffusion-controlled reactions). It is seen that at 80% conversion, mixing of the dispersed phase reduces the reaction volume required by nearly a factor of three. Shain (S18) extended the work with a more efficient computation algorithm. He studied drop interactions on a second-order dimerization reaction and selectivity for two competing parallel reactions occurring in the dispersed phase. Figure 14 shows conversion for the second-order reaction as affected by the reaction modulus and interaction frequency. These results indicate that for high levels of conversion, if one assumes completely mixed dispersed phase, the reactor may be overdesigned by a factor of 2–4 times larger at 80–90% conversion levels.

Hulbert and Akiyama (H18) analyzed the population balance equation to determine coalescence and redispersion effects on extent of conversion, deriving a set of equations for the moments of the distribution. The results agreed with those of Shain (S18). The authors claim the moment method reduces computation effort although there is a loss of information on the distribution.

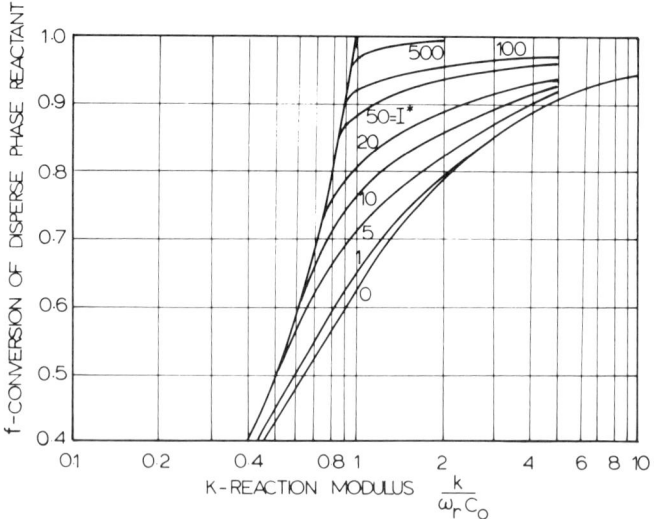

FIG. 13. Effect of dispersed-phase mixing on the conversion for a zero-order reaction in the drops in a single-stage, two-phase, stirred reactor [after Curl (C16)].

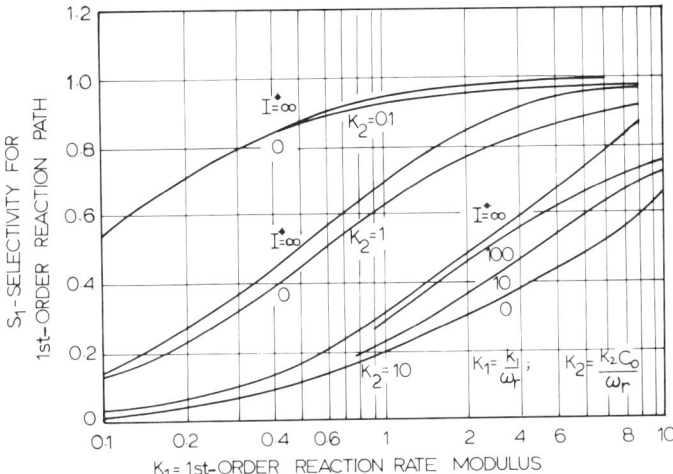

FIG. 14. Effect of dispersed-phase mixing on selectivity for competing reaction paths (A → B first order, 2A → C second order) in the drops in a single-stage, two-phase, stirred reactor [after Shain (S18)].

Evangelista, Katz, and Shinnar (E4) applied the mixing model proposed by Curl (C16) to develop scale-up criteria for stirred tank reactors in which homogeneous fluids undergo reaction. The effects of finite mixing on complex chemical reactions are computed. Turbulence theory is applied to estimate values of the mixing parameters. The model is generalized to include multiple reactions and methods are derived to compute average outlet concentrations for complex systems. A method of moment is employed with the restriction that the rate expressions are polynomials.

Ming-heng et al. (M15) employ a model for liquid–liquid dispersions similar to those mentioned previously. They analyzed droplet mixing on conversion for a second-order reaction occurring in the dispersed phase. The solution was obtained using the method of moment equations.

Valentas and Amundson (V3) studied the performance of continuous flow dispersed phase reactors as affected by droplet breakage processes and size distribution of the droplets. Various reaction cases with and without mass transfer were studied for both completely mixed or completely segregated dispersed phase. Droplet size distribution is shown to have a considerable effect on the efficiency of a segregated reaction system. They indicated that polydispersed drop populations require a larger reactor volume to obtain the same conversion as a monodispersed system for zero-order (or mass-transfer-controlled) reactions in higher conversion regions. As the dispersed phase becomes completely mixed, the distribution of droplet sizes becomes less important. These interactions are underscored in this work. However, the work was limited to breakage only and only the extremes of droplet mixing were analyzed.

Other researchers have attempted to account for mass transfer in polydispersed systems. Baynes and Laurence (B10) present an analysis of mass transfer between two liquid phases for the case of a homogeneous, batch-stirred tank with coalescence and breakup. A laminar-flow spray column in which droplets undergo coalescence only was also examined. Equation (101) was developed for the internal coordinates of drop mass, solute, and age. The resulting expression was cast in terms of moment generating functions. After simplifications, solutions were presented comparing solute concentrations in the dispersed phase with and without droplet interactions. Shah and Ramkrishna (R4, S17) present an analysis of mass transfer in lean liquid–liquid dispersions undergoing breakage only. These authors employ the spatially homogeneous population balance equation (102) in terms of the trivariate droplet number concentration over droplet size, mean solute concentration, and age. Mass transfer in single drops is assumed to be by diffusion, and droplet breakage is assumed to be binary and even. Conditional concentration densities were calculated for various values of hold times and several breakage frequency functions, $g(a)$. The results indicate, as shown by others, that the dynamics of droplet breakage phenomena play an important role in the analysis of mass transfer in dispersions.

It should be pointed out that in the above work one has the problem of defining the average concentration and the rate of change of the average concentration of the solute in the drop. To this end one must solve the conservation of mass equation for the drop for the initial and boundary conditions to which each drop is exposed. The aforementioned workers (B10, S7, S17) considered the case of unsteady diffusion. One can see the possibilities of extending the analysis of mass transfer to account for different modes of mass transfer depending on droplet size. For large droplets, one can employ various models to account for convective effects due to internal circulation within drops. Thus the mass-transfer mechanism can be more accurately described over the various drop size ranges. In theory this approach is all-inclusive. In practice, however, significant computational difficulties can develop when the conservation equations do not have an analytical solution.

An example of the use of the population balance method to predict reaction in particulate systems is presented in the work of Min and Ray (M16, M17). The authors developed a computational algorithm for a batch emulsion polymerization reactor. The model combines general balances, individual particle balances, and particle size distribution balances. The individual particle balances were formulated using the population balance equation based on Eq. (102) with attendant simplifications. Various emulsion properties could be predicted such as total particle size distribution, free radical distribution, monomer and initiator concentrations and conversion. Application of the model to experimental data appears favorable. The authors indicate that computational times were not excessive. Population balance techniques have also found great applications to crystallization modeling (R5, R6).

The work discussed in this section clearly delineates the role of droplet size distribution and coalescence and breakage phenomena in mass transfer with reaction. The population balance equations are shown to be applicable to these problems. However, as the models attempt to be more inclusive, meaningful solutions through these formulations become more elusive. For example, no work exists employing the population balance equations which accounts for the simultaneous affects of coalescence and breakage and size distribution on solute depletion in the dispersed phase when mass transfer accompanied by second-order reaction occurs in a continuous-flow vessel. Nevertheless, the population balance equation approach provides a rational framework to permit analysis of the importance of these individual phenomena.

It is observed, even in the partial solution to the problem, that realistic models of the droplet coalescence and breakage processes as discussed in Section V,D,2 have yet to be employed. A parallel development has occurred. The work is currently at the point where the realistic model of the droplet dynamics can be applied to the pertinent problems of extent of reaction and solute depletion in dispersions. The success of this effort would permit the researcher and designer to predict dispersed-phase reactor performance from fundamental properties of the dispersion, operation conditions of the vessel, and knowledge of the intrinsic kinetics.

E. TYPE II INTERACTION MODELS: MONTE CARLO SIMULATION MODELS

The deterministic population balance equations governing the description of mass transfer with reaction in liquid–liquid dispersions present a framework for analysis. However, significant difficulties exist in obtaining solutions for realistic problems. No analytical solutions are available for even the simplest cases of interest. Extension of the solution to multiple reactants for uniform drops is possible using a method of moments but the solution is limited to rate equations which are polynomials (E3). Solutions to the population balance equations for spatially nonhomogeneous dispersions were only treated for nonreacting dispersions (P4), and only a simple case was solved for a spray column (B19). Treatment of unmixed feeds presents a problem.

Statistical Monte Carlo coalescence redispersion models were developed to bypass the aforementioned difficulties. These statistical models attempt to account for the macromixing and micromixing phenomena. Generally a large number of drops or fluid elements are specified in a reactor, drops are added and removed to simulate macroflow patterns either in a random or prescribed fashion. More detailed models have particles follow the flow field. Micromixing of the dispersed phase is typically simulated by choosing two particles at random from the entire population or from a local population. These particles then coalesce, whereupon their intensive properties are assumed to be averaged instantaneously. The resulting drop may be immediately redispersed, yielding two particles. Separate coalescence and breakage regions may be used. An interaction or coalescence frequency parameter is employed which is assumed to be related to physical parameters and operation conditions. This interaction parameter is a key parameter of these models. Between the coalescence-redispersion and flow steps each drop is assumed to behave as a segregated drop in which reaction occurs with or without mass transfer to the surrounding fluid. Energy exchange can also be included. Average concentrations, conversions, and temperatures of the dispersed and continuous phases can be calculated at any time during the simulation for the entire vessel or the exit stream by summing over all particles and taking mass and energy balances.

Spielman and Levenspiel (S28) first applied such a model to study effects of coalescence and redispersion on the progress of zero- and second-order reactions which occur in the dispersed phase. The model of the coalescence-redispersion process defined by Curl (C16) was used. The dispersed phase consists of a large number of equally sized drops, drops coalesce two at a time and redisperse instantly to form two identical droplets, droplets coalesce at a constant rate with equal probability of coalescence for each drop, concentration is uniform through each droplet, and the continuous phase is homogeneous and chemically inert. Flow is simulated by removing droplets and replacing them in the grid with fresh feed droplets. The authors define a dimensionless interaction parameter I' such that each drop coalesces I' times during its stay in the reactor. Here

$$I' = 2u^*\theta/N_T$$

where u^* is the number of coalescences occurring in the reactor per unit time, N_T is the total number of drops in the reactor, and θ is the mean residence time in the reactor. Performance charts were prepared. The model presents a simplified concept of the mixing processes and gives ranges of the extent of conversion expected for a given level of droplet mixing. However, the interaction parameter was not related to the operating conditions and physical properties.

Komasawa *et al.* (K14) employed the model of Spielman and Levenspiel (S28) to measure interaction rates by comparing the reaction conversions measured with those calculated. The result agreed within experimental error with those values measured by a physical method.

Luss and Amundson (L13) employed this model to analyze reactor stability and control for segregated two-phase systems. The Monte Carlo simulation was employed to model the age distribution of segregated drops in the vessel. Conditions of operation under which heat-transfer effects may control the design of the reactor were given. It was shown that some steady states may be obtained in which the temperature of some drops greatly exceeds the average dispersed-phase temperature. The coalescence-redispersion problem was not considered here because of unreasonable computation times.

Kattan and Adler (K3) presented a stochastic mixing model for homogeneous turbulent tubular reactors. The model was developed for reactions in a homogeneous phase and is based on random coalescence and redispersion of uniformly sized fluid lumps. It has direct applicability to dispersions. The authors define a mixing parameter $\eta(z)$ which is a function of reactor length. The parameter $\eta(z)$ can be estimated from the reaction data of Vassilatos and Toor (V8). Rao and Dunn (R8) extended these concepts to simulate mixing and segregation effects on extent of reaction in a tubular chemical reactor. The model includes axial dispersion on the plug-flow mean velocity in addition to the micromixing details, which are determined by the local coalescence-redispersion of droplets. These models (K3, R8) are similar to those of Curl (C16) and Spielman and Levenspiel (S28). Coalescence-redispersion of a droplet is limited to those adjacent in radial or axial positions. An interaction parameter, constant over the reactor length, similar to that of Spielman and Levenspiel (S28) was defined. The authors left unanswered the question of how to select the model parameters from information on the physical turbulent mixing process.

Zeitlin and Tavlarides (Z2–5) developed a simulation model which attempts to account for the macroflow patterns of the dispersed phase in a turbine-agitated vessel, the droplet mixing via breakage or coalescence, and nonuniform drop size on mass transfer or reaction in dispersions.

Batch, semibatch, or continuous-flow operation can be simulated. The continuous phase is assumed well mixed. Particle movement was either random or followed the flow direction of the sum of the local average fluid velocity and the particle gross terminal velocity. The probability of droplet breakup is assigned based on droplet size. Binary breakage was assumed to form two randomly sized particles whose masses equal the parent drop. The probability of coalescence exists when two drops enter the same grid location. Particles are added and removed to simulate flow.

The drops behave as segregated entities between flow and coalescence-redispersion simulation. The coalescence and breakage frequencies can be varied with vessel position. The computational time was related to coalescence frequency data available in the literature. Figure 15 shows the steady-state dimensionless droplet number size distribution as a function of rotational speed for continuous-flow operation. As expected the model predicts smaller droplet sizes and less variation of the size distribution with increase in rotational speed. Figure 16 is a comparison of the droplet number size distribution with drop size data of Schindler and Treybal (S11).

The model can be employed to predict the effects of droplet size distribution and droplet coalescence-redispersion on conversion and selectivity for reacting dispersions. The reactions can occur in either phase simultaneously with interphase heat and mass transfer.

The utility of the model to predict the effects of interdroplet mixing on extent of reaction was demonstrated for the case of a solute diffusing from the dispersed phase and undergoing second-order reaction in the continuous phase. For this comparison the normalized volumetric dispersed-phase concentration distribution is defined as $f_v(y) \, dy$ equal to the fraction of the total volume of the dispersed phase with dimensionless concentration in the range y to $y + dy$, where $y = c/c_0$ and

$$\int_0^1 f_v(y) \, dy = 1 \tag{109}$$

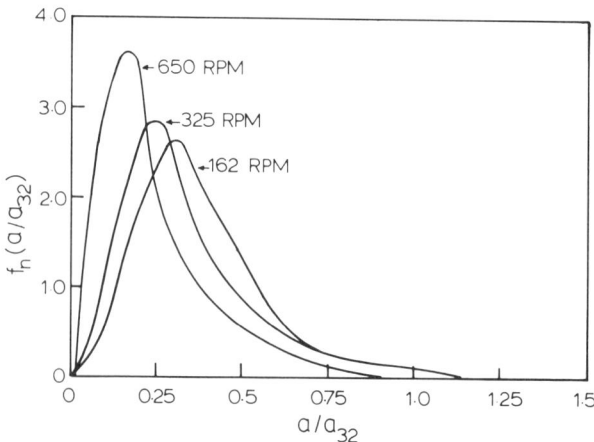

Fig. 15. Steady-state number size distribution versus rpm from Monte Carlo simulation, batch system, ethyl acetate (d)–water (c), $\phi = 0.041$ [after Zeitlin and Tavlarides (Z2)].

Figure 17 compares the dispersed-phase concentration frequency function for dispersed phase of complete segregation, intermediate mixing, and pseudo-complete mixing. Multiple peaks for the frequency function exist for segregated systems due to slow depletion of extractant from large drops in the system. The model permits the incorporation of both size distribution and droplet mixing effects on calculations of conversions. In addition, the simulation permits calculation of reaction time and agitation necessary to insure that all drops have reached a certain concentration range. The model was also employed to follow temperature rises in drops (Z1) from which solute is extracted and undergoes second-order reaction in a continuous-phase film surrounding the drop. The energy was distributed between the phases. The results show that significant temperature rises of some drops are possible over the average dispersed-phase temperature depending upon droplet size and coalescence frequency. These are consistent with other investigators (L13, Y2).

The simulation technique was employed to model hydrocarbon fermentation (S33). The effect of hydrocarbon drop size distribution and droplet interaction on microbial growth rate for pure and two-component feed were delineated.

Jakubowski and Sideman (J1) extended the model of Zeitlin and Tav-

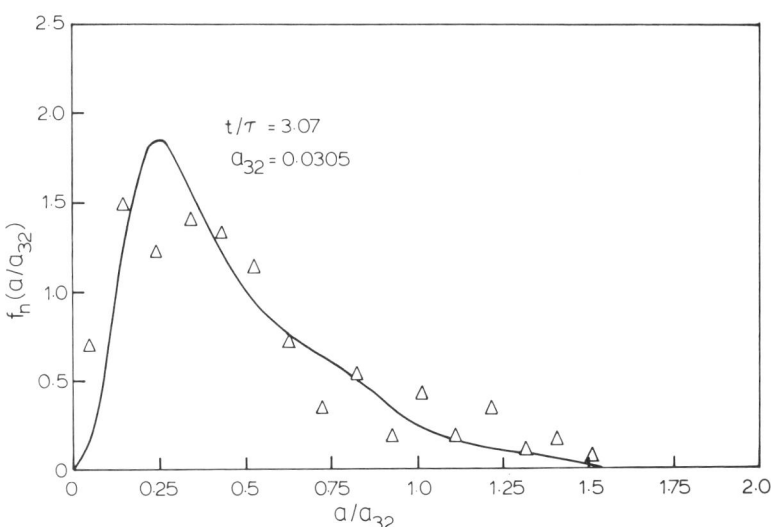

FIG. 16. Comparison of number size distribution from Monte Carlo simulation with data of Schindler and Treybal (S11), ethyl acetate (d)–water (c), continuous-flow, stirred vessel, $\phi = 0.041$, $N^* = 325$ rpm, $a_{32} = 0.313$ mm (experiment) versus 0.305 (simulation) [after Zeitlin and Tavlarides (Z2)].

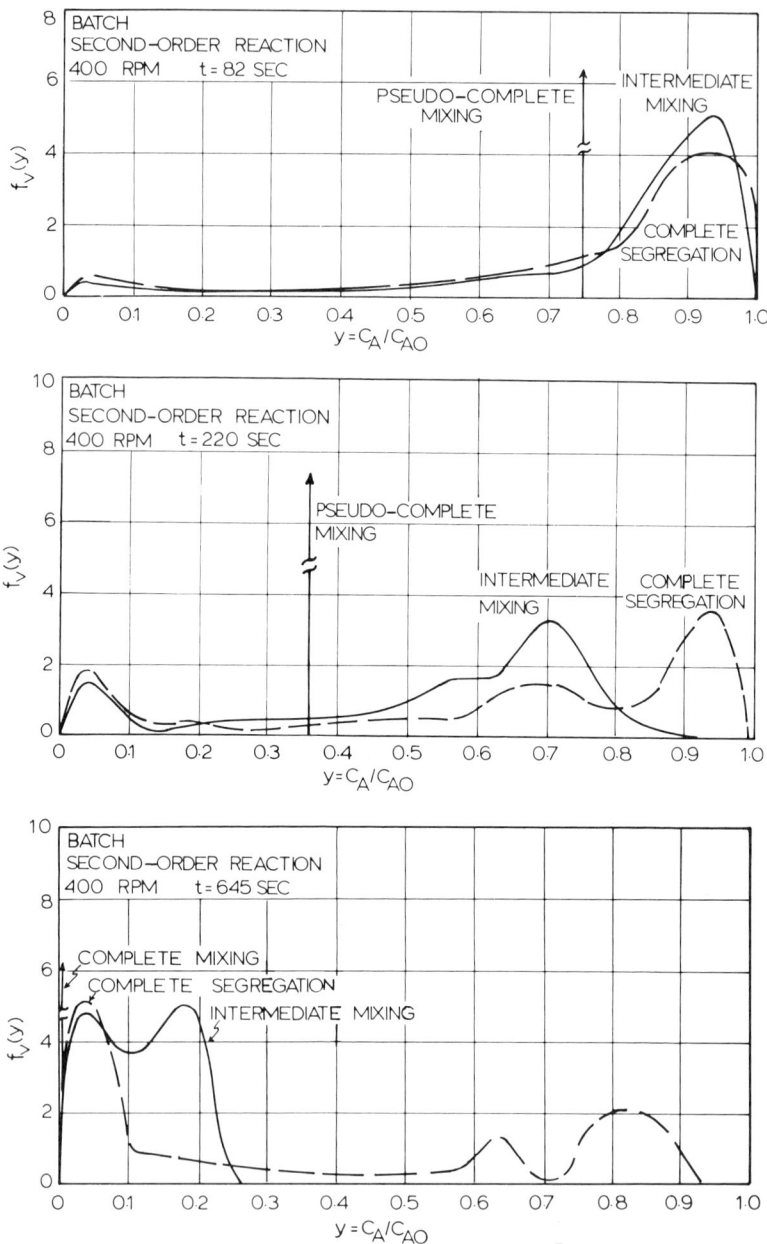

FIG. 17. Comparison of volumetric dispersed-phase concentration distribution for various levels of droplet mixing for second-order reaction in the continuous phase, $\phi = 0.006$ [after Zeitlin and Tavlarides (Z3)].

larides (Z2–5) for the analysis of the hydrodynamics of two- and three-phase agitated systems. The extension of the model to three-phase, liquid–liquid–solid systems enables the hydrodynamics of crystallization in dispersions to be described, while accounting for the kinetics of nucleation and crystal growth.

These simulation methods appear to offer substantial advantages for describing reactions in dispersions over models involving solution of the population balance equations or macro- and micromixing concepts. With simulation models, complex multicomponent reactions can be followed. Effects of droplet coalescence and redispersion for nonuniform drops on transport processes can be accounted for, and effects of unmixed feeds on conversion and selectivity for complex reactions present no problem. Use of the models to predict scale-up of stirred vessels is not yet possible until the macroflow patterns and the local turbulence energy dissipation can be estimated from vessel geometry, operation conditions, and physical properties. Local coalescence and breakage frequencies can then be calculated from available models (C12, D4, D5, R16, S32). These models have yet to incorporate effects of nonideal mixing in the continuous phase.

F. Type III Interaction Models: Use of Macromixing and Micromixing Concepts

There are various models developed to account for effects of droplet coalescence and redispersion on conversion and selectivity in reacting dispersions which directly employ the macromixing and micromixing concepts of Danckwerts (D2) and Zweitering (Z6). These workers described the extreme conditions of complete segregation and maximum mixing for chemical reactors having arbitrary residence time distributions. Considerable attention was devoted to characterizing the intermediate states of mixing and their relation to chemical conversion. One inherent difficulty of these micromixing studies is relating any particular model parameter to the flow patterns existing in the reactor vessel of interest. This approach, accordingly, has more application to establish whether micromixing is of importance in determining the outcome of various reaction types as opposed to describing an actual reactor. In this section various coalescence-redispersion models based on the aforementioned concepts are discussed.

Harada and co-workers (H1) developed two coalescence-redispersion models to describe micromixing in a continuous-flow reactor. In the first model, the incoming dispersed-phase fluid is assumed to consist of uniformly sized droplets. These droplets undergo 0 to n coalescences and redispersions with surrounding droplets of a constant average concentration. The fraction of the droplets in the exit which have encountered a fixed number of coalescences and redispersions are related to the residence time frequency function. A concentration is prescribed for each one of these fractions based on the reaction kinetics. Integration over all fractions yields the average concentration of the effluent. The model in the limits agrees with expected results for simple reaction schemes. However, the model assumptions are less realistic than those developed by Curl (C16) with population balance techniques or Spielman and Levenspiel (S28) with simulation methods.

The second model (H1) is less involved. It has received some use (Y2) because of its simplicity. A representative drop of the dispersed phase is assumed to interact with an ensemble of other drops at a certain frequency defined by an interaction coefficient. The equation governing the temporal variation of concentration in the drop is given by

$$-dc(t)/dt = r(c, t) + (2t_c^*)^{-1}[c(t) - \bar{c}_0] \qquad (110)$$

where $r(c, t)$ is the rate of reaction, t_c^* is the average time interval between interactions, and $\bar{c}_0$ is the average concentration in all drops in the reactor, where $\bar{c}_0$ for a perfectly mixed continuous reactor is the mean value over all ages. The average concentration of reactant in the dispersed phase flowing out of the perfect macromixed reactor is evaluated by integrating the solution of Eq. (110) over the residence frequency function of the drops in the vessel. Experiments were conducted and indicated that the model could predict mixing effects on conversion. This model was employed by Yamazaki and Ichikawa (Y2) to model the effect of coalescence and redispersion on the stability of the temperature control in chemical reactors in which liquid–liquid dispersions are reacting. Villermaux and co-workers (A7, V12) also employed this model. They defined a fictitious transfer coefficient h, equal to $\frac{1}{2}t_c^*$. This parameter was related to the intensity of interaction as proposed by Curl (C16) and Spielman and Levenspiel (S28) for dispersed phases and other workers (E4, K4, T10) for continuous phases. The simplicity of the model permitted application to a variety of cases for reactions in liquid–liquid dispersions. The model was also applied to predict incomplete mixing for homogeneous systems (A7).

Erickson *et al.* (E3) developed a model for batch growth in fermentations with two liquid phases present in which the growth-limiting substrate is dissolved in the dispersed phase. The model accounts for drop size distribution and considers the effect of droplet coalescence and redispersion by an interaction model similar to that of Eq. (110). Droplet interactions were shown to be important if drop size distributions have large variance.

A conceptual framework for describing mixing in continuous-flow chemical reactors was developed by Kattan and Adler (K4) and adapted by Treleaven and Tobgy (T10) to model unmixed feeds whose residence time distributions differ. The conceptual framework (Fig. 18) orders the fluid within the reactor according to the residual lifetime as suggested by Zwietering (Z6). The two feed streams are each represented by a large number of equally sized elements and are assigned residual lifetimes in accordance with theoretically or experimentally determined residence time functions $f_1(t)$ and $f_2(t)$. The reactors fluid is also divided into a large number of uniform fluid elements. Elements within the reactor have a residual lifetime associated with it. The residual lifetime distribution of the elements in the reactor, according to Treleaven and Tobgy, are formed from the residual time function of two feed streams so that

$$\psi(\lambda) = [V_1/(V_1 + V_2)]\psi_1(\lambda) + [V_2/(V_1 + V_2)]\psi_2(\lambda)$$

where

$$\psi_j = (\theta_j)^{-1}[1 - \int_0^\lambda f_j(t')\,dt'], \quad j = 1, 2$$

and where θ_j is the average residence time of the stream j and V_j is the reactor volume due to flow from stream j. Macromixing is accounted for

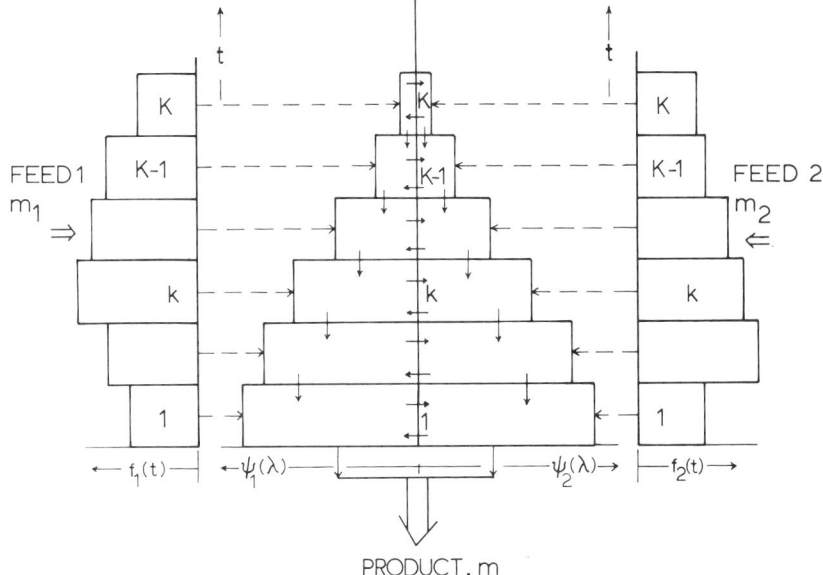

FIG. 18. Diagrammatic representation of stochastic flow simulation (horizontal arrows across center line represent micromixing only) [after Treleaven and Tobgy (T10)].

by grouping fluid elements of the same residual lifetime from each feed stream with reactor fluid elements of the same residual lifetimes. The number of fluid elements of a given residual lifetime is maintained constant by removing an equal number as those which enter a residual lifetime increment. Micromixing is accomplished by prescribing coalescence and redispersion between pairs of fluid elements which have the same residual lifetime. This operation averages the intensive variables of the pair of elements instantaneously. Between collisions fluid elements act as segregated fluid elements in which the chemical species change according to the controlling kinetic expressions. Kattan and Adler developed a deterministic formulation of the model which was shown to be a general case of Curl's single continuous flow stirred tank model. Both workers conducted calculations to quantify micromixing phenomena on conversion employing a Monte Carlo formulation of the conceptual model. Treleaven and Tobgy (T11) measured chemical conversions for a confined jet reactor, for which residence time data have previously been determined. The previously described model was employed to correlate the experiments. The authors suggest that the model may be adaptable to design and operation of industrial reactors.

VI. Conclusions

From fundamental considerations substantial progress has been made in analyzing the hydrodynamics of droplet interactions in dispersions. Effects of flow field, viscosity, holdup fraction, and interfacial surface tension are somewhat delineated. Use of population balance equations with appropriate coalescence and breakage rate functions can permit '*a priori*' prediction of dynamic and steady-state behavior of physically equilibrated liquid–liquid dispersions. Results are confined to particular ranges of operating conditions and physical parameters. Conversely, it is seen that population balance equations present the better modeling approach for extracting information about the coalescence and breakage processes. Further research is needed to account for the effects of mass transfer, high holdup fraction (>20%), and surfactants on droplet coalescence and breakage rates. Experimental data on drop size distributions and coalescence frequencies would be valuable for high-viscosity continuous and dispersed phases over various operation conditions and impeller designs. Limited data exist for size distributions and mixing frequencies for the case of aqueous drops dispersed in organics. Models for predicting behavior of these systems are not tested. Similarly, limited data exist for liquid–liquid systems with large density differences.

Various models were discussed to account for dispersed phase mixing on conversion and selectivity for mass transfer with complex reactions occurring in liquid–liquid dispersions. The researcher must determine what level of sophistication of the modeling is needed. Presently there are few guidelines which clearly delineate the range of operation conditions wherein dispersed-phase mixing processes may be important on conversion and selectivity. The deterministic population balance equations present significant difficulties in obtaining solutions for realistic problems. Simulation methods appear to offer substantial advantages. However, use of the simulation models to predict scale-up of stirred vessels is not possible until macroflow patterns and the local turbulence energy dissipation can be correlated with vessel geometry, operating conditions, and physical properties. Then local coalescence and breakage frequencies could be calculated from available models. In addition, there is need to incorporate effects of nonideal mixing in the continuous phase. Extensions of the simulation models to account for and to delineate the importance of this nonideal mixing effect on reacting liquid–liquid dispersions appears possible and needed.

Nomenclature

a, a' Drop diameter
a_c Characteristic diameter related to the maximum of a distribution function
a^* Effective interfacial area per unit volume of dispersion
a_{max} Maximum drop diameter
a_{min} Minimum drop diameter
a_{21} Surface to size average drop diameter
a_{32} Sauter mean diameter, Eq. (89)
$A(a')da'$ Fraction of drops with diameters between a' and $a' + da'$
$A(h_0)$ Constant, Eq. (20)
AB Initial separation distance
$A'B'(t)$ Separation distance of lumps of fluids, Eq. (39)
c_i Constants
c_0 Concentration of feed droplets
$\bar{c}_0$ Average concentration over all drops
c_r Ratio of variance, Eq. (99)

c_1 Dye concentration in the dye-containing feed stream
C_A^* Solubility of A in the continuous phase
D_I Impeller diameter
D_T Vessel diameter
E_a Adhesion energy, Eq. (20)
E_c Minimum surface energy, Eq. (36)
$E(k)$ Three-dimensional energy spectrum function
E_s Shear energy
E_T Kinetic turbulent energy, Eq. (30)
$f(h_0, h_c)$ Defined by Eq. (70)
$f(t)$ Residence time distribution function
$f_v(y)dy$ Fraction of total dispersed-phase volume with dimensionless concentration in the range y, $y + dy$, Eq. (109)
F Adhesion force, Eq. (23)
$F(a, a')\, da\, da'$ Drop coalescence rate

$g(a')$	Breakage frequency of drop of size a'
g_c	Newton's law-conversion factor
G	Shear rate
G_{eff}	Effective shear rate
h	Fictitious transfer coefficient
$h(a, a')$	Collision frequency between drops of sizes a and a' for a binary collision process based on number concentration
$h^+(\mathbf{x}; t)d\mathbf{x}$	Number of droplets produced per unit volume of dispersion per unit time at time t in the incremental range $\mathbf{x}, \mathbf{x} + d\mathbf{x}$ in phase space
$h^-(\mathbf{x}; t)d\mathbf{x}$	Number of droplets destroyed per unit volume of dispersion per unit time at time t in the incremental range $\mathbf{x}, \mathbf{x} + d\mathbf{x}$ in phase space
I	Amount of light transmitted
I^*	Dispersed-phase mixing modulus, ω_i/ω_r
I'	Dimensionless interaction parameter
I_0	Amount of incident light
$\bar{I}_0, \bar{I}_a, \bar{I}_b$	Average steady-state light transmittance for coalescence frequency tests in Eq. (98)
$I(0), I(t), I(\infty)$	Light transmittance values for times 0, t, and ∞, respectively
k	Wave number
k'	Reaction rate constant
k_d	Kolmogoroff wave number
k_e	Wave number associated with the size of energy-containing eddies
K_i	Constants
K_L	Continuous-phase mass-transfer coefficient
K'	Constant, Eq. (12)
K''	Rate constant for drop decomposition process, Eq. (27)
K^*	Equilibrium constant, Eq. (28)
K_r	Reaction modulus, Eq. (108)
l	Length of optical path
l_e	Linear scale of energy-containing eddies
N	Total number of drops present per unit volume of dispersion
N_T	Total number of drops in the reactor
$n(\mathbf{x}; t)d\mathbf{x}$	Number of particles per unit volume of dispersion at time t in incremental range $\mathbf{x}, \mathbf{x} + d\mathbf{x}$ in phase space
N^*	Rotational speed of impeller
P^*	General function, Eq. (105)
P	Impeller power
$P(E)$	Distribution function for kinetic energy, Eq. (37)
$P(u)$	Probability distribution function for local velocity, Eq. (44)
$P(u, \dot{u})$	Joint probability distribution function of the local velocity and acceleration, Eq. (44)
$P(\dot{u}/u_b)$	Conditional probability of local acceleration when the corresponding velocity difference is u_b
Q_k	Volumetric flow rate of stream k
r	Distance coordinate
$r(a')da'$	Number of drops of diameter between a' and $a' + da'$ breaking per unit volume of dispersion per unit time, Eq. (25)
$r(a', a)da\, da'$	Number of daughter drops of diameter between a and $a + da$ produced by breakage of drops of diameter a' and $a' + da'$ per unit volume of dispersion per unit time
$r(c, t)$	Rate of reaction

R	Rate of reaction per unit area		two colliding drops for coalescence to occur, Eq. (66)
s	Standard deviation		
t	Contact time two drops stay together	$\mathbf{x}$	General coordinate vector in particle phase space
t_b	Breakage time, Eq. (35)	$\dot{\mathbf{x}}$	Coordinate velocity in phase space
t_{br}	Time during which a drop is subjected to shear	$\mathbf{x}_e$	External coordinate vector in particle phase space
t_c	Circulation time	$\mathbf{x}_i$	Internal coordinate vector in particle phase space
t_c^*	Average time interval between interactions	X_β	Internal coordinate velocity of component β
$T(k)$	Scalar associated with the transfer of energy between wave numbers or eddy sizes	y	Dimensionless concentration, c/c_0
u	Velocity parameter characteristic of viscous dissipation range, Eq. (5)	$Z(a, a')da\, da'$	Drop collision rate, number of drops of sizes a and a' colliding per unit volume per unit time
u_b	Velocity necessary to break a drop		
u_c	Cutoff velocity		
u_c^*	Relative velocity of approach along lines of centers before collision		GREEK LETTERS
u^*	Number of coalescences occurring in the reactor per unit time	α	Parameter defined in Eq. (93)
		β	Extinction coefficient of dye
$\dot{u}$	Acceleration	$\beta(a', a)da$	Fraction of drops of diameter between a and $a + da$ produced by breakage of a drop of diameter a'
u_j'	Root-mean-square velocity fluctuation		
$\overline{u^2(r)}$	Mean square of relative velocity fluctuations between two points a distance r apart	ϵ	Energy dissipation
		η	Kolmogoroff length
		$\eta(z)$	Mixing parameter
$\bar{u}(r)$	Average turbulent velocity fluctuation at the distance r apart	θ	Residence time
		λ	Residual lifetime
		$\lambda(a, a')$	Collision efficiency between colliding drops of sizes a and a'
U_c	Modified kinetic energy of collision		
v	Drop volume	μ	Viscosity
$\mathbf{v}_e$	External coordinate velocity vector in phase space	μ_α	αth moment, Eq. (103)
		ν	Kinematic viscosity
V	Volume of control element	$\nu(a')$	Number of daughter droplets produced by breakage of a drop of diameter a'
$V\%$	Cumulative volume percent of particles below diameter a		
		ρ	Density
W_b	Number of crossings of velocity u_b per unit time, Eq. (44)	σ	Interfacial tension
		σ'^2	Variance, Eq. (56)
		τ	Coalescence time
W^*	Critical relative velocity along lines of centers of	ϕ	Dispersed phase holdup fraction

ϕ_f Fraction of dispersed-phase feed stream containing dye
$\psi(\lambda)$ Residual lifetime distribution
ω_i Dispersed-phase mixing frequency
ω_r Residence frequency
ω_v Coalescence frequency on volume basis, Eq. (96)
∇u Local velocity gradient

OVERMARKS

———   Average at a point

SUBSCRIPTS

c   Continuous phase
d   Dispersed phase
i, j, k Unit vectors

SYMBOLS

$\langle \ \rangle$ Mean value

DIMENSIONLESS GROUPS

N_p Power number, $Pg_c/\rho N^{*3}D_I^5$
$(N_{Re})_\lambda$ Microscopic Reynolds number, $u'\lambda g/\nu$
$(N_{Re})_{l_e}$ Macroscopic Reynolds number, $u'l_e/\nu$
$(N_{Re})_T$ Tank Reynolds number, $N^*D_I^2/\nu$
N_{We} Weber number, $D_I^3 N^{*2}\rho/\sigma$

References

A1. Abrahamson, J., *Chem. Eng. Sci.* **30**, 1371 (1975).
A2. Aiba, S., *AIChE J.* **4**, 485 (1958).
A3. Allan, R. S., Charles, G. E., and Mason, S. G., *J. Colloid Sci.* **16**, 150 (1961).
A4. Ambegaonkar, A. S., Dhruv, A. S., and Tavlarides, L. L., *Can. J. Chem. Eng.* **55**, (1977).
A5. Ambegaonkar, A. S., and Tavlarides, L. L., *Proc. DECHEMA Int. Symp. Chem. React. Eng., 4th* II, 76 (1976).
A6. Angelo, J. B., Lightfoot, E. N., and Howard, D. W., *AIChE J.* **12**, 751 (1966).
A7. Aubry, C., and Villermaux, J., *Chem. Eng. Sci.* **30**, 457 (1975).
B1. Bailey, E. D., *Ind. Eng. Chem. Anal. Ed.* **18**, 365 (1946).
B2. Bajpai, R. K., Ramkrishna, D., and Prokop, A., *Chem. Eng. Sci.* **31**, 913 (1976).
B3. Bajpai, R. K., Prokop, A., and Ramkrishna, D., *Biotechnol. Bioeng.* **17**, 541 (1975).
B4. Bartok, W., and Mason, S. G., *J. Colloid Sci.* **14**, 13 (1959).
B5. Batchelor, G. K., "Theory of Homogeneous Turbulence." Cambridge Univ. Press, London and New York, 1953.
B6. Batchelor, G. K., *Proc. Cambridge Philos. Soc.* **48**, 435 (1952).
B7. Batchelor, G. K., *Proc. Cambridge Philos. Soc.* **43**, 533 (1967).
B8. Bates, R. L., Fondy, P. L., and Corpstein, R. R. *Ind. Eng. Chem., Process Des. Dev.* **2**, 310 (1963).
B9. Bates, R. L., Fondy, P. L., and Fenic, J. G., in "Mixing Theory and Practice" (V. W. Uhl and J. B. Grey, eds.), Vol. I. Academic Press, New York, 1966.
B10. Baynes, C. A., and Laurence, R. L., *Ind. Eng. Chem. Fundam.* **8**, 71 (1969).
B11. Baynes, C. A., Ph.D. dissertation, The Johns Hopkins University, Baltimore, Maryland, 1967.
B12. Blanch, H. W., and Fietcher, A., *Biotechnol. Bioeng.* **16**, 539 (1974).
B13. Borzani, W., and Sanchez Podlech, P. A., *Biotechnol. Bioeng.* **13**, 685 (1971).
B14. Bouyatiotis, B. S., and Thornton, J. D., *Int. Chem. Eng. London Symp. Ser.* **26**, 43 (1967).
B15. Brodkey, R. S., "The Phenomena of Fluid Motions." Addison-Wesley, Reading, Massachusetts, 1967.

B16. Brown, A. H., *Br. Chem. Eng.* **13**, 1719 (1968).
B17. Brown, D. E., and Pitt, K., *Chem. Eng. Sci.* **29**, 345 (1974).
B18. Brown, D. E., and Pitt, K., *Proc. Chemeca 1970, Melbourne and Sydney* 83 (1970).
B19. Brown, D. E., and Pitt, K., *Chem. Eng. Sci.* **27**, 577 (1972).
B20. Burrill, K. A., and Woods, D. R., *J. Coll. Inter. Sci.* **30**, 511 (1969).
B21. Burrill, K. A., and Woods, D. T., *J. Colloid Interface Sci.* **42**, 15 (1973).
B22. Burrill, K. A., and Woods, D. R., *J. Colloid Interface Sci.* **42**, 35 (1973).
C1. Calderbank, P. H., *Trans. Inst. Chem. Eng.* **36**, 443 (1958).
C2. Calderbank, P. H., *Trans. Inst. Chem. Eng.* **36**, 443 (1958).
C3. Camp, T. R., and Stein, P. C., *J. Boston Soc. Civil Eng.* **30**, 219 (1943).
C4. Chapman, J. W., Cox, P. R., and Strachan, A. N., *Chem. Eng. Sci.* **29**, 1247 (1974).
C5. Chappelear, D. C., *J. Colloid Sci.* **16**, 186 (1963).
C6. Charles, G. E., and Mason, S. G., *J. Colloid Sci.* **15**, 236 (1960).
C7. Chartres, R. H., and Korchinsky, W. J., *Trans. Inst. Chem. Eng.* **53**, 247 (1975).
C8. Chen, H. T., and Middleman, S., *AIChE J.* **13**, 989 (1967).
C9. Cockbain, E. G., and McRoberts, T. S., *J. Colloid Sci.* **8**, 440 (1953).
C10. Collins, S. B., and Knudsen, J. G., *AIChE J.* **16**, 1072 (1970).
C11. Coulaloglou, C. A., Dispersed Phase Interactions in an Agitated Flow Vessel. Ph.D. thesis, Illinois Institute of Technology, Chicago, 1975.
C12. Coulaloglou, C. A., and Tavlarides, L. L., *Chem. Eng. Sci.* **32**, 1289 (1977).
C13. Coulaloglou, C. A., and Tavlarides, L. L., *AIChE J.* **22**, 289 (1976).
C14. Cox, P. R., and Strachan, A. N., *Chem. Eng. Sci.* **27**, 457 (1972).
C15. Curl, R. L., *AIChE J.* **20**, 184 (1974).
C16. Curl, R. L., *AIChE J.* **9**, 175 (1963).
C17. Cutter, L. A., *AIChE J.* **12**, 35 (1966).
D1. Danckwerts, P. V., *Ind. Eng. Chem.* **43**, 1460 (1951).
D2. Danckwerts, P. V., *Chem. Eng. Sci.* **8**, 93 (1958).
D3. Dang, V., and Steinberg, M., *AIChE J.* **22**, 925 (1976).
D4. Delichatsios, M. A., *Phys. Fluids* **18**, 622 (1975).
D5. Delichatsios, M. A., and Probstein, R. F., *Ind. Eng. Chem. Fundam.* **15**, 134 (1976).
D6. Delichatsios, M. A., and Probstein, R. F., *J. Colloid Interface Sci.* **51**, 394 (1975).
D7. Delichatsios, M. A., and Probstein, R. F., *J. Water Pollut. Control Fed.* **V47**, 941 (1975).
D8. Desouza, A., and Pike, R. W., *Can. J. Chem. Eng.* **50**, 15 (1972).
D9. De Santiago, M., and Trambouze, P., *Chem. Eng. Sci.* **26**, 1803 (1971).
D10. De Santiago, M., and Bidner, M. S., *Chem. Eng. Sci.* **26**, 175 (1971).
D11. Doulah, M. S., *Ind. Eng. Chem. Fundam.* **14**, 137 (1975).
D12. Doulah, M. S., and Thornton, J. D., Paper presented at the Liquid–Liquid Extraction Symp., Inst. Chem. Eng., University of Newcastle, England, 1967.
E1. East, T. W. R., and Marshall, J. S., *Q.J. R. Meteotol. Soc.* **80**, 26 (1954).
E2. Elton, G. A. H., and Picknett, R. G., *Int. Congr. Surface Activity 2nd* **1**, 388. Butterworths, London, 1957.
E3. Erickson, L. E., Fan, L. T., Shah, P. S., and Chen, M. S. K., *Biotechnol. Bioeng.* **12**, 713 (1970).
E4. Evangelista, J. J., Katz, S., and Shinnar, R., *AIChE J.* **15**, 843 (1969).
F1. Fernandes, J. B., and Sharma, M. M., *Chem. Eng. Sci.* **22**, 1267 (1967).
F2. Fernandes, J. B., and Sharma, M. M., *Chem. Eng. Sci.* **23**, 9 (1968).
F3. Fernandes, J. B., *AIChE Symp. Ser. 120* **68**, 124 (1972).
F4. Flumerfelt, R. W., *Ind. Eng. Chem. Fundam.* **11**, 312 (1972).
F5. Fort, I., *Collect. Czech. Chem. Commun.* **36**, 2914 (1971).
G1. Gal-Or, B., Klinzing, G. E., and Tavlarides, L. L., *Ind. Eng. Chem.* **61**, 21 (1969).

G2. Gal-Or, B., and Hoelscher, H. E., *AIChE J.* **12**, 499 (1966).
G3. Gal-Or, B., and Resnick, W., *Chem. Eng. Sci.* **19**, 653 (1964).
G4. Gal-Or, B., and Resnick, W., *Ind. Eng. Chem. Process Des. Dev.* **5**, 15 (1966).
G5. Gal-Or, B., and Walatka, V. V., *AIChE J.* **13**, 650 (1967).
G6. Gibson, C. H., and Massielo, O. J., "Statistical Models and Turbulence" (M. Rosenblatt and C. Van Atta, eds.), p. 427. Springer-Verlag, Berlin and New York, 1972.
G7. Gillespie, T., and Rideal, E., *Trans. Faraday Soc.* **52**, 173 (1956).
G8. Gillespie, T., *J. Colloid Sci.* **18**, 562 (1963).
G9. Giles, J. W., Hanson, C., and Marsland, J. G., *Proc. Int. Solvent Extr. Conf., Hague, Soc. Chem. Ind. (London)* 94 (1971).
G10. Groothuis, H., and Zuiderweg, F. J., *Chem. Eng. Sci.* **19**, 63 (1964).
G11. Grossman, G., *Ind. Eng. Chem. Process Des.* **11**, 537 (1972).
G12. Groves, M. J., and Freshwater, D. C., *J. Pharm. Sci.* **57**, 1273 (1968).
G13. Gunkel, A. A., and Weber, M. E., *AIChE J.* **21**, 931 (1975).
G14. Gunkel, A. A., and Weber, M. E., *AIChE J.* **21**, 939 (1975).
H1. Harada, M., Arima, K., Eguchi, W., and Nagata, S., *Mem. Fac. Eng. Kyoto Univ.* **24**, 431 (1962).
H2. Hartland, S., *Chem. Eng. Sci.* **24**, 987 (1969).
H3. Hartland, S., and Wood, S. M., *AIChE J.* **19**, 810 (1973).
H4. Hartland, S., *Chem. Eng. Sci.* **22**, 1675 (1967).
H5. Hartland, S., *Chem. Eng. J.* **1**, 258 (1970).
H6. Heisenberg, W., *Z. Phys.* **124**, 628 (1948).
H7. Heisenberg, W., *Proc. R. Soc. (London)* **A195**, 402 (1948).
H8. Hillestad, J. G., Ph.D. thesis, Purdue University, West Lafayette, Indiana, 1965.
H9. Hinze, J. O., "Turbulence." McGraw-Hill, New York, 1959.
H10. Hinze, J. O., *AIChE J.* **1**, 289 (1955).
H11. Hodgson, T. D., and Lee, J. C., *J. Colloid Interface Sci.* **30**, 94 (1969).
H12. Hodgson, T. D., and Woods, D. R., *J. Colloid Interface Sci.* **30**, 429 (1969).
H13. Hoffer, M. S., and Resnick, W., *Chem. Eng. Sci.* **30**, 473 (1975).
H14. Holmes, D. B., Vancken, R. M., and Dekker, J. A., *Chem. Eng. Sci.* **19**, 201 (1964).
H15. Howarth, W. J., *Chem. Eng. Sci.* **19**, 33 (1964).
H16. Howarth, W. J., *AIChE J.* **13**, 1007 (1967).
H17. Hulbert, H. M., and Katz, S., *Chem. Eng. Sci.* **19**, 555 (1964).
H18. Hulbert, H. M., and Akiyama, T., *Ind. Eng. Chem. Fundam.* **8**, 319 (1969).
I1. Inoue, H., O'Shima, E., and Mizognihi, K., *Kagaku Kogaku* **37**, 623 (1973).
I2. Ito, S., Urushiyama, S., and Ogawa, K., *J. Chem. Eng. Jpn.* **7**, 462 (1974).
I3. Ito, S., Ogawa, K., and Yoshida, N., *J. Chem. Eng. Jpn.* **8**, 206 (1975).
J1. Jakubowski, S., and Sideman, S., *J. Multiphase Flow* **3**, 171 (1976).
J2. Jeffreys, G. V., and Davies, G. A., in "Recent Advances in Liquid–Liquid Extraction" (C. Hanson, ed.), p. 495. Pergamon, Oxford, 1971.
J3. Jeffreys, G. V., and Hawksley, J. L., *AIChE J.* **11**, 413 (1965).
K1. Kafarov, V. V., and Babanov, B. M., *J. Appl. Chem. USSR* (English trans.) **32**, 810 (1959).
K2. Karam, H. J., and Bellinger, J. C., *Ind. Eng. Chem. Fundam.* **7**, 576 (1968).
K3. Kattan, A., and Adler, R. J., *AIChE J.* **13**, 580 (1967).
K4. Kattan, A., and Adler, R. J., *Chem. Eng. Sci.* **27**, 1013 (1972).
K5. Kawecki, W., Reith, T., Van Heuven, J. W., and Beek, W. J., *Chem. Eng. Sci.* **22**, 1519 (1967).
K6. Keey, R. B., and Glen, J. B., *AIChE J.* **15**, 942 (1969).
K7. Kim, W. J., and Manning, F. S., *AIChE J.* **10**, 747 (1964).

K8. Koetsier, W. J., and Thoenes, D., *Proc. Eur. Symp. Chem. Reactor Eng., Fifth, Amsterdam* B3-15 (1972).
K9. Kolmogoroff, A. N., *Dokl. Akad. Nauk SSSR* **66**, 825 (1949).
K10. Kolmogoroff, A. N., *C. R. Acad. Sci. USSR* **30**, 301 (1941); **32**, 16 (1941).
K11. Komasawa, I., Kuboi, R., and Otake, T., *Chem. Eng. Sci.* **29**, 641 (1974).
K12. Komasawa, I., Morioka, S., and Otake, T., *Kagaku Kogaku* **34**, 538 (1970).
K13. Komasawa, I., Morioka, S., Kuboi, R., and Otake, T., *J. Chem. Eng. Jpn.* **4**, 319 (1971).
K14. Komasawa, I., Susukura, T., and Otake, T., *J. Chem. Eng. Jpn.* **2**, 208 (1969).
K15. Kuboi, R., Komasawa, I., and Otake, T., *Chem. Eng. Sci.* **29**, 651 (1974).
K16. Kuboi, R., Komasawa, I., and Otake, T., *J. Chem. Eng. Jpn.* **5**, 349 (1972).
K17. Kuboi, R., Komasawa, I., and Otake, T., *J. Chem. Eng. Jpn.* **5**, 423 (1972).
K18. Kuo, C. H., and Huang, C. J., *AIChE J.* **16**, 493 (1970).
L1. Laats, M. K., and Frishman, F. A., *Fluid Dyn.* (trans. from Russian) **8**, 304 (1974).
L2. Laddha, S. S., and Sharma, M. M., *Chem. Eng. Sci.* **31**, 843 (1976).
L3. Lang, S. B., and Wilke, C. R., *Ind. Eng. Chem. Fundam.* **10**, 329, 341 (1971).
L4. Langlois, G. E., Gullberg, J. E., and Vermeulen, T., *Rev. Sci. Instr.* **25**, 360 (1954).
L5. Lawson, G. B., *Chem. Proc. Eng.* **48**, 45 (1967).
L6. Lee, J. C., and Hodgson, T. D., *Chem. Eng. Sci.* **23**, 1375 (1968).
L7. Levich, V. G., "Physicochemical Hydrodynamics," p. 458. Prentice Hall, New York, 1962.
L8. Levins, D. M., and Glastonbury, J. R., *Trans. Inst. Chem. Eng.* **50**, 32 (1972).
L9. Liem, A. J. G., and Woods, D. R., *Can. J. Chem. Eng.* **52**, 222 (1974).
L10. Liem, A. J. S., and Woods, D. R., *AIChE Symp. Ser. 144* **70**, 8 (1974).
L11. Linek, V., and Mayrhoferova, J., *Chem. Eng. Sci.* **24**, 481 (1969).
L12. Ludvik, M., and Steidl, H., *Collect. Czech. Chem. Commun.* 35, 1480 (1970).
L13. Luss, D., and Amundson, N. R., *Chem. Eng. Sci.* **22**, 267 (1967).
M1. MacKay, G. D., and Mason, S. G., *J. Colloid Sci.* **16**, 632 (1961).
M2. Madden, A. J., and Damerell, G. L., *AIChE J.* **8**, 233 (1962).
M3. Mashkar, R. D., and Sharma, M. M., *Chem. Eng. Sci.* **30**, 811 (1975).
M4. Matsuzawa, H., and Miyauchi, T., *Kagaku Kogaku* **25**, 582 (1961).
M5. McAvoy, R. M., and Kintner, R. C., *J. Colloid Sci.* **20**, 188 (1965).
M6. McAvoy, R. M., Wiegard, W. A., Tomkin, E. E., and Kintner, R. C., *AIChE Inst. Chem. Eng. Joint Symp. Adv. Sep. Tech.* **1**, 18 (1965).
M7. McCoy, B. J., and Madden, A. J., *Chem. Eng. Sci.* **24**, 416 (1969).
M8. McKay, G. D. M., and Mason, S. G., *Can. J. Chem. Eng.* **41**, 203 (1963).
M9. McLaughlin, C. M., and Rushton, J. H., *AIChE J.* **19**, 817 (1973).
M10. Mehta, V. D., and Sharma, M. M., *Chem. Eng. Sci.* **26**, 461 (1971).
M11. Merchuk, J. C., and Farina, I. H., *Chem. Eng. Sci.* **31**, 645 (1976).
M12. Metzner, A. B., Feehs, R. H., Ramos, H. L., Otto, R. E., and Tuthill, J. D., *AIChE J.* **7**, 3 (1961).
M13. Metzner, A. B., and Taylor, J. S., *AIChE J.* **3**, 3 (1957).
M14. Miller, R. S., Ralph, J. L., Curl, R. L., and Towell, G. D., **9**, 196 (1963).
M15. Ming-Heng, C., Liang-Heng, C., and Wei-Kang, Y., *Sci. Sin.* **15**, 123 (1966).
M16. Min, K. W., and Ray, W. H., *Int. Symp. Chemical Reaction Eng., 4th* Heidelberg April, 1976.
M17. Min, K. W., and Ray, W. H., *J. Appl. Polym. Sci.*, **22**, 89 (1978).
M18. Mizushina, T., Ito, R., Hiraoka, S., and Fujimoto, K., *Kagaku Kogaku* **37**, 409 (1973).
M19. Mlynek, Y., and Resnick, W., *AIChE J.* **18**, 122 (1972).
M20. Mok, Y. I., and Treybal, R. E., *AIChE J.* **17**, 916 (1971).
M21. Moryakov, V. S., Nikolaev, N. A., and Nikolaev, A. M., *Int. Chem. Eng.* **14**, 623 (1974).
M22. Mugele, R. A., and Evans, H. D., *Ind. Eng. Chem.* **43**, 1317 (1951).

M23. Mujumdar, A. S., Huang, B., Wolf, D., Weber, M. E., and Douglas, W. J. M., *Can. J. Chem. Eng.* **48**, 475 (1970).
M24. Murdoch, P. G., and Leng, D. E., *Chem. Eng. Sci.* **26**, 1881 (1971).
N1. Nagata, S., Yamamoto, K., and Hashimato, K., *Mem. Fac. Eng. Kyoto Univ.* **21**, 260 (1959); **22**, 68 (1960).
N2. Nagata, S., and Yamaguchi, I., *Mem. Fac. Eng. Kyoto Univ.* **22**, 249 (1960).
N3. Nagata, S., "Mixing-Principles and Applications." Halsted, New York, 1975.
N4. Nanda, A. K., and Sharma, M. M., *Chem. Eng. Sci.* **21**, 707 (1966).
N5. Neilsen, H. J., Ph.D. thesis, Illinois Institute of Technology, Chicago, 1958.
N6. Nielsen, L. E., Wall, R., and Adams, G., *J. Colloid Sci.* **13**, 441 (1958).
N7. Nienow, A. W., and Miles, D., *Ind. Eng. Chem. Process Des. Dev.* **10**, 41 (1971).
N8. Nishikawa, M., Okamoto, Y., Hashimoto, K., and Nagata, S., *J. Chem. Eng. Jpn.* **9**, 489 (1976).
O1. Oldshue, J. Y., *Biotechnol. Bioeng.* **8**, 3 (1966).
O2. Oldshue, J. Y., *Chemeca 1970* 99 (1970).
O3. Olney, R. B., *AIChE J.* **10**, 827 (1966).
O4. Onda, K., Takeuchi, H., and Takahashi, M., *Kagaku Kogaku* **35**, 221 (1971).
O5. Otake, T., and Komasawa, I., *Kagaku Kogaku* **32**, 475 (1968).
O6. Overbeek, J. Th.G., *in* "Colloid Science," (H. R. Kruyt, ed.), Vol. I, Chap. VII. Elsevier, Amsterdam, 1952.
P1. Padmanaban, L., and Gal-Or, B., *AIChE J.* **14**, 709 (1968).
P2. Padmanaban, L., and Gal-Or, B., *Chem. Eng. Sci.* **23**, 631 (1968).
P3. Pancher, S., "Random Functions and Turbulence," p. 301. Pergamon, Oxford, 1971.
P4. Park, J. Y., and Blair, L. M., *Chem. Eng. Sci.* **30**, 1057 (1975).
P5. Patterson, G. K., *in* "Turbulence in Mixing Operations: Theory and Application to Mixing and Reaction" (R. S. Brodkey, ed.), p. 222. Academic Press, New York, 1975.
P6. Paul, T. H., and Sleicher, C. A., *Chem. Eng. Sci.* **20**, 57 (1965).
P7. Princen, H. M., *J. Colloid Sci.* **18**, 178 (1963).
P8. Puranik, S. A., and Sharma, M. M., *Chem. Eng. Sci.* **25**, 257 (1970).
R1. Ramkrishna, D., *Chem. Eng. Sci.* **29**, 987 (1974).
R2. Ramkrishna, D., *Chem. Eng. Sci.* **20**, 1135 (1971).
R3. Ramkrishna, D., *Chem. Eng. Sci.* **28**, 1362 (1973).
R4. Ramkrishna, D., Paper presented at the 71st National Meeting AIChE., Dallas, Texas, Feb., 1972.
R5. Randolph, A. D., *Ind. Eng. Chem. Fundam.* **8**, 58 (1969).
R6. Randolph, A. D., and Larson, M. A., "Theory of Particulate Processes." Academic Press, New York, 1971.
R7. Rao, M. A., and Brodkey, R. S., *Chem. Eng. Sci.* **27**, 137 (1972).
R8. Rao, D. P., and Dunn, I. J., *Chem. Eng. Sci.* **25**, 1275 (1970).
R9. Reith, T., and Beek, W. J., *Proc. Eur. Symp. Chem. React. Eng., 4th,* Brussels, 1968.
R10. Resnick, W., and Gal-Or, B., *Adv. Chem. Eng.* **7**, 295 (1968).
R11. Rice, S. O., *in* "Selected Papers on Noise and Stochastic Processes" (N. Wax, ed.), p. 189. Dover, New York, 1954.
R12. Rietema, I., *Adv. Chem. Eng.* **5**, 237 (1964).
R13. Rodger, W. A., Trice, V. G., Jr., and Rushton, J. H., *Chem. Eng. Prog.* **52**, 515 (1956).
R14. Rodriguez, F., Grotz, L. C., and Engle, D. L., *AIChE J.* **7**, 663 (1961).
R15. Ross, S. L., Measurements and Models of the Dispersed Phase Mixing Process. Ph.D. thesis, University of Michigan, Ann Arbor, 1971. Ross, S. L., Verhoff, F. H. and Curl, R. L. *Ind. Eng. Chem. Fundl.* **16**, 371 (1977); **17**, 101 (1978).
R16. Ross, S. L., and Curl, R. L., *Paper No. 29b, Joint Chem. Eng. Conf., 4th, Vancouver,* Sept. 1973.

R17. Rotta, J. C., "Turbulente Strömungen," p. 96. Teubner, Stuttgart, 1972.
R18. Rumscheidt, F. P., and Mason, S. G., *J. Colloid Sci.* **16**, 238 (1961).
R19. Rushton, J. H., Costich, E. W., and Everett, H. J., *Chem. Eng. Prog.* **46**, 375, 467 (1950).
R20. Rushton, E., and Davies, G. A., *CHISA Congr., 3rd Prague,* Sept. 1972.
R21. Rushton, E., and Davies, G. A., *Int. Fluid Mech. Symp. McMaster Univ.,* Hamilton, Ontario 1970.
S1. Sachs, J. P., and Rushton, J. H., *Chem. Eng. Prog.* **50**, 597 (1954).
S2. Sachs, J. P., Ph.D. thesis, Illinois Institute of Technology, Chicago, 1952.
S3. Saffman, P. G., and Turner, J. S., *J. Fluid Mech.* **1**, 16 (1956).
S4. Sanchez Podlech, P. A., and Borzani, W., *Biotechnol. Bioeng.* **14**, 43 (1972).
S5. Sankholkar, D. S., and Sharma, M. M., *Chem. Eng. Sci.* **28**, 2089 (1973).
S6. Sankholkar, D. S., and Sharma, M. M., *Chem. Eng. Sci.* **28**, 49 (1973).
S7. Sankholkar, D. S., and Sharma, M. M., *Chem. Eng. Sci.* **30**, 729 (1975).
S8. Sato, Y., Kamimuno, M., and Yamameto, K., *Kagaku Kogaku* **34**, 104 (1970).
S9. Sato, Y., Ishii, K., Horie, Y., Kumiwano, M., and Yamamoto, K., *Kagaku Kogaku* **31**, 275 (1967).
S10. Scheele, G. F., and Leng, D. E., *Chem. Eng. Sci.* **26**, 1867 (1971).
S11. Schindler, H. D., and Treybal, R. E., *AIChE J.* **14**, 790 (1968).
S12. Schwarz, N., and Bezemer, C., *Kolloidzeitschrift* **146**, 139, 145 (1956).
S13. Schwartzberg, H. G., and Treybal, R. E., *Ind. Eng. Chem. Fundam.* **7**, 1 (1968).
S14. Schwartzberg, H. G., and Treybal, R. E., *Ind. Eng. Chem.* **7**, 6 (1968).
S15. Shah, A. K., and Sharma, M. M., *Can. J. Chem. Eng.* **49**, 596 (1971).
S16. Shah, P. S., Fan, L. T., Kao, I. C., and Erickson, L. E., *Adv. Appl. Microbiol.* **15**, 367 (1972).
S17. Shah, B. H., and Ramkrishna, D., *Chem. Eng. Sci.* **28**, 389 (1973).
S18. Shain, S. A., *AIChE J.* **12**, 806 (1966).
S19. Sharma, M. M., and Danckwerts, P. V., *Br. Chem. Eng.* **15**, 522 (1970).
S20. Sharma, M. M., and Nanda, A. K., *Trans. Inst. Chem. Eng.* **46**, T44 (1968).
S21. Shiloh, K., Sideman, S., and Resnick, W., *Can. J. Chem. Eng.* **51**, 542 (1973).
S22. Shinnar, R., *J. Fluid Mech.* **10**, 259 (1961).
S23. Singh, P. N., and Ramkrishna, D., *J. Colloid Interface Sci.* **53**, 214 (1975).
S24. Singh, P. N., and Ramkrishna, D., *Comput. Chem. Eng.* **1**, 23 (1977).
S25. Skelland, A. H. P., "Diffusional Mass Transfer." Wiley, New York, 1974.
S26. Sleicher, C. A., *AIChE J.* **8**, 471 (1962).
S27. Smoluchowski, M., *Z. Phys. Chem.* **92**, 129 (1917).
S28. Spielman, L. A., and Levenspiel, O., *Chem. Eng. Sci.* **20**, 247 (1965).
S29. Sprow, F. B., *AIChE J.* **13**, 995 (1967).
S30. Sprow, F. B., *Chem. Eng. Sci.* **22**, 435 (1967).
S31. Sridharan, K., and Sharma, M. M., *Chem. Eng. Sci.* **31**, 767 (1976).
S32. Stamatoudis, M., Ph.D. thesis, Illinois Institute of Technology Chicago, 1977.
S33. Stamatoudis, M., and Tavlarides, L. L., *Paper No. 24C AIChE Natl. Meet., 75th, Detroit,* June, 1973.
S34. Steidl, H., *Collect. Czech. Chem. Commun.* **33**, 2191 (1968).
S35. Suzuki, K., and Watanabe, T., *Bull. Chem. Soc. Jpn.* **44**, 2039 (1971).
S36. Swartz, J. E., and Kessler, D. P., *AIChE J.* **16**, 254 (1970).
T1. Tavlarides, L. L., Coulaloglou, C. A., Zeitlin, M. A., Klinzing, G. E., and Gal-Or, B., *Ind. Eng. Chem.* **62**, 6 (1970).
T2. Tavlarides, L. L., and Gal-Or, B., *Chem. Eng. Sci.* **24**, 553 (1969).
T3. Tavlarides, L. L., and Gal-Or, B., *Isr. J. Technol.* **7**, 1 (1969).
T4. Taylor, G. I., *Proc. R. Soc.* **A138**, 41 (1932).

T5. Taylor, G. I., "The Scientific Papers of Sir Geofrey Ingram Taylor" (G. K. Batchelor, ed.), p. 457. Cambridge Univ. Press, London and New York, 1963.
T6. Taylor, G. I., *Proc. R. Soc.* **A146**, 501 (1934).
T7. Taylor, G. I., *Proc. R. Soc. (London)* **151A**, 421 (1935).
T8. Thornton, J. D., and Bouyatiotis, B. A., *Ind. Chem.* **39**, 298 (1963).
T9. Torza, S., Cox, R. G., and Mason, S. G., *J. Colloid Sci.* **38**, 395 (1972).
T10. Treleaven, C. R., and Tobgy, A. H., *Chem. Eng. Sci.* **27**, 1497, 1756 (1972).
T11. Treleaven, C. R., and Tobgy, A. H., *Chem. Eng. Sci.* **28**, 413 (1973).
T12. Treybal, R. E. "Mass-Transfer Operations." McGraw-Hill, New York, 1968.
T13. Trice, V. G., Jr., and Rodger, W. A., *AIChE J.* **2**, 205 (1956).
T14. Tsukiyama, S., Takamura, A., Fukuda, Y., and Koishi, M., *Chem. Pharm. Bull. (Tokyo)* **24**, 414 (1976).
T15. Tsukiyama, S., Takahashi, H., Takashima, I., and Hatano, S., *Yakugaku Zaschi* **91**, 305 (1971).
U1. Uhl, V. W., and Gray, J. B., eds., "Mixing Theory and Practice," Vol. I. Academic Press, New York, 1966.
V1. Valentas, K. J., and Amundson, N. R., *Ind. Eng. Chem. Fundam.* **5**, 533 (1966).
V2. Valentas, K. J., Bilous, D., and Amundson, N. R., *Ind. Eng. Chem. Fundam.* **5**, 271 (1966).
V3. Valentas, K. J., and Amundson, N. R., *Ind. Eng. Chem. Fundam.* **7**, 66 (1968).
V4. Van Atta, C., and Park, T., *in* "Statistical Models and Turbulence" (M. Rosenblatt and C. Van Atta, eds.). Springer-Verberg, Berlin and New York, 1972.
V5. Vanderveen, J. H., UCRL-8733, M. S. thesis, University of California, December, 1960.
V6. Van Heuven, J. W., and Hoevenaar, J. C., *Proc. Eur. Symp. Chem. React. Eng., 4th, Brussels* (1968).
V7. Van Heuven, J. W., and Beek, W. J., *Proc. Int. Solvent Extr. Conf., Hague, Soc. Chem. Ind.* Paper 51, 70 (1971).
V8. Vassilatos, G., and Toor, H. L., *AIChE J.* **11**, 66 (1965).
V9. Van't Riet, K., Bruijn, W., and Smith, J. M., *Chem. Eng. Sci.* **31**, 407 (1976).
V10. Van't Riet, K., and Smith, J. M., *Chem. Eng. Sci.* **30**, 1093 (1975).
V11. Verhoff, F. H., A Study of the Bivariate Analysis of Dispersed Phase Mixing. Ph.D. thesis, University of Michigan, Ann Arbor (1969).
V12. Vermeulen, T., Williams, G. M., and Langlois, G. E., *Chem. Eng. Prog.* **51**, 85F (1955).
V13. Vijayan, S., and Ponter, A. B., *Chem.-Ing.-Tech.* **47**, 748 (1975).
V14. Vijayan, S., Ponter, A. B., and Cao Thi, A. T., *Tenside Detergents* **12**, 271 (1975).
V15. Vijayan, S., and Ponter, A. B., *Tenside Detergents* **11**, 241 (1974).
V16. Villermaux, J., and Devillon, J. C., *Proc. Eur. Symp. Chem. React. Eng., 5th, Amsterdam* B1-13 (1972).
W1. Ward, J. P., and Knudsen, J. G., *AIChE J.* **13**, 356 (1967).
W2. Waslo, S., and Gal-Or, B., *Chem. Eng. Sci.* **26**, 829 (1971).
W3. Weinstein, B., and Treybal, R. E., *AIChE J.* **19**, 304 (1973).
W4. Westerterp, K. P., Van Dierendonck, L., and Dekraa, J. A., *Chem. Eng. Sci.* **18**, 157 (1963).
W5. Woods, D. R., and Burill, K. A., *Electroanal. Chem.* **37**, 191 (1972).
Y1. Yamguchi, I., Yabuta, S., and Nagata, S., *Kagaku Kogaku* **27**, 576 (1963).
Y2. Yamazaki, H., Ichikawa, A., *Kagaku Kogaku* **34**, 219 (1970); *Int. Chem. Eng.* **10**, 471 (1970).
Y3. Yoshida, F., and Yamada, T., *J. Ferment. Technol. (Jpn)* **49**, 235 (1971).

Z1. Zavoluk, R. J., M.S. thesis, Illinois Institute of Technology, Chicago, 1972.
Z2. Zeitlin, M. A., and Tavlarides, L. L., *Can. J. Chem. Eng.* **50,** 207 (1972).
Z3. Zeitlin, M. A., and Tavlarides, L. L., *AIChE J.* **18,** 1268 (1972).
Z4. Zeitlin, M. A., and Tavlarides, L. L., *Ind. Eng. Chem. Process Des. Dev.* **11,** 532 (1972).
Z5. Zeitlin, M. A., and Tavlarides, L. L., *Proc. 5th Eur./2nd Int. Symp. Chem. React. Eng.* B1-10 (1972).
Z6. Zweitering, T. N., *Chem. Eng. Sci.* **11,** 1 (1959).

TRANSPORT PHENOMENA AND REACTION IN FLUIDIZED CATALYST BEDS

Terukatsu Miyauchi and Shintaro Furusaki

Department of Chemical Engineering
University of Tokyo
Tokyo, Japan

Shigeharu Morooka

Department of Applied Chemistry
Kyushu University
Fukuoka, Japan

and

Yoneichi Ikeda

Fluidization Engineering Laboratory
Tokyo, Japan

I.	Introduction	276
	A. Purpose and Outline of the Review	276
	B. General Properties of Fluidized Beds	277
	C. Historical Development of FCB Studies	281
	D. Quality of Fluidization in Relation to Particle Properties	283
II.	Flow Properties of Fluid Beds	285
	A. Particle Properties in Relation to Fluidity	285
	B. Formation, Splitting, and Coalescence of Bubbles	290
	C. Operation of Fluid Beds	297
	D. Properties of Bulk Flow Inside Fluid Beds	297
	E. Axial Distribution to Bed Structure	305
	F. Effect of Bed Internals	307
III.	Turbulent-Flow Phenomena in Bubble Columns and Fluidized Catalyst Beds	310
	A. Two-Phase Bubble Flow in Recirculation	311
	B. Comparison of Theory with Experiments on Bubble Columns	317

	C. Phenomenological Background of Turbulent Kinematic Viscosity	326
	D. Interpretation of Experiments for FCB	327
IV.	Longitudinal Dispersion Phenomena as Derived from Flow Properties	330
	A. Theory of Longitudinal Dispersion for the Recirculation Flow Regime	331
	B. Longitudinal Dispersion in a Bubble Column	335
	C. Longitudinal Dispersion in a Fluidized Catalyst Bed	338
V.	Bubble Phenomena in Relation to Bed Performance	340
	A. Mean Bubble Velocity without Bulk Recirculation	341
	B. Mean Bubble Velocity with Bulk Recirculation	344
	C. Bubble Splitting in Turbulent Flow	350
VI.	Heat and Mass Transfer in Fluidized Catalyst Beds	360
	A. Bubble Dynamics	360
	B. Heat and Mass Transfer through the Bubble Wall	363
	C. Mass Transfer between Bubble and Emulsion Phase	365
	D. Particle Capacitance Effect	371
	E. Axial and Radial Mixing of Heat and Mass	373
	F. Review of Wall Heat Transfer in Fluid Beds	379
VII.	The Successive Contact Mechanism for Catalytic Reaction	381
	A. Experimental Facts and Reactor Models	381
	B. The Successive Contact Mechanism	390
	C. Overall Rate Constant K_{oR} Based on L_q	399
VIII.	Further Properties of the Successive Contact Mechanism	402
	A. Axial Distribution of Reactivity in a Fluid Bed	402
	B. Axial Distribution of the Contact-Mechanism Contributions	407
IX.	Nonisothermal Effect on the Bed Performance	413
	A. Effect on Steady Reaction	413
	B. Stability of Fluid Bed Reactors	421
X.	Discussion and Summary	425
	A. Applicability of Reactor Models	425
	B. Development of Industrial Fluidized Catalyst Beds	426
	C. Recent Trends in Fluidized Catalyst Beds	429
	D. Technical Problems in FCB Design	430
	Nomenclature	432
	References	437
	Note Added in Proof	448

I. Introduction

A. Purpose and Outline of the Review

Fluidization today seems a very young and active field, even though Charles E. Robinson originated the technique a century ago. It is being applied today in the chemical industry to catalytic reactions, noncatalytic reactions, and many physical operations. Warren K. Lewis and Edwin R. Gilliland can be said to have begun modern engineering on the fluidized catalyst bed, which was started in an elaborate study on hindered settling and developed in a great number of theses at MIT (M1).

The purpose of this review is to consider the uses of fluidized beds for catalytic reactions, using a viewpoint quite different from studies directed toward the physical handling of solid particles or toward gas–solid noncatalytic reaction. At this point it is useful to outline the scope of the review.

First, it will be shown that flow properties of the fluidized catalyst bed (FCB) are clearly different from those of other conventional fluidized beds. The different treatment required is very significant for research and development on fluidized catalytic beds. Next, factors affecting the flow properties are discussed, especially particle size distribution, and also heat and mass transfer, and mixing properties.

Reactions in the FCB are then discussed. The mechanism of successive contact (M25) is presented. Contact efficiency is greater in the dilute phase, i.e., freeboard region, than in the underlying dense phase. In the FCB, the dilute phase plays an important role in advancing the catalytic reaction when reaction rates are high. This factor provides a basis for identifying the appropriate reaction model and clarifying the effect of the dilute phase on selectivity and stability.

To understand particle circulation in the FCB, the simplified theory (M31) for a bubble column is applied so as to analyze flow patterns of bubbles and "emulsion." Finally, applications of all these studies are investigated. Apparatus problems and the development of improved FCBs are given considerable attention.

B. General Properties of Fluidized Beds

In fluidization, a suspension of fine solid particles behaves like a liquid during the upflow of a supportive gas or liquid phase. Thus the bed of fluidized solid itself may be analyzed similarly to liquid systems. The gas-lift effect produces internal recirculation, by providing a descending flow of high particle concentration and an ascending flow of low particle concentration. This effect resembles the circulation in bubble columns. Whereas bubble columns contain dispersed gas and a continuous liquid phase, the fluidized bed comprises the bubble phase and the emulsion phase in which particles have gained fluid-like properties by the interstitial gas flow.

The interrelation of pressure drop and superficial gas velocity appears to determine the various states of fluidization, as shown in Fig. 1 by

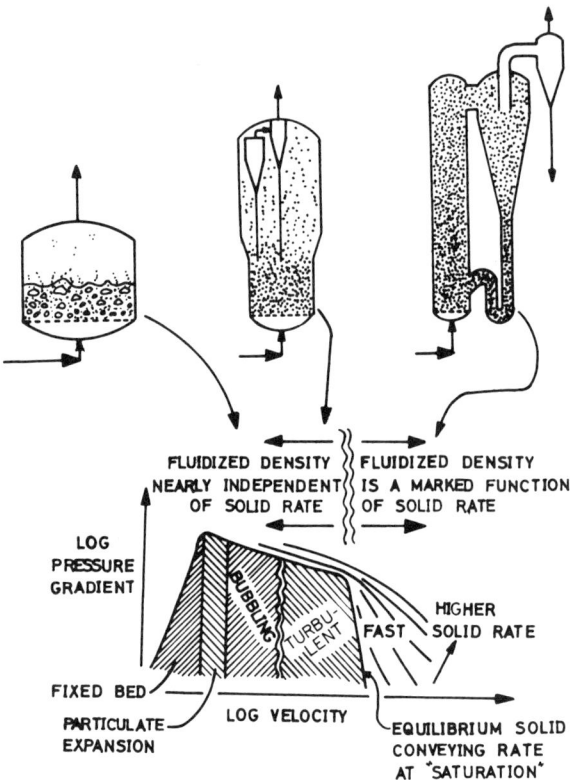

FIG. 1. Various states of fluidization (after Y11, Y12).

Yerushalmi et al. (Y11, Y12). With increasing gas velocity, the pressure drop increases while the bed is in a fixed state; then it stays almost constant, equal to the fluidizing density of the bed after the minimum fluidization velocity. Further increase of gas velocity makes the bed fluidize vigorously; a change occurs from the bubbling bed to the turbulent bed and finally to the fast fluidized bed, and the amount of particles entrained by the gas flow likewise increases. The charge of particles in a fluidized bed will not remain constant unless the particles are separated in cyclones and returned to the bed.

Most industrial fluidized beds using fine powder are found to be in the category of turbulent beds; typical of these is fluidized catalytic cracking. On the other hand, most noncatalytic reactions and physical operations use coarse particles and are operated as bubbling beds. Squires (S14) has pointed out that the state of fluidization varies significantly with fluidized particle diameter, and has identified the following two categories:

(a) *Fluid bed*. The fluidization of fine particles, generally all through 20 mesh; usually with a substantial part smaller than 200 mesh, often smaller than 325 mesh. Superficial velocities generally below 60 cm/sec.

(b) *Teeter bed*. The fluidization of coarse particles, generally all larger than 60 mesh; with a substantial part larger than 10 mesh. Superficial velocities generally above 150 cm/sec.

Considering Squires' criteria and observed industrial performance (G15, K24, M2, O10, V12, Z5), Ikeda (I4) modified the criteria as shown in Table I, stressing that the fluidized catalyst bed should be viewed as the typical fluid bed (I1–I10). The general flow features of a fluid bed operated under relatively high gas velocity are illustrated in Fig. 2 (M25). The dense bed consists of the emulsion phase and bubble phase; the emulsion phase exhibits liquid-like properties, in the case of the fluid bed. Intense circulation of dense phase results from the central rapidly ascending bubble-rich phase and the peripheral descending emulsion phase of low bubble content. The boundary of the dense phase and the dilute phase is rather ambiguous; a transition region exists in which bubbles collapse and particles are blown in from the emulsion phase. In the transition

TABLE I

TYPICAL FLUID BED AND TEETER BED

Characteristics	Fluid bed	Teeter bed
Mean particle size	50–70 μm	0.1–2 mm (mostly ~0.4 mm)
Range of sizes	5–100 μm	0.05–5 mm
U_f [cm/sec]	30–90	30–150
U_f/U_{mf}	50–200	3–10
L_{mf} [cm]	50–500	50–150
L_{mf}/D_T	1–2	0.1–0.5
Constitution	Bed divides into dense phase + dilute phase, although surface of dense phase is ill-defined	Density of bed is almost constant, surface of bed rather well defined
Behavior of bubbles	Most gas flows as uniformly small bubbles	Large bubbles, which may coalesce progressively during rise
Behavior of emulsion phase	Circulative flow is developed (possibly central upward, peripheral downward)	Generally relative movement of particles; partially circulative flow
Stability	Slugging unlikely, even for large L_q/D_T	Frequent slugging for $L_q > 1$–1.5 m

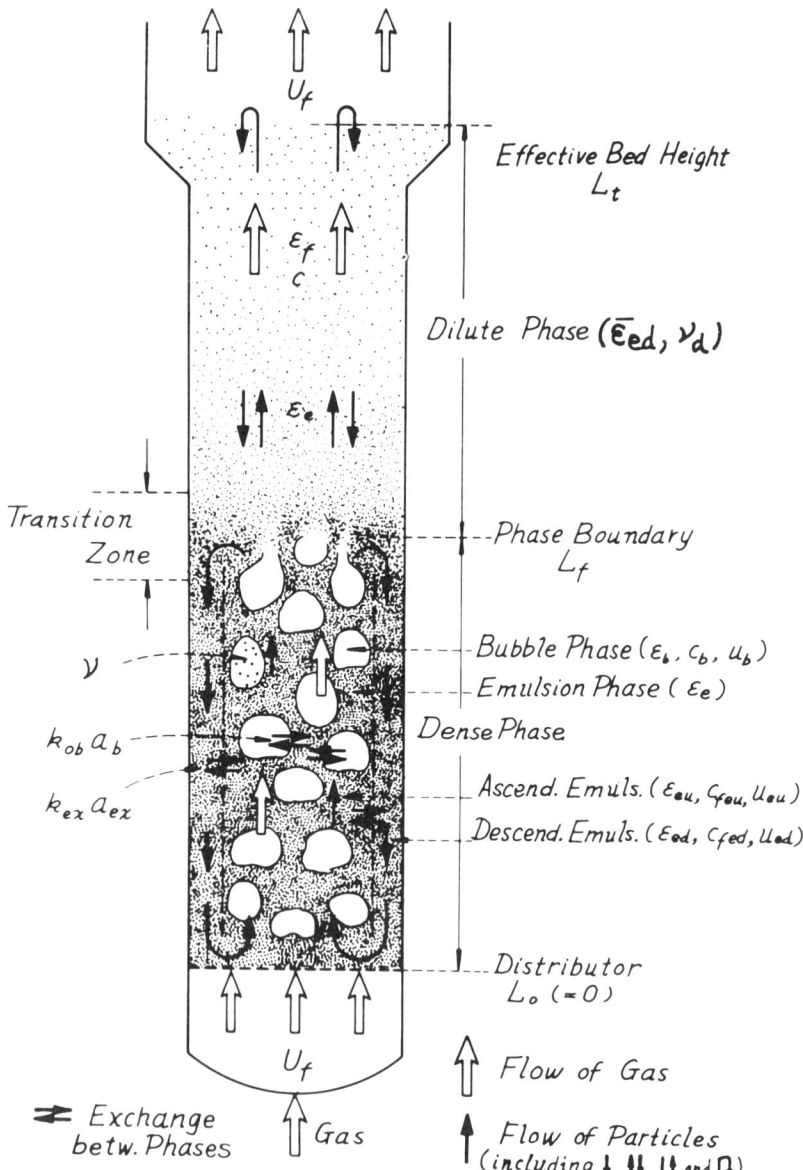

FIG. 2. General flow features of the fluid bed.

region, inversion of the circulating flow can occur. In the dilute phase, particles are suspended particulately in the stream of ascending gas. The content of suspended particles decreases with height, except in the upper part of the dilute phase where it stays constant.

C. HISTORICAL DEVELOPMENT OF FCB STUDIES

As is well known, the fluid catalytic cracking (FCC) process was the first application of the fluidized bed to catalytic reactions. Esso Research and Development Company (now Exxon) developed the process in a very short period (15 months) from the operation of the pilot plant to the start-up of the commercial plant, under the supervision of Warren K. Lewis of MIT. A major reason for the quick success of the development is that Lewis farsightedly chose fluid beds for FCC, instead of teeter beds which had been applied for the Winkler process, and backed up this choice with sustained basic research. Fluid beds are preferable for catalytic reactions, because of: (1) sufficient inventory of catalyst, ensuring adequate contact time, heat-transfer area, and process inertia (fly-wheel effect); (2) good fluidization, with only small perturbations of pressure drop, and consequent stable operation; (3) good fluidity—particles easy to handle, with low rates of erosion of bed apparatus and catalyst attrition; (4) easily manufactured catalyst particles, produced in large quantity in spray dryers. Moreover, the purpose of fluidized-bed operation was to remove coke formed and still operate continuously, rather than to achieve high conversion (conversion of 50–60% was acceptable). Frantz (F6) stated, commenting on the good fortune of matching FCC and FCB, "If some other reaction had been chosen for the commercial fluid unit, fluidization as a commercial operation would have been put back on the shelf for several years."

Carlsmith and Johnson (C2) have reviewed the research and development of the FCC process. Drawing on their data, Ikeda (I4) has recalculated scale-up ratio and operating results, as shown in Table II. The feed rate of oil is roughly proportional to the cross-sectional area of the reactor. Thus a large superficial velocity was chosen for the small-scale reactors, and scale-up was conducted under a constant superficial velocity.

The success of FCC encouraged applications of the fluidized catalyst reactor to other catalytic reactions. Successful applications can be found in fluid catalytic reforming, production of alkyl chloride by oxychlorination, production of phthalic anhydride, acrylonitrile synthesis by ammoxidation, and production of maleic anhydride.

In applying the fluidized bed to catalytic reactions, the catalyst must have the appropriate properties and activity. For production of phthalic

TABLE II
Scale-Up Ratio and Reaction in the Development of FCC

	Small-scale pilot plant	Large-scale pilot plant	Commercial plant
Reactor			
Bed diameter, D_T	2 in.	15 in.	15 ft.
Relative diameter	0.133	1	12
Relative cross section	0.018	1	144
Bed height, L_t [ft]	20	20	28
Relative bed height	1	1	1.4
L_t/D_T	120	16	1.87
Regenerator			
Bed diameter	4 in.	22 in.	19.5 ft.
Relative diameter	0.18	1	10.6
Relative cross section	0.03	1	112
Bed height [ft.]	20	31	37
Relative bed height	0.67	1	1.2
L_t/D_T	60	17	2.3
Capacity [bbl/day]	2	100	15,000
Relative capacity	0.02	1	150
Product			
Gasoline [vol. %]	50.2	49.7	49.5
Gas oil [vol. %]	50.0	50.0	50.0
Dry gas [wt. %]	4.2	5.0	6.2
Carbon [wt. %]	3.2	2.8	2.8

anhydride, Betts (B11) stated that the ground catalyst easily suffers attrition and produces dusty fines; these give rise to heat-exchanger fouling, cyclone or filter blockage, and after-burning. Betts showed that the microspherical catalyst resisted attrition and eliminated the difficulties listed. According to Graham and Way (G14), fluidized beds are ideally suited to a moderately active catalyst because they involve contact times and large masses of catalyst; the more highly active fixed-bed catalysts may give reduced yield due to overoxidation and should be avoided in fluid beds.

The Fischer–Tropsch synthesis produces hydrocarbons from H_2 and CO. Fluid beds were applied to it in the Hydrocol process using ordinary turbulent beds and in the Kellogg process using fast beds (Fig. 1). Development of the Hydrocol process has been reported by Grekel *et al.* (G15) and Hall and Taylor (H1). The main difficulties in scale-up of these fluidized beds were incomplete fluidization and low conversion.

On the basis of this experience with the Hydrocol process, Volk *et al.* (V12) proposed providing internals in fluid beds in order to achieve good fluidization and raise the contact efficiency by performing the following functions: (1) prevention of growth of bubbles; (2) promotion of lateral movement of gas and solids; (3) prevention of formation of dead pockets of solids, (4) prevention of elutriation of fine particles, so as to maintain the original particle size distribution; (5) allowance for periodic removal of the entire bed of solid particles from the reactor.

Volk et al. (V12) used vertical surfaces (tubes or half-rounds) with success in a version of the Hydrocol synthesis known as the HRI process. The conversion of CO gas for several reactor diameters was correlated by the following equivalent diameter:

$$D_{eq} = \frac{4(\text{free cross-sectional area})}{(\text{wetted perimeter of all vertical surface})} \tag{1-1}$$

The HRI process concept has greatly contributed in the application of fluid beds to catalytic reactions.

Perhaps the most successful application since FCC is the Sohio process. The reasons for the success are explained as follows. First, the process achieved high conversion and selectivity in FCB, even in a reaction system involving several parallel and consecutive reactions. Second, the process was industrialized directly by the use of fluid beds, rather than by passing through the stage of fixed beds. Callahan et al. (C1) have reported on development of the initial catalyst, bismuth molybdate, for the Sohio process.

D. Quality of Fluidization in Relation to Particle Properties

In industrial fluidized beds, it is always important to establish good fluidization, and especially so for high-performance catalytic reactions with conversions over 90% and selectivity over 70%. Volk et al. (V12) summarized the variables defining good fluidization and good gas–solid contact, as follows: (1) Gas entry—designed so that the gas entering the bed is well distributed. (2) Gas velocity—high enough to keep the solid in motion, but not so high that gas channeling occurs. (3) Bed height—relatively low. With other factors constant, the greater the bed height, the more difficult it is to obtain good fluidization. (4) Gas and solid densities—relatively high-density gas and low-density solid. The closer the ratio of densities of gas and solid, the easier it is to maintain good fluidization. (5) Particle size—a wide range of sizes gives more stable fluidization than particles of uniform size. (6) Reactor internals—serving the functions described earlier.

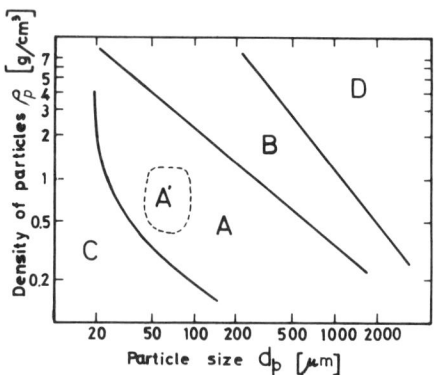

FIG. 3. Classification of particles by Geldart (G4, G5).

Geldart (B1, G4, G5) has classified fluidized particles into four groups, taking account of the fact that the behavior of fluidized beds is considerably affected by properties of the fluidized particles:

Group A. Powders having a particle density less than about 1.4 g/cm^3; in particular, those with porous structure and a mean diameter in the range of 20–100 μm.

Group B. Powders having a particle density in the range of 1.4 to about 4 g/cm^3, and a mean size in the range of 40–500 μm.

TABLE III

DESIRED PROPERTIES OF FLUIDIZED CATALYST PARTICLES

Characteristics	Optimal properties	Main reasons
Particle shape	Spherical, smooth surface	Fluidity; resistance against attrition
Bulk density [g/cm^3]	0.4–1.2	Good fluidization. If larger than this, the bed has a tendency toward slugging. If smaller, attrition and carry-over increase
Particle size distribution	Average 50–70 μm; fraction with ($d_p \leq 44$ μm) should be 10–20% for LPCR, 20–40% for HPCR[a]	Fluidity; good fluidization
Attrition resistance	Attrition rate <0.1%/hr	Minimize catalyst consumption, maintain particle size distribution and particle shape

[a] HPCR = high-performance catalytic reaction, such as ammonia oxidation; LPCR = low-performance catalytic reaction, such as FCC.

Group C. Powders in which surface forces become overwhelmingly important, e.g., all powders having a mean size less than about 30 μm.
Group D. Powders of size usually over 600 μm.

This classification is illustrated in Fig. 3, in terms of size and density of particles. Comparing Geldart's classification with Table I, we find that group A roughly corresponds to the typical fluid bed, and group B (as well as part of group D) corresponds to the typical teeter bed. According to Geldart, the behavior of gas bubbles and the effective viscosity of fluidized beds can be interpreted by a "surface/volume diameter" regardless of the particle size distribution.

Ikeda (I4) paid attention to the effect of the fines fraction ($d_p < 44$ μm), and deduced the optimal size range to obtain good fluidization and good gas–solid contact efficiency from the results of industrial operations (O10, V12, Z5; see Table III). His criterion for good fluidization is indicated as A' in Fig. 3, a target in developing or selecting catalysts for FCB.

II. Flow Properties of Fluid Beds

A. Particle Properties in Relation to Fluidity

1. *Effect of Fines*

Particle size distribution has a great effect on most aspects of fluidization. Zenz (Z4) pointed out that fluid–solid systems could be classified in terms of a bed viscosity, which is probably controlled by the rubbing of coarse particles against one another. According to Trawinski (T25, T26), the fines between coarse particles act as a lubricant, reducing bed viscosity. The minimum concentration of fines is the quantity required to coat each coarse particle with a monolayer of fines. Extending this aproach to multidispense systems, Zenz (Z4) calculated the optimum size distribution of fluidized particles, shown in Fig. 4 and expressed by a skewed normal probability distribution, which agrees with a typical size distribution of FCC particles.

Ikeda *et al.* (I5, I10), investigating the effect of fines on bed expansion, pressure fluctuation, and discharge rate of particles from a 8-cm-i.d. fluid bed, found that bed expansion and discharge rate increase and pressure fluctuations decrease with increasing content of fines (<44 μm). The range of particles smaller than 44 μm is called "fines fraction"; its role has been summarized by Ikeda (I2) as shown in Table IV. The desirable properties of fluidized particles utilized as catalyst were given by Ikeda (I2) as

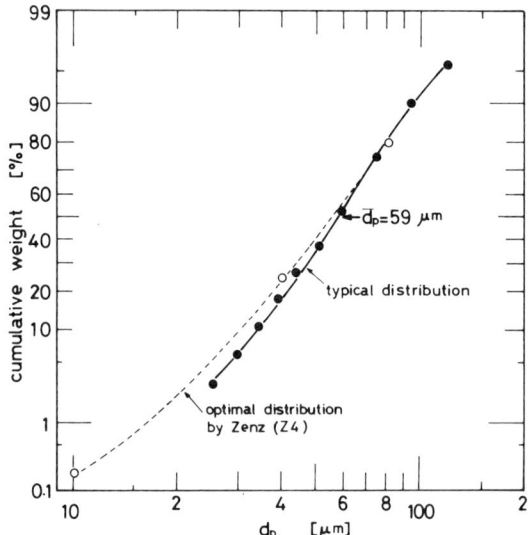

FIG. 4. Typical size distribution of particles used in fluid bed.

shown in Table V. Morse and Ballow (M44), de Groot (D7) and Matsen (M10) have also recognized that the quality of fluidization is improved by using smaller particles with a wider size distribution. Geldart (G4, G5) classified the fluidized beds into four categories characterized by density difference ($\rho_p - \rho_g$) and mean size of particles, as presented in Section I. The terms "fluid bed" and "teeter bed" (Squires, S14) seem to correspond to group A and group B, respectively, although the criteria which distinguish among the groups are not well established.

2. *Expansion of the Emulsion Phase*

The expansion of a gas–solid fluidized bed at gas rates between the minimum-fluidization velocity and the initial bubbling velocity was studied in 1966 by Davies and Richardson (D6). This homogeneous expansion of the bed is considerably larger when finer particles are fluidized. Several studies have been carried out on bed expansion (D6, D15, M41); on particle–particle forces and interactions (D14a, M6, R9), and on the flow rate of gases through beds (D6) under the condition of particulate fluidization.

Figure 5 shows typical expansion data measured by Morooka *et al.* (M41) using unclassified FCC catalyst particles. As soon as the bed reaches minimum fluidization, it starts to expand. After bubbles start to form (i.e., under aggregative fluidization) the bed expands slowly and then

TABLE IV

Effects of the Change of Weight Fraction of Particles Smaller Than 44 μm ("Fines Fraction") (I2)

Change of weight fraction and its causes	Main effect
When (1) attrition resistance of particles is large, (2) catalyst recovery system is less efficient, and (3) fine particles are not supplied and coarse particles are not discharged during the operation, then particles smaller than 44 μm become low in weight fraction.	(1) Slugging prevails (2) Erosion of bed increases due to poor fluidization (3) Load on catalyst recovery system becomes unstable, and instantaneous load on the system increases (4) Pressure and temperature control becomes difficult (5) Yield of products decreases because side reactions come to prevail. Occasionally, afterburning takes place
When (1) attrition resistance of particles is small, (2) catalyst recovery system is efficient, and (3) coarse particles are not supplied and fine particles are not removed during the operation, then particles smaller than 44 μm become high in weight fraction.	(1) Channelling prevails (2) Large clusters of particles grow due to agglomeration (3) Operational troubles occur due to particle adhesion (Decrease of heat transfer, excess reaction at walls in freeboard, and clogging in pipes and cyclones will take place.) (4) Load on catalyst recovery system increases (5) Afterburning takes place (6) Temperature becomes unstable locally and side reactions prevail

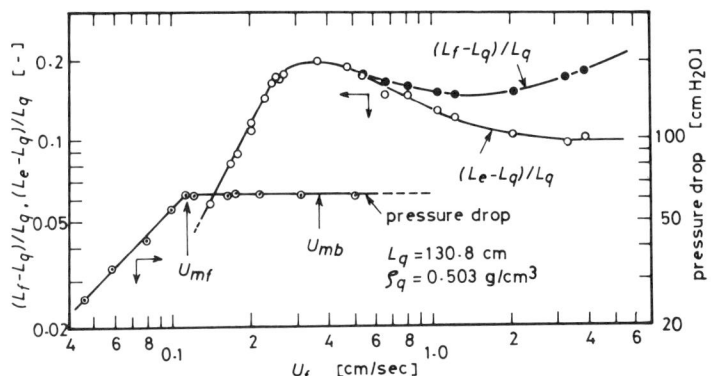

Fig. 5. Expansion and pressure drop of a fluid bed (after M41).

TABLE V

DESIRABLE PROPERTIES OF FLUIDIZED PARTICLES

Property of particles	Optimum values	Main interests of operation	Troubles in commercial-scale fluid bed	Supplementary methods of improvement
Chemical activity	Apparent contact time $3 < \tau < 30$ sec	Stability of operation Homogeneous temperature in bed $L_q/D_T = 1$–2	High activity: excess reaction in dilute phase Low activity: decrease of reaction yield	Dilution with inert particles Multistage fluid bed
Particle size distribution	Weight mean diameter $50 < \overline{dp} < 70\ \mu m$ $+80\ \mu m$, 5–20 wt.% $-40\ \mu m$, 10–30 wt.%	Improved fluidity Good fluidization Decreased catalyst loss and attrition	Severe erosion if particles are coarse Difficulty in recovering particles $<10\ \mu m$	Installation of internals Multistage fluid bed with horizontal baffles
Attrition resistance	Rate of attrition (Forsythe method) <0.1–0.5 wt. %/hr	Avoidance of catalyst loss Maintenance of optimum size distribution	Very fine particles produced by attrition (cf. Table IV)	Reduction of gas velocity injected through distributor
Shape of particles	Microspherical	Low tendency toward attrition Decreased erosion	High erosion rate when irregular hard particles are used	Installation of internals Blending with MS particles
Particle density	Below 2–3 g/cm³	Good fluidization	Unsatisfactory fluidization when high density particles are used	Blending with low-density particles Installation of internals

[a] From 12.

its volume remains constant or even decreases with increasing gas velocity; the volume of the emulsion phase is always smaller than at the bubble point. Thus it is not correct to select the minimum-fluidization bed height as the standard bed height for calculation of gas-bubble holdup. The equivalent height of emulsion phase in the aggregative bed can be determined by sedimentation, extrapolating the sedimentation curve to time zero (M41, R9). The results of Morooka *et al.* (M41) show that, at higher gas velocities, the expansion ratio of the emulsion phase $(L_e - L_q)/L_q$ is independent of gas velocity and bed diameter, as illustrated in Fig. 5; the sedimentation velocity corrected by the deaeration velocity equals the superficial gas velocity in the same manner as in a liquid–solid fluidized bed (R7).

The expansion ratios in the emulsion phase measured by Oltrogge (O5) and Morooka *et al.* (M41) at $U_f \gg U_{mf}$ are correlated in Fig. 6 with the parameter N_p. According to Miyauchi and Yamada (M32), the mean particle diameter in the parameter N_p is

$$\bar{d}_p = \left(\Sigma \Delta n \, d_p^5 / \Sigma \Delta n \, d_p^3\right)^{1/2} \tag{2-1}$$

The ratio $(L_e - L_q)/L_q$ is about 0.1 when typical FCC particles are fluidized.

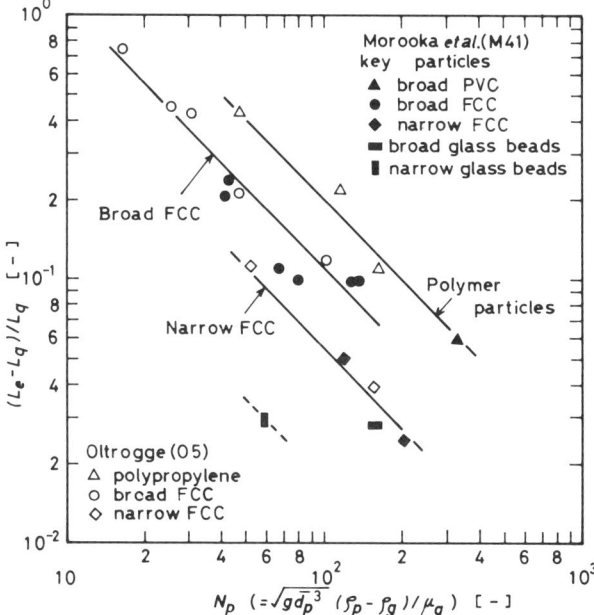

FIG. 6. Expansion ratio of emulsion phase (after M41).

B. Formation, Splitting, and Coalescence of Bubbles

The frequency of gas bubbles which are formed steadily through an orifice in a fluidized bed has been studied by Harrison and Leung (H6). Their results show that the mechanism of chainlike bubble formation is the same as that in an inviscid liquid. If all of the excess gas above the minimum fluidization velocity passes through in the form of gas bubbles, the diameter of a sphere having the same volume as the originated bubble is represented by two equations (in units of cm/sec). Van Krevelen and Hoftijzer (V6) found that

$$d_{bi} = 0.375[A_T(U_f - U_{mf})/n_d]^{2/5} \qquad (2\text{-}2)$$

where A_T is the bed's cross-sectional area, and n_d is the number of orifices. Harrison and Leung (H6) have provided a similar relation:

$$d_{bi} = 0.347[A_T(U_f - U_{mf})/n_d]^{2/5} \qquad (2\text{-}3)$$

where n_d is the number of holes in the perforated plate used as gas distributor. The initial bubble diameter measured by Miwa et al. (M17), Hiraki and Kunii (H13), and Chiba et al. (C4) shows good agreement with Eq. (2-2). The initial equivalent diameter of bubbles formed through a porous plate is given by Miwa et al. (M17):

$$d_{bi} = 0.00376(U_f - U_{mf})^2 \qquad (2\text{-}4)$$

where d_{bi} and $(U_f - U_{mf})$ are in units of cm and cm/sec, respectively. Comparisons between the experimental results and the equations are shown in Fig. 7.

Intensive studies have been carried out on the ascending bubble diameters in free fluidized beds (C5, K27, R14, R16, W9). Various correlations for estimating bubble diameters have appeared (M36, R13, W9). However, the particles utilized in these experiments belong to group B of Geldart's classification. For this type of particle, bubble diameters are expressed as a function of bed diameter, of distance of the bubble above the distributor, of initial bubble diameter, and of physical properties of the fluidized particles. Mori and Wen (M36) emphasized the former three factors and proposed the equation:

$$(d_{bM} - d_b)/(d_{bM} - d_{bi}) = \exp(-0.3z/D_T) \qquad (2\text{-}5)$$

where d_{bM} is a maximum bubble diameter, estimated as

$$d_{bM} = (1.87)(0.347)[A_T(U_f - U_{mf})]^{2/5} \qquad (2\text{-}6)$$

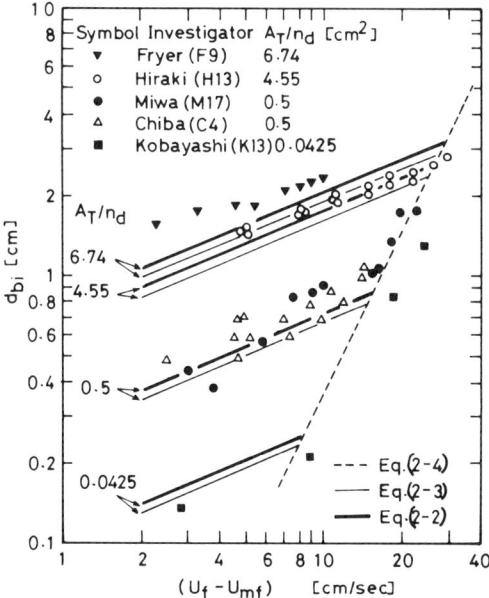

FIG. 7. Comparison of the observed initial bubble diameter with calculated diameter (after M36).

The ranges of data from which their correlation was obtained are

$$0.5 \leq U_{mf} \leq 20 \text{ cm/sec}, \quad U_f - U_{mf} \leq 48 \text{ cm/sec}$$
$$0.006 \leq \bar{d}_p \leq 0.045 \text{ cm}, \quad D_T \leq 130 \text{ cm}$$

The correlation of Werther (W9) gives nearly the same results as Eq. (2-5). Both correlations predict growth of bubble diameter and decrease of bubble frequency due to coalescence. However, experimental data in fluid beds have been excluded from the above correlations.

In contrast to bubbles in teeter beds, naturally occurring bubbles in fluid beds appear to split and recoalesce very frequently, and occasionally a bubble frequency will increase at a height further from the distributor. Rowe (in D5) and Toei *et al.* (T17) studied the mechanism of bubble splitting. Toei *et al.* (T17) found that small downward cusps were generated at the two-dimensional bubble surface near the stagnation point, which grew and thereby caused bubble deformation and, frequently, bubble splitting. Figure 8 shows the effect of particle diameter on the average total frequencies of bubble disturbance and splitting. The results of Toei *et al.* (T17) indicate that the frequency of disturbances or splittings decreases with increasing particle diameter, and is independent of bubble

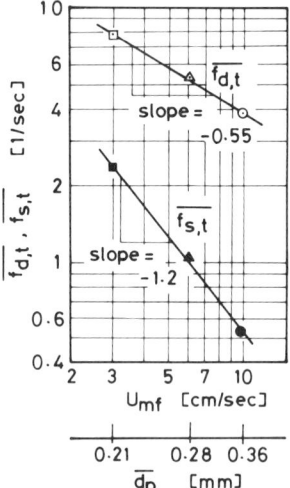

FIG. 8. Effect of the particle diameter on average total frequencies of disturbances $\bar{f}_{d,t}$ and splittings $\bar{f}_{s,t}$ (T17).

diameter. Therefore bubbles which form in a fluid bed with fine particles have more tendency to split than to coalesce, as discussed in Section V,C.

Morooka *et al.* (M43) measured properties of bubbles in fluid beds with FCC particles. The fluid beds were 7.9 and 19.5 cm i.d., and L_f was 100–200 cm. Bubble signals were detected by a hot-wire probe, originally designed by Yamazaki and Miyauchi (Y3–5), which consists of a pair of parallel tungsten wires 10 μm in diameter with a span of about 3 mm and a separation of 12 mm. The electrical resistance of the hot wires changes with the density difference between the bubble and emulsion phase. The disturbance induced by this probe is presumably not greater than that of the miniature capacitance probes of Werther (W6) and Burgess and Calderbank (B17). Figure 9 shows the velocity of bubbles ascending along the central axis of fluid beds. The effects of bubble diameters, gas distributors, and axial position of the probe on the ascending bubble velocity are rather small.

The measured bubble velocity is much greater than predicted from equations derived by Davidson and Harrison (D3) and Taylor (T8), respectively:

$$u_b = U_f - U_{mf} + 0.711(gd_b)^{1/2} \qquad (2\text{-}7)$$

and

$$u_b = 0.711(gd_b)^{1/2} \qquad (2\text{-}8)$$

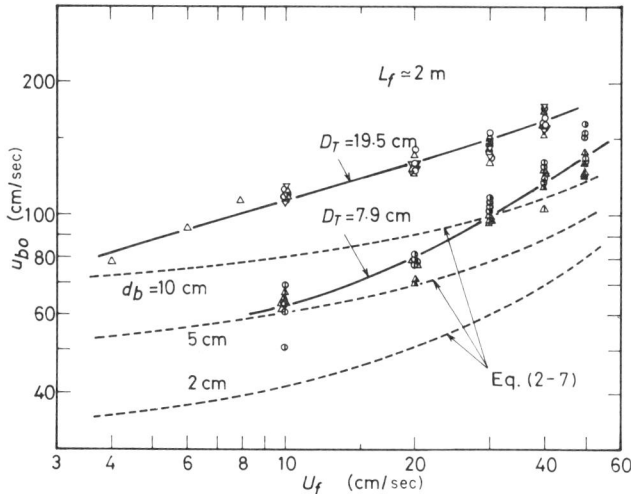

FIG. 9. Velocity of bubbles ascending along the center of a fluid bed (M43).

The reason, pointed out by Turner (T28), is that the bubble velocity in a fluidized bed *without circulation* will obey Eq. (2-8), but that bubbles in fluid beds ascend in an intense circulatory flow of the emulsion phase.

The arithmetic mean of bubble signals measured by Morooka *et al.* (M43) is shown in Figs. 10 and 11. Fig. 10 illustrates the effect of superficial gas velocity on the mean length of bubble signals. Some slugs appear in the range of $U_f < 30$ cm/sec at the upper part of a 7.9-cm-i.d. bed. Figure 11 shows the relationship between the mean length of bubble signals and axial probe positions for operations at higher gas velocity. If an initial bubble diameter is larger than the equilibrium diameter, the bubble has a tendency to split.

Actual bubble diameters can be calculated from the mean bubble signals according to the relationship given by Anderson (A6) and Ueyama and Miyauchi (U1). Assuming (1) the bubbles are spherical, (2) an ascending velocity is independent of a bubble diameter, and (3) the wall effect is negligible, the following equation is obtained:

$$M(l) \, dl = 2l \int_l^\infty \frac{F(d_b) \, d(d_b)}{\int_0^\infty \xi^2 F(\xi) \, d\xi} \, dl \qquad (2\text{-}9)$$

where $M(l)$ is a distribution function of bubble signals measured with a point-shaped probe, and $F(d_b)$ is a distribution function of bubbles actually

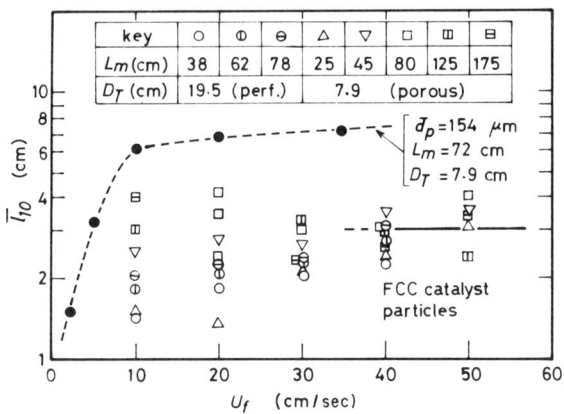

FIG. 10. Mean length of bubble signals in fluid beds (M43); perf. ≡ perforated.

existing in a fluid bed. Integration of Eq. (2-9) yields:

$$\frac{\int_0^\infty d_b^m F(d_b)\, d(d_b)}{\int_0^\infty d_b^n F(d_b)\, d(d_b)} = (m/n) \frac{\int_0^\infty l^{m-2} M(l)\, dl}{\int_0^\infty l^{n-2} M(l)\, dl} \qquad (2\text{-}10)$$

where m and n are positive integers. The actual volume-surface mean of bubbles is calculated with Eq. (2-10) as follows:

$$d_{32} = \frac{3}{2} \frac{\int_0^\infty l M(l)\, dl}{\int_0^\infty M(l)\, dl} = 1.5 \bar{l}_{10} \qquad (2\text{-}11)$$

Under the conditions of Morooka et al., d_{32} is estimated as 4.5–6.5 cm.

Matsen (M10), in studies of bubble diameters in a 61-cm-i.d. fluid bed, recognized that small difference in particle size distribution could produce significant changes in the character of fluidization. With Coke-5 particles ($\bar{d}_p$ at 50 cumulative wt.% = 70 μm, ρ_p = 2.1 g/cm^3), his probe measurements showed initial formation of 25-cm-diameter bubbles, which broke into smaller stable bubbles of ~6–13 cm. The maximum stable bubble size in a 13.8-cm-i.d. fluid bed with 26-μm particles was estimated as 2.5 cm. Massimilla (M3) also reported that smooth fluidization was obtained with fine particles.

Werther (W9) measured bubble diameters in a 100-cm-i.d. bed, using quartz sand (U_{mf} = 1.35 cm/sec) and spent FCC particles (U_{mf} = 0.20 cm/sec). In the quartz sand bed a rapid increase was observed in average bubble size with height above the distributor, whereas in the FCC bed

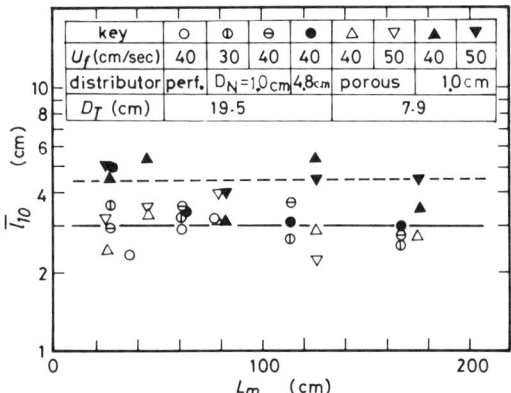

FIG. 11. Mean length of bubble signals in fluid beds operated under relatively high gas velocity (M43); perf. ≡ perforated.

bubble growth decreased with increasing height. When the FCC particles were fluidized at $U_f = 10$ cm/sec, the volume-surface mean of bubble diameter approached the equivalent value (3.5 cm) at 40–50 cm above the perforated plate. In a similar study, Ikeda (I5) calculated the effective bubble diameter to be about 8 cm in a 360-cm-i.d. baffled fluid bed (equivalent bed diameter = 70 cm, $U_f = 45$ cm/sec, $U_{mf} = 0.6$ cm/sec).

Tsusui (T27) studied the effect of particle characteristics in a fluid bed of 6-cm i.d., using a hot-wire probe. The difference of flow characteristics was most evident at high velocity of bubbles and high diameters, as shown in Fig. 12. Fresh FCC particles showed the smoothest fluidization, expected for the high ascending velocity. Removing fine particles ($d_p \leq 44$ μm) from original FCC particles gave a low ascending velocity and larger bubbles. The heavier MS catalyst showed very low ascending velocity and a typical slugging flow. Silica microspheres, which are very light ($\rho_p = 0.3$ g/cm^3) and average around 70 μm in diameter with a very narrow particle size distribution, showed a fluidity similar to that of the FCC particles without fines. The actual bubble size distributions, $F(d_b)$, are shown in Fig. 13.

All these results indicate that bubble diameters in fluid beds are strongly influenced by the phenomenon of bubble splitting. Kehoe and Davidson (K8, K9) observed slugging in a fluid bed with 62-μm catalyst particles, although their results are presumably due to the narrow size distribution of particles. An experiment conducted by Lanneau (L2) showed the presence of large bubbles in a 7.6-cm-i.d. fluid bed with typical fine microspherical alumina catalyst; however, his capacitance tip measured 4.8 × 6.4 × 19 mm. Since many bubbles present were probably smaller than the

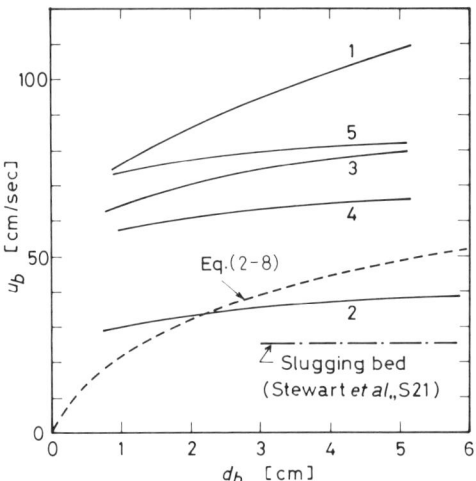

FIG. 12. Ascending velocity of bubbles under the conditions of $U_f = 20$ cm/sec, $D_T = 6$ cm, and $L_q = 40$ cm (T27). (1) FCC particles, (2) MS catalyst, (3) FCC catalyst without fines, (4) silica balloons, (5) heavy fluid bed catalyst.

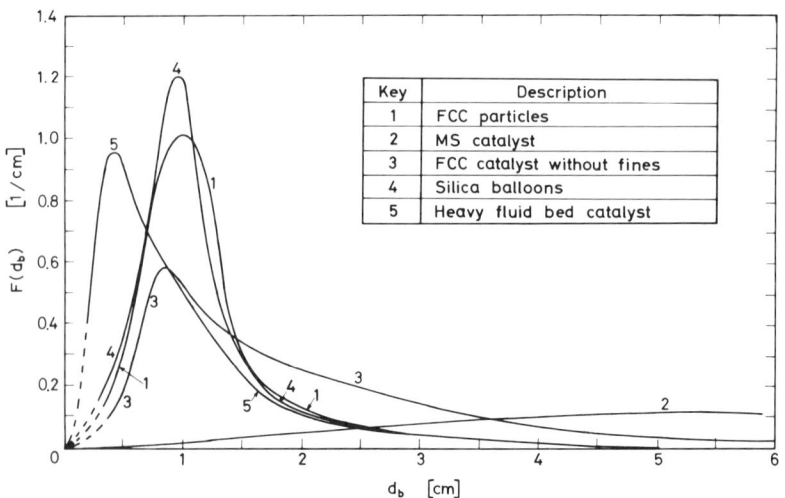

FIG. 13. Frequency distribution functions of actual bubble size (T27). Experimental conditions are the same as in Fig. 12.

probe tip, as shown in Fig. 13, Lanneau's measurement seems to show the maximum size present under given conditions.

C. Operation of Fluid Beds

Operation of the bed is closely related to the properties of particles discussed in Section II,A. Fluid beds are usually operated in the turbulent-flow region to obtain good contact and sufficient throughput of reactant gas. In general, the superficial gas velocity U_f in fluid beds is considerably larger than the minimum fluidizing velocity U_{mf}; the ratio of U_f/U_{mf} is usually more than 100. Also U_f/u_t is more than unity, where u_t is the terminal velocity of an average-size particle. Bubble size does not increase to an unlimited extent, but approaches a certain limiting value determined by the turbulence intensity; when large bubbles are injected into a turbulent bed by a single nozzle, the bubble size reduces to an asymptotic value as bubbles ascend (M43).

In commercial-scale fluid beds, a superficial gas velocity usually more than 30 cm/sec is preferred to provide sufficient turbulence, fluidity, and throughput of reactant. Increase of the contact efficiency by increasing the gas velocity has been observed by Lewis *et al.* (L12) and Gilliland and Knudsen (G7).

However, high gas velocity, may lead to problems in bed operation. It increases the entrainment loss of fluidizing catalyst particles. It also may give rise to excessive reaction in the particle-disengaging space, which sometimes will lead to reactor instability or to decreased selectivity (as discussed in later sections). Attrition or erosion of the reactor is more likely at higher gas velocity. Thus, there is an optimum gas velocity for fluid beds, which for most catalysts is in the range of 20–80 cm/sec, usually 40–60 cm/sec.

Operation of a fluid bed depends on both height and diameter. Slugging properties have been studied extensively by Davidson and co-workers (H17, K8, K9, S21). Generally the ratio L_f/D_T is much larger than unity for small-scale reactors (C2). For solids particles in a teeter bed, this ratio will lie well into the slugging region. However, the fluidity in fluid bed is quite different from that in a teeter bed, as explained in Section II,A.

D. Properties of Bulk Flow Inside Fluid Beds

1. *Average Holdup of Gas Bubbles*

Leva (L7) proposed a correlating equation for expansion in aggregative fluidized beds, but poor agreement between calculated and observed voidage was found for fluidization of FCC particles with atmospheric air (Z5). Bed expansion can be expressed in terms of average bubble holdup, which also enters the calculation of contact efficiency (Section VII). Average bubble holdup is defined by

$$\bar{\epsilon}_b = (L_f - L_e)/L_f \qquad (2\text{-}12)$$

where L_e is the equivalent height of the emulsion phase measured by the flow interruption method (R9). Experimental data by Morooka *et al.* (M40) and van Swaay and Zuiderweg (V8) show (in Fig. 14) that the effect of bed diameter on the bubble holdup is very small. At higher gas velocities, experimental values of bubble holdup in fluid beds agree with those in bubble columns. At $U_f = 50$ cm/sec, $\bar{\epsilon}_b$ does not exceed 0.5, and a fluidized-bed height is at most twice that of a settled bed. In contrast with fluid beds, the expansion of teeter beds is a function of both particle properties and bed diameters (H17, K8, K9).

The data of Wilhelm and Kwauk (W13) plotted in Fig. 14 reveal higher values of $\bar{\epsilon}_b$ because of severe slugging in a small bed. Figure 15, given by de Groot (D7), shows that bubble holdups in beds with narrow-range silica are much smaller than with broad-range silica. Thus, in larger fluidized beds of silica with a narrow size distribution, very large bubbles form with high ascending velocities. The effect of temperature and pressure of the

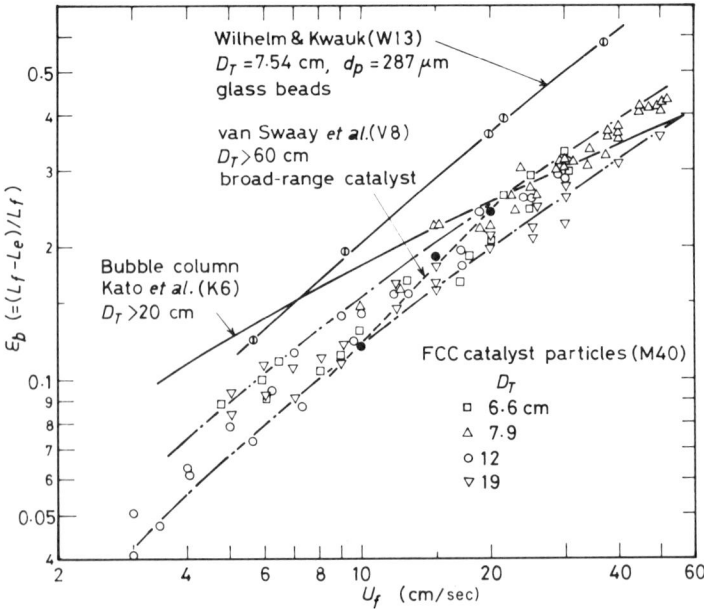

FIG. 14. Holdup of gas bubbles in fluid bed (after M40).

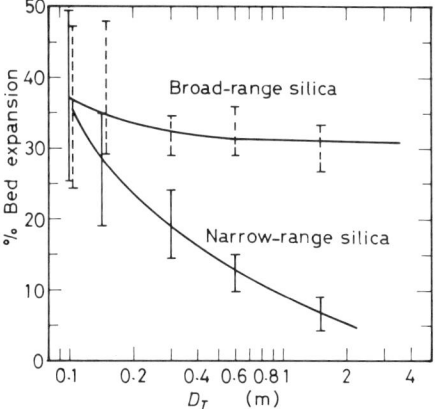

FIG. 15. Bed expansion as a function of bed diameter at $U_f = 20$ cm/sec. (D7).

fluidizing gas on bed expansion has been studied by de Vries *et al.* (D8). As Fig. 16 shows, the influence of the gas pressure is not large for wide-range microspherical silica. These authors mention that emulsion-phase fluidity is quite good at high temperature.

2. *Lateral Distribution of Bubble Holdup*

El Halwagi and Gomezplata (E8), Larroux *et al.* (L3) and Nishinaka *et al.* (N7) studied lateral distributions of bubbles in fluid beds. Their results show that bubbles in a dense phase are likely to ascend in the central part of a bed.

Nishinaka *et al.* (N7) have determined the lateral distribution of gas bubbles in fluid beds with diameters of 6.6, 12, and 19 cm, as plotted in

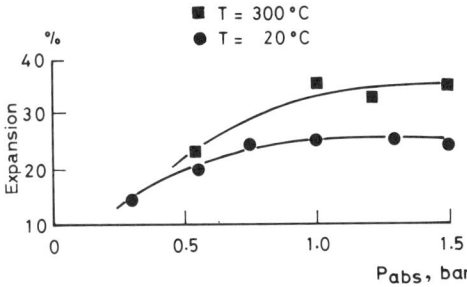

FIG. 16. Fluid bed expansion for SCP catalyst as a function of pressure (D8). Fraction <44 μm $\approx 20\%$, $U_f = 18$ cm/sec, and $D_T = 10$cm.

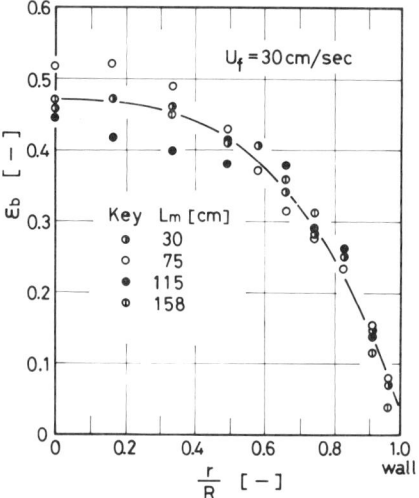

FIG. 17. Lateral holdup distribution of gas bubbles in fluid bed with 12-cm inner diameter (N7).

Fig. 17. The distribution is nearly independent of axial position. The holdup of gas bubbles at radial distance r from the center axis of the bed is expressed as

$$(\epsilon_{b0} - \epsilon_b)/\epsilon_{b0} = [(\epsilon_{b0} - \epsilon_{bw})/\epsilon_{b0}](r/R)^n \qquad (2\text{-}13)$$

where ϵ_{b0} is the holdup of gas bubbles at the center axis of the bed, and ϵ_{bw} is the holdup of gas bubbles extrapolated to the bed wall and may be taken as approximately zero. Parameter n in Eq. (2-13) is variable with gas velocity and bed diameter (Figs. 18 and 19). Their experiments were

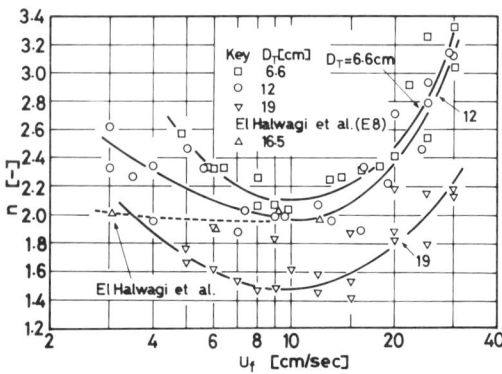

FIG. 18. Effect of superficial gas velocity on n (N7).

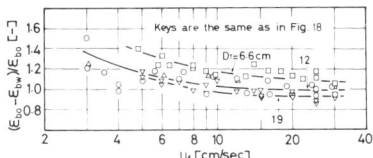

FIG. 19. Effect of superficial gas velocity on $(\epsilon_{bo} - \epsilon_{bw})/\epsilon_{bo}$ (N7).

carried out with FCC particles fluidized in beds of $L_f/D_T \geq 2$. Different distribution behavior has been reported for teeter beds with $U_f/U_{mf} \leq 10$ and $L_f/D_T \leq 1$ (W7,8).

The lateral distributions of bubble holdup in fluid beds are very similar to those of bubble columns for gas–liquid systems. Experimental results by Akehata et al. (A2), Pozin et al. (P6), Ivanov and Bykov (I12), Yamagoshi (Y2), Miyauchi and Shyu (M31), Hills (H9), Kato et al. (K7) and Ueyama and Miyauchi (U3) are reasonably well expressed by Eq. (2-13). When a gas phase is distributed with a perforated plate or a nozzle sparger, the parameters in Eq. (2-13) for the bubble columns are in the range of $n = 1.7–2.5$.

3. *Flow Patterns in Fluid Beds*

The lateral distribution of gas bubbles induces bulk recirculation of the emulsion phase. For teeter beds, the flow pattern of solid particles has been studied by Werther (W7, W8), Burgess and Calderbank (B17), Oki and Shirai (O3) and Whitehead et al. (W12), in experiments carried out with alumina particles, quartz sand, and petroleum coke under conditions of $\bar{d}_p \geq 83$ μm and $(U_f - U_{mf})/U_{mf} \leq 14$. The mode of bulk circulation was centrally descending and peripherally ascending for $L_f/D_T \leq 1$, while the ascending flow moved toward the center with increasing L_f/D_T. Werther (W8) showed that this circulation flow is caused entirely by bubbles which carry solid particles upward in their wakes.

In contrast to the teeter bed, less work has been carried out on the bulk flow pattern in fluid beds (L11, M29). Measurements at the relatively high gas velocities of practical interest ($U_f > 10$ cm/sec, $U_f/U_{mf} \gg 1$) show that the rate of circulation exceeds that of solid particles conveyed by the bubble wake, and results from the buoyant force induced between the centrally ascending bubble-rich phase and the peripherally descending emulsion phase of low bubble content (see Fig. 2). The behavior greatly resembles that in bubble columns, as will be discussed quantitatively in a later section (H9, K15, M31, P3, P6, T23, U3, Y2, Y22, and Kato and Morooka, unpublished data). For a bubble column, Shyu and Miyauchi

(S13) have proposed a semiempirical equation for the circulating liquid velocity at the center axis:

$$u_{l0} = 6.8 U_f^{0.5} D_T^{0.28} \qquad (2\text{-}14)$$

where u_{l0} and U_f are in cm/sec and D_T is in cm.

In fluid beds, the ascending velocity of bubbles at the center axis has been measured by Morooka et al. (M43) and Tsutsui and Miyauchi (T27) with the assumption that the slip velocity of bubbles relative to the adjacent emulsion equals the ascending velocity of a single bubble; the circulation velocity of the emulsion phase along the axis is calculated by

$$u_{e0} = u_{b0} - 0.711(gd_b)^{1/2} \qquad (2\text{-}15)$$

When the fluid bed is operated under relatively high gas velocity, the mean bubble size changes only slightly with axial bed height and depends essentially on the flow intensity. Under these flow conditions, u_{e0} of fluid beds and u_{l0} of bubble columns are seen to be nearly equal, as shown in Fig. 20. This indicates that the mechanism of circulation in the fluid bed agrees well with that in the bubble column. Studies on solid circulation in a fluidized bed with a draft tube have been demonstrated by Yang and Keairns (Y7, Y8) and LaNauze (L1); their models also are based on a density-driving force between annulus and draft tube.

4. *Apparent Emulsion Viscosity*

Fluidized beds have many liquidlike properties. An apparent viscosity is often assigned analogous to Newtonian fluids, measured by means of

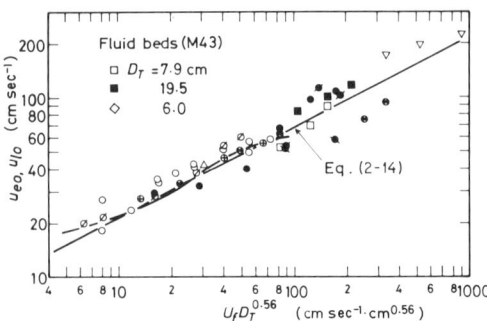

FIG. 20. Circulation velocity ascending at the center of fluid bed and bubble column. Data of bubble columns are measured by: ⊗, Pavlov (P3), D_T = 17.2 cm; ⊕, Pozin et al. (P6), 30.0 cm; △, Yamakoshi (Y2), 25.0 cm; curve, Yoshitome and Shirai (Y22), 15.0 cm; ●, Miyauchi and Shyu (M31), 10.0 cm; ○, Hills (H9), 13.8 cm; ▽, Ueyama and Miyauchi (U3), 60.0 cm; ◕, Koide et al. (K15), 504.0 cm; ⌀, Kato and Morooka (unpublished data), 12.0 cm.

Stormer viscometers, concentric cylinder viscometers, rotating spindles, falling spheres, etc. Because these viscometers expend part of their energy in accelerating the particles, this produces change in their orientation, and because voidage in the bed is affected by the immersed objects, the data on apparent viscosity of fluidized beds have to be carefully examined.

Most previous work has been reviewed by Grace (G13) and Schügerl *et al.* (in D5). Botterill (B12) and Hetzler and Williams (H8) correlated the apparent viscosity of liquid- and gas-fluidized systems, applying a free-volume theory which may be used successfully for glass-forming (polymeric) liquids. Saxton *et al.* (S3) proposed another approach to a free-volume theory. They compared theoretical expectations with the experimental data obtained in liquid-fluidized systems. Their extension to gas-fluidized systems (in cgs units) became, for sand,

$$\mu_b = \frac{2.00 \times 10^4 (1 - \epsilon_f) d_p^{3/2}}{(1 + 304 d_p^{3/2}) U_f^{1/4}} \qquad (2\text{-}16)$$

and for polyvinyl acetate powder,

$$\mu_b = \frac{6.80 \times 10^3 (1 - \epsilon_f) d_p^{3/2}}{(1 + 211 d_p^{3/2}) U_f^{1/4}} \qquad (2\text{-}17)$$

[μ_b (g/cm sec), d_p (cm), and U_f (cm/sec)]. The former expression coincides roughly with the data of Schügerl *et al.* (in D5), but the latter gives values smaller than the data of Furukawa and Ohmae (F20) by a factor of 2. This implies an additional dependence of μ_b on d_p.

Grace (G13) studied the relationship between the included angle of spherical-cap bubbles and the apparent viscosity, and found that apparent viscosity in fluidized beds was between 4 and 13 poise. Hagyard and Sacerdote (H20) measured the bed viscosity by means of the decay of oscillation of a vertical cylinder immersed in the bed. The kinematic viscosity μ_b/ρ_b decreased from about 5 cm²/sec to 1–0.5 cm²/sec.

The general tendencies of the experimental results are as follows: The apparent viscosity in an unfluidized bed is very high (perhaps infinite) but drops rapidly with increasing gas velocity. When particles are fully fluidized, the bed viscosity becomes rather independent of gas velocity and increases with increasing particle size and particle density; i.e., it shows a gradual change from a displacement effect to a collisional momentum transfer.

Furthermore studies on the effect of particle size distribution, by

Matheson *et al.* (M7), show that relatively small amounts of fines added to coarse particles sharply decrease the bed viscosity. Based on this evidence, Zenz (Z4) has calculated an optimum size distribution of fluidized particles (as mentioned in Section II,A,1). The importance of the apparent viscosity in relation to flow characteristics of the bed has been stressed by Matheson *et al.* (M7), Rice and Wilhelm (R5), Finnerty *et al.* (F3), Grace (G13), and others. Matheson *et al.* (M7) have found that the slugging tendencies of a bed can be expressed in terms of the apparent viscosity. Their results for FCC particles containing 20 wt.% of 46-μm-diameter fine particles gave μ_b equal to 6×10^{-2} g/cm · sec; this small bed viscosity agreed with the visible high fluidity of the bed.

Meanwhile, Rice and Wilhelm (R5) have investigated the stability of interfacial surface between the particle-free gas phase and the dense-bed phase. According to their analysis, the lower bed interface (or upper interface of a bubble) is inherently unstable, whereas the upper bed interface (or lower interface of a bubble) is inherently stable. Thus an instability initiated at the upper bubble surface may reduce bubble size; this tendency is presumably stronger in a bed of lower viscosity. The apparent viscosity which applies to the surface dynamics has been studied by Finnerty *et al.* (F3), Takamura (T2) and Morooka and Kato (unpublished data). They have measured wavelengths, frequencies, and attenuation rates for waves generated at the bed surfaces. In the experiments of Takamura (T2) and Morooka and Kato (unpublished data), rectangular fluidized beds of 6×60 cm and 8×80 cm were utilized, respectively, and bed heights were kept at about 30 cm. The attenuation rates of surface waves were calibrated with glycerol–water solutions. Their results, shown in Fig. 21, indicate that the apparent viscosity of FCC particles is the smallest among particles employed in the experiments.

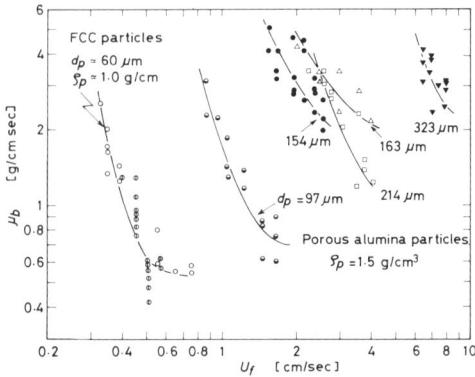

FIG. 21. Apparent viscosity of fluidized bed (T2 and Morooka and Kato, unpublished).

E. Axial Distribution to Bed Structure

1. Bed Density

With increasing gas velocity, the bed surface diffuses gradually, and a dilute phase is formed on the dense bubbling bed owing to entrainment of fluidizing particles. This axial distribution of bed density has been measured by several workers (B4, F2, L13, M25, M33). Fig. 22 shows the axial distributions of bed density measured by Miyauchi *et al.* (M33) in 1969. Their experiment was performed in a 7.9-cm-i.d. fluid bed of FCC particles having a mean diameter of 58 μm.

The influence of the amount of particles in the bed on the axial porosity distribution has been reported by Bakker and Heertjes (B4) for glass beads of 175–210-μm-diameter, fluidized in a 9.0-cm teeter bed. Their results show that the quantity of axially suspended particles in the dilute phase is increased noticeably by increasing the initial bed mass from 200 to 2000 g. In contrast with this behavior, the bed density distribution given by Kaji (K1) for the dilute phase shows less dependence on the initial bed height L_q under constant U_f. He has measured the distribution for a 7.9-cm-diameter bed of FCC catalyst, having a mean size of 58 μm, by varying L_q at 32, 51, and 71 cm, and U_f for 27–58 cm/sec. The mean bed density in a dense fluid bed can be calculated by the following equation knowing the mean bubble holdup in the dense fluid bed given in Section II,D,1:

$$\rho_b = (1 - \bar{\epsilon}_b)\rho_q L_q / L_e \qquad (2\text{-}18)$$

2. Transition Zone and Return Flow of Emulsion

A transition zone exists between the dense and dilute phase. In this region, particle density reduces gradually to the low level of the dilute

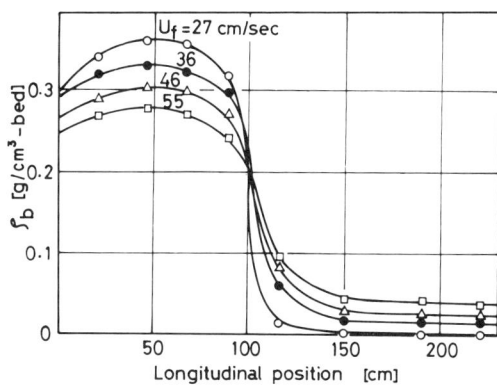

FIG. 22. Longitudinal bulk density distribution (M33).

phase. A circulating flow is formed in the transition zone, changing the direction of flow. When there happens to be an upward flow of the emulsion, the ascending velocity of bubbles is presumably expressed by $u_e + 0.711(gd_b)^{1/2}$, where u_e is the circulating velocity of emulsion. When the circulation returns, the rising velocity of bubbles becomes slowed in the upper part of the dense phase, which gives rise to greater holdup of the bubble phase. This is probably a reason for the large mass-transfer rate in the transition zone (described in Section VIII). Other causes of such good contact in the transition zone may be coalescence and rupture of bubbles at the end of the dense phase, the latter giving great disturbance in the dense phase and pushing particles into the dilute phase. The return flow of the emulsion also occurs at the vicinity of the distributor, and here the holdup of bubbles is reported to be larger than in other parts of the dense phase (Basov et al., B6). Although the flow pattern of bubbles is different in the transition zone, good contact in this zone is experimentally justified as written in Section VIII.

3. *Properties of the Dilute Phase*

The gas leaving the top of the dense phase carries entrained particles with it. Zenz and Weil (Z6) studied the mechanism of particle entrainment from nonslugging fluid beds, observing a transport disengaging height (TDH), above which the rate of decrease in entrainment approaches zero. The TDH, in most cases the design optimum for location of cyclones, is dependent on bed diameter and superficial gas velocity. An empirical correlation for estimating the TDH given by Zenz and Weil (Z6) largely agrees with the results of Tanaka and Shinohara (T6). According to Zenz and Weil (Z6), the entrainment rate can be replaced by the saturation rate for a particle whose diameter equals the geometric mean (50 wt.%) of all particles with terminal velocities less than the superficial gas velocity. Measurements of entrainment from fluid beds have been carried out by Zenz and Weil (Z6), Overcashier et al. (O11), Lewis et al. (L13), Fournol et al. (F5), Hanada (H3), and Kono (K28). Figure 23 shows that their results are strongly affected by particle size and size distribution, particle density, and other factors. Calculations by the procedure of Zenz and Weil (Z6) predict smaller values than the experimental data for mixed-size particles, and larger values for uniform-size particles.

The entrainment rate and the density distribution below the TDH were intensively studied by Lewis et al. (L13) with closely sized particles. Their results on the entrainment rate were correlated by the following equation.

$$\epsilon/U_f = B \exp[-(b/U_f)^2 - aH] \qquad (2\text{-}19)$$

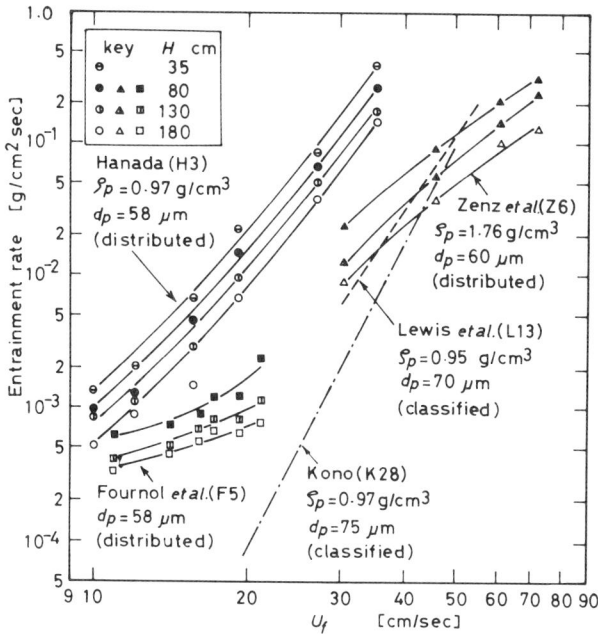

FIG. 23. Entrainment rate from the bed with cracking catalyst particles.

where H is freeboard height, and a, b, and B are parameters dependent on various factors; the effect of particle size distribution on the parameters is not known.

Elutriation—the selective removal of fines by entrainment—has been studied by several workers (K25, L6, M23, O8, T7, W2). Most of this work has been reviewed by Kunii and Levenspiel (K24) and Leva and Wen (in D5). However, Merrick and Highley (M13) point out that early correlations for elutriation rate constant are inaccurate for very fine particles, since they assume u_t proportional to d_p^2 (rather than to d_p) and hence predict that the elutriation rate constant reduces to zero as the particle size reduces to zero. Thus, care must be taken when such empirical equations are applied to a fluid bed containing fine particles.

F. Effect of Bed Internals

The uniformity and contact efficiency of a fluid bed can be improved by immersion of a surface within the bed (G15, I5, V12). Ikeda has defined a generalized equivalent diameter, considering that the commercial reactors have not only vertical internals but also horizontal baffles. Figure 24 indi-

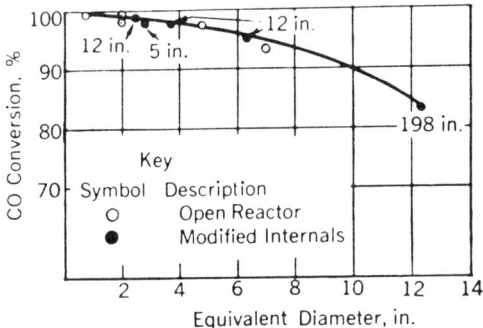

FIG. 24. CO conversion in Hydrocol reaction for several reactor diameters (V12).

cates that the equivalent diameter of a large reactor gives the same conversion as a small reactor of the same actual diameter. The proposed equivalent diameter is

$$(D_{eq})^{-1} = (D_{eV})^{-1} + (D_{eH})^{-1} \tag{2-20}$$

where D_{eV} and D_{eH} are the component equivalent diameters based on the vertical surfaces and the horizontal surfaces, respectively. Ikeda carried out an acrylonitrile synthesis by the Sohio process, in fluid beds with diameters from 8 cm to 3.8 m. Reaction yields were compared under the same reaction conditions and catalyst, and the effective bubble diameter was calculated by modifying the bubbling bed model originally proposed by Kunii and Levenspiel (K22, K23). The effective bubble diameter is given by

$$(d_b)_{eff} = 1.9(D_{eq})^{1/3} \tag{2-21}$$

where $(d_b)_{eff}$ and D_{eq} are both in cm.

Several investigations have been made on the flow characteristics in multistaged fluid beds. Nishinaka et al. (N6, N8, N9) have measured the average bubble holdup, the lateral distribution of bubble holdup, and the longitudinal dispersion of solid particles in four- and eight-stage fluid beds installed with various horizontal baffles. As shown in Fig. 25 the average bubble holdup (except for beds baffled with tube plates) is correlated by the equation of Nishinaka et al., (N8):

$$\bar{\epsilon}_b = 0.08 U_f^{0.75} A_r^{-0.3} N^{0.05} \tag{2-22}$$

where A_r is the free area of baffle in %, and N is a number of horizontal plates; U_f is in cm/sec. The longitudinal-dispersion coefficient of solid particles has been found to decrease with decreasing free area of baffles (N6). The intermixing mass velocity of solid particles between adjacent

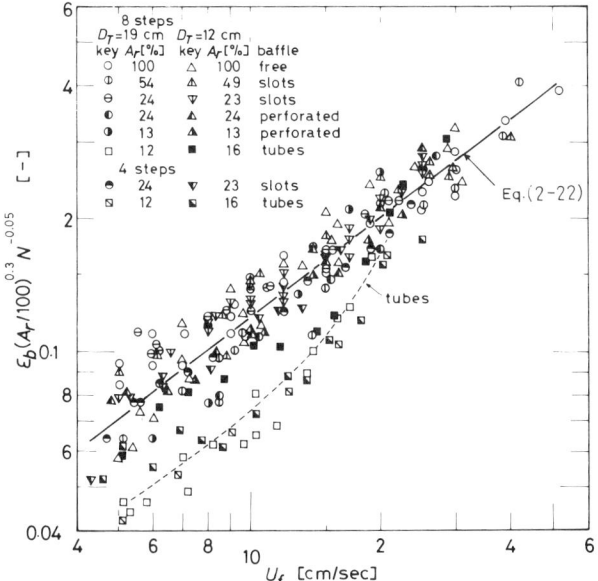

FIG. 25. Correlation of average gas bubble holdup in free and baffled fluid beds (N8).

stages is related to the longitudinal-dispersion coefficient of solid particles:

$$\text{Intermixing mass velocity} = \rho_p U'_s = E_{zs}\rho_b/\Delta L \qquad (2\text{-}23)$$

where U'_s is solid intermixing velocity based on particle density, and ΔL is the length of one stage. This particle-intermixing velocity is correlated in the same manner by Overcashier et al. (O11). In Fig. 26 the solid intermixing velocity is plotted against gz'/U_f^2 and is expressed by the following equations.

For slots and perforated plates

$$U'_s = 11 U_f^2/gz' \qquad (2\text{-}24)$$

and for tube plates

$$U'_s = 2.5 U_f^2/gz' \qquad (2\text{-}25)$$

where z' is the height of the dilute zone under each baffle. Other work has been carried out with regard to holdup of solid particles and axial distribution of bed density (B2, G16, N3), residence-time distribution of gas and solid particles (O11, R1) and entrainment from the bed (O11, H3). Most of these investigations have been reviewed by Harrison and Grace (in D5), Botterill (B12), and Verma (V10). An experimental study on mass-transfer

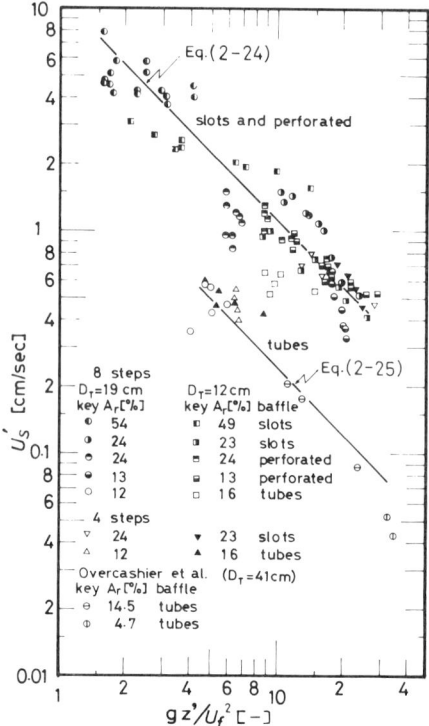

FIG. 26. Relationship between intermixing velocity of solid particles and gz'/U_f^2 (N8).

rate between the bubble and the emulsion phase in baffled fluid beds (M42) is mentioned in Section VI,B.

III. Turbulent-Flow Phenomena in Bubble Columns and Fluidized Catalyst Beds

In the preceding section, the flow properties of fluidized catalyst beds have been clarified mainly on the basis of experimental observations. In the case of FCC catalyst, the apparent viscosity of the emulsion is usually very small, and the emulsion shows good fluidity. Catalyst particles once charged into a fluidized bed reactor are usually in service for several months. Hence it is justifiable to prepare the particles very carefully, so that the fluidized bed shows the best fluidization possible. This kind of careful preparation is usually impractical in the case of single-pass particles such as coal, mineral ores, or grain.

It is interesting to see how far modern technology for fluidized catalyst beds has served to achieve good fluidization. Our criterion of good fluidization is a gas-fluidized bed of a low viscosity liquid (such as water), where the low-viscosity liquid would set the lower limit to the fluidity of the emulsion. Such a gas-fluidized liquid bed is the well-known bubble column, which has been studied extensively. Our objective is to understand the behavior in the recirculation flow regime, since the superficial gas velocity of practical interest is usually more than 30 cm/sec for fluidized catalyst beds and for these conditions intense recirculation of the emulsion has been observed (note Fig. 2 and Section II,D,3).

Currently available data for the flow properties of the fluidized catalyst bed are fragmentary, since the local motion of the emulsion phase is difficult to measure experimentally. Therefore, it is useful to clarify the flow properties of the bed in terms of our knowledge of bubble columns. First, the fluid-dynamic properties of the bubble columns will be explained; then, the available data will be adapted to apply to fluid catalyst beds. The reader will be able to picture an emulsion phase of carefully prepared catalyst particles operating in intense turbulence for fluidized beds under conditions of practical interest. This turbulence distinguishes the flow properties of fluid catalyst beds from those of widely studied teeter beds.

A. Two-Phase Bubble Flow in Recirculation

In a bubble column, as is widely known, the bubble swarm rises uniformly when superficial gas velocity is low (less than about 2 cm/sec) and uniform-size bubbles are released at a bottom gas-distributor. This type of uniform flow, called the bubble flow regime, has been utilized widely for various gas–liquid contacting operations (K21, O9, S1). The bubble-flow regime seems to correspond to the well-known bubbling fluidized bed, where, however, the bubbles grow to larger size due to coalescence as they rise. Upon increasing the gas velocity, the uniform bubble flow becomes unstable, and intense recirculation of mixed gas and liquid sets in due to buoyant forces induced between the centrally ascending bubble-rich phase and the peripherally descending bubble-lean phase. This condition, named the recirculation flow regime (P3, Y2, Y21), resembles the flow in fluid catalyst beds operated at practical gas velocities.

1. Simplified Theory of Recirculation for a Bubble Column

Theories of recirculation have been presented for the bubble column with large L/D_T, without continuous liquid feeding (H9, M31). The studies use the same equation of motion, but different assumptions for the turbulent kinematic viscosity.

The time-averaged flow pattern of liquid in a bubble column may be visualized as shown in Fig. 27, where the column is in recirculation flow, and the ratio L/D_T is large compared to unity. Under steady flow, the basic equation of motion (H9, M31) is:

$$-(1/r)\, d(r\tau)/dr = d\bar{P}/dz + (1 - \epsilon_b)\rho_l g \qquad (3\text{-}1)$$

where the left-hand side of the equation gives the force acting on a fluid element by radial shear-stress difference, and the right-hand side gives the forces acting by axial static-pressure gradient and by gravity. A similar equation of motion has been applied by Wijffels and Rietema (W16) to determine the flow properties of liquid–liquid spray columns.

The shear stress τ is related to the mean axial velocity of liquid through the sum of the molecular and turbulent kinematic viscosities:

$$\tau = -(\nu_M + \nu_t)\rho_l (du/dr) \qquad (3\text{-}2)$$

where ν_M is usually negligible in comparison with ν_t except in the vicinity of the column wall. The turbulent shear stress for the bubble column with $L/D_T \gg 1$ is expressed by the following equation (U2):

$$-\nu_t\, du/dr = (1 - \epsilon_b) u'_r u'_z \qquad (3\text{-}3)$$

where u'_r and u'_z are the radial and axial velocity fluctuations of liquid, respectively. Equation (3-3) is obtained by time-averaging the fluctuation velocity components for two-phase flow.

The boundary conditions to solve the basic equations, Eqs. (3-1) and (3-2), were given earlier (M31). As shown in Fig. 27, the liquid flow undergoes upflow between a and b and downflow between b and c, both with a developed turbulence, and also downflow in the laminar sublayer between c and d near the wall; at the wall, $r = R$ (or $y = R - r = 0$) and $u = 0$. Consistent with the concept of the universal velocity profile (S4, S8), a buffer layer should be present between the turbulent core and the laminar sublayer. However, it is hard to distinguish the laminar-turbulent buffer layer from the turbulent core, since bubbles from the turbulent core enter irregularly into the buffer layer and agitate it. Hence, partly to simplify the mathematical treatment and partly to satisfy a physically sound interpretation of this bubbling buffer layer, the thickness δ used for the laminar sublayer is taken somewhat greater than in single-phase flow and the buffer layer is neglected.

According to the concept of the universal velocity distribution, the velocity profile in the laminar sublayer is given by

$$u/v_* = v_* y/\nu_M \qquad \text{for} \quad v_* y/\nu_M \lesssim 5 \qquad (3\text{-}4)$$

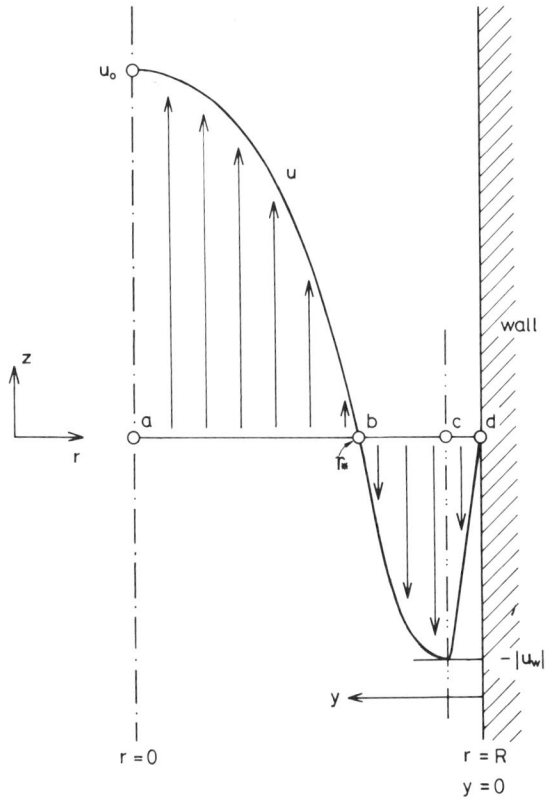

FIG. 27. Recirculation flow pattern in a gas bubble column.

where $v_* = (|\tau_w|/\rho_l)^{1/2}$ is the friction velocity. The velocity distribution in the turbulent core is

$$u/v_* = 5.75 \log(v_* y/\nu_M) + 5.5 \quad \text{for} \quad v_* y/\nu_M \gtrsim 70 \quad (3\text{-}5)$$

By equating Eqs. (3-4) and (3-5), one obtains a value y defined as the thickness δ of the laminar sublayer, for which:

$$v_* \delta/\nu_M = 11.63 \quad (3\text{-}6)$$

The velocity u_δ at $y = \delta$ is given by combining Eqs. (3-4) and (3-6):

$$u_\delta = 11.63 v_* = 11.63(|\tau_w|/\rho_l)^{1/2} \quad (3\text{-}7)$$

According to the analysis of many experimental data (M31, U5), the thickness δ is usually much smaller than the column radius R, so that u_δ essentially equals the peripheral velocity $|u_w|$ of the turbulent core. As a consequence, the boundary condition at $r = R$ for the turbulent core is given by

$$|u_w| \approx u_\delta = 11.63(|\tau_w|/\rho_l)^{1/2} \quad \text{at} \quad r = R \tag{3-8}$$

Another boundary condition is obviously

$$du/dr = 0 \quad \text{at} \quad r = 0 \tag{3-9}$$

2. *Solution of the Simplified Theory*

The solution of Eqs. (3-1) and (3-2) with the boundary conditions of Eqs. (3-8) and (3-9) has been given by Miyauchi and Shyu (M31) for the case without liquid feeding, and by Ueyama and Miyauchi (U5) for the case with continuous liquid feeding. The procedure here follows the latter reference, with extension to a simplified treatment (M27).

Multiplying by $2\pi r\, dr$ and integrating radially from $r = 0$ to $r = R$, Eq. (3-1) is transformed to

$$-d\bar{P}/dz = (2/R)\tau_w + (1 - \bar{\epsilon}_b)\rho_l g \tag{3-10}$$

where $\bar{\epsilon}_b$ is the mean gas holdup, defined by

$$\bar{\epsilon}_b = \int_0^R 2\pi r \epsilon_b\, dr/\pi R^2 \tag{3-11}$$

By introducing Eqs. (3-2) and (3-10) into Eq. (3-1) and neglecting ν_M in comparison with ν_t, the following basic equation is obtained for the turbulent core:

$$-(1/r)d[\nu_t r(du/dr)]/dr = (2/R)(\tau_w/\rho_l) - (\bar{\epsilon}_b - \epsilon_b)g \tag{3-12}$$

This equation is solved with respect to u under the simplifying assumptions (M31, U5) that ν_t is constant in the turbulent flow region, and that bubble gas holdup ϵ_b is distributed radially according to the relation:

$$\epsilon_b/\bar{\epsilon}_b = 2(1 - \phi^2) \tag{3-13}$$

where $\phi = r/R$. The physical significance of the assumption regarding ν_t will be explained later; trial calculations to find a radially variable ν_t giving the best fit to the experimental velocity profile of the bubble–liquid mixed phases have shown that a constant ν_t is quite workable if taken as the radial-cross-sectionally averaged mean value of a distributed ν_t. This assumption simplifies the mathematical treatment, and also is supported by Figs. 17 and 18 (refer to Section II,D,2). Because $n = 2$ the gas holdup ϵ_{b0} at the center axis is twice $\bar{\epsilon}_b$. The empirically adjustable viscosity term ν_t has different values when n is changed, but the flow properties of the mixed phases are insensitive to n. Therefore, $n = 2$ is a satisfactory average for all the experimental data.

With these assumptions, Eq. (3-12) is solved directly, with use of Eqs. (3-8) and (3-9):

$$u + |u_w| = \frac{D_T}{4\nu_t} \cdot \frac{\tau_w}{\rho_l} (1 - \phi^2) + \frac{1}{32} \cdot \frac{gD_T^2 \bar{\epsilon}_b}{\nu_t} (1 - \phi^2)^2 \qquad (3\text{-}14)$$

The numerical value of ν_t will be given in the next section. In Eq. (3-14), the first term of the right-hand side has been shown by Ueyama and Miyauchi (U5) to be essentially negligible for the recirculation flow regime. As a consequence, one has the following simplified quantitative relation:

$$u + |u_w| = \frac{1}{32} \cdot \frac{gD_T^2 \bar{\epsilon}_b}{\nu_t} (1 - \phi^2)^2 \qquad (3\text{-}15)$$

The undetermined term $|u_w|$ is given by applying a mass balance for the liquid phase:

$$\int_0^R 2\pi r(1 - \epsilon_b) u \, dr \pm \pi R^2 U_L = 0 \qquad (3\text{-}16)$$

where the column is continuously supplied by a liquid feed with superficial velocity $\pm U_L$ (+ indicates countercurrent gas–liquid flow and − concurrent) ($U_L = 0$ if the liquid is stationary).

By introducing Eq. (3-13) for ϵ_b, $|u_w|$ is obtained from Eqs. (3-15) and (3-16):

$$|u_w| = \frac{1}{192} \cdot \frac{gD_T^2 \bar{\epsilon}_b}{\nu_t} \cdot \frac{2 - 3\bar{\epsilon}_b}{(1 - \bar{\epsilon}_b)} \pm \frac{U_L}{1 - \bar{\epsilon}_b} \qquad (3\text{-}17)$$

where the $\pm$ sign has the same meaning as that in Eq. (3-16). With this $|u_w|$, the radial distribution of linear liquid velocity u is given by Eq. (3-15).

Another simplified liquid velocity profile is also obtained directly from Eq. (3-15):

$$(u + |u_w|)/(u_0 + |u_w|) = (1 - \phi^2)^2 \qquad (3\text{-}18)$$

where u_0 is the linear liquid velocity at the column axis, $\phi = 0$, given from Eqs. (3-15) and (3-17) as:

$$u_0 = \frac{1}{192} \cdot \frac{gD_T^2 \bar{\epsilon}_b}{\nu_t} \cdot \frac{4 - 3\bar{\epsilon}_b}{(1 - \bar{\epsilon}_b)} - \left(\pm \frac{U_L}{1 - \bar{\epsilon}_b} \right) \qquad (3\text{-}19)$$

The radial position r_* (or $\phi_* = r_*/R$) where the liquid velocity u relative to the column wall is zero (point b in Fig. 27) is obtained from Eqs. (3-15) and

(3-17) by setting $u = 0$. For the simple case of $U_L = 0$ which is usually utilized (batch operation for liquid), the above procedure gives the relation:

$$\phi_* = (1 - e_b)^{1/2}, \quad \text{where} \quad e_b \equiv \{(2 - 3\bar{\epsilon}_b)/[6(1 - \bar{\epsilon}_b)]\}^{1/2} \quad (3\text{-}20)$$

The shear stress at the wall τ_w is given from Eq. (3-8):

$$\tau_w/\rho_l = -(|u_w|/11.63)^2 \quad (3\text{-}21)$$

where $|u_w|$ is calculated adequately from Eq. (3-17) (or still more accurately from U5).

Another useful relation results from the mass balance for the gas phase of the bubble column:

$$\int_0^R 2\pi r(u + u_s)\epsilon_b \, dr = \pi R^2 U_G \quad (3\text{-}22)$$

where u_s is the slip velocity of a bubble relative to the liquid surrounding it. According to experimental measurements of u_s in the recirculation flow regime (U3, Y2), u_s is approximately constant radially, and hence a mean constant value $\bar{u}_s$ may be used. With Eqs. (3-13) for ϵ_b, Eq. (3-18) for u, and Eq. (3-19) for u_0, Eq. (3-22) is integrated to

$$U_G/\bar{\epsilon}_b \pm U_L/(1 - \bar{\epsilon}_b) = \bar{u}_s + (u_0 + |u_w|)/6(1 - \bar{\epsilon}_b) \quad (3\text{-}23\text{a})$$

With $u_0 + |u_w|$ given by setting $\phi = 0$ in Eq. (3-15), Eq. (3-23a) becomes

$$U_G/\bar{\epsilon}_b \pm U_L/(1 - \bar{\epsilon}_b) = \bar{u}_s + (gD_T^2/192\nu_t)\bar{\epsilon}_b/(1 - \bar{\epsilon}_b) \quad (3\text{-}23\text{b})$$

where the $\pm$ sign has the same meaning as in Eq. (3-16). The second term on the right-hand side of the equation is the apparent increase of slip velocity due to recirculation flow. [Calculation shows that this term equals $U^*/\bar{\epsilon}_b$, where U^* is the net linear velocity of the recirculating liquid averaged over the column cross section.[1]] This term is a natural consequence of recirculation flow, and does not appear in the bubble-flow regime where the following well-known relationship applies (N4, Y20):

$$U_G/\bar{\epsilon}_b \pm U_L/(1 - \bar{\epsilon}_b) = \bar{u}_s \quad (3\text{-}24)$$

The mean gas holdup $\bar{\epsilon}_b$ in the recirculation flow regime is obtained at $U_L = 0$ by solving Eq. (3-23b) with respect to $\bar{\epsilon}_b$:

$$\bar{\epsilon}_b = \frac{1}{2}\left(\frac{1 + \alpha}{1 - \gamma}\right)\left[1 - \left(1 - 4\frac{\alpha}{1 + \alpha} \cdot \frac{1 - \gamma}{1 + \alpha}\right)^{1/2}\right] \quad (3\text{-}25)$$

---

[1] $U^* = (\pi R^2)^{-1} \int_0^R 2\pi r u \, dr$.

where

$$\alpha = U_G/\bar{u}_s \quad \text{and} \quad \gamma = gD_T^2/(192\,\nu_t\bar{u}_s)$$

The reason $\bar{\epsilon}_b$ is approximately proportional to $U_G^{1/2}$ and decreases only slightly with column scale-up in recirculation flow (cf. Fig. 14) is explained from Eq. (3-25) by including the behavior of $\bar{u}_s$ and ν_t, which are discussed in the next section.

The mean flow rate of the recirculating liquid is useful when known. For upflow, we take the superficial value:

$$\bar{U} = (\pi R^2)^{-1} \int_0^{r_*} 2\pi r(1 - \epsilon_b) u \, dr = 2 \int_0^{\phi_*} (1 - \epsilon_b) u \phi \, d\phi \quad (3\text{-}26)$$

With integration limits in Eq. (3-26) from r_* to R with $U_L = 0$ (no continuous liquid feed), the mean superficial downflow velocity results. By using Eq. (3-13) for ϵ_b and Eqs. (3-15) and (3-17) for u, with $U_L = 0$, Eq. (3-26), when integrated, is

$$\bar{U} = (gD_T^2/192\nu_t) \cdot \bar{\epsilon}_b e_b^3 (4 - 3\bar{\epsilon}_b e_b) \quad (3\text{-}27)$$

with e_b as given in Eq. (3-20). Once $\bar{U}$ is known, one can easily obtain the interstitial mean velocities of upflow and downflow.

B. COMPARISON OF THEORY WITH EXPERIMENTS ON BUBBLE COLUMNS

Experimental facts which support the recirculation flow and its turbulence properties are given here, mainly in relation to the simplified theory developed in the preceding section.

1. *Interstitial Velocity Profile of Recirculating Liquid*

The available data were correlated by Ueyama and Miyauchi (U5), who found that the normalized velocity profile Eq. (3-18) represents the data reasonably well. Figure 28 illustrates the experimentally measured velocity profile of water containing bubbles from a single-nozzle gas-distributor (Y2). The bubble column was 25 cm in diameter operating in the recirculation flow regime at a superficial gas velocity U_G of 5.2 cm/sec. Estimated values of $u_0 = 45$ cm/sec and $|u_w| = 26$ cm/sec, arrived at by trial, give the best fit of the normalized profile data, and Eq. (3-18) provides a reasonable approximation.

The radial distance ϕ_* at which the mean liquid velocity is zero, given by Eq. (3-20), is illustrated in Fig. 29 as a function of $\bar{\epsilon}_b$, where $\bar{\epsilon}_b$ is taken from the following correlation:

$$\bar{\epsilon}_b = 0.054 U_G^{1/2} \quad (3\text{-}28)$$

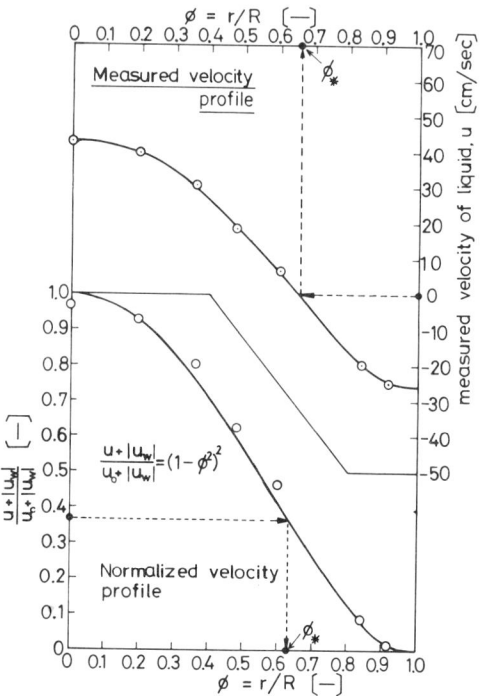

FIG. 28. Interstitial liquid velocity profile and normalized velocity profile [data by Yamagoshi (Y2)], $D_T = 25$ cm, $U_G = 5.2$ cm/sec, air–water system at room conditions. Adjusted u_0 and $|u_w|$ are 45 and 26 cm/sec, respectively.

with U_G in centimeters per second. This equation is a good approximation for $\bar{\varepsilon}_b$ of a bubble column operated in the recirculation flow regime at $U_L \simeq 0$ and is equipped with a single-nozzle gas sparger. The columns tested are in the range of $D_T = 10$–60 cm and aspect ratio $L_q/D_T \gtrsim 2$ (H11, U3).

In Fig. 28, the radial position ϕ_* at $u = 0$ is 0.66 from the measured profile and 0.63 from the normalized profile, where the term $(u + |u_w|)/(u_0 + |u_w|) = 26/(26 + 45) = 0.366$. These two values of ϕ_* are shown in Fig. 29 with some additional bubble-column data (U3) and some fluidized-bed data (M29). Equation (3-20) seems applicable to the bubble-column data.

When a perforated plate is used as a gas sparger in a bubble column, the mean gas holdup $\bar{\varepsilon}_b$ is influenced by the size and arrangement of holes as well as column diameter as a result of change in both of bubble size and the flow pattern of the bubble swarm. These effects have been reviewed and correlated by Kato and Nishiwaki (K6) for air–water systems where a sparger is perforated uniformly. When the hole diameter δ is smaller than 1.0–1.4 mm and the gas velocity U_G is low, $\bar{\varepsilon}_b$ for a given diameter column

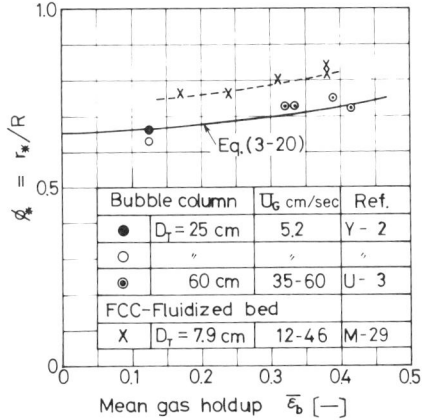

FIG. 29. Radial position ϕ_* as a function of $\bar{\epsilon}_b$.

changes with δ, U_G, and the number of holes. However, $\bar{\epsilon}_b$ tends to be variable only with U_G, for U_G above 10–15 cm/sec, where it approaches the values observed for a column with $\delta \gtrsim 2$ mm.

Essentially the same performance is observed for a column equipped with a porous sintered metal or glass plate as a gas distributor.

When the hole size δ is equal to or greater than about 2 mm, $\bar{\epsilon}_b$ increases in proportion to U_G in the low-gas-flow range ($U_G \lesssim 5$–6 cm/sec) for different diameters of columns. As U_G increases beyond this range, $\bar{\epsilon}_b$ increases approximately in proportion to $(U_G)^{1/2}$, decreasing slightly at constant gas velocity as the column diameter D_T increases. Here, the extent of decrease seems to depend on the type of sparger, which depends slightly on D_T for a perforated plate, and is essentially independent of D_T for a single-nozzle sparger.

2. *Turbulent Kinematic Viscosity*

This term may be determined in various ways from experimental data. A frequently used method for single-phase flow (S4) is to differentiate an experimental velocity profile graphically, and to obtain the kinematic viscosity ν_t from the velocity gradient with basic relations [e.g., Eqs. (3-1) and (3-2)]. The currently available data for velocity profile do not appear accurate enough to yield ν_t as a function of radial position. As an alternative, Hills (H9) has assumed a relation between the local ν_t and the radial position:

$$\nu_t \propto \epsilon_b(1 - \epsilon_b) \propto D_T^2/u \qquad (3\text{-}29)$$

The last proportionality comes from his experimental test of the relation. Along the column axis, u equals u_0 and is approximately proportional to $(U_G D_T)^{1/4}$, as is seen in Fig. 30. Hence, $\nu_t \propto D_T^{7/4}/U_G^{1/4}$ along the column axis.

The constant effective ν_t assumed in the present simplified treatment is obtained by knowing u_0 and $|u_w|$. Along the column axis ($\phi = 0$), Eq. (3-15) gives

$$u_0 + |u_w| = gD_T^2 \bar{\epsilon}_b/(32\nu_t) \tag{3-30}$$

With u_0 and $|u_w|$ obtained by trial and $\bar{\epsilon}_b$ from Eq. (3-28), ν_t is given by Eq. (3-30). In the case of Fig. 28, $\bar{\epsilon}_b$ is calculated to be 0.123 (the measured value is 0.127) and $\nu_t = 33.2$ cm²/sec.

In the same manner, ν_t has been calculated by Ueyama and Miyauchi (U5) from a group of available experimental velocity profiles. Fig. 31 shows the recalculated ν_t as a function of D_T, matched empirically by the power function:

$$\nu_t = 0.128 D_T^{1.70} \tag{3-31}$$

where ν_t is in cm²/sec and D_T in cm. Earlier work of Miyauchi and Shyu (M31) also gives Eq. (3-31) with a different coefficient. Recent experience with a plant-size bubble column ($D_T = 550$ cm) suggests that the ν_t dependence on D_T may be extrapolated to larger-size columns for estimating the flow properties (K15).

The dependence of ν_t on U_G is (U5) approximately proportional to $U_G^{1/6}$, but not always consistently. This dependence is confirmed when ν_t is calculated theoretically from u_0 (in the next section). Figure 32 shows a

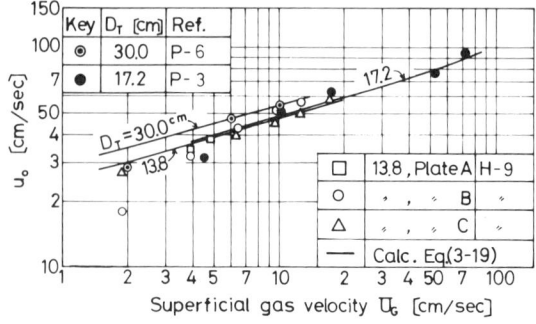

FIG. 30. Dependence of liquid velocity at the center axis, u_0, on the superficial gas velocity U_G and the column diameter D_T; ν_t is taken from Fig. 33 and $\bar{\epsilon}_b$ from Eq. (3-28) for the calculated curves. Data are taken from Hills (H9).

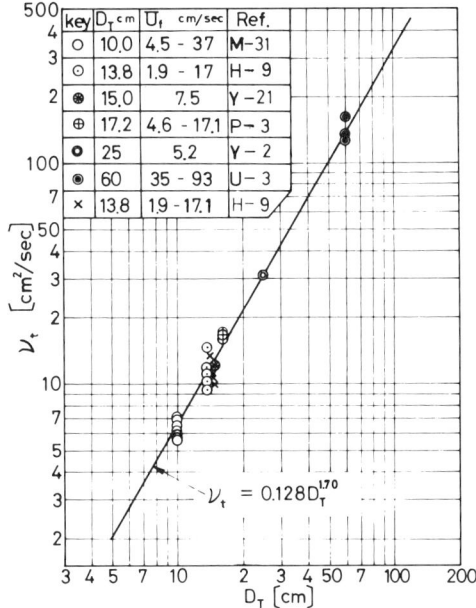

FIG. 31. Turbulent kinematic viscosity ν_t as a function of column diameter D_T for bubble columns; ν_t increases rapidly with D_T, but is little influenced by U_G. Note that $U_G = 1.9$–93 cm/sec.

correlation based on this dependence. The line B–B is for the bubble columns and is expressed (in cm-sec) by

$$\nu_t = 0.160 U_G^{1/6} D_T^{3/2} \qquad (3\text{-}32)$$

The correlation shown in Figs. 32 and 33 as curve A–A is the turbulent kinematic viscosity of the emulsion phase of fluidized cracking catalyst beds estimated by an indirect method explained in Section IV,C in relation to axial dispersion of the emulsion. The correlation is given in cm-sec units by

$$\nu_t = 0.079 U_G^{1/6} D_T^{1.70} \qquad (3\text{-}31\text{a})$$

The exponent 1.70 on D_T is different from 1.50 of Eq. 3-32, but Eq. 3-31a gives a better correlation for axial dispersion of the emulsion.

We have seen that ν_t can be obtained from $|u_w|$ or u_0. In the case of Fig. 28, $\nu_t = 28.1$ cm²/sec from Eq. (3-17), and $\nu_t = 36.1$ cm²/sec from Eq. (3-19). Some data are available for u_0 (H9, P3, P6) from which ν_t are calculated and shown in Fig. 33. In general, ν_t as calculated from u_0 tends to give a larger value than from $u_0 + |u_w|$, perhaps due to lack of suffi-

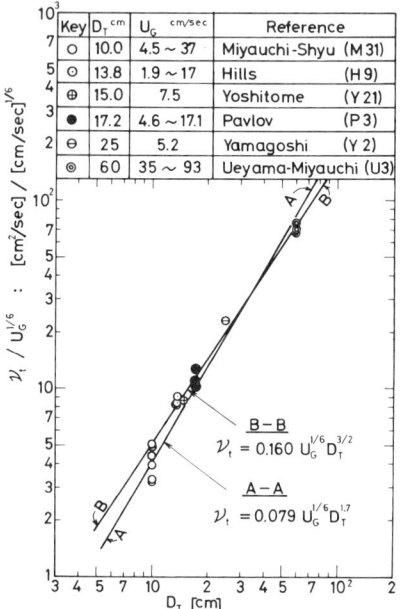

FIG. 32. Turbulent kinematic viscosity ν_t as a function of D_T and U_G. Line A–A is for FCC-catalyst beds and line B–B for bubble columns of low-viscosity liquid.

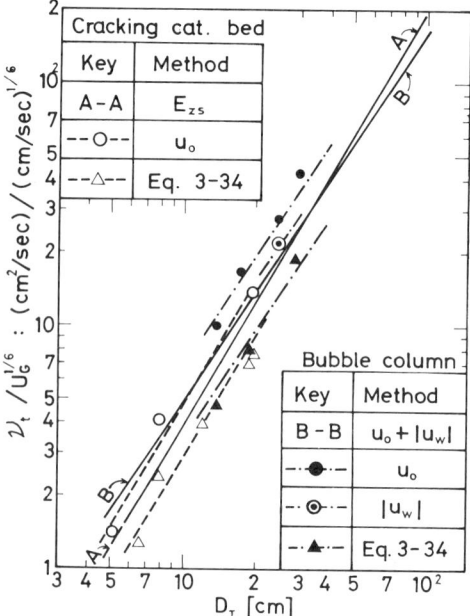

FIG. 33. Turbulent kinematic viscosity ν_t as determined by different methods; the lines A–A and B–B are the same as those given in Fig. 32.

ciently accurate measurements. Once a mean ν_t is obtained, Eq. (3-19) accounts well for the performance of u_0, as illustrated in Fig. 30 for the recirculation-flow regime with $U_G \geq 3.8$ cm/sec. Figures 28–33 show definitely that the flow field in the bubble columns is intensely turbulent in the recirculation-flow regime, since ν_t is much greater than ν_M.

3. *Mean Slip Velocity of Bubbles Relative to Liquid*

Equation (3-23a) gives $\bar{u}_s$ from experimental knowledge of $u_0 + |u_w|$ and $\bar{\epsilon}_b$. In the case of Fig. 28, $u_0 + |u_w| = 71$ cm/sec and $\bar{\epsilon}_b = 0.123$, so that $\bar{u}_s = 28.8$ cm/sec at $U_G = 5.2$ cm/sec with $D_T = 25$ cm. Figure 34 (from U5) summarizes $\bar{u}_s$ so obtained as a function of U_G. New data are added for a commercial-scale column (K15, $D_T = 550$ cm). The mean slip velocity $\bar{u}_s$ remains essentially unchanged at 50 cm/sec for $U_G \geq 20$ cm/sec, but drops gradually as U_G decreases below this value. Also the column diameter has essentially no effect, for $D_T \geq 10$ cm. The intense turbulent field induced in the column seems to keep a steady bubble size, by dynamic balance between coalescence and redispersion of bubbles.

Experimental $\bar{u}_s$ values determined as the difference between the local mean velocities of bubbles and liquid are also shown in Fig. 34 as "open keys." For these bubbles, the volume-surface mean diameters d_{32} have been measured experimentally. The free-rise velocity of the bubbles (T1, V6) is

$$\bar{u}_s = (g d_{32}/2)^{1/2} \qquad (3\text{-}33)$$

The $\bar{u}_s$ thus obtained, shown in the figure as half-open circles (U3), is in

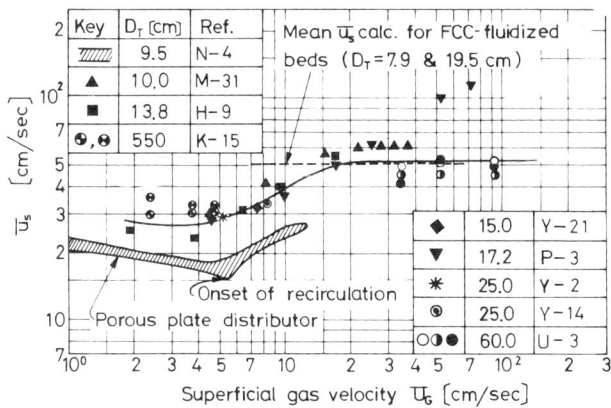

FIG. 34. Dependence of average slip velocity $\bar{u}_s$ on the superficial gas velocity U_G; column diameter D_T has little effect on $\bar{u}_s$. A porous plate distributor seems to give smaller $\bar{u}_s$.

reasonable agreement with other data, suggesting that bubbles in highly turbulent flow have little interaction with one another. Data from Nicklin (N4) show smaller $\bar{u}_s$, perhaps due to use of a porous plate gas-distributor instead of a sieve plate or a single nozzle.

4. *Bubble Volume Fraction*

So far ν_t and $\bar{u}_s$ have been given numerically. Figure 36 (upper part) shows $\bar{\epsilon}_b$ as calculated by Eq. (3-25) for a bubble column with $D_T = 19$ cm using values of these two variables. (The fluidized-bed data in the figure will be discussed in Section III,D,4.) Curve C–C expresses Eq. (3-28). Curve BQB is from Eq. (3-25) with $\bar{u}_s$ from Fig. 34 and ν_t from Eq. (3-32); BQP, the same, except with ν_t from Eq. (3-31). Curve BQP matches better than BQB to curve C–C.

When a bubble column is scaled-up to a larger D_T, calculations similar to the above show that $\bar{\epsilon}_b$ decreases only slightly with increasing D_T, and retains the same functional dependence on U_G because of the properties of ν_t and $\bar{u}_s$.

5. *Solution for ν_t and $\bar{u}_s$*

By reversing the procedure of obtaining $\bar{\epsilon}_b$ from Eq. (3-25) after determining ν_t and $\bar{u}_s$, these two parameters can be determined if $\bar{\epsilon}_b$ is known as a function of U_G for a given D_T. This step involves the use of Eq. (3-23b) or, equivalently, of Eq. (3-25).

For simplicity, we take $U_L = 0$. Also, for a bubble column, $\bar{u}_s$ is constant for $U_G \geq 20$ cm/sec. We designate this value as $\bar{u}_{s0}$. Eq. (3-23b) is then rewritten:

$$\frac{U_G}{\bar{\epsilon}_b} = \bar{u}_{s0} + \left(\frac{1}{\nu_t}\right)\left(\frac{gD_T^2\bar{\epsilon}_b}{192(1 - \bar{\epsilon}_b)}\right) \quad (3\text{-}34)$$

By taking $\bar{\epsilon}_b$ from Eq. (3-28) and plotting $U_G/\bar{\epsilon}_b$ as ordinate and the co-factor of $(1/\nu_t)$ as abscissa, $\bar{u}_{s0}$ and $1/\nu_t$ are obtained simultaneously as shown in Fig. 35. Here ν_t is assumed as being variable only with D_T (as for curve BQP in Fig. 36), with $D_T = 19.0$ cm and $U_G \geq 20$ cm/sec. Figure 35 gives $\nu_t = 14.1$ cm^2/sec and $\bar{u}_{s0} = 46$ cm/sec. On the other hand, $\nu_t = 19.1$ cm^2/sec from Eq. (3-31) and $\bar{u}_{s0} = 50$ cm/sec from Fig. 34.

As is seen from Fig. 36, Eq. (3-28) (curve C–C) is lower than the calculated curve BQP. Lower $\bar{\epsilon}_b$ means that the liquid is more easily recirculated, so that a lower ν_t is obtained. According to several test calculations, the above method tends to give ν_t 25–35% smaller than by Eq. (3-31) (cf. Fig. 33), but gives a reasonable value for $\bar{u}_{s0}$.

Essentially the same procedure is applicable to a fluidized catalyst bed,

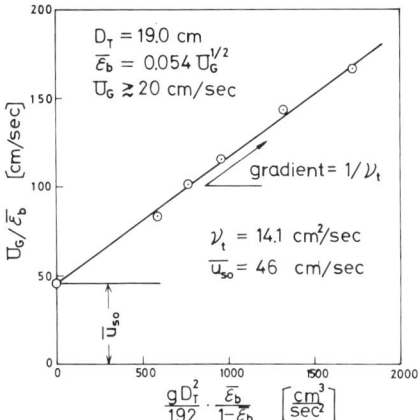

FIG. 35. Simultaneous determination of ν_t and $\bar{u}_{s0}$ from $\bar{\epsilon}_b$ based on Eq. (3-34).

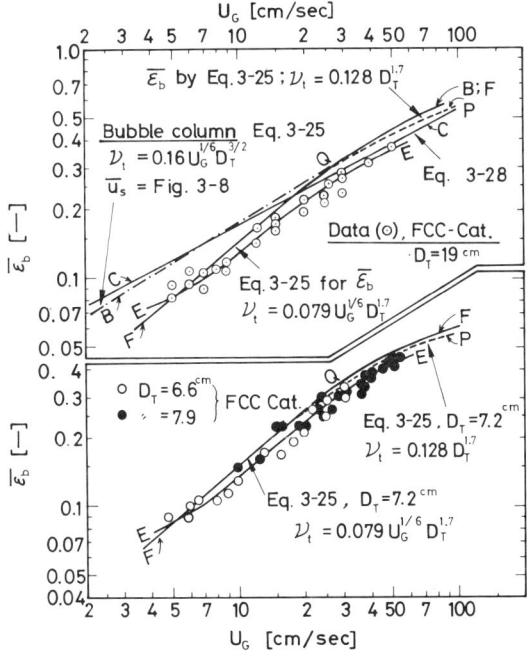

FIG. 36. Comparison of the calculated $\bar{\epsilon}_b$ by Eq. (3-25) with the data given in Fig. 14. Unless otherwise mentioned, $\bar{u}_s = 49.5$ cm/sec. Data are for FCC-catalyst beds of different bed diameters. Curve E–E is an average of the data. The calculated curves BQB and BQP are for bubble columns, FQF and FQP are for FCC-catalyst beds.

where $\bar{u}_s$ remains nearly unchanged for $U_G \geq 7\text{–}8$ cm/sec as will be shown. This procedure, although approximate, provides a convenient tool for data analysis, since the dependence of $\bar{\epsilon}_b$ on U_G is easy to measure and not many data are yet available on turbulence in fluidized catalyst beds.

C. Phenomenological Background of Turbulent Kinematic Viscosity

The turbulent kinematic viscosity ν_t has been introduced in the basic relation, Eq. (3-2), and is correlated with the operational variables through the experimental data (cf. Figs. 31–33) with a simplifying assumption that ν_t is constant radially as an experimentally adjustable parameter.

It is desirable to establish a phenomenologically sound basis by relating ν_t to well-established fluid-dynamic concepts, since we need to know the possibility of scale-up or of applying the concept of ν_t to a fluidized catalyst bed, as discussed in what follows (M27).

1. Bubble Column of Low-Viscosity Liquid

The radial velocity distribution inside a bubble column has been given by Eq. (3-15). Differentiating this with respect to ϕ, one finds the velocity gradient at the column wall, $(du/d\phi)_{\phi=1}$, to be zero, which means that peripherally descending liquid flow slides freely along the column wall. In other words, the column wall keeps the downflow in a cylindrical form at the periphery, but exerts only a negligible frictional force.

As a consequence, the peripherally descending liquid flows freely past the centrally ascending liquid. This counterflow is inherently unstable; a time-varying vortex-like mixing zone should be induced between the counterflowing two streams, leading to intense turbulence. Takahashi (T5) observed such vortex-like motion in measuring a two-point velocity correlation along the column axis in a bubble column of 25 cm in diameter.

The time-varying turbulent motion of the bubble column seems to resemble the free turbulence of open jets (S4) or in the mixing zone of two parallel streams of different velocities (D13).

2. Turbulent Viscosity on the Basis of Prandtl's Hypothesis

Prandtl (P7, S4) established a simplified equation for the apparent kinematic viscosity of free turbulent flow based on experiments with free-turbulent flow by Reichardt (R2). He assumed that the dimensions of the lumps of fluid moving traversally during turbulent mixing are comparable in magnitude to the width of the mixing zone. The virtual kinematic viscosity ν_t is now formed by multiplying the maximum difference in time-mean flow velocity by a length proportional to the width of the zone. In the case of the bubble column, the maximum difference in the velocity is $u_0 + |u_w|$ (see Fig. 27) and the length is the column radius R. Correcting the effect of the liquid-volume fraction $\bar{\epsilon}_L$, ν_t is approximately expressed as:

$$\nu_t = k\bar{\epsilon}_L R(u_0 + |u_w|) \tag{3-35}$$

where k denotes a dimensionless number to be determined experimentally. It follows from Eq. (3-35) that ν_t may be considered constant over the whole width at every cross section of the column.

Introducing Eq. (3-30) with $\phi = 0$ for $u_0 + |u_w|$ in the above expression and solving Eq. (3-35) with respect to ν_t, we obtain the relation

$$\nu_t = (k^{1/2}/8)[gD_T^3\bar{\epsilon}_b(1 - \bar{\epsilon}_b)]^{1/2} \tag{3-36}$$

Introducing Eq. (3-28) for $\bar{\epsilon}_b$, ν_t is approximated by

$$\nu_t = [0.0318(kg)^{1/2}]U_G^{1/6}D_T^{3/2} \tag{3-37}$$

With $k = 0.0259$, Eq. (3-37) reduces to Eq. (3-32) (in cm-sec), a correlation based on experiment. This k value is the same order of magnitude as for free turbulence flow (S4).

For liquid–liquid spray columns Wijffels and Rietema (W17) have studied the combined effect of bulk turbulent flow and the drop motion suspended in it on the intrinsic turbulent viscosity ν_t^*. For the contribution of turbulent flow alone they assume that ν_t^* is given by $k^*R\,|u_w|$, with k^* adjusted at 0.0167 from experiment. Numerically, k^* is the same order of magnitude as k, although smaller, with $u_0 + |u_w|$ taken here instead of $|u_w|$.

For bubble columns the contribution of bubble wake motion has been included in the bulk turbulent motion, since the recirculation flow is induced primarily by the buoyant force of bubbles. The bubble-wake motion in the intense turbulent flow seems quite unstable, and is difficult to distinguish from the motion of the turbulent bulk liquid.

Based on the interpretation developed here, the assumption of constant ν_t and its empirical expression by Eq. (3-32), seem to be sound physically for bubble columns. As shown in later sections, the flow properties of a fluid catalyst bed of good fluidity (e.g., cracking catalyst) are in many respects similar; hence the same interpretation can be applied to the fluid catalyst bed (FCB).

D. INTERPRETATION OF EXPERIMENTS FOR FCB

Thus far the experiments considered have been confined to bubble columns, which provide a fair number of data over a wide range of conditions and are therefore useful for testing the simplified theory. Equivalent data are often not available for fluidized beds of carefully prepared particles, especially with solids other than FCC catalyst.

The main uncertainty in applying the simplified theory to a fluidized catalyst bed probably lies in the question as to whether the shear-stress term in Eq. (3-10) is negligible in comparison with the buoyant-force term. In what follows, although we still lack direct experimental proof regarding the shear-stress term, the simplified theory will be seen to account well for the bed performance (M27).

1. *Interstitial Velocity Profile of the Emulsion Phase*

As stated in Section II,D,3, the pioneer work of Lewis *et al.* (L11) showed intense recirculation of the emulsion phase in an air-fluidized MS-catalyst bed. The time-averaged velocity profile in the recirculation-flow regime is still lacking, although Yamazaki (Y3a) has revealed the profile at a low gas velocity ingeniously by utilizing thermal response. The radial position ϕ_* where the mean velocity is zero relative to the bed wall, measured by Morooka (M43) for a FCC bed of 7.9-cm diameter, is shown in Fig. 29. The deviation from prediction by Eq. (3-20) is about 0.3 cm and is comparable with the resolution power of the strain-gauge probe.

2. *Turbulent Kinematic Viscosity*

Several data for mean ascending velocity u_{b0} of bubbles along the bed axis, in FCC-catalyst beds, are shown in Fig. 37. Velocity u_{b0} equals $\bar{u}_s + u_0$, with $\bar{u}_s$ from Eq. (3-33) for a given d_{32}, and u_0 from Eq. (3-19). With $u_s = 49.5$ cm/sec for $d_{32} = 5$ cm (cf. Fig. 34), $\bar{\epsilon}_b$ from Fig. 36, and ν_t

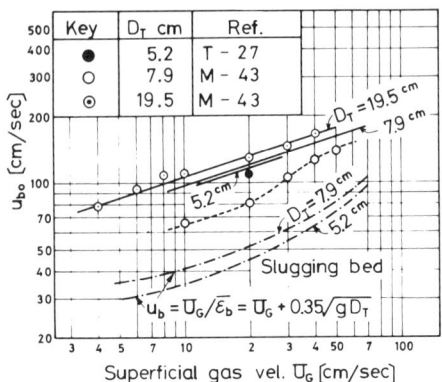

FIG. 37. Mean bubble velocity u_{b0} along the column axis; experimental data are for FCC-catalyst beds. Full curves are calculated by $u_{b0} = \bar{u}_s + u_0$ with $\bar{u}_s = 49.5$ cm/sec and u_0 from Eq. (3-19). U_{mf} is neglected due to $U_{mf} \ll U_G$.

from Eq. (3-31a), the u_{b0} which results is shown in Fig. 37. The calculation matches the data for columns with diameters of 5.2 and 19.5 cm, but not for those with diameters of 7.9 cm, for reasons which are not obvious. In Fig. 37 the bubble velocity in slugging beds [cf. Eq. (5-4b)] is also shown. The FCC-catalyst bed of 5.2-cm diameter is fluidizing smoothly in the recirculation-flow regime with no indication of slugging when L_q is smaller than 50–60 cm.

By reversing the above procedure of obtaining u_{b0} by knowing $\bar{u}_s$, $\bar{\epsilon}_b$ and v_t, one can calculate v_t by knowing $\bar{u}_s$, $\bar{\epsilon}_b$ and u_{b0}. These v_t are shown in Fig. 33 as open circles. They show reasonable agreement with the correlation A–A for fluidized catalyst beds, where the data for 7.9-cm diameter with $u_G < 30$ cm/sec are omitted.

3. Mean Slip Velocity of Bubbles

The mean bubble size in a fluidized bed $\bar{u}_s$ has been discussed in Section II,B. As discussed, d_{32} for a fluidized catalyst bed of good fluidity may be taken as approximately 5.0 cm [cf. Figs. 10 and 11, and Eq. (2-11)] for $U_G \gtrsim 10$ cm/sec. With Eq. (3-33), this d_{32} gives $\bar{u}_s = 49.5$ cm/sec, which is shown in Fig. 34 as a dashed line. It is interesting that the mean slip velocity is essentially the same as for a bubble column, when $U_G \gtrsim 20$ cm/sec. As noted in Section II,B, d_{32} and $\bar{u}_s$ are very sensitive to change in particle size, size distribution, shape, and density.

4. Bubble Volume Fraction

The averaged volume fraction $\bar{\epsilon}_b$, calculated by Eq. (3-25), is shown in Fig. 36, for bubble columns and also for FCC-catalyst beds (M40). The mean slip velocity of bubbles is again taken as $\bar{u}_s = 49.5$ cm/sec. Also, v_t is calculated by Eq. (3-31a) for curve FQF and by Eq. (3-31) for curve FQP, while curve EE is an empirical fit of the data. As in the case of a bubble column, curve FQP matches better with curve EE, although FQP is consistently higher. Curve EE tends to decrease the slope for $U_G \lesssim 7$–8 cm/sec, perhaps due to the decrease in $\bar{u}_s$; the scatter of data makes the behavior unclear. This is explained by the difference in the region of $U_G < 20$ cm/sec. The bubble column shows higher $\bar{\epsilon}_b$ values than those for the fluidized bed, which is due to the bubble column's lower $\bar{u}_s$ (cf. Fig. 34).

5. Solution for v_t and $\bar{u}_s$

As shown above, $\bar{u}_s$ is constant at about 49.5 cm/sec for a fluidized catalyst bed of good fluidity when $U_G > 7$–8 cm/sec. Hence v_t and $\bar{u}_s$ are obtained in this range by plotting the variables of Eq. (3-34) as determined from the experimental data. Figure 38 shows the calculation, with gas holdup data taken from curve EE for $D_T = 19.0$ cm in Fig. 36. The resulting turbulent kinematic viscosity, lower than would be obtained from curve FQP in Fig. 36 or curve AA in Fig. 33, is shown in Fig. 33 as open triangles.

In summary, the calculations show convincingly that modern fluidized-catalyst-bed technology has attained an emulsion fluidity nearly equivalent to that of low-viscosity liquids. With such fluidity, data obtained for a bubble column shed light on the performance of a fluidized catalyst bed, and vice versa.

IV. Longitudinal Dispersion Phenomena as Derived from Flow Properties

Longitudinal dispersion in the continuous phase (the liquid phase for a bubble column, and the emulsion phase for a fluidized catalyst bed) is closely related to flow properties of the equipment. Here, we wish to describe the longitudinal dispersion phenomena in terms of the fluid-dynamic properties of the equipment. The prime purpose is to test whether the fluid-dynamic analysis developed earlier is sound, but lon-

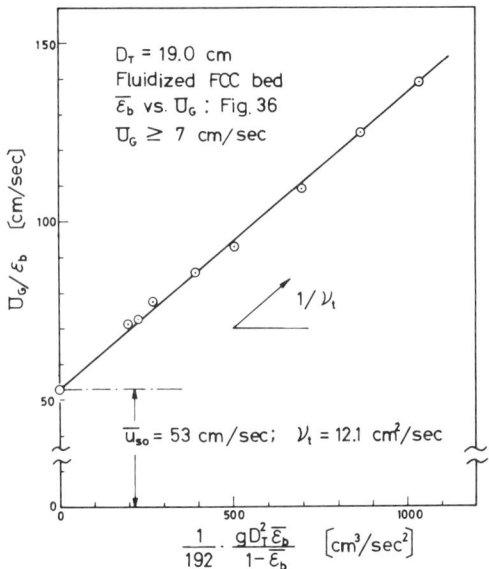

FIG. 38. Simultaneous determination of ν_t and $\bar{u}_{s0}$ from mean gas holdup data by using Eq. (3-34).

gitudinal dispersion itself is important for estimating the reactor performance (D2, G17, K11, L9, M20, P4, S8, V11, Y1).

A. Theory of Longitudinal Dispersion for the Recirculation Flow Regime

A gas bubble column is taken here as a model equipment undergoing longitudinal dispersion of the continuous phase. The theory obtained is equally applicable to a fluidized catalyst bed of good fluidity exhibiting similar flow properties. The following procedure is from Miyauchi (M27).

1. *Underlying Concept*

In the bubble column the velocity profile of recirculating liquid is shown in Fig. 27, where the momentum of the mixed gas and liquid phases diffuses radially, controlled by the turbulent kinematic viscosity ν_t. When $U_L = 0$ (essentially no liquid feed), there is still an intense recirculation flow inside the column. If a tracer solution is introduced at a given cross section of the column, the solution diffuses radially with the radial diffusion coefficient E_r and axially with the axial diffusion coefficient E_z. At the same time the tracer solution is transported axially by the recirculating liquid flow. Thus, the tracer material disperses axially by virtue of both the axial diffusivity E_z and the combined effect of radial diffusion and the radial velocity profile.

The latter mechanism assumed is the well-known Taylor dispersion (T9, T10, T11), which has been studied extensively (A11, G6, L9, T14, S2). High-speed motion pictures taken by Towell *et al.* (T23) in a 40-cm bubble column (R3) have shown the presence of turbulent eddies, on a scale roughly equal to the column diameter, with systematic large-scale circulation patterns superimposed. Their pictures showed that liquid near the wall flowed downward, while liquid near the center of the column flowed upward, consistent with the flow theory developed earlier and with the Taylor dispersion mechanism.

Axial dispersion in the bubble column is usually well expressed by the following one-dimensional diffusion model with respect to liquid concentration c:

$$\partial c/\partial \theta = E_{zT}\, \partial^2 c/\partial z^2 - \bar{u}\, \partial c/\partial z - \Phi(c) \qquad (4\text{-}1)$$

where E_{zT} is the total axial dispersion coefficient (interstitial value), $\bar{u}$ is defined by $\bar{u} = U_L/\bar{\epsilon}_L$, the mean interstitial liquid velocity, and $\Phi(c)$ is the reaction or mass-transfer term. The total dispersion coefficient E_{zT} is split into the contributions of two mechanisms:

$$E_{zT} = E_{zr} + \bar{E}_z \tag{4-2}$$

where E_{zr} is the contribution of Taylor dispersion, and $\bar{E}_z$ that of the usual axial diffusion averaged over the entire column cross-sectional area.

The total dispersion coefficient is usually determined by measuring the concentration distribution of tracer material steadily back-mixed upstream, or by knowing the impulse response of the column for tracer liquid (see Section IV,A,3).

2. *Basic Relations*

The equation of continuity for the liquid phase of the bubble column is obtained from a mass balance on tracer material for differential gas–liquid mixed phases. With the same procedure as for a homogeneous flow, the following equation is obtained when $\Phi(c) = 0$ and the aspect ratio $L/D_T \gg 1$:

$$-\frac{1}{r}\frac{\partial}{\partial r}\left[r\left(-E_r\epsilon_L\frac{\partial c}{\partial r}\right)\right] - u\epsilon_L\frac{\partial c}{\partial z} = \epsilon_L\frac{\partial c}{\partial \theta} \tag{4-3}$$

Here, the axial diffusion term, $\epsilon_L E_z\, \partial^2 c/\partial z^2$, is tentatively omitted for simplicity, and the radial mass flux N_r (with $\nu_m = \epsilon_L E_r$ for simplicity) is defined by

$$N_r = -\epsilon_L E_r\, \partial c/\partial r = -\nu_m\, \partial c/\partial r \tag{4-4}$$

When the reference plane moves with mean net liquid velocity of $\pm U_L/(1 - \bar{\epsilon}_b)$ (+ indicates cocurrent and − countercurrent), Eq. (4-3) remains valid; u, the liquid velocity relative to the moving reference plane, is given from Eqs. (3-15) and (3-17) by setting $U_L = 0$. This is justified because the net flow term, which defines the moving reference plane, cancels with the net flow term included in the linear velocity term u, [see Eqs. (3-15) and (3-17)]. In what follows in this section we take u as velocity relative to the moving plane.

We now consider the case of steady back-mixing experiment to measure the total axial diffusivity E_{zT}. A small amount of fresh liquid is continuously introduced into the column cocurrently or countercurrently with the gas flow. A still smaller amount of tracer liquid is steadily introduced at a downstream section. Under such circumstances the concentration distribution of the tracer at a given cross section is steady, and the axial concentration gradient $\partial c/\partial z$ is constant radially. As a consequence, Eq. (4-3) simplifies to

$$-\frac{1}{r}\frac{\partial}{\partial r}\left(-r\nu_m\frac{\partial c}{\partial r}\right) - u\epsilon_L\frac{\partial c}{\partial z} = 0 \tag{4-5}$$

with $\partial c/\partial z$ almost independent of r at the steady state.

The axial dispersion coefficient E_{zr} given by the Taylor mechanism for the above case is given by the procedure of Tichacek *et al.* (T14) for turbulent pipe flow. Integrating Eq. (4-5) twice with respect to r, one has

$$c_r - c_R = \frac{\partial c}{\partial z} \int_R^r \frac{dr}{\nu_m r} \int_0^r u\epsilon_L r \, dr \tag{4-6}$$

The net transport Q past any reference plane moving with liquid at the mean velocity $\pm U_L/(1 - \bar{\epsilon}_b)$ is

$$Q = 2\pi \int_0^R (c_r - c_R) u\epsilon_L r \, dr \tag{4-7}$$

Q is seen to be proportional to the axial concentration gradient $\partial c/\partial z$ when Eqs. (4-6) and (4-7) are combined. Hence the mean axial dispersion coefficient E_{zr} is defined by

$$Q = -\pi R^2 \bar{\epsilon}_L E_{zr}(\partial c/\partial z) \tag{4-8}$$

Combining Eqs. (4-6)–(4-8), E_{zr} is given by

$$E_{zr} = -\frac{D_T^2}{2\bar{\epsilon}_L} \int_0^1 \epsilon_L u\phi \, d\phi \int_1^\phi \frac{d\phi}{\nu_m \phi} \int_0^\phi \epsilon_L u\phi \, d\phi \tag{4-9}$$

where $\phi = r/R$, and u is obtained from Eqs. (3-15) and (3-17) by setting $U_L = 0$.

3. *Radial and Axial Diffusivity; Simplification of Eq.* (4-9)

For integrating Eq. (4-9), $\nu_m(= \epsilon_L E_r)$ should be known as a function of ϕ and operating variables. However, the momentum diffusivity ν_t is the only term we know, with essentially no systematic data for ν_m. In the case of free turbulence of a homogeneous fluid, the diffusivity of a scalar quantity like heat and mass is estimated to be about two times that of momentum (S4) and the two diffusivities are not far apart for turbulent pipe flow (S8). However, such a relation is not available yet for gas–liquid bubble flow in bubble columns. Generally the local radial mass diffusivity ν_m may be expressed by $\alpha\nu_t$, with α being a numerical coefficient of order unity.

In the recirculation flow regime Eissa *et al.* (E6) have reported radial diffusivities $\bar{E}_r$ of 1.21 cm²/sec at $U_G = 6.3$ cm/sec and 1.26 cm²/sec at 7.8 cm/sec for a 5-cm-diam bubble column equipped with a multiple-orifice-plate gas distributor. These diffusivities amount to 60% of the eddy kinematic viscosity; $\nu_t = 2$ cm²/sec for $D_T = 5$ cm (Fig. 31). Reith *et al.* (R3) have measured a radial diffusivity of 10 cm²/sec at $U_G = 9.3$ cm/sec for a 14-cm-diam bubble column equipped with a perforated plate orifice having one 2-mm-diam hole per cm²; this is slightly lower than the average kinematic viscosity for the same size column ($\nu_t = 11.4$ cm²/sec). At U_G of 6.0 and 10 cm/sec, Pozin *et al.* (P6) report a radial diffusivity of 35–40 cm²/sec in the core region of a 30-cm-diam bubble column with a gas distributor plate having 1.5% free area with 2.5-mm hole diam. The turbulent kinematic viscosity read from Fig. 31 is 42 cm²/sec for a 30-cm-diam column.

As for axial diffusivity $\bar{E}_z$ included in Eq. (4-2), the situation is not much different from that for radial diffusivity. Pozin et al. (P6) report that the mean value of axial diffusivity $\bar{E}_z$ is 2.5 times the value of $\bar{E}_r$, when radial and axial diffusion are assumed to be homogeneous and nonisotropic throughout the bubble column. Hence it is reasonable to assume for the local axial diffusion coefficient $E_z = \zeta E_r$, with ζ of order unity. On the other hand, one has the relations that $E_r = \nu_m/\epsilon_L$ [see Eq. (4-4)] and $\nu_m = \alpha \nu_t$, so that E_z equals the product $(\alpha\zeta/\epsilon_L)\nu_t$. Accordingly, the cross-sectionally averaged axial diffusivity $\bar{E}_z$ is expressed by the relation $\bar{E}_z = \overline{(\alpha\zeta/\epsilon_L)}\nu_t$, where the overbar shows an effective mean value.

Two approximations are introduced, for simplicity, to perform the integration of Eq. (4-9) for the Taylor mechanism dispersion coefficient E_{zT}. First, the term ν_m included in the second integral of Eq. (4-9) is eliminated from the integral by taking an effective mean value $\bar{\nu}_m$ $(= \bar{\alpha}\nu_t)$, with $\bar{\alpha}$ being a mean of α. Second, the local liquid holdup ϵ_L is eliminated from the multiple integral by using the mean liquid holdup $\bar{\epsilon}_L$, with a correction factor f of order unity ($f \gtrsim 1$).

With these simplifications, Eq. (4-2) is rewritten for subsequent use:

$$E_{zT} = -\frac{D_T^2}{2\nu_t} q_r \int_0^1 u\phi \, d\phi \int_1^\phi \frac{d\phi}{\phi} \int_0^\phi u\phi \, d\phi + q_z \nu_t \qquad (4\text{-}10)$$

where q_r and q_z are defined by $q_r = f\bar{\epsilon}_L/\bar{\alpha}$ and $q_z = \overline{\alpha\zeta/\epsilon_L}$, respectively. Properties of q_r and q_z are not yet evident from experiment, but they are of order unity and should be weak functions of the superficial gas velocity U_G because they include the liquid holdup ϵ_L. Also, q_z increases and q_r just as ϵ_L decreases, with increasing U_G.

In regard to axial dispersion in unbaffled bubble-flow equipment like liquid–liquid spray columns, gas bubble columns, or fluidized catalyst beds, a close similarity has been supposed as a result of bubble flow and of turbulence induced by bubbles (B3, M33). Baird and Rice (B3) have assumed that the Kolmogoroff concept for eddy viscosity in isotropic turbulence is applicable to evaluate E_{zT} in the unbaffled bubble column under turbulent conditions, concluding that E_{zT} is 0.35 $D_T^{4/3}\epsilon_m^{1/3}$ in cm-sec units, with ϵ_m the rate of energy dissipation per unit mass of continuous phase. Experimental observations indicate that axial dispersion follows the Taylor dispersion mechanism.

The eddy viscosity ν_t as given earlier for bubble columns has also been correlated (M27) according to Kolmogoroff's local-isotropy concept, to relate ν_t for the columns with data for atmospheric turbulence taken by Richardson (R6, O1). This approach gives ν_t approximately equal to 0.12 $l^{4/3}\epsilon_m^{1/3}$ in cm-sec units; for atmospheric turbulence ϵ_m is assumed to be 5 cm^2/sec^3 (O1) and l one-half the distance between pairs of observational points; for the bubble column l is assumed to be one-half the column radius. Physically the Prandtl hypothesis may be a sounder approach.

B. Longitudinal Dispersion in a Bubble Column

Many data are available for the longitudinal dispersion coefficient of the bubble column in the recirculation flow regime. Most are summarized in Fig. 39. Of these, Kato and Nishiwaki (K6, ⊕ and ○), Ohki and Inoue

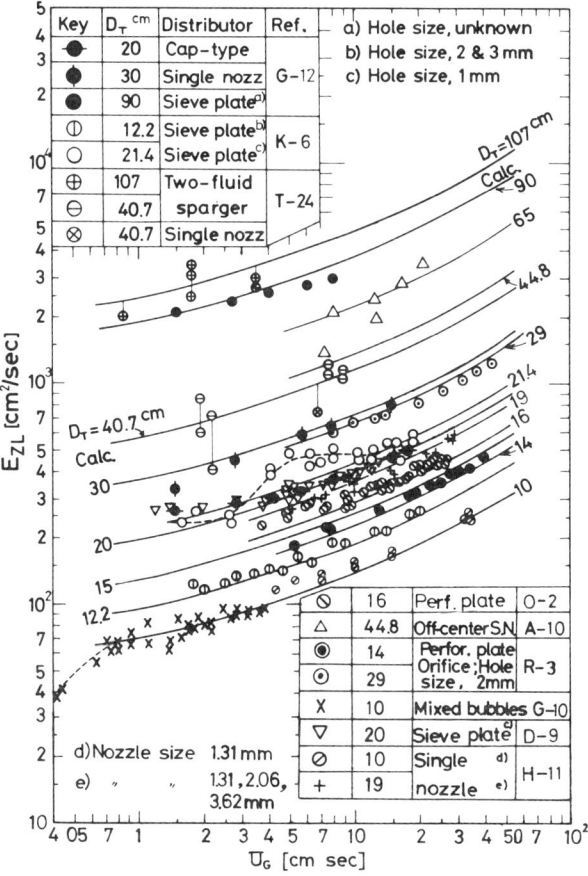

Fig. 39. Longitudinal dispersion coefficients of liquid in bubble columns (SN ≡ single nozzle; PP ≡ perforated plate). The full circles are calculated with the use of Eq. (4-12).

(O2), Towell (T24, ⊗), and Hikita and Kikukawa (H11) have utilized a pulse response technique or step response; and the rest of the data have been obtained by the steady back-mixing method. These are correlated as follows (M27).

1. Basic Relations for Longitudinal Dispersion Coefficient

Equation (4-10) is solved for the bubble column by utilizing Eqs. (3-15) and (3-17) for u; the u included in Eq. (4-10) is obtained by setting $U_L = 0$ in the two earlier equations. The final expression, in cm-sec units, is

$$E_{zT} = \frac{1}{1.18 \times 10^6}\left[\frac{11}{15} + \frac{1}{(1-\bar{\epsilon}_b)^2}\right] q_r \left(\frac{g^2 D_T^6 \bar{\epsilon}_b^2}{\nu_t^3}\right) + q_z \nu_t \quad (4\text{-}11)$$

For bubble columns of relatively small diameter the first right-hand-side term of Eq. (4-11) is much larger than the second term, so that E_{zT} is nearly proportional to D_T^6/ν_t^3 at a given U_G. As a consequence, the functional dependence of ν_t on D_T can be tested accurately by inspecting the dependence of E_{zT} on D_T.

Introducing Eq. (3-32) for ν_t and Eq. (3-28) for $\bar{\epsilon}_b$, Eq. (4-11) is transformed in cm-sec units to

$$E_{zL} = (E_{zT})_{\text{Liquid}} = \left[0.922 + \frac{1.080}{(1 - 0.054 U_G^{1/2})^2}\right] U_G^{1/4} D_T^{3/2} \quad (4\text{-}12)$$

where empirically $q_r = 1.864/U_G^{1/4}$ and $q_z = 0.813 U_G^{1/12}$, with U_G in cm/sec, to give Eq. (4-11) the best match with experimental data. The given dependence of q_r and q_z on U_G looks qualitatively sound, but note that a slight change in the dependence of ν_t and $\bar{\epsilon}_b$ on U_G and D_T results in a large influence on q_r and q_z.

2. Comparison of Theory with Experiment

The total axial dispersion coefficient E_{zL} calculated by Eq. (4-12) is shown in Fig. 39 as a function of D_T and U_G. Agreement with experiment seems satisfactory except for data from Argo and Cova (A10), who used an off-center single-pipe gas distributor tending to create tangential swirling motion, which may have modified the liquid recirculation flow pattern. They also found a higher gas holdup $\bar{\epsilon}_b$ than is given by the correlation utilized in Eq. (4-12), which could also lead to higher E_{zL}. Their axial-dispersion coefficients correspond to $D_T = 65$ cm, about 1.5 times that of their column. Data of Aoyama et al. (A9) for 5-, 10-, and 20-cm-diam columns equipped using a sieve plate as a gas distributor fall slightly above the calculated curves.

The superficial liquid velocity varies over the range of 0–2.18 cm/sec for the data available; this seems to have very little effect except at very low gas velocity. Also, the gas distributor construction has little effect in the recirculation flow regime, provided the distributor holes are placed symmetric around the column axis.

Eq. (4-12) is simplified further by approximating the term in parentheses as a function of U_G. In cm-sec units this leads to

$$E_{zL} = 1.35 U_G^{1/2} D_T^{3/2} \tag{4-13}$$

for $U_G = 5\text{--}60$ cm/sec. For $D_T = 10\text{--}30$ cm Eq. (4-13) is modified to

$$U_G D_T / E_{zL} = 2.0 (U_G/(gD_T)^{1/2})^{1/2} \tag{4-14}$$

which is nearly the same as the dimensionless correlation by Kato and Nishiwaki (K6) for $U_G/(gD_T)^{1/2} \gtrsim 0.03$. Also, Aoyama et al. (A9) report that E_{zL} is proportional to $D_T^{3/2}$ for the columns with a sieve plate as a gas distributor. An empirical correlation by Bedura et al. (B7) gives E_{zL} equal to $2.4 U_G^{0.33} D_T^{1.4}$.

Some correlations are available for the apparent axial diffusivity $\bar{\epsilon}_L E_{zL}$. Towell and Ackerman (T24) have given $\bar{\epsilon}_L E_{zL} = 1.23\ U_G^{1/2} D_T^{3/2}$. Deckwer et al. (D14) obtain the same relation with 2.0 in place of 1.23. Towell and Ackerman's correlation is reduced to $E_{zL} \approx 1.5\ U_G^{3/8} D_T^{3/2}$ in cm-sec units.

Equation (4-12) explains the axial dispersion coefficient reasonably well, as shown in Fig. 39, where the range of experimental variables is $D_T \geq 10$ cm, the perforation hole size ≥ 2 mm and $U_L \leq 2.18$ cm/sec, and the liquid is water or aqueous solutions with the liquid viscosity between 0.86 and 14.5 cp. and the surface tension $= 33\text{--}75.5$ dyne/cm.

At low gas rate, the size of gas bubbles and the gas hold-up are fairly sensitive to the gas-distributor structure (A9, F7, K17, O2, T15). Mixing and bulk recirculation of liquid are induced by the interaction of liquid and bubble motion, so that axial dispersion is influenced by the structure of gas distributor and may show strange behavior (A9, O2). For example, Ohki and Inoue (O2, Fig. 9) show that E_{zL} increases rapidly with U_G to a maximum at $U_G = 5$ cm/sec, then decreases gradually, and approaches typical recirculation-flow behavior regime at around $U_G = 16$ cm/sec. In this example a sieve plate with 91 holes of 0.4-mm diam is utilized for a 16-cm-diam column. Aoyama et al. (A9) found maximum and minimum E_{zL} in the low-gas-flow rate region of their 5.0-cm-diam bubble column using a sintered porous plate for gas distribution. These anomalies are closely related to the properties of bubble holdup in the transition from bubble flow to the ultimate circulation flow with increasing U_G. The properties of axial dispersion in these flow regimes may be explained by the Taylor dispersion mechanism (T9, A11); thus,

$$E_{zL} = E_{zr} + \bar{E}_z = \frac{\bar{u}^2 D_T^2}{k_0 E_r} + \bar{E}_z \tag{4-15}$$

where k_0 is a constant characteristic of the liquid flow pattern and $\bar{u}$ is an effective mean liquid velocity, and measures the axial spread of liquid due to velocity distribution. When many small bubbles of uniform size are generated and rise up through the liquid rather homogeneously, the liquid flow pattern and velocity are modified and radial mixing is suppressed. E_{zr} changes as a result of relative magnitude $\bar{u}^2$ and $k_0 E_r$, sometimes increasing or decreasing quite rapidly as U_G increases.

For organic liquid aeration, Hikita and Kikukawa (H11) show that $E_{zL} \propto 1/\mu_L^{0.12}$ with negligible effect of surface tension. Alexander and Shah (A5) obtain essentially no effect of liquid viscosity μ_L or surface tension on E_{zL}.

For recirculation flow the Taylor dispersion mechanism was introduced by Shyu and Miyauchi (S13). Equation (4-12) is a revised result for it. For this flow regime, Ohki and Inoue (O2) developed an expansion model with parameters adjusted to the data available, and also introduced the Taylor dispersion mechanism for the low-gas-velocity region of uniform bubble flow.

C. Longitudinal Dispersion in a Fluidized Catalyst Bed

Most data available from past studies are summarized in Figs. 40 and 41 for the longitudinal dispersion coefficient of the emulsion phase in fluidized catalyst beds of good fluidity. Such coefficients were first measured in the pioneer work of Bart (1950, B20) for a wide range of gas velocity (U_G = 7.5–90 cm/sec). His values with a 3.2-cm-diam column are approximately one-half those by Morooka *et al.* (M40) in the higher flow rate region. The basis for their correlation follows (M27).

1. *Underlying Concept for Longitudinal Dispersion*

When the velocity profile of the emulsion phase is similar to that of the liquid phase in a bubble column, Eq. (4-11) will apply to the fluidized catalyst bed. This similarity seems to be well justified as mentioned in Sections III,A,4–5, although there is no direct calculation of the turbulent kinematic viscosity ν_t from the measurement of velocity profile in the fluidized catalyst bed.

For bubble columns E_{zT} is estimated from Eq. (4-11) by determining ν_t and $\bar{\epsilon}_b$. For the fluidized bed, reversing the process utilized for the bubble column, ν_t can be calculated from $\bar{\epsilon}_b$ and an experimental E_{zT}. As discussed in Section IV,B,1, E_{zT} is approximately proportional to D_T^6/ν_t^3 at a given U_G, so that the dependence of ν_t on D_T is fairly accurately determined. Also, the ν_t dependence on U_G should be weak as mentioned in Section III,A,4 in relation to the Prandtl hypothesis. Systematic data for ν_t of the fluidized catalyst bed are not yet available from flow properties.

For the FCB we adopt the notation E_{zs} instead of E_{zT}. Experimental data for E_{zs} have been processed to know how they relate to D_T under a given U_G and to U_G under a given D_T. With these relations in hand, Eq. (4-11) is utilized to find the turbulent kinematic viscosity ν_t of the fluid catalyst bed as a function of D_T and U_G, applying the q_r and q_z utilized for the bubble columns [Eq. (4-12)]. The ν_t thus obtained has been given in Fig. 32 as curve A–A, matching Eq. (3-31a). As is evident from Fig. 32, the turbulent kinematic viscosity for a fluidized catalyst bed with good fluidity is approximately the same as for the bubble column, and this suggests strongly that the two systems behave similarly. In the recirculation flow, for example, the emulsion phase should be in intense turbulence. The bubble sizes in this flow regime should be similar for the two cases as is found in Fig. 34.

Introducing Eq. (3-31a) for ν_t into Eq. (4-11), again taking q_r and q_z from Eq. (4-12), the longitudinal dispersion coefficient E_{zs} of the emulsion phase is given in cm-sec units as follows:

$$E_{zs} = (E_{zT})_{\text{emulsion}}$$
$$= \left\{ 9.05 \left[\frac{11}{15} + \frac{1}{(1 - \bar{\epsilon}_b)^2} \right] D_T^{0.90} + 0.064 D_T^{1.70} \right\} U_G^{1/4} \quad (4\text{-}16)$$

The equation applies for to $U_G \geq 10$ cm/sec, utilizing the approximation taken for $\bar{\epsilon}_b$. The equation simplifies further, with a slight change in coefficient to fit best the data available (in cm-sec units):

$$E_{zs} = 12.0 U_G^{1/2} D_T^{0.9} \quad (4\text{-}17)$$

where the range of parameters is $U_G = 10$–50 cm/sec and $D_T = 3.2$–200 cm.

2. Comparison of Eq. (4-16) with Experimental Data

The total axial dispersion coefficient E_{zs} of the emulsion phase calculated from Eq. (4-16) is shown in Figs. 40 and 41 as a function of D_T and U_G, in reasonable agreement with experimental data. The E_{zs} shown here are for particles of good fluidity. Also, note in Figs. 40 and 41 that the second term in the right-hand side of Eq. (4-16) tends to be significant as D_T increases. At $U_G = 20$ cm/sec the contribution of this term to E_{zs} is 10% for $D_T = 100$ cm and 22% for $D_T = 300$ cm. At $U_G = 50$ cm/sec the contribution decreases down to about 0.75 times the mentioned levels.

Figure 41 illustrates the influence of D_T and the properties of bed particles on E_{zs} at $U_G = 20$ cm/sec. Although data from de Vries *et al.* (D8, V8,

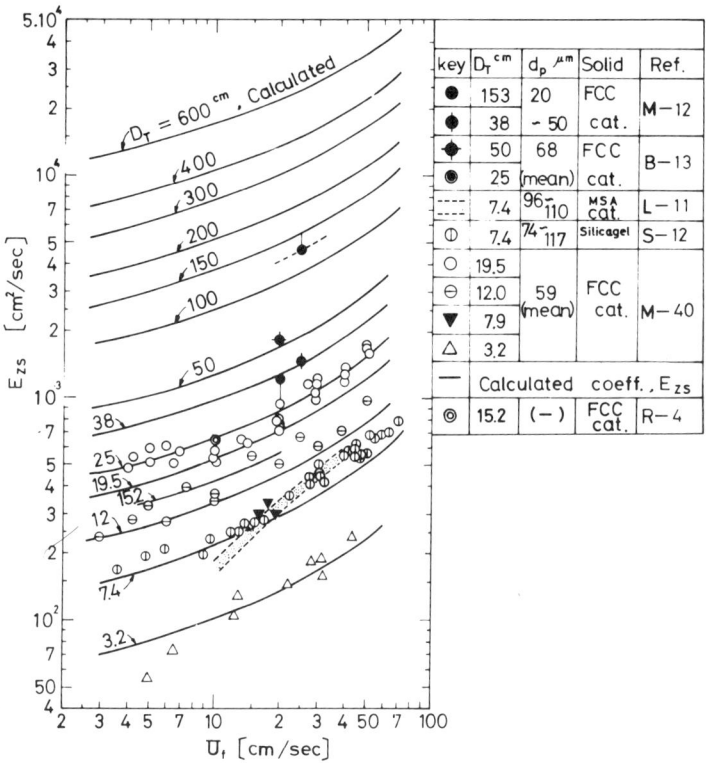

FIG. 40. Longitudinal dispersion coefficient E_{zs} of emulsion phase in fluidized catalyst beds as a function of U_G and D_T.

◎) and those from de Groot (D7, V8, ⊕) deviate from those for FCC particles at smaller or larger bed diameters, respectively, the general trend is nearly parallel.

The turbulent kinematic viscosity ν_t of the fluidized catalyst bed has been determined, as Eq. (3-31a), from the use of axial dispersion coefficient E_{zs}. This is a natural consequence of the analogy between the bubble column and the fluidized catalyst bed of good fluidity (such as in fluidized catalytic cracking). The mean gas holdup (Fig. 36) and the mean bubble velocity along the bed axis (Fig. 37) are reasonably well predicted by applying Eq. (3-31a) for the fluidized cracking catalyst bed.

V. Bubble Phenomena in Relation to Bed Performance

The bubble holdup $\bar{\epsilon}_b$ for a bubble column or for a fluidized bed is easily obtained as a function of superficial gas velocity U_G. Useful information

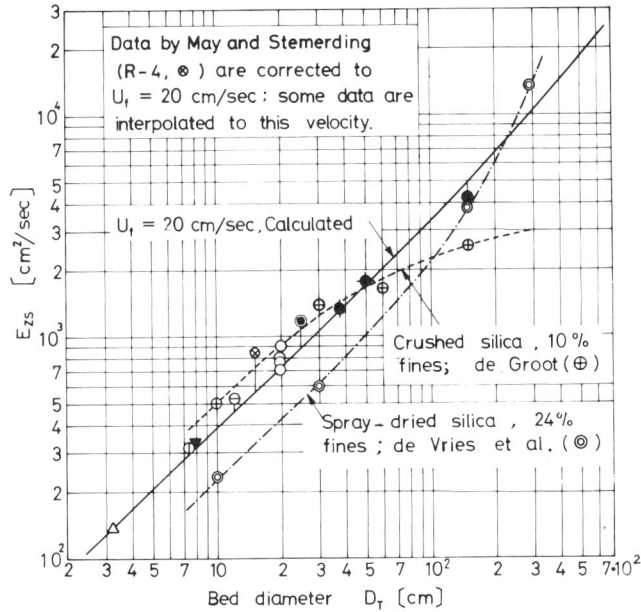

FIG. 41. Influence of particle properties on axial dispersion of the emulsion phase; $U_G = 20$ cm/sec with variable bed diameter D_T; keys are given in Fig. 40. Data by May ($U_G = 25$ cm/sec) and those by Stemerding ($U_G = 10$ cm/sec) are extrapolated to $U_G = 20$ cm/sec according to Fig. 40.

has been extracted from the $\bar{\epsilon}_b$ versus U_G relation for the fluidized bed (D3, N4, P8, T29). In recirculation flow an additional term appears in the apparent mean slip velocity shown on the right-hand side of Eq. (3-23b). This equation has been utilized to determine the turbulent kinematic viscosity ν_t and the mean slip velocity $\bar{u}_s$ (cf. Sections III,B,2, III,B,5, and III,D,5). In this chapter, the equation is further examined in relation to bed performance, since the turbulence properties of the bed result from interaction between bubbles and the continuous phase. As shown in Fig. 34, the mean slip velocity of bubbles in a fluidized catalyst bed of good fluidity is essentially the same as that for a bubble column when $U_G \gtrsim 20$ cm/sec. A criterion under which bubble size approaches a dynamic equilibrium is obviously needed for predicting or evaluating the performance of scaled-up beds.

A. Mean Bubble Velocity without Bulk Recirculation

A brief survey is needed here of bubble flow, with or without a bubble wake. Bubbles of a uniform size are generated by a gas distributor and rise uniformly through the continuous liquid phase in a bubble column, or through the emulsion phase for a fluidized catalyst bed. This one-dimensional bubble flow has been discussed extensively (B14, D4, N4, Z7). A bubble column without continuous liquid feed ($U_L = 0$) is taken for simplicity.

1. *Velocity of the Bubbles*

Case a of Fig. 42 shows the steady bubbling of gas through stagnant liquid in a bubble column, according to Nicklin (N4), where $U_L = 0$. Superficial gas velocity U_G is related to the ascending velocity $\bar{u}_b$ of bubbles relative to the wall:

$$U_G/\bar{\epsilon}_b = \bar{u}_b \qquad (5\text{-}1)$$

where $\bar{\epsilon}_b$ is the gas holdup under continuous bubbling.

Case b of Fig. 42 shows bubbles in exactly the same configuration, but rising relative to the stagnant liquid above in a swarm of finite size. In this case the bubble swarm has the same gas holdup $\bar{\epsilon}_b$ as case a, but rises at a velocity $\bar{u}_{b0}$ (relative to the wall) different from $\bar{u}_b$. In case a there is no net flow of liquid across the section A–A'. In case b there must be a net downward flow of liquid across the section A–A' to cancel the upward flow of gas.

The net downward flow of liquid is the amount necessary to fill in the bubble cavities left behind the swarm per unit time, i.e., $\bar{\epsilon}_b \bar{u}_{b0}$. Hence the interstitial velocity of the downward flow of liquid through the bubble swarm is $\bar{\epsilon}_b \bar{u}_{b0}/(1 - \bar{\epsilon}_b)$, and the ascending velocity of the swarm is reduced from $\bar{u}_b$ by this amount:

$$\bar{u}_{b0} = \bar{u}_b - \bar{\epsilon}_b \bar{u}_{b0}/(1 - \bar{\epsilon}_b) \qquad (5\text{-}2)$$

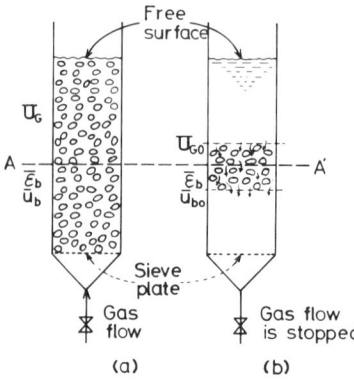

FIG. 42. Example of the rise of swarm of bubbles: (a) for continuous bubbling; (b) for a rising swarm of bubbles.

Equations (5-1) and (5-2) give the following relation (N4):

$$U_G/\bar{\epsilon}_b = \bar{u}_b = \bar{u}_{b0} + U_G \tag{5-3}$$

Taking $U_G - U_{mf}$ instead of U_G, Eq. (5-3) is equally applicable to the fluidized bed, where U_{mf} is the superficial velocity at incipient fluidization and is usually negligible in comparison with U_G for the fluidized catalyst bed.

Nicklin *et al.* (N5) have shown that an equation similar to Eq. (5-3) is applicable to slug flow. A single slug rises at a velocity equal to $0.35\sqrt{gD_T}$. If Eq. (5-3) is modified to allow for a nonuniform velocity profile in the liquid regions between the slugs, an empirical equation results:

$$U_G/\bar{\epsilon}_b = 1.2U_G + 0.35(gD_T)^{1/2} \tag{5-4a}$$

For fluidized beds Eq. (5-4a) is modified (D3, G19, O7, P10, S19):

$$(U_G - U_{mf})/\bar{\epsilon}_b = (U_G - U_{mf}) + 0.35(gD_T)^{1/2} \tag{5-4b}$$

The onset of slugging represents the terminal stage of bubble coalescence when a bubble spans the entire cross section of the bubble column (B15, S21).

If the bubbles in case a of Fig. 42 do not interfere with each other (D3), then the velocity of the bubbles should equal the free-rise velocity $\bar{u}_{s0}$ of a single bubble in a cross-sectional area of liquid large enough for the liquid phase to have no net vertical velocity. This has been indicated by Turner (T28) in regard to the fluidized bed, for which he modified Eq. (5-1) to the form

$$\bar{u}_b = (U_G - U_{mf})/\bar{\epsilon}_b = \bar{u}_{s0} = 0.71g^{1/2}V_b^{1/6} \tag{5-5}$$

where V_b is the volume of a spherical-cap bubble. This relation has been once applied to a bubbling bed (L5), but later abandoned (F12).

2. *Influence of the Bubble Wake*

So far, the influence of bubble wake on mean bubble velocity $\bar{u}_b$ relative to the column wall has not been mentioned, since Eq. (5-3) has been formulated on the basis of $\bar{u}_{b0}$, which already includes the effect of the wake (although it lacks a correction for wake fraction). In bubbling-bed models (F10, F12, K24, L5, S18) an upward flow of solid carried by the bubbles and bubble wakes leads to a downflow of solid (that has been assumed uniform) in the remainder of the bed. Then the bubble velocity $\bar{u}_b$ relative to bed wall should be smaller than the slip velocity of the bubble $\bar{u}_s$ relative to emulsion, since the bubble phase is retarded by downflow of solid. The rate of downflow of solid will be a maximum when the wake is formed at the gas inlet and released at the free surface without being renewed during the travel; also, the rate of downflow will be essentially zero when renewal is very fast during the travel. The actual situation lies between these two extremes (R17); and essentially the same situation applies in the gas bubble column.

We now consider the extreme case of negligible renewal of bubble wake for a continuously bubbling bed. The wake fraction in the bed is $f_w \bar{\epsilon}_b$. The bubble phase is composed of a bubble and a wake, and the fraction of this phase is $(1 + f_w)\bar{\epsilon}_b$. In the bubbling-bed model (F12, K24) the bubble phase is assumed to ascend at velocity $\bar{u}_s$ relative to the emulsion phase. The rate of release of wake solid at the free surface of the bed is then $f_w \bar{\epsilon}_b \bar{u}_b$, where $\bar{u}_b$ is the rising velocity of the bubble phase relative to the bed wall. Released solid constitutes the downward return flow through the emulsion, its velocity being given by $f_w \bar{\epsilon}_b \bar{u}_b / [1 - (1 + f_w)\bar{\epsilon}_b]$. This return flow retards the ascending motion of the bubble phase, and leads to the following relations:

$$\bar{u}_b = \bar{u}_s - f_w \bar{\epsilon}_b \bar{u}_b / [1 - (1 + f_w)\bar{\epsilon}_b] \tag{5-6}$$

or

$$\bar{u}_b = (U_G - U_{mf})/\bar{\epsilon}_b = [1 - f_w \bar{\epsilon}_b/(1 - \bar{\epsilon}_b)]\bar{u}_s \tag{5-7}$$

As a consequence $\bar{u}_b$ is always smaller than the slip velocity $\bar{u}_s$ for the extreme case of negligible renewal of the wake and equals $\bar{u}_s$ when $f_w = 0$ or when the rate of renewal is very rapid for finite f_w. In the latter example the wake loses its identity relative to the emulsion.

In conclusion, for a homogeneously bubbling bed, the velocity of bubble rise $\bar{u}_b$ relative to the bed wall is equal to or lower than the bubble slip velocity $\bar{u}_s$ relative to the continuous phase.

B. Mean Bubble Velocity with Bulk Recirculation

This section is concerned with the behavior of Eq. (3-23b) when applied to all bubble-flow equipment (M27).

In the preceding section the bubble velocity $\bar{u}_b$ in a homogeneously bubbling bed has been shown to be no greater than the slip velocity $\bar{u}_s$. However, numerous measurements do show $\bar{u}_b$ exceeding $\bar{u}_s$; they are generally expressed by the following empirical formula, with β an empirical coefficient characteristic of the flow properties:

$$\bar{u}_b = (U_G - U_{mf})/\bar{\epsilon}_b = \bar{u}_s + \beta(U_G - U_{mf}) \tag{5-8}$$

where U_{mf} is the superficial gas velocity at incipient fluidization and is zero for the bubble column; $\bar{u}_s$ is the mean slip velocity of bubbles, equal to the free-rise velocity $\bar{u}_{s0}$ when the interaction between bubbles is negligible. Towell *et al.* (T23) give $\beta = 2$ for their 40-cm-diam. bubble column under fairly high speed aeration. In Davidson and Harrison's well-known equation (D4), β equals unity with $\bar{u}_s = \bar{u}_{s0}$, although their equation is derived on the basis of homogeneous bubble flow.

1. *Simplified Recirculation-Flow Model*

The relation given by Eq. (5-8) is obtained easily for a bulk recirculation flow of the continuous phase. We consider a bubble column of radius R in which the column liquid is in upflow centrally with a constant interstitial velocity $\bar{u}_{lu}$. Peripherally the liquid descends, forming a recirculation flow. For simplicity, it is assumed[2] that gas bubbles distribute uniformly in the central upflowing liquid and ascend with it, and that no bubbles rise through the peripheral downflow. Here, the mean gas holdup of the upflow is $\bar{\epsilon}_{b*}$ and that averaged over the total column cross section is $\bar{\epsilon}_b$.

The bubble wake released at the free surface, joined with the upflow, constitutes the downflow; so that there is no downward return flow through the bubble layer. Under these circumstances, the mean bubble velocity $\bar{u}_b$ relative to the column wall is

$$\bar{u}_b = \bar{u}_s + \bar{u}_{lu} \tag{5-9}$$

The mass balance for the bubble phase gives

$$\pi R^2 U_G = \pi r_*^2 \bar{\epsilon}_{b*} \bar{u}_b \tag{5-10}$$

where r_* is the radius of the central upflow stream. In Eq. (5-10) $\pi r_*^2 \bar{\epsilon}_{b*}$ is obviously equal to $\pi R^2 \bar{\epsilon}_b$. Hence

$$\bar{u}_b = U_G/\bar{\epsilon}_b = \bar{u}_s + \bar{u}_{lu} \tag{5-11}$$

In this simple example $\bar{u}_{lu}$ corresponds to $\beta(U_G - U_{mf})$, with $U_{mf} = 0$, in Eq. (5-8), and the bubble wake shows no influence on $\bar{u}_b$. Here the simple recirculation flow gives the additional term $\bar{u}_{lu}$ even if there is no turbulent motion.

[2] When a gas bubble column of $D_T = 20$ cm is operated at $U_G = 20$ cm/sec, the fraction of gas throughput which rises through the central upflow zone amounts to 89.3% of the total gas, where the fraction φ_b is given as follows:

$$\varphi_b = (\pi R^2 U_G)^{-1} \int_0^{r_*} 2\pi r(u + u_s)\epsilon_b \, dr$$

$$= 1 - \left(1 - \frac{u_0/2}{U_G/\bar{\epsilon}_b}\right) \cdot \frac{1 - 1.5\bar{\epsilon}_b}{3(1 - \bar{\epsilon}_b)}$$

Hence, the simplification of 100% is not unreasonable.

2. Properties of the Bubble Wake

For one-dimensional homogeneous bubble flow, the wake usually reduces the ascending velocity $\bar{u}_b$ of the bubbles. The extent of reduction, however, depends on the rate of wake renewal, as shown in Section V,A,2. For the case of Section V,B,1, the wake has no influence on $\bar{u}_b$. Thus the way by which the wake influences $\bar{u}_b$ depends strongly on the physical situation encountered. In what follows the typical recirculation flow under intense turbulence will be treated, since this is the focus of our interest. We now show that the influence of the wake can be neglected in this flow regime.

Several investigators have observed the size of wakes in the turbulent flow field. Ahlborn (A1) has observed that the wake formed behind a solid cylinder gets much smaller as the cylinder approaches a turbulence grid. Kojima *et al.* (K16) observe for a spherical-cap bubble that the size of wake when the main stream is disturbed by a turbulence grid is not much different from when the main turbulent stream is undisturbed. They indicate that the volume of wake is approximately 4.7 times greater than the bubble volume. However, their bubbles are located downstream from a turbulence-promoting bar (a single glass cylinder of 1-cm diameter) a distance 8.3–38 times the bubble diameter, e.g., these bubbles would be located 15 cm downstream from the bar. Accordingly the bubbles do not seem unstable; the shape of the bubbles changes irregularly, but the wake is not clear whether washed out frequently or not.

When a spherical-cap bubble rises through a uniform swarm of smaller bubbles in water, velocity enhancement of the cap bubble is observed, the enhancement being associated with a change in its shape to an axially elongated form (H10, Fig. 4). In this case the wake will probably be reduced considerably in volume.

We now take a frequently encountered case of a bubble column of 20-cm diameter, operated at $U_G = 20$ cm/sec, with water. The bubble holdup is about 24% [Eq. (3-28)], and many bubbles of fair sizes ascend irregularly through the turbulent water phase. If a wake with 4.7 times the bubble volume accompanied each bubble, the continuous water phase would be essentially composed of wakes. Hence we have no way of distinguishing wakes from turbulent bulk liquid. Also, each bubble should be relatively spherical although of irregular shape, as suggested from Hills' observations (H10) and also observed experimentally. Washing out of wakes seems to occur frequently, due to the irregular motion of bubbles in the intense turbulence. At one moment a bubble may have wake-like flow behind it, and at the next moment the flow may be washed out. As a consequence it is hard to distinguish the wakes from the turbulent bulk liquid.

When the wakes are renewed rapidly, they have little influence on bubble motion (Section V,A,2). A relatively small influence of wakes is particularly the case when bubble motion is in the recirculation-flow regime (Section V,B,1).

The influence of wake motion on bulk turbulence induced in the liquid is understood more clearly by inspecting oscillograms which show the fluctuation of local liquid velocity. Figure 43 shows such oscillograms taken by Kikuchi (K30) with a hot-wire probe. The bubble column is 8.0 cm in diameter, water-filled to a 170-cm height, with the probe 115 cm above the bottom gas distributor. In the column bubbles of constant volume (100 cc) are injected successively at constant time intervals, either at 2.1 sec (case a) or 0.50 sec (case b). Case c is an example of continuous bubbling at $U_G = 6.45$ cm/sec.

In case a, wake motion is obviously identified for each bubble passing over the probe. The local velocity of liquid changes periodically in resonance with the bubble frequency. The resonance, however, is considerably disturbed as the bubble frequency increases. The oscillogram shows the onset of irregular liquid motions, which have different frequencies from those of bubbles. The relation between wake motion and local liquid motion becomes quite dubious. This tendency is increased in case b, where liquid motion exhibits fairly low frequencies. These flow properties are succeeded by continuous bubbling (case c). Here the local liquid motion is governed by the properties of bulk liquid motion, rather than by the bubble wakes.

Based on these discussions, the influence of bubble wakes has been omitted in what follows, since the wake motion is difficult to distinguish from bulk turbulence. However, the question of wake effects has not yet been settled experimentally.

3. *Mean Bubble Velocity in Recirculation Flow*

Under negligible influence of bubble wakes, the gas balance for the bubble phase is:

$$\pi R^2 U_G = \int_0^R 2\pi r(u + u_s)\epsilon_b \, dr \tag{5-12}$$

In the same manner as for Eq. (3-23), a radially constant mean $\bar{u}_s$ is taken for the slip velocity u_s. Also, with an obvious relation that $\epsilon_b = 1 - (1 - \epsilon_b)$, Eq. (5-12) is modified to

$$\pi R^2 U_G = \int_0^R 2\pi r u \, dr - \int_0^R 2\pi r u(1 - \epsilon_b) \, dr + \pi R^2 \bar{\epsilon}_b \bar{u}_s \tag{5-13}$$

The second integral is the net liquid flow through the column and is zero

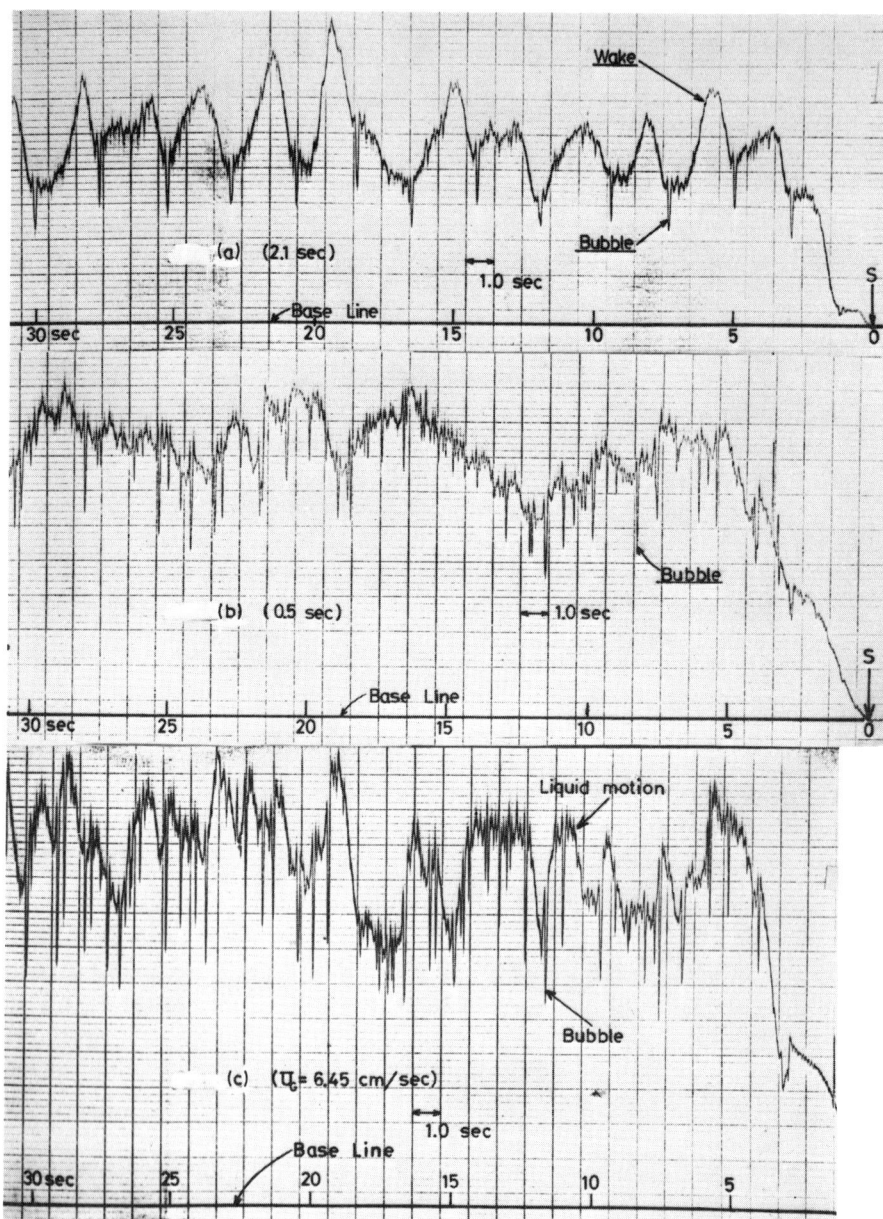

FIG. 43. Hot wire oscillogram showing the fluctuation of local liquid velocity; air–water system, $D_T = 8.0$ cm, S = start of bubbling; data by T. Kikuchi. (a) Time interval is 2.1 sec for bubble injection. (b) Time interval is 0.5 sec. (c) Continuous bubbling at $U_G = 6.45$ cm/sec.

when $U_L = 0$. Hence the equation simplifies to

$$\bar{u}_b \equiv U_G/\bar{\epsilon}_b = \bar{u}_s + U^*/\bar{\epsilon}_b \tag{5-14}$$

where U^* equals the first integral of Eq. (5-13), given in a footnote to Eq. (3-23b). Equation (5-14) is general for conditions of constant $\bar{u}_s$ and no liquid feed, since it assumes no particular radial-velocity distribution. For the fluidized catalyst bed, $(U_G - U_{mf})$ is taken for U_G in Eqs. (5-12)–(5-14).

From Eqs. (5-8) and (5-14) (with $U_{mf} = 0$), one has for the bubble column

$$\beta = U^*/\bar{\epsilon}_b U_G \tag{5-15}$$

In the case of recirculation flow, as developed in Section III,A, $U^*/\bar{\epsilon}_b$ is given by the second term on the right-hand side of Eq. (3-23b). Hence

$$\beta = (v_0/U_G)\bar{\epsilon}_b/(1 - \bar{\epsilon}_b) \tag{5-16}$$

where $v_0 = gD_T^2/192\nu_t$. When the quadratic equation with respect to $\bar{\epsilon}_b$, which is obtained from Eq. (3-23b), is modified into the form of Eq. (5-8), β is given as follows:

$$\beta = 1 + (v_0 - \bar{u}_s)\bar{\epsilon}_b/U_G. \tag{5-17}$$

The data necessary for calculating β using the above equations have already been given in Section III.

Figure 44 shows β as calculated by Eq. (5-16) as a function of U_G and D_T. Generally, β changes with these two parameters. For $D_T = 40$ cm, β is

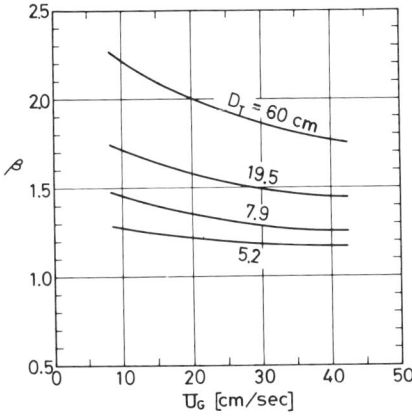

FIG. 44. β as a function of U_G and D_T for a bubble column, where $U_G/\bar{\epsilon}_b = \bar{u}_s + \beta U_G$.

approximately 2 in conformity with Towell (T23). Also, $\beta \gtrsim 1.1$ for the usual bubble columns, operated in the recirculation-flow regime.

Figure 45 illustrates the plot of $(U_G - U_{mf})/\bar{\epsilon}_b$ versus $(U_G - U_{mf})$ according to Eq. (5-8) for fluidized beds. In the figure, Eq. (5-4) is also shown for slugging beds. The mean gas holdup for the FCC-catalyst bed is taken from Fig. 36. The plot for the FCC bed shows clearly that the bed is fluidized smoothly, without slugging. The plot also shows data for a bed of fluidized glass beads of mean diameter of 287 μm. (Data are taken from W13; cf. Fig. 14). The bed behavior is seen to approach that of the slugging bed as $(U_G - U_{mf})$ increases beyond 20 cm/sec. Also, the average β is approximately 0.8 for $U_G < 20$ cm/sec, showing the possibility of bulk recirculation of the emulsion.

The plot shown in Fig. 45 is useful in understanding the flow properties of bubble-flow equipment. Also, it is quite probable that the Davidson and Harrison relation (D4, P11) with $\beta = 1$ is for a fluidized bed with bulk recirculation, although the recirculation pattern may be different.

C. BUBBLE SPLITTING IN TURBULENT FLOW

The mean slip velocity $\bar{u}_s$ has been shown in Fig. 34 for the bubbles observed in bubble columns or in fluidized FCC beds. Here the mean

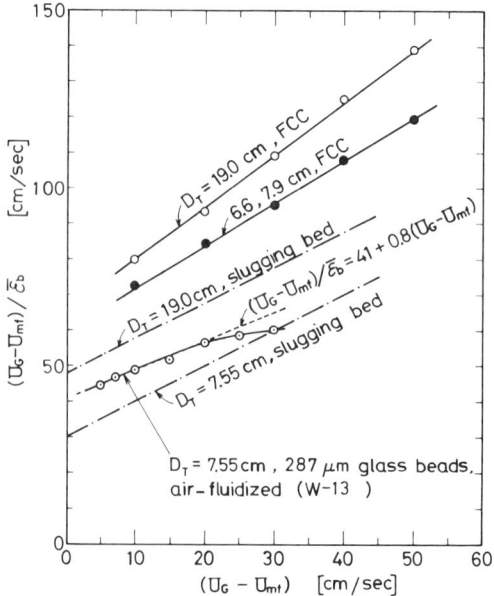

FIG. 45. $(U_G - U_{mf})/\bar{\epsilon}_b$ as a function of $(U_G - U_{mf})$ for fluidized beds. FCC-catalyst beds show no indication of slugging.

bubble size remains essentially unchanged as U_G increases beyond 20 cm/sec, suggesting a dynamic balance of coalescence and splitting of bubbles in the turbulent flow. Recently Yamazaki et al. (Y6) have observed the bubble-size distribution in a FCC-catalyst bed (8.1 cm in diameter with a quiescent bed height of 83 cm). They used a multichannel hot-wire system with a resolving power of 0.5 cm in bubble size. Figure 46a shows how the bubble size distribution at a given bed section changes with the height of the section z above the gas distributor (a canvas-coated sieve plate with 1-mm hole size). At the lower bed section, the distribution is unimodal, but with increasing z it shifts gradually to bimodal distribution as a result of bubble coalescence. Figure 46b shows the volume-

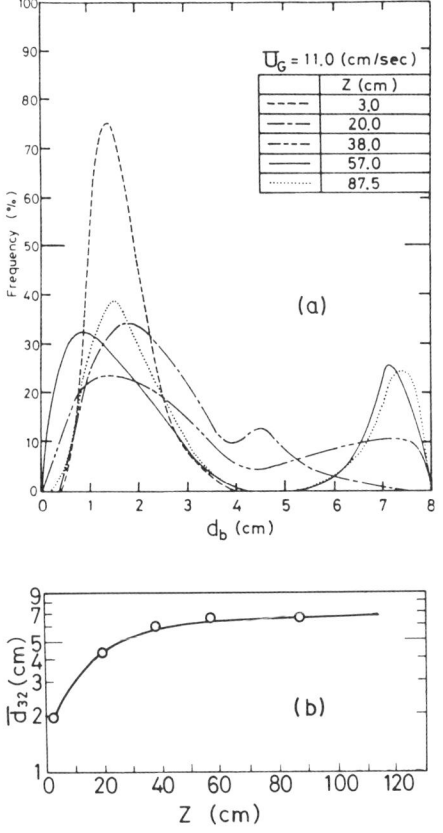

FIG. 46. (a) Bubble size distribution at various heights z from the gas distributor. (b) Volume-surface mean bubble diameter d_{32} as a function of z. $D_T = 8.1$ cm, $U_G = 11$ cm/sec, $L_q = 83$ cm; air-fluidized FCC-catalyst bed (Y6).

surface mean bubble size calculated from the distributions; the mean size becomes constant with $z \gtrsim 3D_T$.

Clift *et al.* (C6), in their study of bubbles in fluidized beds, indicate that instead of having a discrete "maximum stable bubble size" we can expect bubble splitting to occur over a relatively broad and continuous range of bubble sizes. Whether or not a particular bubble splits will depend not only on size but also on angular position, wavelength, and amplitude of disturbances of the bubble interface. It seems likely that measured maximum stable bubble diameters correspond to mean diameters for systems in which dynamic equilibrium has been achieved between coalescence and splitting.

There are three main mechanisms which determine the bubble sizes: coalescence of bubbles, splitting of a bubble under a given disturbance, and the occurrence of disturbances in the bubble-flow equipment. The latter two will be discussed in what follows. (Coalescence is discussed in C7, M10, T16, and W10).

1. *Splitting of a Bubble*

In 1950 G. I. Taylor (T8) investigated whether a small disturbance superimposed on the interface between two immiscible fluids of different densities was amplified or damped away when the fluids were accelerated at a constant rate. Four years later Bellman and Pennington (B10) developed this concept further to include the influence of liquid viscosity and surface tension. Rice and Wilhelm (R5) applied Taylor instability to the mechanism of bubble formation in a fluidized bed. Recently, Clift and Grace (C7) have shown that the first stage of bubble splitting in a fluidized bed is to develop an indentation in the bubble roof, which then moves around the bubble periphery while growing to form a curtain of particles. Splitting occurs if the lower edge of the curtain reaches the base of the bubble before the top passes the equator. It was suggested that this phenomenon results from instability of the type discussed by Taylor, where a heavy fluid overlies a lighter one.

For a gas bubble in liquid, Henriksen and Ostergard (H7) have observed that the bubble is broken up by disturbances created by the downward jet of liquid. According to their observation, a finger of liquid projects down from the roof and eventually divides the bubble in two. The reason that bubble breakup in methanol is easier than in water is due to lower surface tension. Based on these observations, they support the hypothesis of Clift and Grace that the bubble is broken up as a result of the Taylor instability.

Linearized stability analysis has been performed (B10) by superposing a sinusoidal disturbance of wavelength λ (which equals $2\pi/k$, with k the wave number) on the interface. The disturbance is assumed to grow with time θ according to exp $(n\theta)$, where n is the growth factor. Clift *et al.* (C6) extend the analysis to the fluidized bed, where the particles and fluid comprising the dense phase are considered separately by treating the dense phase as two superimposed, continuous fluids. If the viscosity and density of the fluidizing fluid are very much smaller than the corresponding quantities for the particulate phase (i.e., for gas fluidized beds) the stability analysis is found to be identical with that by Bellman and Pennington (B10) for real fluids with negligible interfacial tension and with the upper fluid much more dense and viscous than the lower.

Predictions from these theories for a two-dimensional interface are presented in Fig. 47 for $\nu_M = 0.01$–10 cm^2/sec, $\sigma = 0$ and 60 dyn/cm for the upper dense phase, where the lower light phase is air at room condition. The curves for zero surface tension are for a fluidized bed, and the curves for $\sigma = 60$ dyn/cm are for a bubble column of an aqueous solution.

In the case of zero surface tension, the analysis shows that the growth factor n is always positive for all wavelength λ, hence the disturbance is always amplified. Increase in dense-phase kinematic viscosity results in a

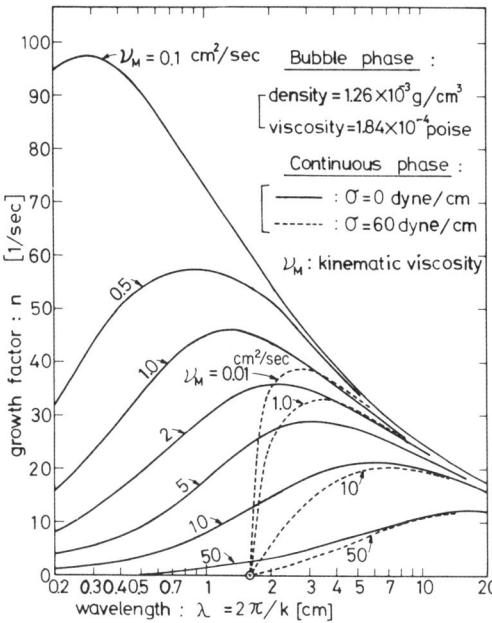

FIG. 47. Growth factor n as a function of wavelength of disturbance, λ, and the kinematic viscosity of the continuous phase for zero surface tension (full curves) and for $\sigma = 60$ dyne/cm (dotted curves) (B10, C6).

reduction in the growth factor and an increase in the most sensitive wavelength, at which n is a maximum. For the gas-fluidized bed, ν_M is the apparent kinematic viscosity of the particulate phase defined (C6) by $\nu_M = \mu_p/\rho_p\epsilon_{se}$, where μ_p is the effective Newtonian shear viscosity of particulate fluid, ρ_p the density of single particle, and ϵ_{se} the volume fraction of particles in the emulsion phase. Clift *et al.* (C6) have shown that the interstitial gas velocity has virtually no effect on the stability of a bubble roof. In their analysis the velocity is varied in the range of 1–100 cm/sec.

When the surface tension is finite, the disturbance tends to be suppressed by the surface tension, the influence of which is noticeable due to smaller curvature of the interface as the wavelength decreases. Hence there is a lower limit for wavelength, below which the growth factor takes negative values. The wavelength λ_{min} where the instability is neutral is (H7):

$$\lambda_{min} = 2\pi(\sigma/\rho_l g)^{1/2} \qquad (5\text{-}18)$$

For an air–water system at room temperature λ_{min} equals 1.70 cm, and for air–methanol λ_{min} is 1.04 cm. The influence of surface tension on growth factor is shown in Fig. 47, where the behavior discussed previously is observed clearly.

As noted previously, disturbances initiated on the roof of a bubble are swept around the periphery, so that in practice a bubble does not split unless the disturbance has grown sufficiently before the tip of the growing spike reaches the side of the bubble. Clift *et al.* (C6) have estimated the likelihood of splitting by comparing the time required for a disturbance to grow by a given factor with the time available for growth.

The required growth time may be provided by $\tau_e = 1/n$, where τ_e is the time required for a small-amplitude disturbance to grow by a factor e. Although the τ_e approach is no longer accurate for disturbances which have grown beyond the scale described by the linearized analysis, the estimate is retained in the absence of a better alternative.

An estimate of the time available for growth, τ_a, may be obtained from the dense-phase tangential velocity $U\psi$ (equal to $r_b \, d\psi/d\theta$, with ψ the angle measured from the vertical) at the bubble–emulsion interface:

$$\tau_a = r_b \int_{\psi_1}^{\pi/2} (1/U_\psi) \, d\psi \qquad (5\text{-}19)$$

where the disturbance originates at $\psi = \psi_1$. Thus τ_a becomes large if the disturbance originates very close to the nose of the bubble. Observations of splitting bubbles suggest that disturbances usually develop on either side of the nose.

Two cases have been studied. Case A has the bubble nose a node when the disturbance originates, and case B has a node located $\lambda/4$ from the bubble nose (so that the nose is an antinode in the initial disturbance). With these the maximum time available for growth, τ_{am}, is calculated from Eq. (5-19). A bubble is liable to be split by a disturbance for which $\tau_{am} > \tau_e$; here, only disturbances with wavelength less than the arc length from the nose to the equator, $\lambda \leq \lambda_{max} = \pi r_b/2$, are considered, since a disturbance with wavelength greater than λ_{max} represents a gross deformation of the bubble rather than a perturbation on the interface.

Figure 48 shows the calculation for a bubble with a radius of 2 or 5 cm in a gas-fluidized bed. In the bed with $\nu_M = 4$ cm^2/sec, the bubble with $r_b = 2$ cm shows $\tau_{am} > \tau_e$ for λ in the range 0.35–3.1 cm, so that a disturbance in this range may cause splitting. However, the same bubble in the bed with $\nu_M = 10$ cm^2/sec always shows $\tau_{am} < \tau_e$, and therefore should not split whatever the wavelength of the disturbance. By contrast, the bubble of 5-cm radius in either bed is liable to be split by a broad range of λ. The most sensitive wavelength (minimum τ_e; point m in Fig. 48) does not correspond to the λ most likely to cause splitting; in some cases λ_{max} is shorter than the most sensitive wavelength. When the most sensitive wavelength is within the range of possible disturbance, the ratio τ_{am}/τ_e is a maximum for a wavelength less than the most sensitive.

An important conclusion of the above approximate analysis is that the effective kinematic viscosity of the emulsion is the dominant factor determining both the initial growth of instabilities and the most sensitive wave number. Thus, prediction of the effect of system properties on bubble stability depends on prediction of the effect on the effective kinematic

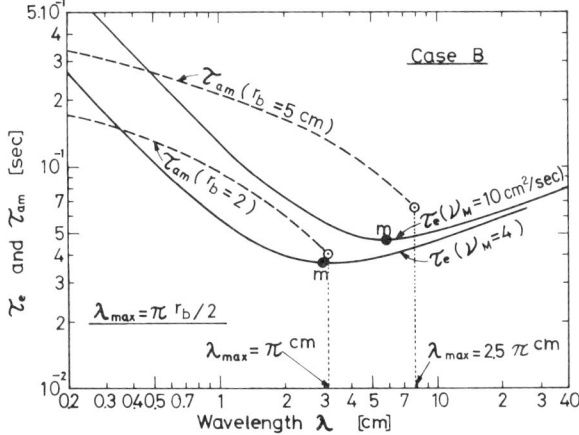

FIG. 48. Comparison of maximum time available for growth, τ_{am}, with time for a disturbance to grow by a factor e, τ_e, for a gas-fluidized bed; disturbance is case B (after C6).

viscosity. Based on this concept, Clift *et al.* (C6) have suggested that bubbles split more readily in a small-particle system and that fluidization is more aggregative as the ratio ρ_p/ρ_g is increased. As for particle size distribution, Tsutsui (T27) has shown that the wide size distribution of a standard FCC catalyst gives better fluidization than the FCC catalyst having a very narrow size cut with the same equivalent mean diameter. Geldart (G4, G5) indicates that fine particles of Group A (see Fig. 3) show good fluidity. Better fluidity of given particles is primarily determined by the volume-surface mean diameter of the particles and is not much influenced by their size distribution, according to Geldart.

Basic understanding of the effective kinematic viscosity of the emulsion is potentially important from the viewpoint of planning and operating industrial fluidized beds. There are mechanisms which lead to the formation of very small bubbles; these are omitted here since the performance of larger bubbles usually dominates the rate processes under high rates of aeration.

2. *Onset of Disturbances Effective in Bubble Splitting*

As explained previously Henricksen and Østergaard (H7) have observed that a bubble in water or methanol is broken up by the disturbances created by the downward jet of the liquid. Plate A in Fig. 49 shows a bubble that is about to break up because of the disturbances created by two liquid jets.

For a bubble to break up, disturbances should take place in the continuous phase, induced by the interaction between bubbles and the continuous phase. If the disturbances induced by an ascending bubble are almost completely damped away before the arrival of the next bubble, the succeeding bubble will hardly be broken up by the Taylor instability. If not, the bubble is liable to be split by the residual disturbances. As the extent of disturbance increases, bubbles split to smaller sizes and finally attain a steady mean size and size distribution as a result of the dynamic balance between splitting and coalescence (cf. Fig. 49). The decay of disturbances, however, depends on factors such as the strength of disturbances, viscosity of the continuous phase, frequency of bubble passage, and column dimensions.

When bubbles are successively fed to the bubble column filled with quiescent liquid, the first bubble will rarely be broken up due to the absence of disturbances on the bubble roof. The bubble rises as a slug or a spherical-cap bubble, depending on the gas flow rate and the column diameter. However, the first bubble leaves disturbances behind so that the succeeding bubbles will be broken up when the disturbances persist at

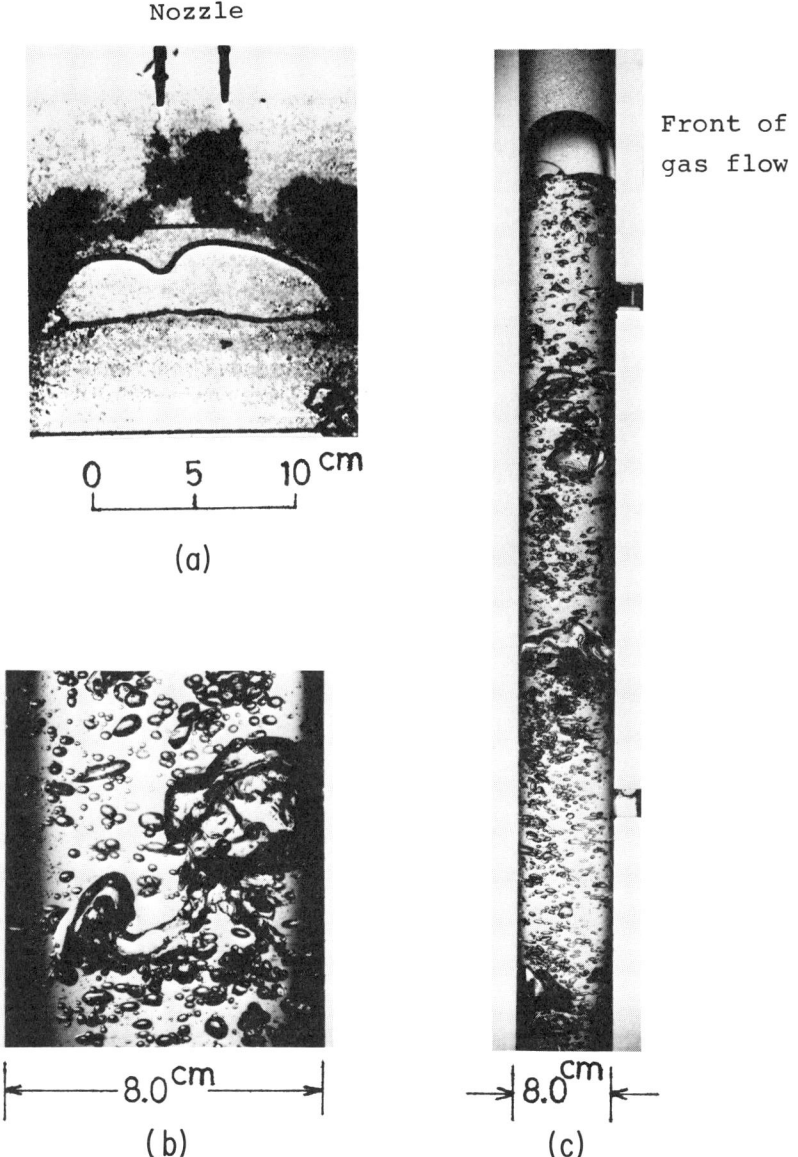

FIG. 49. Splitting of bubble by fluid-dynamic disturbances: (a) from Henricksen and Østergaard (H7); (b) and (c) from T. Kikuchi. (b) Air–water system, $D_T = 8$ cm, $U_G = 2.15$ cm/sec, continuous bubbling. (c) Air–aqueous glycerol system, $D_T = 8$ cm, $U_G = 4$ cm/sec, continuous bubbling.

a certain high level; this is clearly observed in plate C of Fig. 49. Plate B in the figure shows a large bubble about to be split by the turbulence induced by continuous bubbling.

Figure 50 shows the influence of molecular kinematic viscosity ν_M of liquid on the mean bubble size, when a 15-cm-diam. bubble column is operated at the superficial air velocity of $U_G = 6.7–12.7$ cm/sec (see data in K10). The mean bubble size is seen to increase only slightly as ν_M increases from 0.04 to 0.5–0.7 cm²/sec (aqueous glycerol solution); the flow properties here are well into the recirculation-flow regime. As ν_M increases beyond 0.7 cm²/sec the mean bubble size starts to increase rapidly, transferring finally to the high-viscosity branch $\overline{PQ}$ of insufficient breakup due to disturbances of low level. A similar phenomenon has also been observed for a larger-diameter bubble column by Ueyama and Miyauchi (U4), where a different viscosity is obtained for the flow transition effective to bubble splitting.

The mean bubble size that concerns us here is on the order of 5 cm, so that the Eötvos number E_0 (equal to $d_b^2 g \rho_l / \sigma$) is well over 40 for usual bubble-column liquids. The bubbles are of spherical-cap type under this condition, which is essentially equivalent to a Weber-number criterion We (equal to $d_b \rho_l \bar{u}_s^2 / \sigma) > 20$, since $\bar{u}_s = \sqrt{g d_b / 2}$ (H4, H5). The bubbles in a fluidized catalyst bed satisfy the above criterion, since $\sigma \sim 0$. Consequently, surface tension has relatively little effect, and instead the splitting is closely related to disturbances induced by the bulk turbulence, the intensity and the scale of which are mainly governed by the fluidity of the continuous phase and the operating gas velocity.

Figure 51 shows the observed bubble diameter d_b in fluidized beds as a

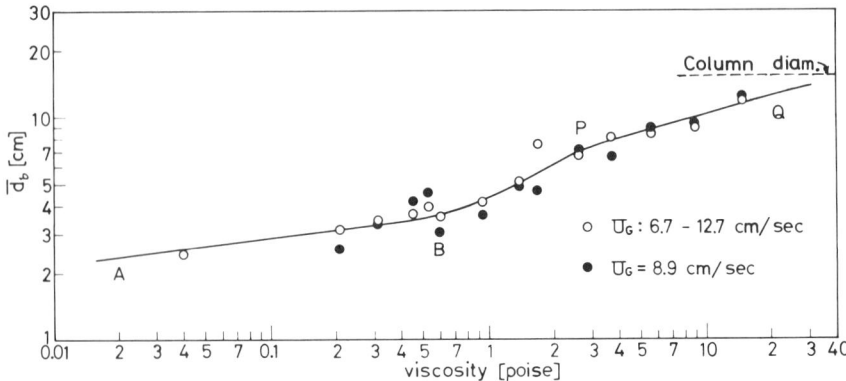

FIG. 50. Change of mean bubble size with liquid viscosity under constant aeration. Aqueous glycerol solution, $D_T = 15$ cm, $U_G = 6.7–12.7$ cm/sec. Data by Kimura (K10).

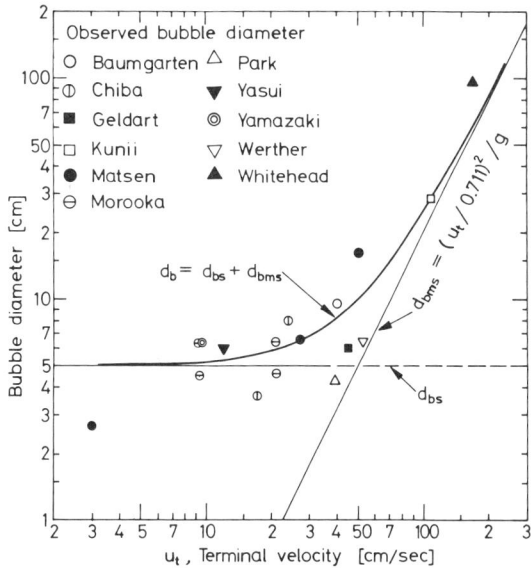

FIG. 51. Observed bubble diameter d_b versus particle terminal velocity u_t; data from Horio and Wen (H21). Mean bubble size by Morooka *et al.* (M43) and that by Yamazaki *et al.* (Y6) are added.

function of the particle terminal velocity u_t. The data are mostly taken from Horio and Wen (H21). Values of mean steady bubble sizes under turbulent aeration are from Morooka *et al.* (M43) and Yamazaki *et al.* (Y6). As the size of particles decreases while good fluidity is retained, d_b tends to approach the steady bubble diameter d_{bs} observed for bubble columns of low-viscosity liquids (Fig. 34). In this domain where $d_b \approx d_{bs}$, dynamic equilibrium exists between bubbles of various sizes as a result of turbulent motion prevailing in the emulsion; there is a steady mean bubble size here rather than the maximum stable bubble size.

The maximum stable bubble diameter d_{bms} defined by Davidson and Harrison (D3) seems to apply for an emulsion of large particles, where the turbulent motion is weak. Bubbles tend to coalesce with each other to grow ultimately to d_{bms}. The observed bubble diameter d_b is closely related to bubble splitting and coalescence as a result of turbulence. The data in Fig. 51 are expressed by an approximation that d_b is the sum of d_{bs} and d_{bms} or

$$d_b = d_{bs} + d_{bms}$$
$$= d_{bs} + (u_t/0.711)^2/g \qquad (5\text{-}20)$$

Scientific approaches to improve bed fluidity are potentially important for fluidized bed technology. Also, further quantitative relations between bubble splitting and bed properties would be very helpful in planning and scaling-up fluidized catalyst beds.

VI. Heat and Mass Transfer in Fluidized Catalyst Beds

Heat and mass transfer constitute fundamentally important transport properties for design of a fluidized catalyst bed. Intense mixing of emulsion phase with a large heat capacity results in uniform temperature at a level determined by the balance between the rates of heat generation from reaction and heat removal through wall heat transfer, and by the heat capacity of feed gas. However, thermal stability of the dilute phase depends also on the heat-diffusive power of the phase (Section IX). The mechanism by which a reactant gas is transferred from the bubble phase to the emulsion phase is part of the basic information needed to formulate the design equation for the bed (Sections VII–IX). These properties are closely related to the flow behavior of the bed (Sections II–V) and to the bubble dynamics.

A. Bubble Dynamics

Although gas bubbles ascending through the emulsion of fine catalyst particles are constantly splitting and coalescing (Sections II, III, and V), they are largely free of particles (H14, K13, T19). Such a bubble, may be pictured as essentially spherical, with the lower $\frac{1}{3}-\frac{1}{4}$ of its volume occupied by particles (Fig. 52). As the bubble rises, it displaces solid particles around it and carries some particles upward in its wake. Its rise velocity $\bar{u}_s$ is proportional to the square root of its frontal diameter. If $\bar{u}_s$ is lower than the velocity of the interstitial gas $u_{mf} = U_{mf}/\epsilon_{mf}$, the gas from the emulsion enters the lower part of the bubble and leaves through the roof; however, for small particles, $\bar{u}_s$ is usually larger than u_{mf}, and the circulation pattern as shown in Fig. 52 is encountered. Gas circulates up the central core of the bubble and, upon emerging from the roof, encounters drag from the particles streaming past. Consequently this gas is swept downward relative to the bubble and is re-entrained at the bottom because of the prevailing pressure gradient there.

It follows that as the bubble rises more quickly, the gas which emerges from its roof will penetrate outward a smaller distance before being swept downward; i.e., the limit of penetration will be nearer to the bubble wall. The total volume enclosed by the limit of penetration is called the cloud,

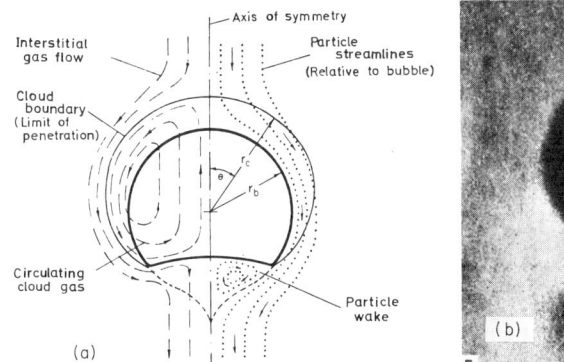
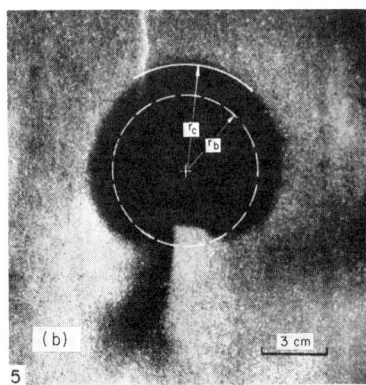

FIG. 52. (a) Gas and particle flow pattern near a bubble, $\alpha \approx 2.5$ and (b) a photograph showing cloud size r_c and bubble size r_b by using NO_2-containing-bubble technique, $\alpha = 2.5$, 230 μm ballotini (after R17).

and the region between the bubble-cavity wall and the limit of penetration is called the cloud overlap. This region and part of the wake are the only regions where the gas in the bubble can contact particles.

These physical pictures of bubble dynamics have been developed by Davidson and Harrison (D3), Murray (M46, M47), Pyle and Rose (P9), Rowe *et al.* (R17) and others. Jackson (J1) considered that a mantle of bed with increased voidage exists near the roof of the bubble. Most of the work on fluid-mechanical theories of aggregative fluidization have been reviewed by Jackson (in D5).

The radius of the cloud r_c has been given by Davidson and Harrison (D3) for $\alpha = \bar{u}_s/\bar{u}_{mf} > 1$:

$$r_c/r_b = [(\alpha + 2)/(\alpha - 1)]^{1/3} \tag{6-1}$$

Alternatively, Murray (M47) proposes for $\alpha > 1$:

$$(\alpha - 1)(r_c/r_b)^4 - \alpha(r_c/r_b) = \cos \theta \tag{6-2}$$

The experiments of Stewart (S20) show (Fig. 53) that the radius of an observed bubble cloud at $\theta = 0$ falls between Eqs. (6-1) and (6-2). Figure 54 shows the thickness of the cloud-overlap region estimated by Eqs. (6-1) and (6-2) for a fine microspherical catalyst fluidized by ambient air, where the minimum fluidization velocity U_{mf} is calculated by Wen and Yu's equation (W4). The cloud-overlap region is seen as being limited to at most a few layers of fine particles (M30). Rieke and Pigford (R8) observed experimentally for a two-dimensional fluidized bed that the gas in the

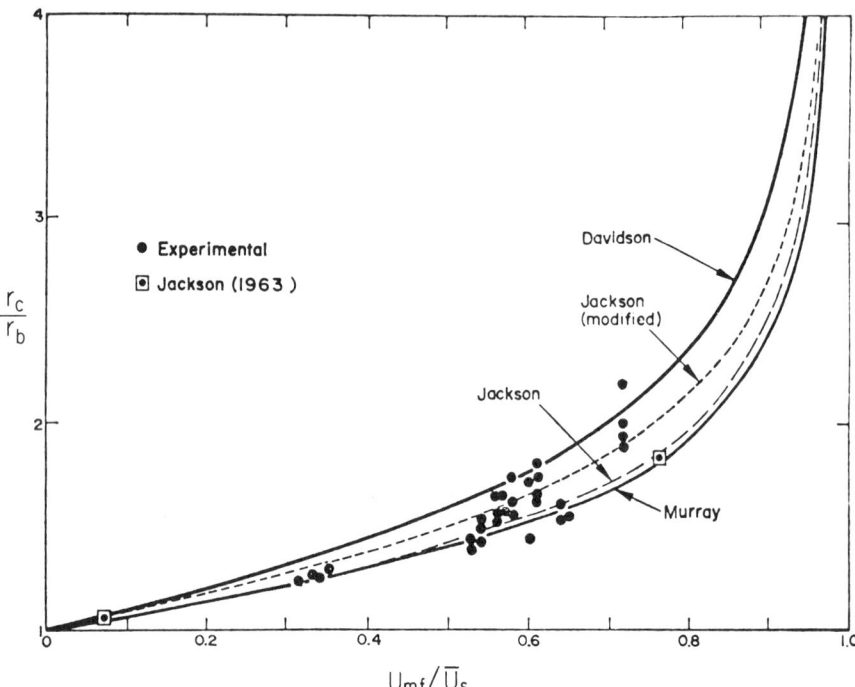

FIG. 53. Ratio of cloud radius r_c to bubble radius r_b on vertical axis above three-dimensional bubbles. Comparison of experimental values with various theoretical predictions. The curve "Jackson (modified)" refers to results obtained by Jackson's type of analysis, but with the Davies–Taylor type rise velocity $k\sqrt{gr_b}$, where k is chosen to give the best fit to the experimental rise velocity (from S20 and D5).

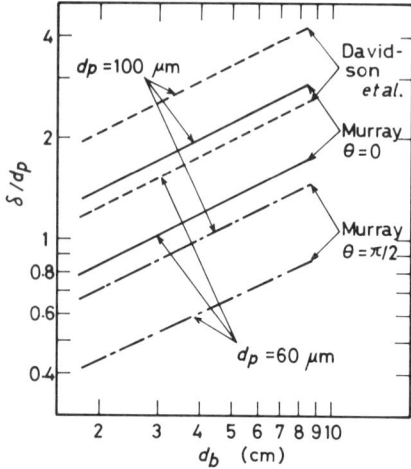

FIG. 54. Thickness of gas cloud. $\rho_p = 1.0$ g/cm³, $\epsilon_{mf} = 0.5$, fluidized by ambient air.

bubble wake does not return to the bubble cavity while circulating. This is in conformity with the observation by Rowe *et al.* (R17); see Fig. 52. As a consequence, the observed gas-flow patterns suggest that there is no direct gas communication between the wake and the cloud-overlap region.

Progress in understanding the bubble dynamics has made it possible to formulate transport equations for heat and mass transfer through the bubble wall, which we now consider.

B. HEAT AND MASS TRANSFER THROUGH THE BUBBLE WALL

Most of the experimental work so far has been concerned with the performance of an isolated bubble; few publications are available on fluidized catalyst beds under a high aeration rate. In 1961 van Deemter (V1) analyzed the transient and steady-state gas mixing data by Mason (given in V1) taken in a 7.6-cm-diam. cracking catalyst bed. Van Deemter based his calculation of the overall mass transfer coefficient $k_{ob}a_b$ on the two-phase theory of fluidization. Further, axial mixing of gas and solid particles has been studied for fluidized catalyst beds by de Groot (D7), Botton (B13), van Swaay and Zuiderweg (V8), and Morooka *et al.* (M42).

For crushed silica fluidized at ambient temperature and pressure in beds with diameter 0.3–1.5 m, van Swaay and Zuiderweg (V8) have reported the apparent HTU:

$$(H_{ob})_{app} = \frac{U_G}{k_{ob}a_b} = \left(1.8 - \frac{1.06}{D_T^{0.25}}\right)\left(3.5 - \frac{2.5}{L_f^{0.25}}\right) \quad (6\text{-}3)$$

where $(H_{ob})_{app}$, D_T, and L_f are in meters. This relation is expected to apply for D_T and L_f up to 10 m, with 15% fines content of the solid ($d_p < 44\ \mu$m). The use of microspherical particles with a larger fraction of fines showed somewhat better mass transfer; but the dependence of HTU or of $k_{ob}a_b/\bar{\epsilon}_b$ on bed dimensions was the same, as shown in Fig. 55. The coefficients are less dependent on gas velocity and bed height for particles with good fluidity. The data by de Groot (D7) indicate a doubtful effect of bed height, in contrast to Eq. (6-3).

Morooka *et al.* (M42) studied the effect of gas adsorption on mass transfer between bubbles and emulsion phase, while measuring the residence time distributions of tracer gases for helium, carbon dioxide, Freon 12, and Freon 22 in 12- and 19-cm-diam. free and baffled beds of cracking catalyst. The $k_{ob}a_b$ values are calculated from the distributions for different gases according to the two-phase model, providing for the adsorption equilibrium.

The quantity $k_{ob}a_b$ increases with increasing superficial gas velocity U_G; it shows a higher value when gas is more strongly adsorbed by the parti-

364 MIYAUCHI, FURUSAKI, MOROOKA, AND IKEDA

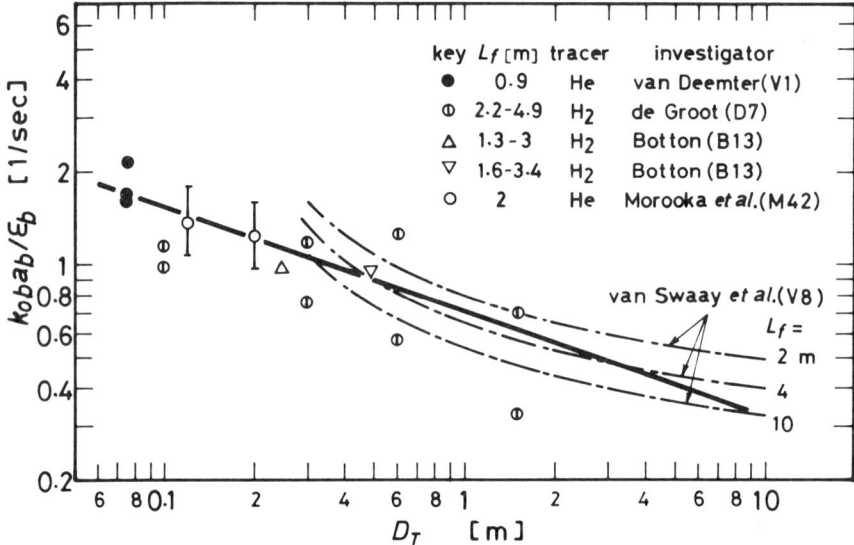

FIG. 55. Effect of bed diameter on $k_{ob}a_b/\bar{\epsilon}_b$.

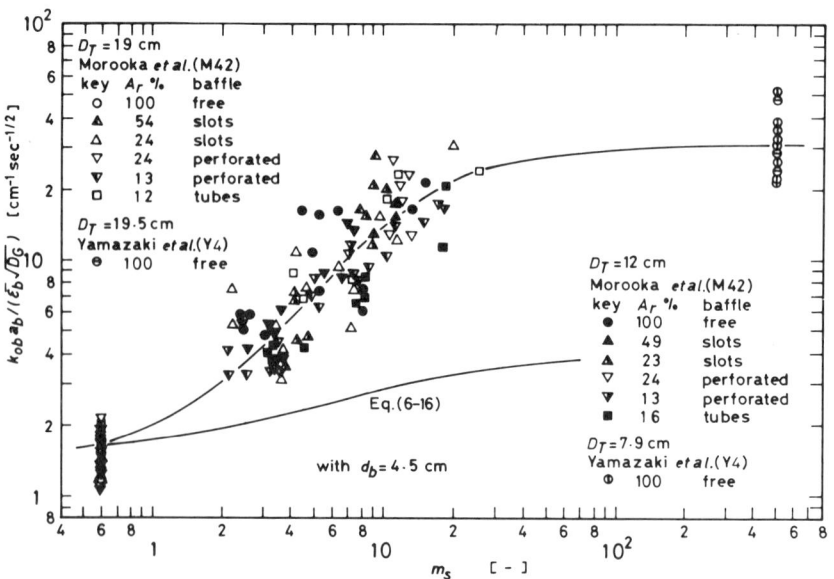

FIG. 56. Effect of adsorption equilibrium constant on $k_{ob}a_b/\bar{\epsilon}_b \sqrt{D_G}$ in free and multistep fluid beds (after M42).

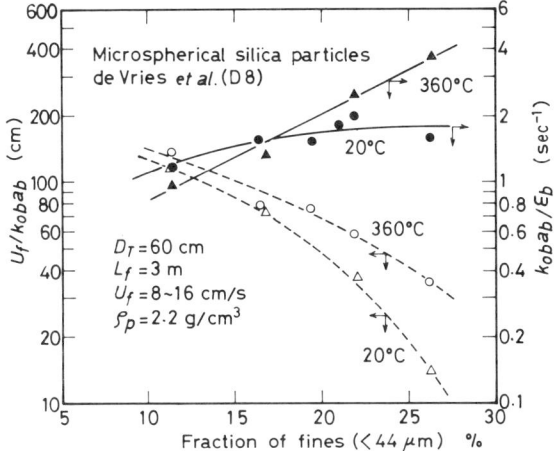

FIG. 57. Effect of fines on $k_{ob}a_b$ in fluid bed (data taken from D8).

cles (m larger). The effect of increasing gas velocity or adding horizontal baffles is mainly to increase the mean bubble holdup in the bed.

The group $k_{ob}a_b/\bar{\epsilon}_b\sqrt{D_G}$ is plotted in Fig. 56 as a function of the gas partition ratio m_s. Obviously k_{ob} increases with m_s, owing to decreased particle-side mass-transfer resistance. Data of Yamazaki and Miyauchi (Y4) are plotted in Fig. 56 and match well as an extension of the Morooka data. Using a quick-response thermoelement, Yamazaki measured the temperature distributions of the bubble and emulsion phases, from which the overall heat-transfer coefficient is calculated for 7.9- and 19-cm-diam. cracking catalyst beds. Since the heat capacity of the emulsion phase is much higher than that of the bubble phase, the overall coefficient is nearly equal to the film coefficient $h_b a_b$ for the bubble side. The heat-transfer coefficient reduces to the mass-transfer coefficient according to the equality $h_b a_b / c_{pg} \rho_g = k_b a_b$.

The effect of added fines content in increasing $k_{ob}a_b$, measured by de Vries *et al.* (D8), is shown in Fig. 57. Other work on heat and mass transfer for teeter-type beds is reviewed by Kunii and Levenspiel (K24) and in Davidson and Harrison's book (D5).

C. Mass Transfer between Bubble and Emulsion Phase

Gas exchange during catalytic reaction between bubbles and the emulsion of catalyst particles is essentially a diffusional phenomenon, with simultaneous adsorption and reaction in the emulsion. Based on this concept, the overall mass transfer has been analyzed by Miyauchi and

Morooka (M30), Chiba and Kobayashi (C3), and Drinkenburg and Rietema (D17). According to this approach, the bubble-side film coefficient $k_b a_b$ tends to provide the main resistance to mass transfer when gas adsorption onto the particles is strong and the reaction rate constant is high. The following theory is by Miyauchi and Morooka (M30).

1. *The Underlying Concept and Its Formulation*

As stated in Section VI,A, bubbles ascending through the usual fluid catalyst bed show a sufficiently large α (equal to $\bar{u}_s/U_{mf}$) to make the thickness of the cloud-overlap region very thin, as illustrated in Fig. 54. Consequently the streamlines for gas and particles are essentially parallel to the bubble surface. Convective bulk flow of "cloud" gas across the bubble interface is negligible. Hence it is assumed that only the portion of cloud gas which ascends along the bubble axis region flows into the roof of the bubble at a volumetric flow rate q_f, then descends through the cloud-overlap region and flows out of the bottom of the region to recirculate again into the bubble void (see Fig. 58). The recirculating gas flowing through the cloud-overlap region exchanges the reactant gas by diffusion from the bubble void to the bulk emulsion phase.

With a simplifying assumption that local adsorption equilibrium is instantaneously attained between gas and particles, the mass-transfer process is expressed by Fig. 58 for the cloud-overlap region. Here the influence of bubble wall curvature is neglected, since the region is very thin. When no catalyst is suspended in the bubble void, the equations of continuity for the reactant gas are as follows (M30):

$$\partial c_b/\partial \theta = D_G \, \partial^2 c_b/\partial x^2 \qquad (6\text{-}4)$$

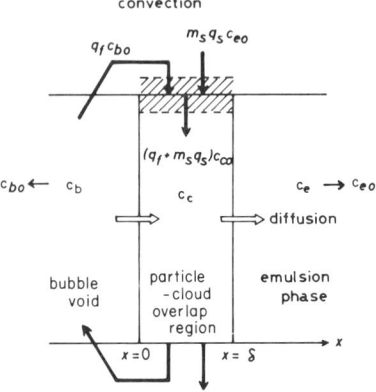

FIG. 58. Schematic diagram of mass transfer near a bubble in fluid bed (M30).

for the bubble void;

$$m\, \partial c_c/\partial \theta = D_{\text{eff}}\, \partial^2 c_c/\partial x^2 - k_r c_c \qquad (6\text{-}5)$$

for the cloud-overlap region;

$$m\, \partial c_e/\partial \theta = D_{\text{eff}}\, \partial^2 c_e/\partial x^2 - k_r c_e \qquad (6\text{-}6)$$

for the emulsion phase, where $m = \epsilon_{\text{fe}} + m_s \epsilon_{\text{se}}$, with $m_s = (c_s/c_f)_{\text{equilibrium}}$. c_s is moles of adsorbed gas per unit volume of particles including intraparticle pore volume. Also, a first-order irreversible reaction is assumed to take place in the particulate phase.

The initial and boundary conditions are as follows:

for $\theta = 0$,

$$c_b = c_{b0}, \qquad c_c = c_{c0}, \qquad c_e = c_{e0}$$

for $\theta > 0$,

$$\begin{aligned}
x &= -\infty: & c_b &= c_{b0} \\
x &= 0: & -D_G\, \partial c_b/\partial x &= -D_{\text{eff}}\, \partial c_c/\partial x,\quad c_b = c_c \\
x &= \delta: & -D_{\text{eff}}\, \partial c_c/\partial x &= -D_{\text{eff}}\, \partial c_e/\partial x,\quad c_c = c_e \\
x &= +\infty: & c_e &= c_{e0}
\end{aligned} \qquad (6\text{-}7)$$

The initial concentration of the cloud-overlap region c_{c0} is obtained from the reactant gas balance taken at the roof of the bubble, where the catalyst particles are fed to the cloud-overlap region from the emulsion above at volumetric flow rate q_s:

$$c_{b0} q_f + m_s q_s c_{e0} = (q_f + m_s q_s) c_{c0} \qquad (6\text{-}8)$$

When $\alpha \gg 1$, the slip velocity of particles relative to the gas is negligible in comparison with the particle velocity, so that q_f/ϵ_{fe} is nearly equal to q_s/ϵ_{se}. With this equality Eq. (6-8) is modified to the form:

$$c_{c0} = (\epsilon_{\text{fe}} c_{b0} + m_s \epsilon_{\text{se}} c_{e0})/m \qquad (6\text{-}9)$$

The concentration distributions in each phase and the mass flux across the bubble surface have been solved to satisfy Eqs. (6-4)–(6-6) under the restrictions of Eqs. (6-7) and (6-9) (M26, M30). The overall mass-transfer coefficient between the bubble cavity and the emulsion phase k_{ob} has been given as follows:

$$1/k_{\text{ob}} = 1/k_b + 1/\beta_r k_e \qquad (6\text{-}10)$$

where

$$\beta_r = \beta - (\epsilon_{\text{fe}}/m) J. \qquad (6\text{-}11)$$

The film mass-transfer coefficients for the bubble and the emulsion phase (k_b and k_e, respectively) are given by

$$k_b = (2/\pi^{1/2})(D_G/\tau_b)^{1/2} \tag{6-12}$$

$$k_e = (2/\pi^{1/2})(mD_{\text{eff}}/\tau_b)^{1/2} \tag{6-13}$$

where τ_b is the time necessary for local surface renewal; β is the Hatta number (or reaction factor) for unsteady gas absorption with a first-order irreversible reaction, as given by Danckwerts (D1):

$$\beta = \left(m_H + \frac{\pi}{8m_H}\right) \text{erf}\left(\frac{2m_H}{\pi^{1/2}}\right) + \frac{1}{2}\exp\left(-\frac{4m_H^2}{\pi}\right) \tag{6-14}$$

with $m_H = (k_r D_{\text{eff}})^{1/2}/k_e$. J in Eq. (6-11), associated with the cloud-overlap region, is plotted in Fig. 59 and conforms to the relation:

$$J = \frac{\pi}{4m_H}\text{erf}\left(\frac{2m_H}{\pi^{1/2}}\right) - \int_0^1 \frac{1}{2(x^{1/2})}\exp\left[-\left(\frac{4m_H^2}{\pi}x + \frac{\text{Pe}}{x}\right)\right] dx \tag{6-15}$$

where $\text{Pe} = m\delta^2 \bar{u}_s/4d_b D_{\text{eff}}$. J is approximated accurately enough by the following equation for $0 \leq \text{Pe} \leq 0.1$ and $0 \leq m_H \leq 2$, the usually encountered domain for fluidized catalyst beds of fine particles:

$$J = (\pi \text{Pe})^{1/2} - \text{Pe} - 0.454 m_H^{1.40} \text{Pe}^{0.84} \tag{6-15a}$$

The physical data needed to compute k_{ob} are D_G, m, k_r, U_{mf}, d_b, and ϵ_{fe} (ϵ_{fe} can be approximated by ϵ_{mf}). With the Higbie penetration theory for bubbles, τ_b included in Eqs. (6-12) and (6-13) is approximated by $d_b/\bar{u}_s$, so that Eq. (6-10) is modified to

$$k_{ob} = \frac{2/\pi^{1/2}}{1 + (\beta_r)^{-1}(\chi/m\epsilon_{fe})^{1/2}}\left(\frac{D_G \bar{u}_s}{d_b}\right)^{1/2} \tag{6-16}$$

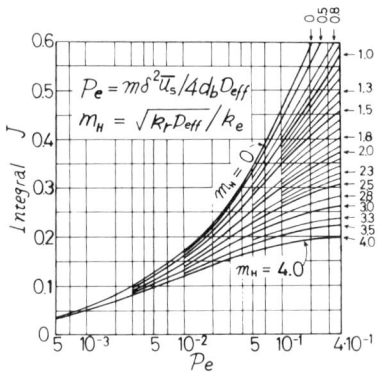

FIG. 59. Integral J as a function of Pe and m_H (M26).

Here, the tortuosity factor χ is approximately 1.5, as given by Hoogschagen (M23). When $\text{Pe} \leq 0.01$ and $m_H \geq 3$, Eq. (6-11) reduces to $\beta_r \approx \beta \approx m_H$. Consequently, Eq. (6-10) is modified to

$$(k_{ob})^{-1} = (k_b)^{-1} + (k_r D_{eff})^{-1/2} \tag{6-17}$$

k_{ob} is independent of k_e under these circumstances.

The gas flow velocity through the emulsion phase is close to the minimum fluidization velocity U_{mf}. When the particles are spherical and have diameters of several tens of microns, this flow condition gives a quite small particle Péclet number, $d_p U_{mf}/D_G$. For example, the Péclet number is estimated as 0.1–0.01 when 122-μm-diam. cracking catalyst is fluidized by gas, with $U_{mf} = 0.73$ cm/sec and $D_G = 0.09$ cm²/sec; and it is estimated as 0.001–0.01 for 58-μm-diam. particles, with $U_{mf} = 0.16$ cm/sec. The mechanism of mass transfer between fluid and particles in packed beds is controlled by molecular diffusion under such low Péclet numbers, and the particle Sherwood number $k_f d_p/D_G$ is well over 10 (M24). Consequently with intraparticle diffusion shown to be negligible (M21), instantaneous equilibrium is established to be a good approximation [see Eq. (6-24)].

2. Comparison with Other Theories

Davidson and Harrison (D3) have assumed that both convective and diffusional flows contribute to gas exchange between bubble and emulsion phases. Partridge and Rowe (P2) have developed a boundary-layer equation for the transfer from cloud to emulsion phase. Kunii and Levenspiel (K22), in their model for estimating the gas-exchange coefficient, assume two transfer steps—between bubble void and cloud-particle overlap region, and between cloud-particle overlap region and emulsion phase. The mass-transfer coefficient for the former step is assumed of the form given Davidson and Harrison, and for the latter step of the form of the Higbie penetration model. In terms of mass-transfer coefficients, the equation of Kunii and Levenspiel is:

$$(k_{ob})^{-1} = (k_{bc})^{-1} + (k_{ce})^{-1} \tag{6-18}$$

where

$$k_{bc} = \tfrac{3}{4} U_{mf} + 0.975 D_G^{1/2}(g/d_b)^{1/4} \tag{6-19}$$

$$k_{ce} = (2/\pi^{1/2})(\epsilon_{mf}^2 D_G \bar{u}_b/d_b)^{1/2} \tag{6-20}$$

where $\bar{u}_b$ is given by Eq. (5-8), with $\beta = 1$ and $\bar{u}_s = \bar{u}_{s0}$.

Chiba and Kobayashi (C3) have presented the following equation, based on the Murray stream function (M47):

$$k_{ob} = \frac{2}{\pi^{1/2}} \left(\frac{\epsilon_{mf}^2 D_G \bar{u}_s}{d_b} \right)^{1/2} \left(1 - \frac{1}{\alpha} \right)^{1/2} \tag{6-21}$$

They further introduce the equilibrium influence of physical adsorption on to the particles. In terms of the present notation and with replacement of ϵ_{mf} by the nearly equal term ϵ_{fe}, their relation is written:

$$(k_{ob})_{m_s \neq 0}/(k_{ob})_{m_s=0} = \eta(m/\epsilon_{fe})^{1/2} \qquad (6\text{-}22)$$

where

$$\eta = [1 + \tfrac{2}{3}[1 - (\epsilon_{fe}/m)](\alpha - 1)^{-1}]^{1/2} \qquad (6\text{-}22\text{a})$$

The relation between proposed equations (6-10) and (6-18) merits discussion. When $\beta_r = 1$ (no reaction) and $m = \epsilon_{mf}$ (no partition to particles), k_{ob} as calculated by Eq. (6-10) or (6-16) is smaller than that by Eq. (6-18) for fine particles, since the former takes $\bar{u}_s$ for bubble velocity and the latter uses $\bar{u}_b$, which is larger than $\bar{u}_s$. Equations (6-20) and (6-13) are generally different because m and ϵ_{mf} are not necessarily equal. The bubble-void resistance to mass transfer has been assumed negligible in Eq. (6-21). This equation is rendered applicable to the case of arbitrary m by utilizing Eq. (6-22). Equation (6-13) can be rewritten in a form similar to Eq. (6-22), with the observations that $m = \epsilon_{fe} + m_s \epsilon_{se}$, and $\epsilon_{fe} \approx \epsilon_{mf}$:

$$(k_e)_{m_s \neq 0}/(k_e)_{m_s=0} = (m/\epsilon_{fe})^{1/2} \qquad (6\text{-}23)$$

The correction factor η to Eq. (6-22) is nearly equal to unity for $\alpha > 10$, so that Eq. (6-22) is almost equivalent to Eq. (6-23) for smaller particles. Equation (6-21) is independent of m_H or rate of chemical reaction.

Drinkenburg and Rietema (D17) have presented a numerical computation of k_{ob} based on the stream functions given by Davidson and Harrison (D3) and by Murray (M47). The bubble-void resistance to mass transfer has been neglected. Enhancement of gas transfer rate by diffusion with simultaneous chemical reaction (Fig. 5 of D17) is reasonably well expressed by Eq. (6-11), the enhancement being expressed as the Hatta number. Enhancement by physical adsorption (Fig. 2 of D17) is also approximated by Eqs. (6-22) or (6-23) for smaller particles.

When the reaction data by Lewis *et al.* (L12) are analyzed by utilizing Eq. (6-10), it turns out that β_r amounts to 1.45 (free-flow bubbles) for the highest reaction rate constant. Hence it seems reasonable to take into consideration the influence of β_r on k_{ob}. Also, it has been shown experimentally that k_b is not necessarily negligible for fluid beds (Y4).

If the adsorption equilibrium is not attained instantaneously, a different analysis is needed. Toei *et al.* (T18) studied the mechanism of heat and mass transfer between bubbles and emulsion phase under such circumstances. The dependence of diffusion rate on bulk flow across the bubble interface also becomes important when coarse particles are fluidized (H16). For two-dimensional bubbles Chavarie and Grace (C7a) compared various interphase mass-transfer models.

3. Applicability of the Theories

Many experimental tests have been made on the theories previously given, although most workers have fluidized particles coarser than the typical cracking catalyst under relatively low gas velocities. When the gas component is not adsorbed, the observed mass-transfer coefficients [see Kunii and Levenspiel (K22, K24), Chiba and Kobayashi (C3), and Drinkenburg and Rietema (D18)] agree with those obtained by the theories. Miyauchi (M26, M30) has shown that the influence of chemical reaction on k_{ob} is expressed adequately by Eq. (6-16). Under relatively high aeration rates Morooka *et al.* (M42) measured the bubble diameter (cf. Section II,B) and calculated k_{ob} and $k_{ob} a_b / \bar{\epsilon}_b \sqrt{D_G}$ from Eq. (6-16). As shown in Fig. 56, the prediction is in good agreement with the experimental results, when helium is used as the tracer gas. However, the prediction for $m_s > 1$ is much smaller than the observed data, though the adsorption equilibrium is taken into account in Eq. (6-16). Nguyen and Potter (N2) mention the same effect, but details are not reported.

The reason that $k_b a_b$ is higher than calculated from Eq. (6-12) may be explained qualitatively by three effects: (1) splitting, coalescence, and rupture of bubbles (T18, T20); (2) direct contact of gas and particles in the transition zone from dense phase to dilute phase (F18); (3) the influence of the particle capacitance effect (M21, M22) as a result of a small steady interchange of particles between the bubble void and the emulsion. An example of this is the case where particles are raining through the bubble (D18, R8, W1) and (4) asphericity of the bubbles (D18). If the particle capacitance effect (discussed in the next section) is responsible for high experimental values for $k_b a_b$, such values should not be applied to the usual catalytic reactions, where m is on the order of unity and particle capacitance has little effect on $k_b a_b$. For design purposes it is normally better to use experimental mass-transfer coefficients obtained by a properly sized fluid bed for the reaction system of interest.

D. Particle Capacitance Effect

As is well known, the temperature in a fluidized catalyst bed is nearly uniform even when considerable heat is generated in the bed (G8, L11). This comes from the bed's large content of solid particles, which disperse the heat axially as a result of particle heat capacity. Similarly, the gas concentration in the bed becomes uniform if the partition of transferring gas component is favorable to the particle phase. The circulating particles, holding a large amount of transferring component, locally exchange the component with the surrounding gas phase; the larger the partition ratio is of the component, the higher the rate of concentration equalization in the gas phase of the bed. Enhancement of heat and mass transport in the bed as a result of particle movement is named the particle capacitance effect; a general formulation has been presented by Miyauchi (M21, M22).

When each catalyst particle is in adsorption equilibrium locally with the gas phase, the axial-dispersion term of the emulsion phase gas in the equations of continuity is $(E_{ef} + m_s E_{es})\, \partial^2 c_{fe}/\partial z^2$, where E_{ef} and E_{es} are the dispersion coefficients in the z direction for gas phase and particles, respectively. The emulsion of fine particles entrains interstitial gas without appreciable relative motion, giving $E_{ef} = \epsilon_{fe} E_{zs}$ and $E_{es} = \epsilon_{se} E_{zs}$, where E_{zs} is the dispersion coefficient of the emulsion in the z direction. As a consequence, the axial-dispersion term is expressed by $(\epsilon_{fe} + m_s \epsilon_{se}) E_{zs}\, \partial^2 c_{fe}/\partial z^2$ or $m E_{zs}\, \partial^2 c_{fe}/\partial z$. In other words, as a result of the capacitance effect, the axial dispersion coefficient of the adsorptive gas is now $m E_{zs}$, which is much larger than E_{ef} when $m \gg 1$.

Essentially the same enhancement takes place for the transport term by convection. That term is ultimately given by $-m u_e\, \partial c_{fe}/\partial z$, where u_e is the mean velocity of the emulsion in the z direction.

When the particles move relative to the gas phase, as when they rain or shower through a gas bubble, they can transport the adsorptive gas rapidly to or from the emulsion phase. The same is true for heat transport (see Section VI,C,3).

The time constant τ_p for a particle of radius r_p placed in the emulsion phase approaching the adsorption equilibrium with the surrounding gas has been estimated (M21) to be (in cm · sec):

$$\tau_p \approx r_p^2 m / 20 D_e \qquad (6\text{-}24)$$

where D_e is the effective diffusivity of adsorptive gas in the particle. For $r_p = 30\,\mu\text{m}$, when $m \approx 10$, τ_p is on the order of 10^{-3}–10^{-4} sec, so that local partition equilibrium is well established in the usual fluidized catalyst beds. This has been shown experimentally by measuring the mean residence times of adsorptive tracer gases in a fluidized cracking catalyst bed (M39). In the work of Nguyen and Potter (N1) with coarser particles, adsorption equilibrium was not attained. Gilliland and Mason (G9) mention the influence of adsorption on the bed transient response early in the 1950s.

Few studies have been performed on the physical adsorption of reactant gases at higher temperatures. Eberly (E1, E2), Eberly and Kimberlin (E3), and Tam and Miyauchi (T4) measured the adsorption equilibria of hydrocarbons on catalysts. The last-named authors have obtained high-temperature equilibria of benzene and nine normal paraffins (C_6H_{14} to $C_{14}H_{30}$) for a silica–alumina cracking catalyst at 150–450°C. Their results suggest that a fair amount of hydrocarbon is physically adsorbed on the cracking catalyst in high-temperature operations.

The particle capacitance effect in bubble mass transfer was shown earlier (see Section VI,C) for the streaming emulsion outside a bubble, so that m is included in the emulsion-side film coefficient k_e. Chiba and Kobayashi (C3) and Drinkenburg and Rietema (D17) introduce the same effect. Yokota et al. (Y13), in studies of mass transfer from submerged surfaces in a gas fludized bed, find that the rate is strongly enhanced by adsorption of the transferring gas component on fluidized particles. The role of the particle capacitance effect in heat transfer has been discussed by Mickley and Fairbanks (M14), Mickley et al. (M16), Drinkenburg (D16), Yoshida et al. (Y18), and Kunii and Levenspiel (K24).

E. Axial and Radial Mixing of Heat and Mass

1. *Overall Mixing of Nonadsorptive Gas*

The extent of longitudinal mixing of nonadsorptive gas has been measured by Gilliland and Mason (G8, G9), Mason (in R4), Stemerding (in R4), May (M12), Overcashier et al. (O11), De Maria and Longfield (D10), Miyauchi et al. (M33), Muchi et al. (M49), and Ogasawara et al. (O12). All these experiments were carried out under relatively high gas velocities of practical interest. The mixing process in the beds is approximated by the one-dimensional diffusion model:

$$\partial c/\partial \theta = E_z \, \partial^2 c/\partial z^2 - U_G \, \partial c/\partial z \qquad (6\text{-}25)$$

where c is the mean concentration of gas, E_z is the superficial longitudinal dispersion coefficient of gas, and U_G is the superficial gas velocity.

Experimental values for E_z available so far for fluidized catalyst beds are summarized in Fig. 60, where the full curve has been calculated by combining Eqs. (6-31) and (6-34) (see Section VI,E,3). In obtaining E_z experimentally there have been two kinds of methods, one of which utilizes a step response and the other a steady backmixing of a nonadsorptive tracer gas. In Fig. 60, the data obtained by the step-response method (solid keys) are seen generally to agree with the calculated curve, where the response curves by Gilliland and Mason (G9) and that by May (M12) are treated by the step-response theory of Yagi and Miyauchi (Y1) to obtain E_z/U_G. The same quantity obtained by the steady back-mixing method (open keys) is seen to scatter considerably, giving lower E_z/U_G for smaller D_T and higher values for larger D_T when compared with the calculated curve. Gilliland and Mason (G9) have indicated the inadequacy of the steady back-mixing method for measuring axial-mixing of gas, although the method suggests information as to flow properties of the beds.

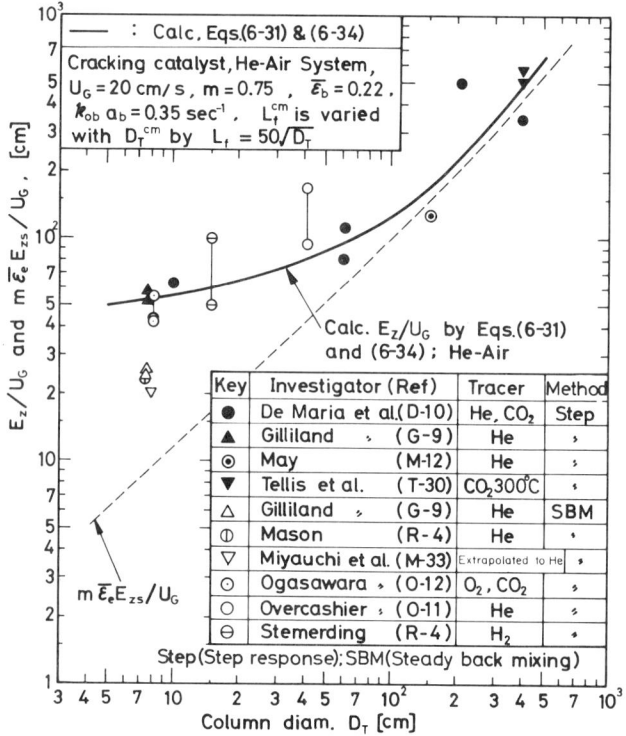

FIG. 60. Longitudinal dispersion of gas in fluid beds (after Miyauchi). The full curve is calculated from Eqs. (6-31) and (6-34) for He–air system.

Other works on overall axial and radial mixing of nonadsorptive gases in fluidized beds have been reviewed by Kunii and Levenspiel (K24) and Potter (in D5).

2. *Mixing of Solid Particles*

Axial mixing in fluidized catalyst beds has been explained in Section IV, where bed particles of good fluidity are shown to give a fair amount of dispersion as a result of the Taylor dispersion mechanism. Porous particles with diameter larger than about 150 μm or heavy particles such as glass beds show considerably different flow properties from the fluidized beds of good fluidity. Morooka *et al.* (M40) find that particle-caused dispersion approaches total dispersion, as size and density decrease and as the size distribution broadens. They explain the deviations of data from Shrikhande (S12) and de Groot (D7) from their correlation, which applies mainly to broad size distributions of porous particles smaller than 100 μm in mean diameter and to equal-sized glass beds smaller than about 20–30 μm.

De Vries et al. (D8) have studied the effect of fines content on the axial dispersion coefficient of the particles. As shown in Fig. 61, good fluidization is obtained when at least 10% of fines (-44 μm) are present. Axial dispersion of particles has also been studied by other workers (L8, M5, P5, S6, T3, W14). Lateral dispersion of particles is covered by studies performed under relatively limited conditions (B16, G1, H15, L11, M37).

3. *Axial Dispersion of Fluid with Particle Capacitance Effect*

The overall dispersion coefficient of gas has been observed to increase when a more adsorptive gas is utilized as a tracer for transient response under constant aeration (Stemerding in R4; M33, M42). Figure 62 shows such examples taken for a fluidized cracking catalyst bed (M33), where the partition ratios of adsorptive gas m (equal to $\epsilon_{fe} + m_s\epsilon_{se}$) are measured as 0.6, 4.5 and 10 for hydrogen, carbon dioxide, and Freon 12, respectively. Enhancement of gas dispersion by adsorption has also been studied by Yoshida and Kunii (Y15), Yoshida et al. (Y17), Yates and Constants (Y9), Zalewski and Hanesian (Z3), and Nguyen and Potter (N2).

Overall longitudinal dispersion of an adsorptive gas is obtained by solving the interaction between several rate processes and the particle capacitance effect. Morooka et al. (M39) have analyzed the behavior of the fluidized catalyst bed with recirculation flow as given in Fig. 2. The simplifying assumptions for modeling the bed behavior are essentially the same as in Section VII,B,2, where the influence of directly contacting

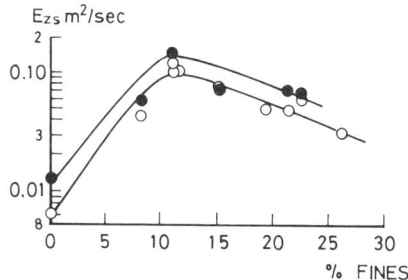

FIG. 61. Longitudinal dispersion coefficient of solid particles as a function of size fraction <44 μm. $D_T = 30$ cm/sec, $L_f = 2$–2.5 m, and superficial gas velocities are: (○) 10 cm/sec, (●) 20 cm/sec (D8).

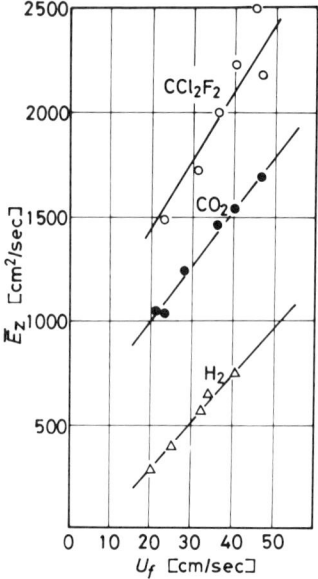

FIG. 62. Influence of particle capacitance effect on overall longitudinal dispersion coefficient of gas (M33).

particles with bed gas has been neglected, since the influence is usually small for transient response of tracer gas at a moderate partition ratio.

The simplifications taken here are, first, that the catalyst particles are usually several tens of μm in mean diameter, so that the interstitial gas velocity through the emulsion phase is about u_{mf} (equal to U_{mf}/ϵ_{mf}) and is neglected in comparison with U_G and emulsion-phase recirculation velocity. Also, the particles and the gas in the emulsion are in local adsorption equilibrium the partition ratio m (equal to $\epsilon_{fe} + m_s\epsilon_{se}$) applying for unit volume of the emulsion phase. Second, the central upflow and the peripheral downflow recirculate in the bed as ideal plug flows at approximately the same linear velocities. Furthermore, the bubble phase ascends only through the central upflow emulsion.

With these simplifying assumptions, the equations of continuity are given for transient response of a tracer gas as follows (M38):

$$\bar{\epsilon}_b \, \partial c_b/\partial \theta = -U_G \, \partial c_b/\partial z - k_{ob}a_b(c_b - c_{feu}) \quad (6\text{-}26\text{a})$$

for the bubble phase;

$$m\bar{\epsilon}_{eu} \, \partial c_{feu}/\partial \theta = -mU_e \, \partial c_{feu}/\partial z + k_{ob}a_b(c_b - c_{feu})$$
$$- mk_{ex}a_{ex}(c_{feu} - c_{fed}) \quad (6\text{-}26\text{b})$$

for the centrally ascending emulsion phase;

$$m\bar{\epsilon}_{\text{ed}}\, \partial c_{\text{fed}}/\partial\theta = mU_e\, \partial c_{\text{fed}}/\partial z + mk_{\text{ex}}a_{\text{ex}}(c_{\text{feu}} - c_{\text{fed}}) \quad (6\text{-}26\text{c})$$

for the peripherally descending emulsion phase, where $k_{\text{ob}}a_{\text{b}}$ is the overall coefficient of mass transfer between bubbles and emulsion, and $k_{\text{ex}}a_{\text{ex}}$ is the exchange rate coefficient between upflow and downflow portions of emulsion phase as a result of radial mixing (Fig. 2). The $\bar{\epsilon}_{\text{eu}}$ and $\bar{\epsilon}_{\text{ed}}$ are the volume fractions of ascending and descending emulsion phases, related by the equation $\bar{\epsilon}_{\text{eu}} + \bar{\epsilon}_{\text{ed}} = \bar{\epsilon}_e$. $U_e = \bar{\epsilon}_{\text{eu}}u_{\text{eu}}$ is the superficial circulation velocity of emulsion phase, where u_{eu} is the interstitial velocity of the ascending emulsion. Also $U_e = \bar{\epsilon}_{\text{ed}}u_{\text{ed}}$, as a result of continuity of the recirculating emulsion [see Eq. (3-26)].

The initial and boundary conditions to solve Eqs. (6-26) for an impulse response of the adsorptive gas are:

$$\begin{aligned} \theta < 0: &\quad c_b = c_{\text{feu}} = c_{\text{fed}} = 0 \\ \theta \geq 0: &\quad x = 0; \quad c_b = \delta(\theta), \quad c_{\text{feu}} = c_{\text{fed}} \\ &\quad x = L_f; \quad c_{\text{feu}} = c_{\text{fed}} \end{aligned} \quad (6\text{-}27)$$

The variance of the residence times of the tracer gas with respect to mean residence time is obtained from Eqs. (6-26) and (6-27), according to van der Laan (V3), as follows:

$$\begin{aligned}\frac{\sigma^2}{2} = &\frac{1}{N_{\text{ob}}}\left(\frac{m\bar{\epsilon}_e}{\bar{\epsilon}_b + m\bar{\epsilon}_e}\right)^2 + \frac{1}{N_{\text{ex}}}\left(N_v + \frac{m\bar{\epsilon}_e}{2(\bar{\epsilon}_b + m\bar{\epsilon}_e)}\right)^2 \\ &\times \left[1 - \frac{N_v^2}{N_{\text{ob}}N_{\text{ex}}}\frac{(\lambda_1 - \lambda_2)(1 - e^{-\lambda_1})(1 - e^{-\lambda_2})}{(e^{-\lambda_1} - e^{-\lambda_2})}\right]\end{aligned} \quad (6\text{-}28)$$

where $N_{\text{ex}} = mk_{\text{ex}}a_{\text{ex}}L_f/U_G$, $N_{\text{ob}} = k_{\text{ob}}a_bL_f/U_G$, $N_v = mU_e/U_G$, and

$$\lambda_1 = N_{\text{ob}}(1 + 1/N_v)(1 + \beta)/2, \quad \lambda_2 = N_{\text{ob}}(1 + 1/N_v)(1 - \beta)/2$$

$$\beta = [1 + 4N_{\text{ex}}/(1 + N_v^2)N_{\text{ob}}]^{1/2}$$

Here, we note that the influence of particle motion on an impulse response has been treated by van Deemter (V2), Yoshida and Kunii (Y15), Fryer and Potter (F10, F11), and Nguyen and Fryer (N2), but their models are all different from the one treated above.

Morooka *et al.* (M42) further solved the impulse response of a tracer gas for a fluidized catalyst bed according to the one-dimensional two-phase diffusion model (V1), where the influence of the particle capacitance effect was considered under the assumption of local adsorption equilibrium. The equations of continuity for the tracer gas are:

$$\bar{\epsilon}_b\, \partial c_b/\partial\theta = -U_G\, \partial c_b/\partial z - k_{\text{ob}}a_b(c_b - c_e) \quad (6\text{-}29\text{a})$$

for the bubble phase;

$$m\bar{\epsilon}_e \, \partial c_e/\partial \theta = E_e \, \partial^2 c_e/\partial z^2 + k_{ob}a_b(c_b - c_e) \tag{6-29b}$$

for the emulsion phase, where E_e is the apparent longitudinal dispersion coefficient of emulsion phase. Since the slip velocity of gas through the particles is negligible for the emulsion phase of small particles, E_e is connected with E_{zs} [cf. Eqs. (4-16) and (4-17) and Section VI,D] with reasonable accuracy:

$$E_e = m\bar{\epsilon}_e E_{zs} \tag{6-30}$$

Under the same boundary conditions as given by van Deemter (V1) the variance of residence times is given by

$$\frac{\sigma^2}{2} = \frac{1}{N_{ob}} \left(\frac{m\bar{\epsilon}_e}{\bar{\epsilon}_b + m\bar{\epsilon}_e}\right)^2 + \frac{1}{(\text{PeB})_s}\left[1 - \frac{q}{(\text{PeB})_s}\left(\frac{\cosh \lambda - \cosh(N_{ob}/2)}{\sinh \lambda}\right)\right] \tag{6-31}$$

where $N_{ob} = k_{ob}a_b L_f/U_G$, $(\text{PeB})_s = U_G L_f/m\bar{\epsilon}_e E_{zs}$, and

$$q = [1 + 4(\text{PeB})_s/N_{ob}]^{1/2}, \quad \lambda = qN_{ob}/2$$

For the present simplified recirculation model, the axial dispersion coefficient is given by the following approximate equation for arbitrary m:

$$(mU_e)^2/mk_{ex}a_{ex} = E_e \tag{6-32}$$

This relation is derived by considering a steady-state tracer exchange between the ascending and descending emulsions and the bubbles, and comparing the resulting equation with that of the diffusion model. Since $E_e = m\bar{\epsilon}_e E_{zs}$, the above equation is equivalent to the following relation from van Deemter (V2):

$$U_e^2/k_{ex}a_{ex} = \bar{\epsilon}_e E_{zs} \tag{6-33}$$

Once the variance of the residence time distribution has been obtained, the axial dispersion coefficient of the tracer gas E_z is calculated by combining the following equation for the one-dimensional diffusion model (V3) with Eq. (6-28) or Eq. (6-31):

$$\frac{\sigma^2}{2} = \frac{1}{(\text{PeB})_f}\left[1 - \frac{1}{(\text{PeB})_f}\{1 - \exp[-(\text{PeB})_f]\}\right] \tag{6-34}$$

where $(\text{PeB})_f = U_G L_f/E_z$, the column Peclet number for the tracer gas [cf. Eq. (6-25)]. The full curve given in Fig. 60 is calculated by combining Eqs. (6-31) and (6-34) for a helium–air system, where the numerical values necessary to the calculation are given in the figure and the dispersion coefficient of the emulsion E_{zs} has been taken from Fig. 41. Calculation gives good mean values for the data obtained by step response (solid keys), but deviates considerably from those by steady back-mixing (open keys). The quantity E_z/U_G is reported to be little influenced by changing U_G widely (G9, R4).

The apparent axial dispersion coefficient of the emulsion phase, $m\bar{\epsilon}_e E_{zs}/U_G$, is also shown in Fig. 60 as a dotted curve. Axial dispersion of gas is seen to be larger than that of emulsion for smaller D_T, but is not necessarily larger for larger-diameter beds. With the operating variables chosen here, E_z/U_G seems to approach $m\bar{\epsilon}_e E_{zs}/U_G$ for $D_T \gtrsim 200$ cm. De Maria and Longfield (D10) suppose that E_z/U_G can be considerably larger than $m\bar{\epsilon}_e E_{zs}/U_G$ at increasing D_T.

Figure 60 indicates that the overall mass-transfer coefficient $k_{ob}a_b$ is obtained from the use of transient-response methods, and this approach has been utilized extensively. However, the approach seems to be more appropriate to small-diameter beds. In fact, E_z/U_G for large-diameter beds is mostly governed by the term $m\bar{\epsilon}_e E_{zs}/U_G$, but only little influenced by $k_{ob}a_b$.

The influence of longitudinal dispersion on the extent of a first-order catalytic reaction has been studied by Kobayashi and Arai (K14), Furusaki (F13), van Swaay and Zuiderweg (V8), and others. They use the one-dimensional two-phase diffusion model, and show that longitudinal dispersion of the emulsion has little effect when the reaction rate is low. Based on the circulation flow model (Fig. 2) Miyauchi and Morooka (M29) have shown that the mechanism of longitudinal dispersion in a fluidized catalyst bed is a kind of Taylor dispersion (G6, T9). The influence of the emulsion-phase recirculation on the extent of reaction disappears when the term φ defined by Eq. (7-18) (see Section VII) is greater than about 10. For large-diameter beds, where φ does not satisfy this restriction, their general treatment includes the contribution of Taylor dispersion for both the reactant gas and the emulsion (M29).

Mixing and dispersion of gas and particles in the transition zone and in the freeboard region are important to know, in relation to the mechanisms of the reactions taking place (see Sections VII–IX). However, little work has thus far been done in these areas.

F. Review of Wall Heat Transfer in Fluid Beds

Numerous measurements of heat transfer between the fluid bed and immersed surface have been carried out; these have been reviewed by Kunii and Levenspiel (K24), Davidson and Harrison (D5), Botterill (B12), and Zabrodsky et al. (Z1, Z2). Wen and Leva (W3) and Wender and Cooper (W5) correlated heat-transfer data at the walls of fluidized beds. However, the data utilized for their correlation were obtained with a fluidized bed less than 10.2 cm in diameter, and measurements under fluid bed conditions were not included. Heat transfer between fluidized beds and immersed vertical tubes was correlated by Wender and Cooper (W5). This correlation includes data with fine porous catalyst particles by Olin and Dean (O4) and Mickley and Fairbanks (M14):

$$\frac{h_w d_p}{k_g} = 0.01844 C_R (1 - \bar{\epsilon}_f) \left(\frac{c_{pg} \rho_g}{k_g}\right)^{0.43}$$
$$\times \left(\frac{d_p \rho_g U_G}{\mu_g}\right)^{0.23} \left(\frac{c_{ps}}{c_{pg}}\right)^{0.8} \left(\frac{\rho_s}{\rho_g}\right)^{0.66} \quad (6\text{-}35)$$

for $d_p \rho_g U_G / \mu_g = 10^{-2}$–$10^2$, where $c_{pg} \rho_g / k_g$ is in sec/cm². The quantity C_R represents the correlation factor for nonaxial tube location, given in Fig. 63.

The time-averaged heat-transfer coefficient h_w between the vertical wall and fluidized beds, using fine particles, is controlled by the unsteady heat transfer due to packet renewal at the heat-transfer surface (M14, M16, Y18). According to this mechanism, h_w is expressed as:

$$h_w = (k_{\text{eff}} \rho_e c_{pe} / \pi \hat{t})^{1/2} \quad (6\text{-}36)$$

where $\hat{t}$ is the proper characteristic contact time of packets. Therefore, $h_w / \sqrt{k_{\text{eff}} \rho_e c_{pe}}$ is a function of $\hat{t}$ or of U_G, as shown in Fig. 64; data on wall-to-bed heat transfer in bubble columns are given in the figure. In the case of bubble columns, corresponding values for the water–air system are used in calculating $(k_{\text{eff}} \rho_l c_{pl})^{1/2}$. The correlation by Kato (K5) includes data by Fair *et al.* (F1) and Kölbel *et al.* (K18, K19). The rate of the packet

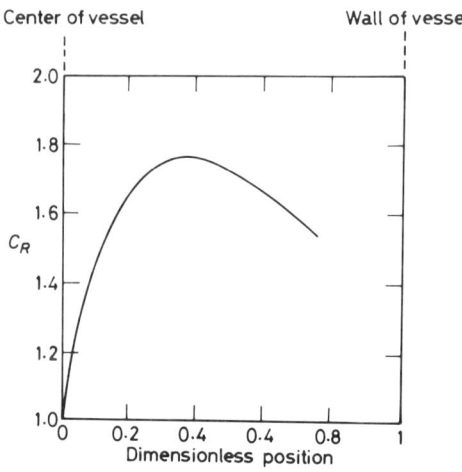

FIG. 63. Correction factor C_R for nonaxial location of immersed tubes (V14).

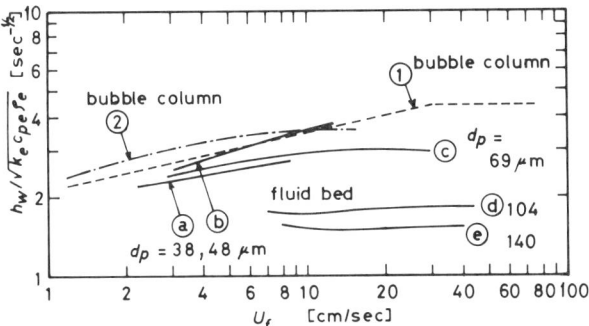

FIG. 64. Comparison between bed-to-surface heat transfer coefficient of fluid bed and bubble column. (a) and (b) Olin and Dean (O4); (c) Mickley and Fairbanks (M14), (d) and (e) Mickley et al. (M16); (1) Kato (K5); (2) Yoshitome et al. (Y23).

renewal on the vertical heat-transfer surface in fluid beds agrees with the data in bubble columns. The larger the particle size that is used in fluidized beds, the smaller the rate of the packet renewal becomes compared with that in bubble columns. Vreedenburg (V13) correlated data on heat transfer between horizontal tubes and the fluidized bed, including data for fluid beds; the data are widely scattered, and further study appears to be needed.

The bed-to-wall heat-transfer coefficient in a fluidized bed at high temperatures is larger than at room temperature (Y19). Questions have been raised about the effect of radiative heat transfer at high temperatures (B12, Y19), and more studies are necessary on this problem.

In the transition zone and the freeboard region, heat transfer between bed and wall is a function of bed density. Shirai et al. (S9, S11) studied heat transfer from a sphere immersed in the fluidized bed and showed the trend of decreasing heat-transfer coefficient with decreasing bed density.

VII. The Successive Contact Mechanism for Catalytic Reaction

A. Experimental Facts and Reactor Models

Early attempts to approximate gas–solid contacting in fluid catalyst beds were based on the assumption either of isothermal plug flow of the fluidizing gas through the bed with the catalyst uniformly distributed or of isothermal complete mixing of the gas within the bed. The simple dispersion model, falling between the above two cases, was also used (G8, R4). Evidence from both large-scale and laboratory observations (G9a, L12), however, indicated conversions substantially below those predicted by the completely mixed case. Due to this, a number of models to describe fluidized reactor behavior based on the two-phase theory of fluidization (G9b, J2, S7) have been proposed over the last 20 years. Most assume with this theory that all gas in excess of that necessary to produce minimum fluidization passes through the bed as the bubble phase. For the fluid catalyst beds, the net flow of gas through the emulsion phase may be negligible because of its relatively small magnitude in comparison with the bubble-phase gas. In what follows this simplification has been made consistently.

1. *Overall Rate Constant of Reaction*

When a fixed bed of catalyst is operated isothermally as an ideal *piston-flow* reactor without radial concentration gradients and without change in the gas density, the differential mass balance for reactant along the bed is given by

$$-U_G \, dc/dz - k_r c = 0 \tag{7-1}$$

where U_G is the superficial gas velocity, c the reactant concentration, z the axial distance from the reactor inlet, and a first-order irreversible reaction with the rate constant k_r is assumed to take place in the bed. This equation is integrated to give the bed outlet concentration c_0 at $z = L_q$, the fixed bed height:

$$(c_0/c^0)_{\text{fixed bed}} = \exp(-k_r L_q/U_G) \tag{7-2}$$

When the bed is not fixed, but is fluidized freely at the same gas velocity as above, with U_G much greater than the minimum fluidization velocity U_{mf}, the bed becomes *aggregative* and deviates from the piston-flow reactor. As a consequence, the apparent overall rate constant of reaction K_{oR} for the fluidized bed is smaller than k_r, and the extent of reaction is expressed by

$$(c_0/c^0)_{\text{fluid bed}} = \exp(-K_{oR} L_q/U_G) \tag{7-3}$$

K_{oR} is the same physical quantity as the specific converting power Q of Lewis *et al.* (L12).

Usually, the fluidized catalyst bed constitutes an aggregative dense phase for z between 0 and L_f, and a dilute phase for $z > L_f$. Hence, another apparent overall rate constant k_{oR} is defined on the basis of L_f. Taking L_f instead of L_q, Eq. (7-3) is rewritten:

$$(c_0/c^0)_{\text{fluid bed}} = \exp(-k_{oR} L_f/U_G) \tag{7-4}$$

where obviously the relation

$$K_{oR}L_q = k_{oR}L_f \tag{7-5}$$

holds.

The object of the following treatment is to establish a physically sound reactor model to obtain k_{oR}, based on the flow and transport properties of fluidized catalyst beds. Bed performance for chemical kinetics other than the first-order reaction may be computed after a sound bed performance has been established.

2. *General Properties of Reactor Models*

Many investigations have been performed to make clear the nature of gas flow through fluidized particles by studying gas mixing, residence time distribution, and gas bypassing. Various kinetic reactor models have been proposed to describe gas–solid contacting in fluidized reactors (C9, D3, F10, F12, G7–9, G9a, H16, I13, K4, K14, K24, L5, L12, M8, M12, M29, M48, O6, P1, P2, P9, S7, V1, V2). Essentially these models are variations of the two-phase concept (J2, S7) of a gas bubble phase and an emulsion phase, constituting together a dense phase, with gas being interchanged between the bubble and emulsion phases. Some models take into consideration the gas-cloud region associated with each bubble (K4, K12, K24, P2, P9).

The main differences between the models lies in whether or not some fraction of the catalyst is in direct contact with the bubble gas, and in the extent of axial mixing in each phase. Properties of various models have been discussed extensively by Gilliland and Knudsen (G7) in relation to the extent of reaction in experimental fluidized bed reactors, considering that allowance for direct contact between bubble gas and a certain amount of catalyst in it is the sole way to account for the contact efficiency. Unless a fraction of the catalyst particles is assumed to be entrained in the bubble gas, the bubble size calculated to fit the reaction data is found to decrease with increasing catalyst activity at otherwise identical fluidization conditions, in which the bubble size should remain constant. Essentially the same decrease in bubble size was observed by Miyauchi and Morooka (M29) in their analysis of the data by Lewis *et al.* (L12), and by Furusaki (F14) in his fluidized bed data for the Deacon reaction.

The general two-phase model of fluidization contains such parameters (see Fig. 65) as the volume fraction of bubbles $\bar{\epsilon}_b$, the volume fraction of particulate phase $\bar{\epsilon}_e$ (equal to $1 - \bar{\epsilon}_b$), the overall mass transfer coefficient $k_{ob}a_b$ (sometimes also the bubble-side and emulsion-side film coefficients $k_b a_b$ and $k_e a_b$, respectively), the fraction of catalyst directly contacting

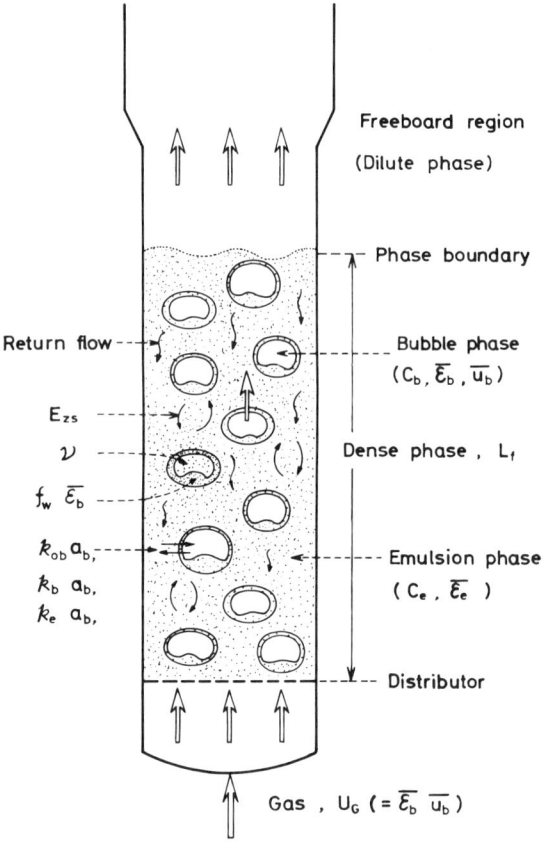

FIG. 65. Schematic expression of general two-phase dense bed model and the parameters included (homogeneous return flow is by F10–12, K24, L5).

with bubble gas ν, the cloud-wake fraction $f_w \bar{\epsilon}_b$, and the axial dispersion coefficient of the emulsion E_{zs}.

In the case of fluid catalyst beds, the mean bubble size stays approximately constant axially, so that the above parameters are treated as remaining unchanged axially. Also, the interstitial gas flow through the emulsion is neglected in comparison with the feed gas flow U_G, so as to considerably simplify the treatment. Usually the bubble phase is assumed to be unmixed axially.

Under the above simplifications, the two-phase consecutive model by Shen and Johnston (S7) with vertically unmixed emulsion is written for a differential bed height:

$$-U_G \, dc_b/dz - k_{ob}a_b(c_b - c_e) = 0 \qquad (7\text{-}6a)$$

with

$$k_{ob}a_b(c_b - c_e) - \bar{\epsilon}_e k_r c_e = 0 \tag{7-6b}$$

where the original capacity coefficient is expressed as the product of the overall coefficient k_{ob} and the specific surface area of the bubbles a_b included in the bubble–emulsion mixed phases. A first-order irreversible reaction takes place in the emulsion. Solving the equations, the overall rate constant as defined by Eq. (7-4) is given by

$$k_{oR} = [(k_{ob}a_b)^{-1} + (\bar{\epsilon}_e k_r)^{-1}]^{-1} \tag{7-7}$$

When a part of the emulsion volume ν is included in bubbles and contacts freely with the bubble gas, the effective fraction of the emulsion phase is $\bar{\epsilon}_e - \nu$, and Eq. (7-6) is rewritten:

$$-U_G \, dc_b/dz - k_{ob}a_b(c_b - c_e) - \nu k_r c_b = 0 \tag{7-8a}$$

$$k_{ob}a_b(c_b - c_e) - (\bar{\epsilon}_e - \nu)k_r c_e = 0 \tag{7-8b}$$

This case is the direct-contact model corresponding to the vertically unmixed emulsion (VUE) by Lewis *et al.* (L12), a special case of the model by Mathis and Watson (M8). Solving the equations, the overall rate constant is

$$\{(k_{ob}a_b)^{-1} + [(\bar{\epsilon}_e - \nu)k_r]^{-1}\}^{-1} \tag{7-9}$$

The bubble flow model by Orcutt *et al.* (C9, O6) uses the film coefficient $k_b a_b$ between the bubble gas and the emulsion. In their vertically unmixed emulsion k_{oR} is given, with negligible gas flow through the emulsion, by Eq. (7-7) with $k_b a_b$ in place of $k_{ob}a_b$. When the emulsion is mixed perfectly, the extent of reaction is given by

$$c_o/c^o = e^{-N_b} + (1 - e^{-N_b})/[1 + \bar{\epsilon}_e N_r/(1 - e^{-N_b})] \tag{7-10}$$

where $N_b = k_b a_b L_f/U_G$ and $N_r = k_r L_f/U_G$. Hence, k_{oR} is obtained by equating Eqs. (7-10) and (7-4).

In the bubbling bed model of Kunii and Levenspiel (K24), there are two transfer steps for the bubble mass transfer, namely, the transfer between bubble void and cloud-particle overlap region $k_b a_b$ and that between the cloud-particle overlap region and the emulsion phase $k_e a_b$. They further assume that the cloud-particle overlap region and the bubble wake are mixed perfectly, and contact freely with the cloud gas. Their basic equations in the present notation are (for their case 2):

$$-U_G \, dc_b/dz - k_b a_b(c_b - c_c) = 0 \tag{7-11a}$$

$$k_b a_b(c_b - c_c) - f_w \bar{\epsilon}_b k_r c_c - k_e a_b(c_c - c_e) = 0 \tag{7-11b}$$

$$k_e a_b(c_c - c_e) - (\bar{\epsilon}_e - f_w \bar{\epsilon}_b) k_r c_e = 0 \tag{7-11c}$$

where $f_w \bar{\epsilon}_b$ is the fraction of catalyst located in the cloud-wake region, and the catalyst directly contacts gas of concentration c_c. Equation (7-11b) applies for the cloud-wake region, and Eq. (7-11c) for the emulsion not including $f_w \bar{\epsilon}_b$. These equations lead to

$$1/k_{oR} = 1/k_b a_b + 1/k_{or} \qquad (7-12)$$

with

$$k_{or} = \{(k_e a_b)^{-1} + [(\bar{\epsilon}_e - f_w \bar{\epsilon}_b) k_r]^{-1}\}^{-1} + f_w \bar{\epsilon}_b k_r \qquad (7-13)$$

Here k_b is given by Davidson and Harrison (D3), and k_e by the Higbie penetration mechanism (see Section VI where k_{bc} and k_{ce} are utilized in place of k_b and k_e, respectively).

It is interesting to observe the limiting cases where one particular mechanism dominates the whole reactor performance. One extreme is the case where $k_r \ll k_{ob} a_b$, leading to (for all models):

$$k_{oR} = \bar{\epsilon}_e k_r \quad \text{for} \quad k_r \ll k_{ob} a_b \qquad (7-14)$$

Obviously, model identification is not possible under this condition. Another extreme is k_r much larger than the bubble mass-transfer coefficient. The extremes differ from one another, as follows:

$$k_{oR} = k_{ob} a_b \qquad (7-15)$$

for the two-phase consecutive model, Eq. (7-7);

$$k_{oR} = k_b a_b \qquad (7-16)$$

for the bubble flow model, Eq. (7-10), and bubbling bed model, Eq. (7-13);

$$k_{oR} = \nu k_r \qquad (7-17)$$

for the direct contact model (VUE), Eq. (7-9). Here we note that k_{oR} for the models not including directly contacting catalyst are upper-end limited by the mass-transfer coefficient $k_{ob} a_b$ or $k_b a_b$. Only the direct-contact model shows k_{oR} increasing linearly with k_r under a given flow condition. These tendencies are more clearly visualized from Fig. 66, where k_{oR} is shown as a function of k_r at constant gas velocity. Numerical values given to various parameters are those for an ethylene hydrogenation run by Lewis et al. (L12).

The models show considerably different k_{oR} when $k_r \geq 1$ sec^{-1}. The bubble flow model (BFM) is for perfectly mixed emulsion, but a k_{oR} only slightly larger is obtained for BFM when the emulsion is vertically unmixed. For the bubbling bed model (BBM) k_{oR} is fairly sensitive to different assumptions for f_w; an f_w of about unity is recommended (F11, K24) to fit the reaction data available. The value $f_w = 0.35$ is taken from Rowe (D5) for 75-μm-diam. spherical particles.

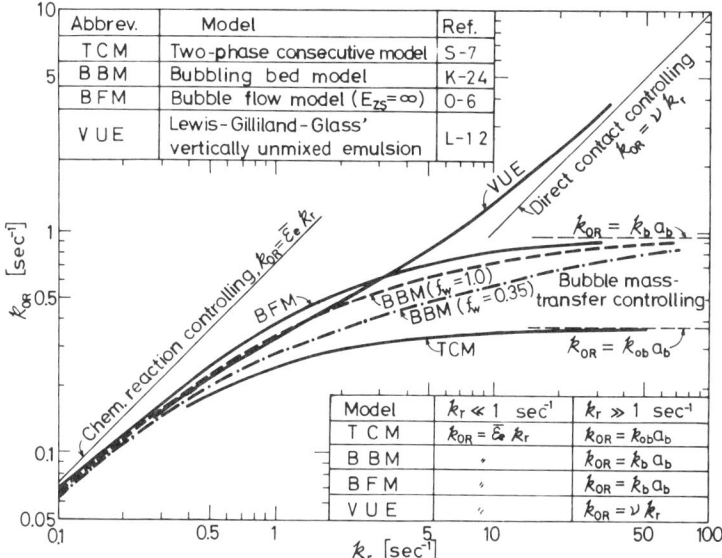

FIG. 66. Overall reaction rate constant k_{oR} as a function of reaction rate constant k_r for various reactor models; $U_G = 25$ cm/sec, $\bar{\epsilon}_b = 0.27$, $k_{ob}a_b = 0.37$ sec^{-1}, $k_b a_b = 0.94$ sec^{-1}, $k_e a_b = 0.61$ sec^{-1}, $\nu = 0.10$, $f_w = 1.0$ and 0.35; data are for a fluidized catalyst bed.

The direct-contact model with vertically unmixed emulsion (VUE) shows considerably different behavior from other models for $k_r \gtrsim 5$ sec^{-1}; k_{oR} increases approximately linearly with k_r, but the rest of the models are upper-end limited by $k_b a_b$ or $k_{ob} a_b$. As a consequence, the models are best identified with one another by comparing their performance with the data obtained for $k_r \gtrsim 5$ sec^{-1}.

3. *Comparison of Models with Experiments*

Catalytic hydrogenation of ethylene by nickel- or copper-impregnated cracking catalyst is taken here for comparison. Figure 67 shows typical experimental k_{oR} values taken under a constant superficial gas velocity U_G by varying k_r. Curve LGG is based on the data by Lewis *et al.* (L12), GK by Gilliland and Knudsen (G7), and FKM those by Furusaki *et al.* (F18). The calculation of k_{oR} will be explained later. The dotted curves are calculated by the two-phase consecutive model (TCM) and by the bubbling bed model (BBM) for $U_G = 20$ and 25 cm/sec, where the mean bubble size is 4.5 cm and the wake fraction $f_w = 1.0$.

Comparing Fig. 67 with Fig. 66 shows that only a direct contact model such as VUE accounts for the experimental data. For other models to

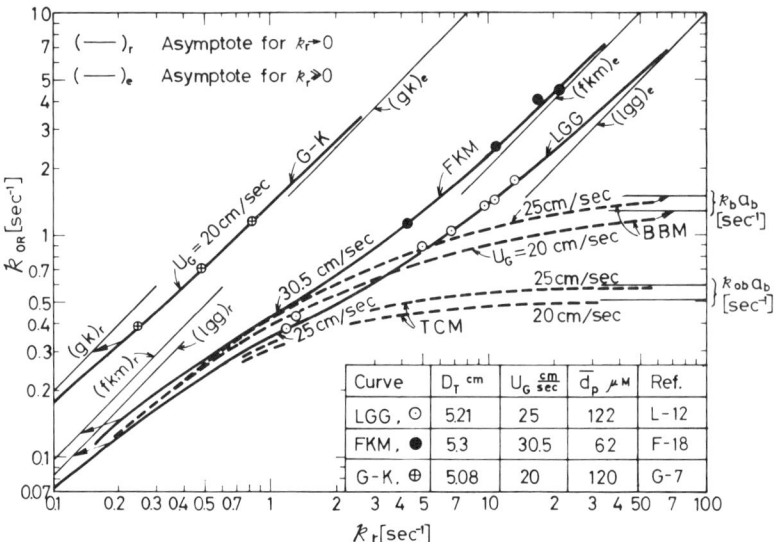

FIG. 67. Overall rate constant k_{oR} for experimental ethylene hydrogenation runs by Ni- or Cu-impregnated cracking catalyst beds. Dotted curves were calculated by the two-phase consecutive model (TCM) and by the bubbling bed model (BBM). Full curves were calculated by the successive contact model.

account for the data, $k_b a_b$ or $k_{ob} a_b$ should be increased with increasing k_r; this would imply that the bubble size decreases with increasing catalyst activity at otherwise identical fluidization conditions, whereas the bubble size should remain constant. A similar tendency is also observed for catalytic ozone decomposition by a fluidized catalyst bed (see Fig. 68), although not so decisively.

The direct contact model has some difficulties, however. In fluidized beds, gas bubbles of very low solid content are usually considered to exist in the dense phase (H14, K13, T19). Also, the cloud layer is negligibly thin, due to small U_{mf} for the usual fluid catalyst beds, according to equations of Davidson and Harrison (D3) and Murray (M47). The streamlines of gas phase through a bubble have been observed to pass through the cloud, but not through the bubble wake (R17). Thus there seems little possibility of believing that the bubble gas is in direct contact with a substantial amount of catalyst in the bubble phase (see also Section VI,A). Furthermore, the direct contact model is applied to the data by Gilliland and Knudsen, and ν in Eq. (7-9) is calculated to fit the data. Calculation (M26) shows that the volume of catalyst, with an apparent density the same as for the emulsion, which contacts the bubble gas freely exceeds the volume of bubble gas itself ($\nu/\bar{\varepsilon}_b$ = 3.3, 2.0, and 1.5, respectively, for U_G = 10, 20, and 30 cm/sec). This seems to be unsound physically.

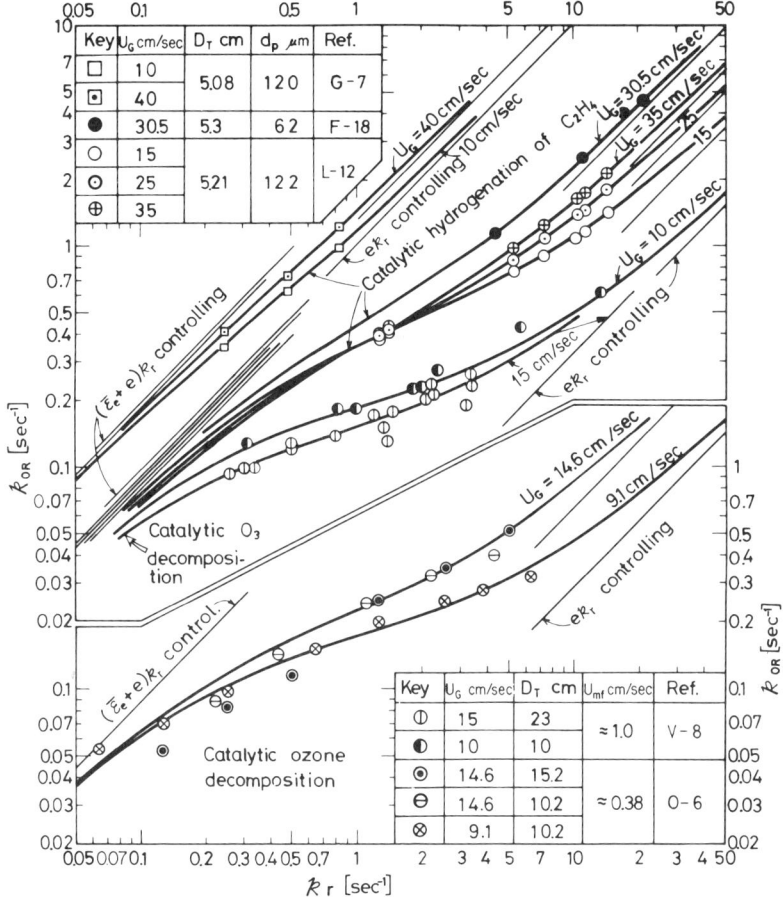

FIG. 68. Summary of catalytic reaction data by fluid beds for ethylene hydrogenation and ozone decomposition; curves were calculated by the successive contact model.

Similarly the bubbling bed model may be visualized as a kind of direct contact model when $k_b a_b$ is much greater than k_{or} in Eq. (7-12), since the equation is now reduced to Eq. (7-9) with an apparent equality of $f_w \bar{\epsilon}_b = \nu$ under this extreme. Actually, however, the transfer of bubble gas to the cloud-wake region is limited by $k_b a_b$, so that the amount of catalyst $f_w \bar{\epsilon}_b$ should be greater than ν to account for the above data; such a large $f_w \bar{\epsilon}_b$ again seems unreasonable.

So far, the models explained here view the fluidized catalyst bed as being composed only of dense phase, so that there is difficulty in locating directly contacting catalyst. This difficulty may be avoided by placing the catalyst somewhere outside the dense phase. Based on the axial distribution of bed density, Miyauchi (M25) located the catalyst in the dilute phase and developed the concept of the successive-contact mechanism. A part of the catalyst may be placed in the jetting zone above the gas distributor, as explained in Section VIII. Catalyst erosion could be excessive with too much jetting, however. Also, such a large amount of directly contacting catalyst as observed for the Gilliland and Knudsen data is difficult to locate solely in the jetting zone.

Perhaps the influence of dilute phase on the progress of reaction must have appeared in the minds of many investigators, but it was never formulated. It was stated (M33) that "Visually the particles in the dilute phase are dispersed rather uniformly, so that under these flow conditions this portion of the particles may take part in increasing the catalytic conversion." Interesting discussions by Riley (R12) are another example.

4. *Other Reactions*

Fluidized catalytic reactions have been industrially operated in the "fluid bed" conditions, but most of the research has been carried out for the "teeter bed." Several studies of fluidized catalytic reaction are listed in Table VI, which are of interest in considering transport phenomena in fluidized catalyst beds.

B. THE SUCCESSIVE CONTACT MECHANISM

A reactor model is developed to include reaction taking place in the dilute phase, and to be reasonably consistent with the known flow properties of fluidized catalyst beds operated under relatively high gas velocity. According to this model, reaction proceeds successively in the dense phase and in the dilute phase.

1. *Flow Properties of Fluidized Catalyst Beds*

As stated previously, fluid beds are operated under high gas velocities with small particles ($U_G/U_{mf} \geq 100$, $U_G/u_t = 1-3$). The apparent density of particles is rather small, i.e., usually less than 1.5 g/cm^3 (and preferably less than 1.0 g/cm^3).

The bed consists of the dense phase and the dilute phase, and the former constitutes the emulsion and bubble phases (Fig. 2). In many respects the flow behavior resembles that in gas bubble columns operated at about equivalent gas velocity (Sections II–V). Intense circulation of the dense phase takes place by the buoyant force induced between the centrally ascending bubble-rich phase and the peripherally descending

TABLE VI. Catalytic Reactions in Fluidized Beds[a]

Author	Reaction	d_p [μm]	D_T [cm]	U_G [cm/sec]	U_{mf} [cm/sec]	Reaction rate constant [sec^{-1}]	References
Pansing	Regeneration of cracking catalyst	44.5–79 (4 types)	25	3–18	—	28.5 Cr at $U_G = 18$[b]	P1
Askins et al.	Regeneration of cracking catalyst	—	1220	47	—	—	A12
Mathis and Watson	Dealkylation of cumene by SiO_2–Al_2O_3 catalyst	74–149	5–10	1.5–2.5	0.6	~1	M8
Lewis et al.	Hydrogenation of ethylene Ni on MS catalyst	80–188 (av. 122)	5	6–45	0.73	1.1–14.3	L12
May	Catalytic cracking	80–150	5–150	3–60	—	—	M12
Gometzplata and Shuster	Catalytic cracking of cumene, SiO_2–Al_2O_3 catalyst	74–149	7.6	1.8–7.3	—	~0.77	G18
Massimilla and Johnstone	Oxidation of ammonia by alumina catalyst	44–149 (av. 105)	11.4	2–16	0.27	0.086	M4
Orcutt et al.	Ozone cracking by Fe_2O_3 on FCC catalyst	20–60	10.2–15.2	3.6–15	0.3	0.05–7	O6
Ishii and Osberg	Isomerization of cyclopropane	74–105	2.3–15	0.9–20	0.3	—	I11
Echigoya et al.	Dealkylation of cumene Hydrogenation of benzene	105–149	5.3	1.2–14.2	0.25	—	E4, E5
Gilliland and Knudsen	Hydrogenation of ethylene Cu on FCC catalyst	44–149 (av. 52, 120)	5.2	3–49	—	0.06–0.95	G7
de Vries et al.	Oxidation of HCl		30–150	~20	—	~0.5	D8
van Swaaij and Zuiderweg	Ozone cracking by Fe_2O_3		5–60	~20	1	<1	V7
Furusaki	HCl oxidation by $CuCl_2$–KCl–$SnCl_2$ on MS SiO_2	Av. 84	5.5	10–60	~0.6	—	F14
Tone et al.	Cracking of methyl cyclohexane by FCC catalyst	88–105	4.5–5.2	1–3	0.25–0.3	—	T21
Furusaki et al.	Hydrogenation of ethylene	62	5.3	10 ~ 30	0.2	4.3 ~ 21	F18

[a] Selected for fluid beds. [b] Cr = gm-coke/gm-catalyst.

bubble-lean phase. Mean bubble size stays relatively unchanged axially; the size seems to be characteristic of flow intensity in the bed. The upflow and downflow emulsions exchange with each other by radial eddy mixing. Thus the general flow features may be visualized as in Fig. 2.

In the dilute phase the catalyst population decreases with increasing axial height from the dense-bed surface (Section II). It has been shown (M26) that the amount of directly contacting catalyst is approximately equal to "a" of Lewis *et al.* (L12), if one takes the amount as the volume of catalyst existing above L_f. The data are compared in Fig. 69. Here, L_f is taken as the height above the distributor where the volume fraction of the emulsion equals that of the bubbles, i.e., $\bar{\epsilon}_e = \bar{\epsilon}_b$; $\bar{\epsilon}_e$ becomes meaningless in the dilute phase or in the transition zone, since $\bar{\epsilon}_b$ becomes dubious. However, $\bar{\epsilon}_{ed}$ (instead of $\bar{\epsilon}_e$) is used as the hypothetical volume fraction if the dispersed particles in the dilute phase are concentrated into the same density as the emulsion. This treatment is convenient in calculating conversion in the dilute phase.

One important point is that the phase boundary which separates the dilute phase from the dense phase is somewhat arbitrary to define and is located with difficulty. Instead of defining the phase boundary, it might be more reasonable to place a transition zone between the above two phases. In what follows, however, the concept of phase boundary has been

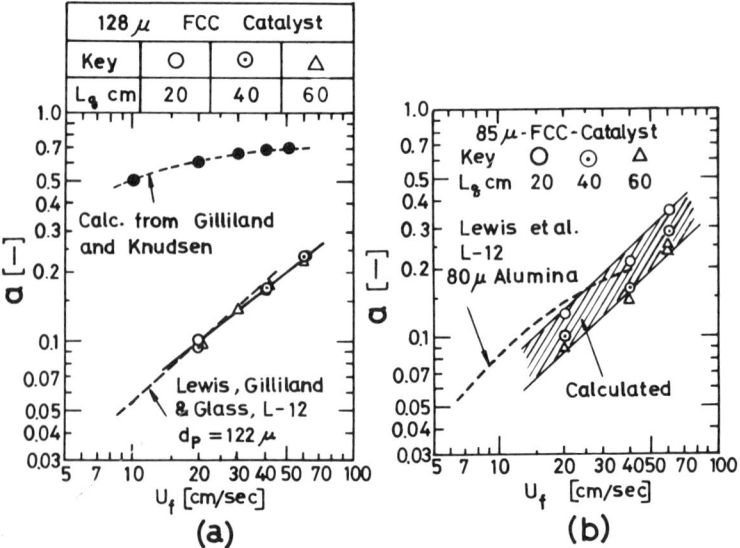

FIG. 69. Comparison of experimental "a" with calculated values based on the successive contact model (M26).

utilized, owing partly to the simplicity of developing a reactor model and partly to the lack of sufficient information to define a transition zone. When the dense phase height L_f is taken at $\bar{\epsilon}_e = 0.50$, the bed density distributions show that L_f remains approximately unchanged for different gas velocities (M25).

2. *Formulation of the Model*

The previously mentioned flow properties of the beds lead to the following formulation for the successive-contact model (M25). Simplifying assumptions additional to those given in Section VII,A,2 are, for the dense phase (M29):

(a) Ascending or descending velocity constant for bubble phase or for circulating emulsion phase, respectively.

(b) Bubbles ascend mostly with the upflowing dense phase, so that they are included only in this phase (see footnote on p. 345).

(c) The upflow and downflow emulsions exchange a part of the emulsion with each other at the exchange rate coefficient $k_{ex}a_{ex}$ (Fig. 2).

(d) A small quantity of freely suspended particles has been observed in the bubble phase (H14, K13, T19). These particles, with the volume fraction v, are allowed to directly contact the bubble gas.

(e) Physical adsorption equilibrium is almost instantaneously attained between particles and fluid in the emulsion phase with the partition ratio $m = \epsilon_{fe} + m_s\epsilon_{se}$, with $m_s = c_{se}/c_{fe}$.

(f) An isothermal and irreversible first-order reaction takes place steadily in the bed.

With these premises, analytical solutions have been presented for the dense phase (M25, M29). The solutions show essential features of Taylor dispersion with chemical reaction, similar to the case treated by Cleland and Wilhelm and others (C8, S2, W15) for cylindrical-pipe flow (see M29). Here, the axial dispersion coefficient of gas affecting the steady reaction diminishes as the rate of chemical reaction increases, compared with the dispersion coefficient affecting axial mixing without accompanying chemical reaction. These solutions are general, but inconvenient for computation. Several simplifications have been discussed; one simplification potentially applicable to the usual fluid-bed operations is to adopt the restriction:

$$\varphi = \frac{U_G}{mU_e}\left(1 + 4m\frac{k_{ex}a_{ex}}{k_{ob}a_b} + \frac{\bar{\epsilon}_e k_r}{k_{ob}a_b}\right) \gtrsim 10 \qquad (7-18)$$

With this behavior of φ the influence of the emulsion-phase circulation apparently disappears (U_e is the superficial circulation velocity of the emulsion phase) and the computation is greatly simplified.

With $\varphi \gtrsim 10$, the following equations of continuity are obtained for the dense phase:

$$-U_G \, dc_b/dz - k_{ob}a_b(c_b - c_{feu}) - \nu k_r c_b = 0 \quad (7\text{-}19)$$

$$k_{ob}a_b(c_b - c_{feu}) - mk_{ex}a_{ex}(c_{feu} - c_{fed}) - (\tfrac{1}{2}\bar{\epsilon}_e - \nu)k_r c_{feu} = 0 \quad (7\text{-}20)$$

$$mk_{ex}a_{ex}(c_{feu} - c_{fed}) - \tfrac{1}{2}\bar{\epsilon}_e k_r c_{fed} = 0 \quad (7\text{-}21)$$

where half of the emulsion phase is in ascending flow and the other half is in descending flow. Equation (7-19) is for the bubble phase; Eqs. (7-20) and (7-21) are for the ascending and descending emulsion phase, respectively. The term $(\tfrac{1}{2}\bar{\epsilon}_e - \nu)$ is the fraction of the ascending emulsion and $\tfrac{1}{2}\bar{\epsilon}_e$ is that of the descending emulsion.

When the exchange rate of emulsion between the upflow and the downflow is high enough, the emulsion gas concentrations c_{eu} and c_{ed} become nearly equal, so that Eqs. (7-20) and (7-21) are combined and reduced to Lewis' VUE model, Eq. (7-8.b).

Equations (7-19)–(7-21) are combined by eliminating c_{eu} and c_{ed}, yielding

$$-U_G \, dc_b/dz - (k_{or} + \nu k_r)c_b = 0 \quad (7\text{-}22)$$

where $1/k_{or} = 1/k_{ob}a_b + 1/\xi\bar{\epsilon}_e k_r$, and ξ is given in Eq. (7-28). For large-diameter beds, φ could be smaller than 10 (see M29).

The bubble gas, with initial concentration c^0, leaves the dense phase at $z = L_f$ with concentration c_{L_f}, and enters the dilute phase. For the dilute phase the diffusion-model approach is more general, since a fair amount of gas mixing is observed in a large-diameter bed (H19). Here, however, the usual piston-flow assumption is made for simplicity, since in small-scale beds the particles in the dilute phase are suspended rather uniformly. Actually the observed flow properties of the phase, including the transition zone, seem to be more complicated.

The equation of continuity with the piston-flow assumption is

$$-U_G \, dc/dz - \bar{\epsilon}_{ed} k_r c = 0 \quad (7\text{-}23)$$

where the physical meaning of $\bar{\epsilon}_{ed}$ has been explained in Section VII,B,1.

The reactant concentration c_0 of the gas leaving the dilute phase at $z = L_t$ is obtained by integrating Eqs. (7-22) and (7-23) as follows:

$$c_0/c^0 = \exp\{-[k_{or} + (\nu + e)k_r]L_f/U_G\} \quad (7\text{-}24)$$

where

$$e = (L_f)^{-1} \int_{L_f}^{L_t} \bar{\epsilon}_{ed} \, dz \quad (7\text{-}25)$$

Accordingly, eL_f is the total fraction of catalyst (with the same density as the emulsion) suspended in the dilute phase. The total fraction of catalyst in the dense phase is $\bar{\epsilon}_e L_f$, so that the following equality exists for the successive contact model:

$$L_q = (e + \bar{\epsilon}_e)L_f \qquad (7\text{-}26)$$

Comparing Eqs. (7-24) and (7-4), k_{oR} can be given by

$$k_{oR} = [(k_{ob}a_b)^{-1} + (\xi\bar{\epsilon}_e k_r)^{-1}]^{-1} + (\nu + e)k_r \qquad (7\text{-}27)$$

where ξ, defined by Eq. (7-28) is the fraction of catalyst in the dense phase effectively utilized for reaction:

$$\xi = \tfrac{1}{2} - (\nu/\bar{\epsilon}_e) + [2 + (\bar{\epsilon}_e k_r/mk_{ex}a_{ex})]^{-1} \qquad (7\text{-}28)$$

Equation (7-27) is reduced to the VUE model by Lewis *et al.*, Eq. (7-9), when $e = 0$ and $\bar{\epsilon}_e k_r/mk_{ex}a_{ex} \ll 2$; ν is usually small in comparison with e (Section VIII). In this case, with $\xi = 1$, Eq. (7-27) is reduced to the simplest expression of the model:

$$k_{oR} = [(k_{ob}a_b)^{-1} + (\bar{\epsilon}_e k_r)^{-1}]^{-1} + ek_r \qquad (7\text{-}29)$$

3. *Experimental Evidence*

Ethylene hydrogenation runs are summarized in Fig. 68, where the data by Lewis *et al.* are taken from their smoothed curves (Figs. 7 and 8 of L12). For these, approximate numerical values for φ and ξ are $\varphi = 25\text{--}38$ and $\xi = 0.95\text{--}0.73$, for $U_G = 10\,\text{cm/sec}$, and $\varphi = 52\text{--}63$ and $\xi = 0.98\text{--}0.83$ for $U_G = 40\,\text{cm/sec}$. Also, the simplest expression [Eq. (7-29)] is utilized to calculate k_{oR}. The Lewis–Gilliland–Glass plot (see Section VII,C,2) has been applied to obtain $k_{ob}a_b$ and e from the experimental data; the other data are similarly processed.

The data by Orcutt *et al.* (O6) for catalytic ozone decomposition have been processed in the same manner. Those by van Swaay and Zuiderweg (V7, V8) for ozone decomposition are treated as follows. Their data are presented in terms of an apparent HTU, $(H_{ob})_{app}$, as defined by

$$(H_{ob})_{app} = U_G/k_{oR} - U_G/\bar{\epsilon}_e k_r \qquad (7\text{-}30)$$

where U_G/k_{oR} equals H_{oR}, the overall height of a reaction unit by Hurt (H18). Because $(H_{ob})_{app}$ equals $U_G/(k_{ob}a_b)_{app}$, they presume Eq. (7-7) holds. Now k_{oR} is calculated using Eq. (7-30) by knowing the experimentally given $(H_{ob})_{app}$; see Fig. 70.

In processing the above data, ν is consistently assumed to be negligible in comparison with e, and Eq. (7-29) is utilized in calculating k_{oR}, including the full curves given in Fig. 68.

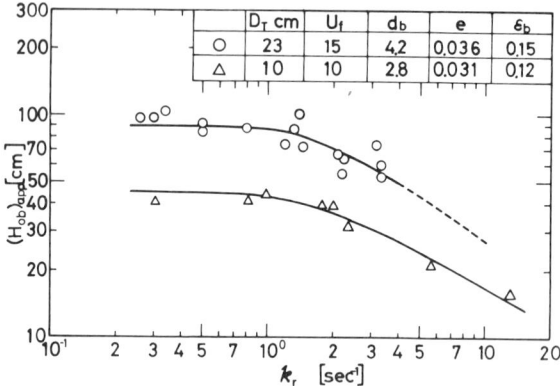

FIG. 70. Comparison of the calculated $(H_{ob})_{app}$ with the data of van Swaay and Zuiderweg (V8).

In the figure the data obtained by Gilliland and Knudsen from the use of copper-impregnated cracking catalyst (quiescent bed density $\rho_{mf} = 0.33$ g/cm^3) show considerably higher k_{oR} than those by Lewis *et al.*, who used nickel-impregnated cracking catalyst ($\rho_{mf} = 0.52$ g/cm^3). Since the experimental runs have been performed under nearly isothermal conditions, the higher k_{oR} by Gilliland seems mostly to come from their using less dense catalyst than Lewis *et al.* used. In other words, at least qualitatively, a fair amount of catalyst is suspended in the dilute phase in conformity with higher e (or a).

Furusaki *et al.* use a standard cracking catalyst of wide size distribution with smaller mean particle size, higher e, and higher U_G than those of Lewis *et al.* (cf. Fig. 68).

The successive contact model seems to be sound in accounting for the fluid-bed performance given in Fig. 68, since the model satisfies the flow and transport properties of the beds as well. The model is also applied successfully to catalytic oxidation of HCl (the Deacon reaction) in a fluid bed (F14).

Recently, Yates and Rowe (Y10) have observed, on the basis of their model for catalyst distribution in the freeboard region, that this region can usually exert a considerable influence on the course of the reaction. Their observation is essentially parallel with the concept of the successive contact mechanism. However, they use the bubbling bed model in calculating the reaction in the dense phase, so that the effect of directly contacting catalyst seems to be corrected two times, first partially in the dense phase and then in the freeboard region (see Section VII,A,3).

4. Relative Contribution to the Overall Extent of Reaction

As explained in Section VI, the bubble mass-transfer coefficient $k_{ob}a_b$ is given by

$$(k_{ob}a_b)^{-1} = (k_b a_b)^{-1} + (\beta_r k_e a_b)^{-1} \quad (7\text{-}31)$$

where β_r is the modified Hatta number. Also, the variables affecting the reaction in a fluid bed have been formulated as Eq. (7-27). It is important to know their relative contributions, i.e., β_r, the amount of catalyst in the dilute phase e, the fraction of catalyst directly contacting with the bubble gas ν, and the jetting zone. Of these, ν is difficult to measure directly by a physical method, but is usually small in comparison with e (Section VIII).

For a typical example we take the data from Lewis *et al.* for the system C_2H_4–H_2, where $U_G = 16$ cm/sec, $d_p = 122$ μm and $D_G = 0.891$ cm²/sec. Then, the mean bubble size is approximately 5.0 cm (Fig. 51), so that $k_{ob}a_b$ is obtained from Eq. (7-31) (Section VI). With these and $\xi \approx 1$, k_{oR} is calculated from Eq. (7-29) by taking k_r and e as variables.

The results (M28) are given in Fig. 71, where $H_{oR} = U_G/k_{oR}$. From this it

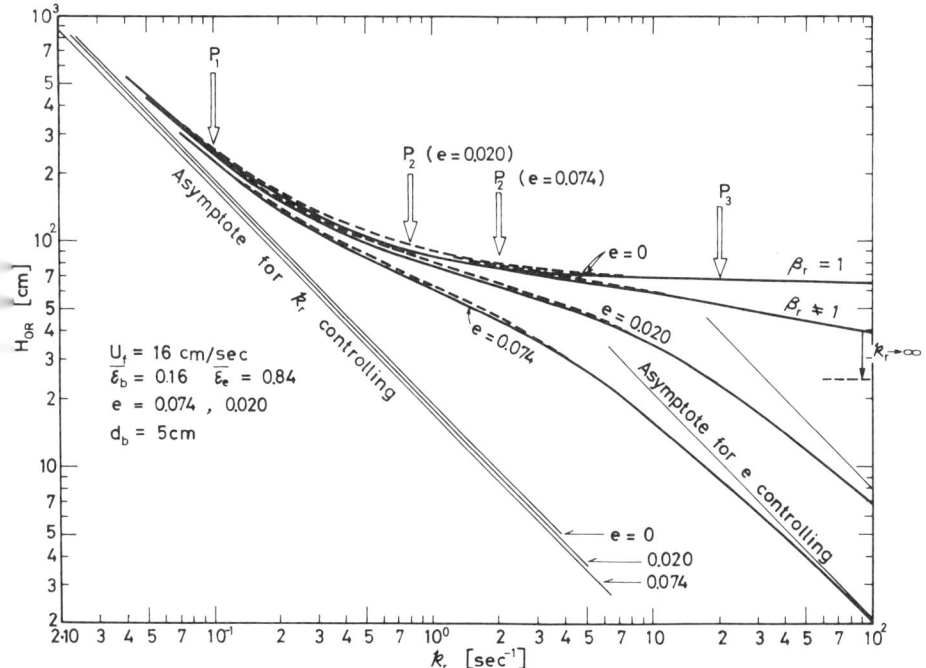

FIG. 71. Relative contribution of parameters to $H_{oR} = U_G/k_{oR}$); P_1, P_2, and P_3 are the respective locations where $k_{ob}a_b$, e, and β_r start to be significant (M28).

is obvious that e is most significant; β_r is important only for very large k_r (≥ 20 sec^{-1}, point P_3). In the region of very small k_r (≤ 0.1 sec^{-1}) the reaction is the rate-determining step. Points where $k_{ob}a_b$ and e begin to affect H_{oR} are shown as points P_1 and P_2, respectively, for an approximate estimate. Point P_2 for an e which is starting to be significant changes considerably as e changes.

The properties of point P_2 are more clearly observed from Fig. 70, for the experimental apparent HTU by van Swaay and Zwiderweg (V7, V8). The HTU decreases as k_r increases beyond about 1.5 sec^{-1}, showing an apparent enhancement of bubble mass transfer due to ek_r (M28). This k_r corresponds to point P_2. The full curves are calculated by Eqs. (7-29) and (7-30) from the use of numerical values given in Fig. 70. The equations give two extremes:

$$(H_{ob})_{app} = U_G/k_{ob}a_b \tag{7-32a}$$

for negligible ek_r; and

$$(H_{ob})_{app} = U_G/ek_r \tag{7-32b}$$

for $k_r \gg k_{ob}a_b$.

Axial distribution of $k_{ob}a_b$ has been shown to have only a minor effect on the performance of fluid catalyst reactors (K14, M28). It has been shown in Section II that (a) bubbles from a single nozzle break up in rising a certain distance to attain a final size; (b) bubbles from a perforated plate associate together when rising; and (c) d_b stays fairly constant axially thereafter.

Basov et al. (B6) measure the longitudinal-bed density distribution for crushed cracking catalyst (average $d_p = 120$–130 μm). They observe that near the distributor (0–20 cm above it), there is a region of varying density with height, then stays unchanged. The density is low near the distributor, since dense-phase circulation is low there and particles may not be easy to supply from the emulsion above. This lower region is called the jetting zone. Here d_b may be smaller or larger than the average d_b of the bed, depending on the type of distributor and on energy input to the jet.

To see the influence of the jetting zone on the extent of reaction, H_{oR} (equal to U_G/k_{oR}) has been calculated under the assumptions that (a) $k_{ob}a_b$ is five times greater in the jetting zone than in the dense phase above, (b) the height of the jetting zone is one-fifth of the bed height L_f, and (c) the other four-fifths has $k_{ob}a_b$ such that the mean $k_{ob}a_b$ averaged over L_f equals that in the example given in Fig. 71. Calculation gives the dotted curves given in Fig. 71. The effect of the jetting zone appears insignificant, as far as $k_{ob}a_b$ is concerned. However, a fraction of directly contacting catalyst seems to be located in the jetting zone, although it is not very significant (see Section VIII).

Behie and Kehoe (B9) state that the effect of the jetting zone is salient, especially for the case of very fast reaction. They use a very high mass-transfer coefficient in their calculation ($k_{oJ} = 57$ cm/sec, $u_J = 30$ m/sec) (B8). On the other hand, $\bar{\epsilon}_b$ could be very small in the jetting zone if the dense phase holds only the jets of high speed and not bubbles, since $\bar{\epsilon}_b = U_G/u_b = U_G/u_j \ll 1$ due to $u_j \gg 0$. Actually, the fraction of bubble phase is high in this zone (B6, E8). In case of fluid beds, the spout above the nozzle of the distributor is quite unsteady and splits into bubbles within a very short distance. As for this, the observation by van Krevelen and Hoftizer (V6) for a gas jet submerged in liquid is quite suggestive for practical values of U_G.

C. Overall Rate Constant K_{oR} Based on L_q

So far, the overall rate constant has been based on the fluidized bed height L_f. This definition of k_{oR} has some advantages, since the parameters which appear in formulating the reactor theory are directly related to flow properties of the bed and are thus easier to understand physically. However, K_{oR}, based on quiescent bed height L_q, is more convenient for design purposes, since L_q is known once the catalyst inventory is known. In contrast, L_f changes with U_G, and may be difficult to define when the bed is aerated with large gas velocity. In such a case L_f may even be zero when the catalyst inventory is small. Also, K_{oR} has the definite advantage that, as stated in what follows, the Lewis–Gilliland–Glass plot is applicable to the successive contact model without any trial calculations.

1. Reactor Models Based on K_{oR}

Equation (7-5) gives the relation that K_{oR} equals $k_{oR}L_f/L_q$. For the successive contact model the ratio L_q/L_f equals $(\bar{\epsilon}_e + e)$ from Eq. (7-26), so that we have:

$$K_{oR} = k_{oR}/(\bar{\epsilon}_e + e) \qquad (7\text{-}33)$$

For the rest of the models there is no dilute phase. Hence $e = 0$ and $L_q = \bar{\epsilon}_e L_f$, or

$$K_{oR} = k_{oR}/\bar{\epsilon}_e \qquad (7\text{-}34)$$

Equation (7-33) is combined with Eq. (7-27), so that a K_{oR} is obtained for the successive contact model:

$$K_{oR} = (1 - a)[F^{-1} + k_r^{-1}]^{-1} + ak_r \qquad (7\text{-}35)$$

with $F = k_{ob}a_b/(\bar{\epsilon}_e - \nu)$ and $a = (\nu + e)/(\bar{\epsilon}_e + e)$.

When $e = 0$ in Eq. (7-35), the above model is reduced to the direct contact model (VUE) of Lewis *et al.* This relation is also obtained by combining Eqs. (7-34) and (7-9). In an essentially similar manner the other models are reduced to the form of Eq. (7-35) and these are summarized in Table VII. Various extreme cases of chemical reaction, bubble mass transfer, or directly contacting catalyst being controlling are easily obtained from this table.

Figure 72 shows a numerical example taken from the catalytic hydrogenation of ethylene by Gilliland and Knudsen (G7). Reaction data are shown at $U_G = 30$ cm/sec. The rest of the parameters are given in Fig. 72, and these make it possible to calculate K_{oR} for different reactor models. The experimental data are first processed by the Lewis–Gilliland–Glass plot to obtain F and a, from which e and $k_{ob}a_b$ are calculated for the successive contact model (SCM, $\nu \approx 0$). This example perhaps constitutes one of the most severe tests of different reactor models. In Fig. 72 DCM(VUE) approximates the data equally well.

Except for the mechanistic concepts, there is little difference between the two models as far as the capability of correlating data is concerned. The differences are that (a) L_f/L_q is different for the two models due to existence of the dilute phase, and (b) $k_{ob}a_b$ as calculated from F is 0.166

TABLE VII

K_{oR} FOR VARIOUS REACTOR MODELS[a]

Reactor model	$K_{oR} = (1 - a)/(1/F + 1/k_r) + ak_r$		k_{oR}/K_{oR} $(=L_q/L_f)$
	F	a	
Two-phase consecutive model (S7)	$k_{ob}a_b/\bar{\epsilon}_e$	0	$\bar{\epsilon}_e$
Bubble-flow model (VUE)[b] (O6)	$k_b a_b/\bar{\epsilon}_e$	0	$\bar{\epsilon}_e$
Direct contact model (VUE)[b] (L12)	$k_{ob}a_b/(\bar{\epsilon}_e - \nu)$	$\nu/\bar{\epsilon}_e$	$\bar{\epsilon}_e$
Successive contact model (usually $\nu \ll e$) (M25, M26)	$\dfrac{k_{ob}a_b}{\bar{\epsilon}_e - \nu} \approx \dfrac{k_{ob}a_b}{\bar{\epsilon}_e}$	$\dfrac{\nu + e}{\bar{\epsilon}_e + e} \approx \dfrac{e}{\bar{\epsilon}_e + e}$	$\bar{\epsilon}_e + e$
Bubbling bed model (K22–24)	$1/K_{oR} = \bar{\epsilon}_e/k_b a_b + 1/K_{or}$ $K_{or} = (1 - a)/(1/F + 1/k_r) + ak_r$ $F = k_e a_b/(\bar{\epsilon}_e - f_w \bar{\epsilon}_b), a = f_w \bar{\epsilon}_b/\bar{\epsilon}_e$		$\bar{\epsilon}_e$

[a] $U_{mf} \ll U_G$; $\bar{\epsilon}_e = (1 - \bar{\epsilon}_b)$; $e = (1/L_f) \int_{L_f}^{L_t} \bar{\epsilon}_{ed} \, dz$

[b] VUE, vertically unmixed emulsion.

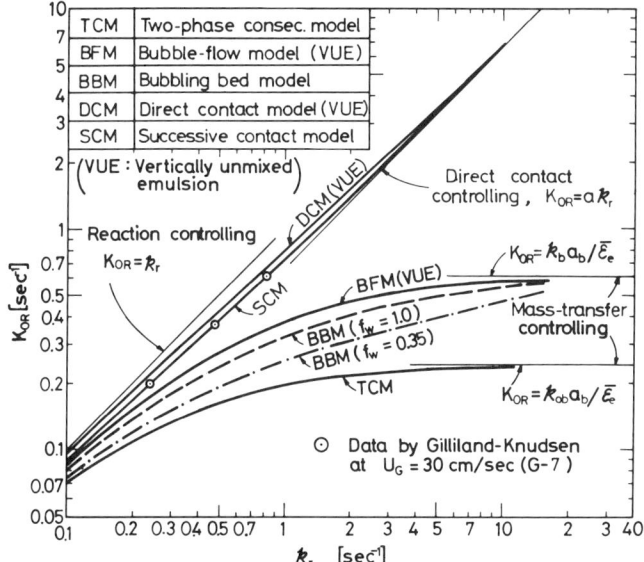

FIG. 72. Overall rate constant K_{oR} based on the quiescent bed height L_q calculated by various reactor models; $U_G = 30$ cm/sec, $\bar{\epsilon}_b = 0.31$, $k_{ob}a_b = 0.17$ sec^{-1}, $k_b a_b = 0.42$ sec^{-1}, $k_e a_b = 0.27$ sec^{-1}, $a = 0.66$, $m = 1$, $f_w = 1.0$ and 0.35.

sec^{-1} for SCM and 0.057 sec^{-1} for DCM; the latter value seems too small. From Table VII one has

$$F = [k_{ob}a_b/(\bar{\epsilon}_e - \nu)]_{DCM(VUE)} = (k_{ob}a_b/\bar{\epsilon}_e)_{SCM(\nu \approx 0)}$$

so that

$$(k_{ob}a_b)_{DCM(VUE)}/(k_{ob}a_b)_{SCM(\nu \approx 0)} = (\bar{\epsilon}_e - \nu)/\bar{\epsilon}_e = 1 - a \quad (7\text{-}36)$$

In the present example $a = 0.657$ and $1 - a \approx \frac{1}{3}$ (see also discussion given in Section VII,A,3).

2. *Evaluation of e and $k_{ob}a_b$—the Lewis–Gilliland–Glass Plot*

In evaluating the transport properties in fluid catalyst beds during reaction, it is necessary to utilize reaction data obtained at relatively high reaction rate. The reactor models of different mechanisms have been reduced to the form of Eq. (7-35), as shown in Table VII, including the bubbling bed model when $k_b a_b \gg K_{or}$. Eq. (7-35) is equivalent to one developed by Lewis *et al.* (L12) for their direct contact model of vertically unmixed emulsion (VUE). As a consequence, Eq. (7-35) is transformed (L12) to:

$$k_r K_{oR}/(k_r - K_{oR}) = k_r^2 a/(k_r - k_{oR}) + F \quad (7\text{-}37)$$

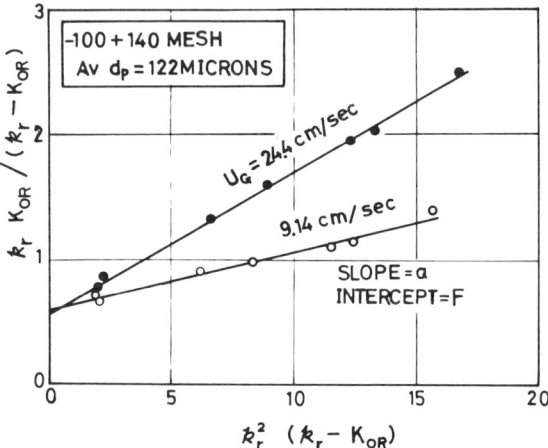

FIG. 73. Typical Lewis–Gilliland–Glass plot (after L12).

The plot of $k_r K_{oR}/(k_r - K_{oR})$ versus $k_r^2/(k_r - K_{oR})$ for the data taken at constant U_G should give a straight line of slope a and intercept F, as shown in Fig. 73 (L12). In this way, one can estimate a and F by reaction data for several catalysts of different activity. This evaluation method, named the Lewis–Gilliland–Glass plot (F18), gives reliable F and a during reaction, since the original data are obtained from the reaction itself. For the successive contact model, $k_{ob}a_b$ and e are calculated from F and a, since v is usually small in comparison with e (Section VIII).

VIII. Further Properties of the Successive Contact Mechanism

The concept of the successive contact mechanism has been given its simplest form by dividing the fluidized catalyst bed into two parts—dense phase and dilute phase. The concept has been found to apply to bed performance, as shown in the preceding section. The reactor model has been developed on the basis of several simplifying assumptions, partly to retain mathematical simplicity as a workable design equation accounting for the relative effects of the variables, and partly due to a relative lack of information about bed performance. Further properties of the mechanism are examined here, particularly as to axial distribution of reactivity inside the bed.

A. Axial Distribution of Reactivity in a Fluid Bed

The dilute phase as defined in Section VII,B,1 partly includes the transition zone. Gas–solid contacting there is much more complicated than is indicated by the simplifying assumptions for the successive contact model.

Tsutsui (T29) measured the axial bed density distribution inside a 2-inch-diam. fluid bed with the purpose of reproducing the runs by Gilliland and Knudsen (G7). The fluid-flow experiment was intended to determine whether or not so large an amount of particles was suspended in the dilute phase. Since the catalyst utilized by Gilliland was not available, Tsutsui used microspherical carbon and, glass "baloons" (*ballotini*) and mixtures of them with cracking catalyst as the fluidizing particles to adjust the quiescent bed density. He found that the quantity of particles suspended in the dilute phase was qualitatively parallel with those observed for the runs by Gilliland, but not always quantitatively.

Later, Furusaki *et al.* (F17) studied the hydrogenation of ethylene by fluidized Ni catalyst to obtain the axial reactivity distribution. Here the samples of bed gas were removed by a traveling sampler placed at the center of the bed during steady reaction, so that the sample taken in the dense phase shows an average of the concentration in the bubble and emulsion phase. Figure 74 shows an example of the axial concentration profile.

Apparently the reaction seems to have almost ended near the distributor; this is because the sample has been mostly taken from the emulsion phase. The calculated concentration profile, assuming $(\bar{\epsilon}_b)_{sample} = 0.2$, is close to the observed profile. However, the axial reactivity distribution inside the bed is not always clear, although this kind of experiment does give useful information. A similar experimental approach has been utilized by other investigators (C7a, F12).

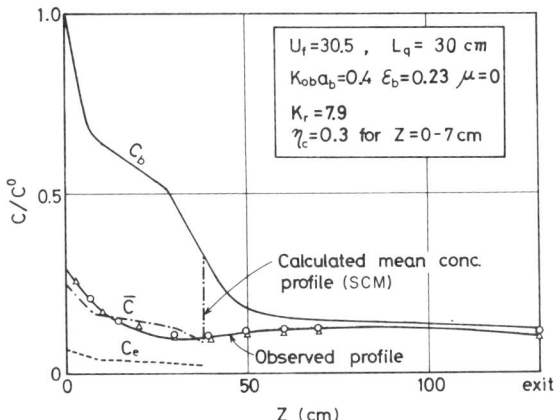

FIG. 74. Axial concentration profile of bed gas, observed by taking out the gas by an axially traveling sampler (F17).

An alternative to the above method is to introduce one of the reactants at a given vertical bed height and to measure steady-state conversion at the bed outlet by varying the height. This method, named the differential reactivity test, was suggested by T. Kikuchi and applied experimentally to catalytic hydrogenation of ethylene by Furusaki et al. (F18).

1. The Differential Reactivity Test

When a fluid bed is operated as an ideal piston-flow reactor as in the simplest case of the successive contact model [Eq. (7-29)], the equation of continuity is given by

$$-U_G \, dc/dz - r(c) = 0 \tag{8-1}$$

If the reaction rate $r(c)$ is the product of a function of vertical position $k_{oR}(z)$ and that of concentration $f(c)$, Eq. (8-1) is integrated to

$$\int_z^{L_t} (k_{oR}(z)/U_G) \, dz = -\int_{c^0}^{c_0} dc/f(c) \tag{8-2}$$

where one of the reactants at initial concentration c^0, is introduced steadily into the bed at height z measured from the distributor, and leaves the bed at $z = L_t$ with the concentration c_0. Differentiating Eq. (8-2) by z, we have:

$$k_{oR}(z)/U_G = (d/dz) \int_{c^0}^{c_0} dc/f(c) \tag{8-3}$$

This is the general relation to obtain the local overall rate constant $k_{oR}(z)$ from experimental runs. When the reaction is of first order, $f(c) = c$, then

$$k_{oR}(z)/U_G = d[\ln(c_0/c^0)]/dz \tag{8-4}$$

Thus, the local slope of the plot of $\ln(c_0/c^0)$ versus z for the data taken under a constant U_G will give $k_{oR}(z)/U_G$, and from this $k_{oR}(z)$.

2. Application of the Test to a Fluidized Catalyst Bed

Distribution of the local reactivity $k_{oR}(z)$ was tested by hydrogenation of ethylene (F18). Reaction proceeded in a fluid bed 5.3 cm in diameter and 130 cm in height. Nickel-impregnated cracking catalyst was fluidized by hydrogen, introduced through the gas distributor at a chosen superficial velocity. Ethylene was injected into the fluidized bed through the injection nozzle at several vertical positions. Conversions thus obtained are plotted against vertical position z in Figs. 75 and 76. In these figures, the slope is large near the distributor and at the transition zone; point f corresponds to the dense bed height L_f at $\bar{\epsilon}_e = 0.5$ (M25). Thus, local reactivity is large at these regions. The dense bed may be divided into three sections: the

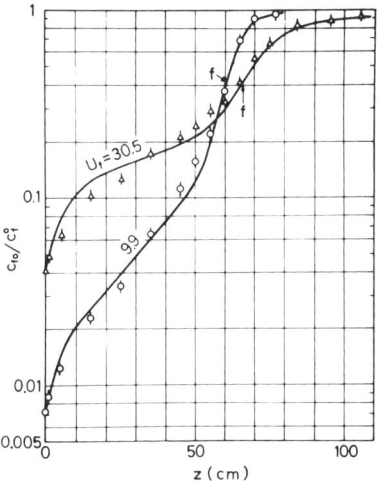

FIG. 75. Longitudinal distribution of reactivity for ethylene hydrogenation by Ni-impregnated cracking catalyst, $k_r = 10$ sec^{-1}, catalyst 1000 cm^3 (F18).

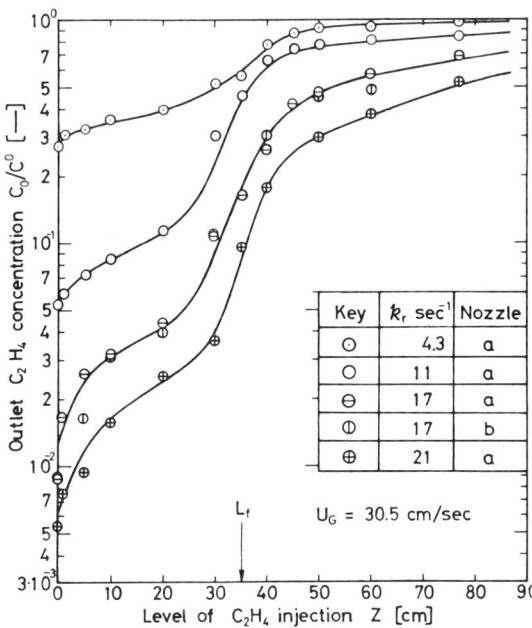

FIG. 76. Effect of the reaction rate constant k_r on the longitudinal distribution of reactivity, ethylene hydrogenation, catalyst 600 cm^3, $U_G = 30.5$ cm/sec. Nozzle designs a and b refer to Fig. 3 of ref. F18.

jetting zone near the distributor, the main dense phase, and the transition zone between the dense and dilute phases. Besides these three zones, the dilute phase must also be considered.

From the differential reactivity test, k_{oR} is obtained as a measure of reactivity. Another measure is to define the local contact efficiency η_c by the following relation:

$$\eta_c = k_{oR}(z)/\bar{\epsilon}_e k_r \tag{8-5}$$

where $\bar{\epsilon}_{ed}$ is taken instead of $\bar{\epsilon}_e$ in applying η_c to the dilute phase. In the case of the dense phase, the fraction of catalyst ν is directly contacting with the bubble gas, so that $k_{oR}(z)$ equals $k_{or} + \nu k_r$ from Eq. (7-22); then η_c for this case is given by

$$\eta_c = \bar{\epsilon}_e^{-1}([k_r(k_{ob}a_b)^{-1} + (\xi \bar{\epsilon}_e)^{-1}]^{-1} + \nu) \tag{8-6}$$

Values of η_c are given in Fig. 77 for the data of Fig. 76. The line of the dense phase in the figure is calculated by Eq. (8-6) assuming $\xi = 1$ and $\nu = 0$. The term $k_{ob}a_b$ is 0.4 sec^{-1} at $U_G = 30.5$ cm/sec and is obtained from the use of the Lewis–Gilliland–Glass plot (Section VII,C,2).

The contact in the dilute phase and the transition zone is different from that in the dense phase. The former is related to mass transfer between the gas phase and an agglomerate of solid particles, whereas the latter is mainly related to mass transfer between bubbles and emulsion including a certain amount of directly contacting catalyst. Upon consideration of hindered settling of the swarm of particles (S16, Z7), we also find contact efficiency to be a function of the population density of particles. Normalizing by the $\bar{\epsilon}_e$ of the main dense phase, $\bar{\epsilon}_{e,de}$, $\bar{\epsilon}_e/\bar{\epsilon}_{e,de}$ is chosen as a variable

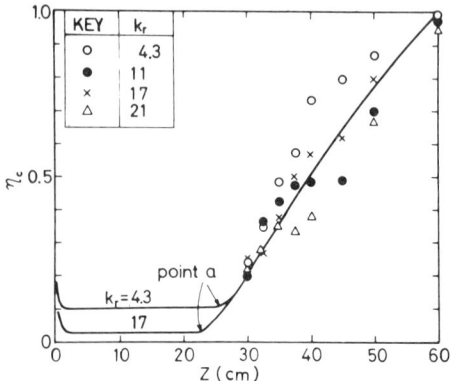

FIG. 77. Dependence of η_c on the height from the distributor, catalyst 600 cm^3, $U_G = 30.5$ cm/sec (F18). Point a indicates the upper limit of the dense phase.

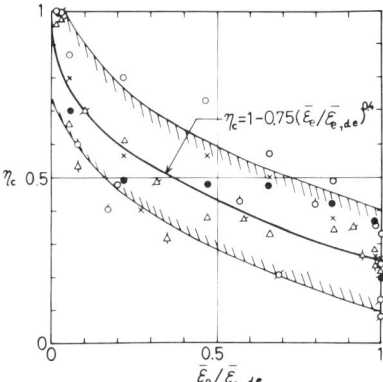

FIG. 78. Correlation of the contact efficiency with $\bar{\epsilon}_e/\bar{\epsilon}_{e,de}$ for the dilute phase including the transition zone (F18).

($\bar{\epsilon}_{ed}/\bar{\epsilon}_{e,de}$ for the dilute phase) to affect η_c because these are primary data which can be obtained by measuring the solid density in the bed.

Figure 78 shows this relation. The data are rather scattered, but most of the efficiencies are inside the shaded area of the figure. The most probable value is given by the solid line, expressed by:

$$\eta_c = 1 - 0.75(\bar{\epsilon}_e/\bar{\epsilon}_{e,de})^{0.4} \tag{8-7}$$

The equation shows that the strong interaction of particles affects the contact efficiency. The interaction seems to increase rapidly with increasing particle population. Figure 78 can be used to estimate local contact efficiency when the density distribution of solid catalyst is known.

B. Axial Distribution of Contact-Mechanism Contributions

So far, the local reactivity distribution has been explained in terms of overall rate constant k_{oR} and overall contact efficiency η_c, following Furusaki et al. (F18). The overall rate constant is further split into two parts, the local mass-transfer term and the local fraction of directly contacting catalyst, by developing a concept of the local Lewis–Gilliland–Glass plot (M27a). The relative contribution of the variables will be explained here according to (M27a).

1. Axial Distribution of Local Overall Rate Constant

Figure 76 has shown the distribution of local reactivity. To see the distribution more clearly, the smoothed curves which connect the data in the figure are drawn and differentiated graphically with respect to the axial bed height z at which one of the reactants, ethylene, has been injected. The differentiation gives, by Eq. (8-4), the local overall rate constant k_{oR}:

$$k_{oR} = 2.303 U_G \, d[\log_{10}(c_0/c^0)]/dz \qquad (8\text{-}8)$$

Figure 79 shows the axial distribution of k_{oR} obtained in this manner. The local reactivity is high in the distributor region and in the transition region ($z \approx L_f$) as well. Also, the reaction proceeds in the freeboard region, indicating the progress of reaction by freely suspended catalyst. The reactivity is rather low in the main dense phase, in contrast to that anticipated from the direct-contact model (L12) or the bubbling-bed model (K24). More reaction takes place in the transition zone than in the distributor region.

The local reactivity distribution, Fig. 79, is further processed in the following sections to determine the relative contribution of directly and indirectly contacting catalyst to the progress of the reaction.

2. *Local Lewis–Gilliland–Glass Plot*

As stated already in Section VII,B,2, Eqs. (7-19)–(7-21) hold for a given cross section of the dense phase (see Fig. 2). These equations have been

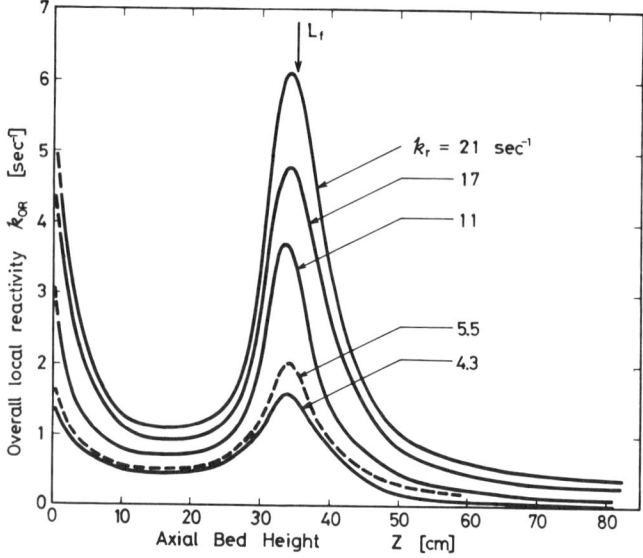

FIG. 79. Axial distribution of the overall rate constant k_{oR} for different catalyst activities, catalyst 600 cm³, $U_G = 30.5$ cm/sec (M27a).

combined, leading ultimately to Eq. (7-22) or

$$-U_G \, dc_b/dz - k_{oR}c_b = 0 \qquad (8\text{-}9)$$

where

$$k_{oR} = k_{or} + \nu k_r = \{(k_{ob}a_b)^{-1} + [(\bar{\epsilon}_e - \nu)k_r]^{-1}\}^{-1} + \nu k_r \qquad (8\text{-}10)$$

For the usual experimental beds, $\bar{\epsilon}_e k_r/mk_{ex}a_{ex} \ll 2$, so that ξ as given by Eq. (7-28) is assumed here to be

$$\xi = 1 - (\nu/\bar{\epsilon}_e) \qquad (8\text{-}11)$$

For a given cross section of the dilute phase (this phase includes a part of the transition zone) the particles contact the ascending gas in a complicated way. Swarms of suspended particles may show a mass-transfer resistance for gas–particle contact, whereas a very dilute suspension of catalyst will not involve any appreciable mass-transfer resistance. Accordingly, one can assume that the fraction ν_d directly contacts the reactant gas; the rest of the catalyst $\bar{\epsilon}_{ed} - \nu_d$, in dense clusters, indirectly contacts the gas with the overall mass-transfer coefficient $k_{oc}a_c$. Under the simplifying assumption of ideal plug flow of gas through the dilute phase, the equations of continuity are given as follows:

$$-U_G \, dc/dz - k_{oc}a_c(c - c_c) - \nu_d k_r c = 0 \qquad (8\text{-}12a)$$

for free-flowing gas at concentration c, and

$$k_{oc}a_c(c - c_c) - (\bar{\epsilon}_{ed} - \nu_d)k_r c_c = 0 \qquad (8\text{-}12b)$$

for dense-cluster gas at concentration c_c.

Eliminating the concentration c_c for Eqs. (8-12) we obtain

$$-U_G \, dc/dz - k_{oRc}c = 0 \qquad (8\text{-}13)$$

where

$$k_{oRc} = \{(k_{oc}a_c)^{-1} + [(\bar{\epsilon}_{ed} - \nu_d)k_r]^{-1}\}^{-1} + \nu_d k_r \qquad (8\text{-}14)$$

Equations (8-10) and (8-14) are essentially the same type of equations. Also, they are each equivalent to Eq. (7-9), the direct contact model of VUE by Lewis et al. (L12). As a consequence, the local mass transfer term ($k_{ob}a_b$ or $k_{oc}a_c$) and the local fraction of directly contacting catalyst (ν or ν_d) should be obtained from the use of the Lewis–Gilliland–Glass plot.

At a given bed section, k_{oR} or k_{oRc} is obtained from Fig. 79 for each run of different catalyst activity under constant aeration. Consequently, the local parameters are determined by this plot. In contrast to Eq. (7-9), which is for the overall bed performance, Eqs. (8-10) and (8-14) are for the local performance. In this sense the plot as applied to the latter equations is called the local Lewis–Gilliland–Glass plot (M27a).

Combining Eq. (7-37) with a and F as given in Table VII for the direct contact model (VUE) or directly modifying Eq. (8-10), one has the following local relation for the dense phase section (with $\xi\bar{\epsilon}_e = \bar{\epsilon}_e - \nu$):

$$k_{oR} = \nu k_r + \beta_r^*(1 - k_{oR}/\bar{\epsilon}_e k_r)(k_{ob}a_b)_{\beta_r=1}/(1 - \nu/\bar{\epsilon}_e) \tag{8-15}$$

where β_r^* is defined [see also Eq. (7-31)] by:

$$\beta_r^* = \frac{k_{ob}a_b}{(k_{ob}a_b)_{\beta_r=1}} = \frac{1/k_b + 1/k_e}{1/k_b + 1/\beta_r k_e} \tag{8-16}$$

Calculation shows for the data of Fig. 79 that $\beta_r^* = 1.21$ and 1.35 for $k_r = 10$ and 20 sec^{-1}, respectively.

In processing data by Eq. (8-15), first assume $\beta_r^* = 1$ and determine $(k_{ob}a_b)_{\beta_r=1}$, from which k_b and k_e are obtained (Section VI). With the k_b and k_e, β_r^* is calculated as a function of k_r. Usually, further iteration is not needed, because $\beta_r^* = 1$ is a good approximation.

The plot $k_{oR}/(1 - k_{oR}/\bar{\epsilon}_e k_r)\beta_r^*$ versus $k_r/(1 - k_{oR}/\bar{\epsilon}_e k_r)\beta_r^*$ for data such as those in Fig. 79 at a given cross section should give a straight line of slope ν and intercept $(k_{ob}a_b)_{\beta_r=1}/(1 - \nu/\bar{\epsilon}_e)$ as shown in Fig. 80a. In this way, one can estimate the local ν and $(k_{ob}a_b)_{\beta_r=1}$ from the reaction data for the dense phase.

Similarly, the relation for the dilute phase is obtained by modifying Eq. (8-14). Equation (8-15) again applies by taking k_{oRc}, $k_{oc}a_c$, $\bar{\epsilon}_{ed}$, and ν_d, respectively, in place of k_{oR}, $k_{ob}a_b$, $\bar{\epsilon}_e$, and ν. Also, simply assume $\beta_r^* = 1$ for the dilute phase, since the mass transfer term is usually negligible here.

3. Axial Distribution of Local Mass-Transfer Coefficient and Directly Contacting Catalyst

Figure 80a is a plot of Eq. (8-15) for the mean k_{oR} of the main dense phase ($z = 10$–20 cm) given in Fig. 79. Figure 80b is the same plot for the dilute phase data ($z = 50$ cm). For the dilute phase the latter plot shows that the straight line passes through the origin, indicating that the mass transfer term is negligible; for the dense phase the intercept remains always finite. Similar plots give the axial distribution of the mass-transfer term $(k_{ob}a_b)_{\beta_r=1}$ and the fraction of directly contacting catalyst ν (Fig. 81).

Having in mind the possibility of inaccuracy inherent to data smoothing and graphical differentiation, it is still quite interesting to see the characteristic distributions of $(k_{ob}a_b)_{\beta_r=1}$ and ν. Here the dense phase height L_f is approximately 35 cm. The term $(k_{ob}a_b)_{\beta_r=1}$ remains approximately constant in the dense phase, and drops rapidly to zero above L_f. Mass transfer seems somewhat better in the transition zone and in the jetting zone than

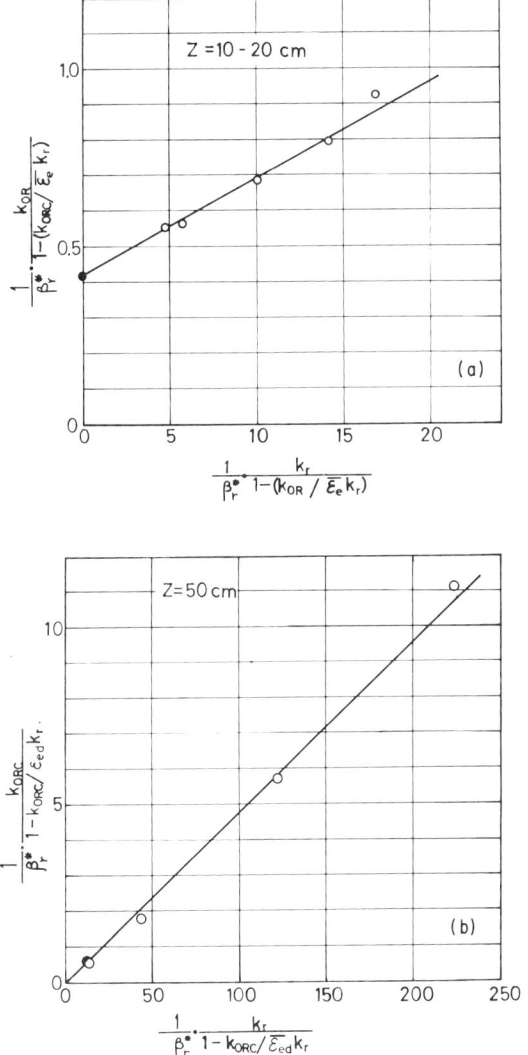

FIG. 80. Local Lewis–Gilliland–Glass plot for the data of Fig. 79: (a) at $z = 10$–20 cm; (b) at $z = 50$ cm (M27a).

in the main dense phase, but not very different. As a consequence, mass transfer in the dense phase can be effectively expressed by a mean value for $(k_{ob}a_b)_{\beta_r=1}$.

In contrast to this, the fraction ν distributes in a complicated manner. Of the total directly contacting catalyst (the area enveloped by the ν

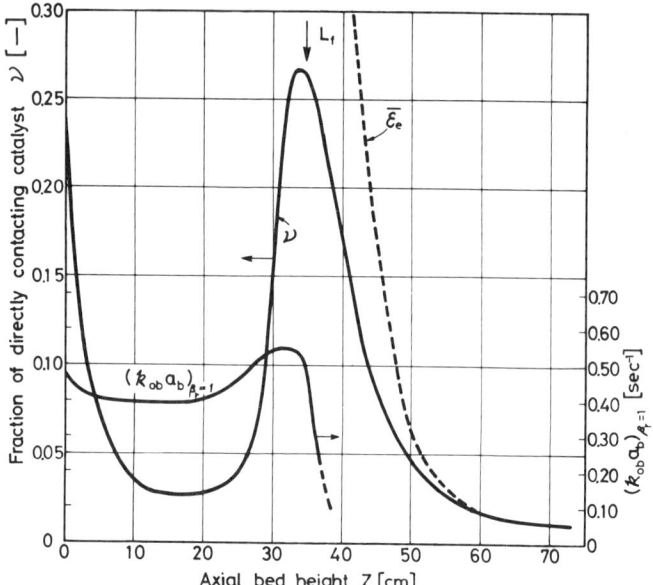

FIG. 81. Axial distribution of local mass transfer term $k_{ob}a_b$ and fraction of direct contact catalyst ν for the data of Fig. 79 (M27a).

versus z curve, Fig. 81), about 15% is located in the jetting zone ($z = 0$–7.5 cm), another 15% in the main dense phase ($z = 7.5$–30 cm), about 40% in the transition zone ($z = 30$–40 cm), and the remaining 30% is in the dilute phase ($z \geq 40$ cm). The ν distribution curve implies that direct contacting of catalyst with the reactant gases is most noticeable in the transition zone and in the dilute phase, less so in the jetting zone, and rather lacking in the main dense phase. We note that the relative contribution of variables given here is based on only one experimental example that would allow separation of ν and $k_{ob}a_b$. More data are needed for further progress.

The kinetic approach so far presented as the successive contact mechanism, in its simplest form, has been applied to very complicated fluid-bed behavior. Nevertheless the model has consistently paralleled the observations. A fair amount of directly contacting catalyst is shown to exist in the transition zone and in the freeboard region, with less in the jetting zone, under aeration with relatively high gas velocity. The relative location of the directly contacting catalyst will certainly influence the selectivity of complex reactions. Directly contacting catalyst in the freeboard region may sometimes produce a drastic effect on thermal stability of the bed (R12; also Section IX).

The successive contact mechanism seems to provide a reasonable approach to fluid catalyst bed design, by making it feasible to take into consideration the combined influence of bed design and reaction kinetics of a given chemical system on both optimum selectivity and bed stability.

Fluid beds with internals are frequently utilized for exothermic reactions. Qualitative and quantitative effects of the internals design have been discussed extensively (see Section II; I2, I5, L12). The internals are believed to increase $k_{ob}a_b$ and perhaps v (not e), and also to decrease the catalyst content in the dilute phase by reducing both the circulation rate of the dense phase and the ascending velocity of bubbles (M26). These effects are still under investigation.

IX. Nonisothermal Effect on the Bed Performance

In the dilute phase the contact efficiency is rather high but the rate of heat transfer is low, and hence the temperature may not remain uniform in this region. Thus, the temperature effect on conversion and selectivity becomes important when a substantial part of the entire reaction occurs in the dilute phase. This chapter discusses salient features of the temperature effect on fluid bed reactions.

A. EFFECT ON STEADY REACTION

1. *Bed Design and Local Reactivity*

Several types of internals have been proposed so far. Some studies are presented by Volk *et al.* (V12) and Grekel *et al.* (G15), who studied several arrangements of tubular internals and baffle trays. Volk suggested vertical tubes to increase the contact efficiency. For fluidized beds with diameter of 5–198 inches, conversions for beds of the same equivalent diameter (cf. Section I,C) were found equal. A large heat-exchange surface in the fluid bed is necessary for non-isothermal reactions, so the use of vertical tubes is quite practical.

Grekel *et al.* (G15) pursued the idea of vertical tube internals, but also recognized the effect of the horizontal grid structure. They increased contacting efficiency by the use of spaced and staggered horizontal tubes, which could have the effect of increased direct contact below and just above the grid. However, care must be taken in applying such a design, because an unstable hot spot may form in a region of good gas–solid contact and poor thermal conduction. This instability in the dilute phase, which is discussed later, is more extreme in the vicinity of a grid. Horizontal internals tend to decrease the flow circulation rate and to produce underlying freeboard space, thus sometimes permitting occurrence of a temperature increase.

Grekel *et al.* also discussed the effect of the region of vicinity of the distributor. This region is important, since there is some catalyst directly contacting the entering gas phase (cf. Section VIII). Designs for the distributor region are proposed in many patents. However, the effect of the distributor disappears within a few decimeters because of the violent turbulence within the dense phase.

Furusaki *et al.* (F18) presented an increase of temperature in the bottom part of the dilute phase for the reaction by activating fine particles. Results for homogeneous catalysts are shown in Fig. 82. Here, the temperature rise is not so significant as the case of activated fine particles (shown in F18) because of the effective cooling through the wall of the reactor. However, if the cooling surface area is small compared with the reactor volume, the temperature rise will be more significant, especially in the upper part of the dilute phase. In the lower part, the intensive longitudinal mixing of solid particles prevents temperature from ascending significantly. The necessity of cooling devices in the dilute phase will be shown later. In some cases the internal cooler or an emergency shot of coolant is provided. The effects of internals are also discussed by Overcashier *et al.* (O11), Morooka *et al.* (M42), and Nishinaka *et al.* (N6, N8, N9).

Heat transfer between the bed and wall is important in considering temperature rise in the dilute phase (F16). This is discussed briefly in Section VI,G.

2. *Overall Conversion*

The effect of the dilute phase on overall conversion of the reactant is straightforward. Conversion in the bed increases if the contact efficiency increases due to suspended catalyst in the dilute phase. If temperature

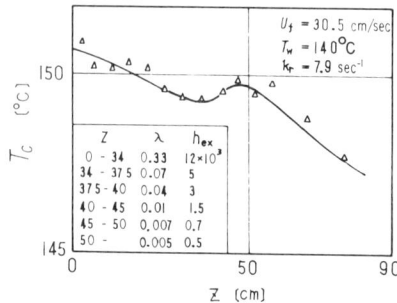

FIG. 82. Temperature profile in a fluid bed. Standard catalyst 600 cm³. Bed diameter 5.5 cm. Solid line shows calculated value using Eqs. (9-7) and (9-8). T_c = temperature at the axis.

increases in the dilute phase, conversion will further increase. Thus, from the viewpoint of conversion of the reactant, it is recommended to use the direct contact effect in the dilute phase and transition zone. Stability and selectivity must now be discussed more thoroughly.

3. *Effect on Selectivity*

Industrially, selectivity is often as important as conversion in considering the efficiency of the reactor. In isothermal reactions, the dilute phase and transition zone may cause better selectivity due to better contact in that region. But in nonisothermal reactions, the effect will be different because of the temperature effect. Mixing of gas and solids in the dilute phase is not sufficient, and this may cause a temperature distribution for exo- or endothermic reactions.

As an illustration, a simplified model will be considered here to show the importance of the nonisothermal effect in the dilute phase. We assume that $\varphi \geq 10$, reaction in the bubble phase is negligible, and temperature in the dense phase is uniform. Then the material balance equation for the dense phase is:

$$U_f(dc_b/dz) + k_{ob}a_b(c_b - c_e) = 0 \tag{9-1}$$

$$k_{ob}a_b(c_b - c_e) - \epsilon_e k_r c_e = 0 \tag{9-2}$$

and for the dilute phase,

$$U_f(dc/dz) + \eta_c \epsilon_{ed} A_r e^{-E/RT} c = 0 \tag{9-3}$$

where the temperature is a radially averaged value. The above equations are rearranged into dimensionless forms:

$$(dC_b/dZ) + N_{ob}(C_b - C_e) = 0 \tag{9-4}$$

$$N_{ob}(C_b - C_e) - \epsilon_e N_r C_e = 0 \tag{9-5}$$

$$(dC/dZ) - \eta_c \epsilon_{ed} N_{Ar} e^{-\Gamma/\theta} C = 0 \tag{9-6}$$

In calculating temperature distribution in the dilute phase, solid motion must be accounted for. Solid motion in the dilute phase is shown in Sections II and VI. Laboratory-scale fluid beds exhibit a circulating flow of solid particles with ascending central core and descending peripheral region. The enthalpy balances for both ascending and descending zones for the steady state are

$$-c_{pg}\rho_g U_f(1 + \lambda m)(dT_u/dz)$$
$$+ \tfrac{1}{2}\eta_c \epsilon_{ed}(-\Delta H)A_r e^{-E/RT_u} c - h_{ex}a_{ex}(T_u - T_d) = 0 \tag{9-7}$$

$$\lambda mc_{pg}\rho_g U_f(dT_d/dz) + \tfrac{1}{2}\eta_c\epsilon_{ed}(-\Delta H)A_r e^{-E/RT_d}c$$
$$+ h_{ex}a_{ex}(T_u - T_d) - h_w a_w(T_d - T_w) = 0 \qquad (9\text{-}8)$$

where λU_f is the superficial flow of solid particles; $h_{ex}a_{ex}$ is the volumetric heat exchange coefficient between the ascending and descending emulsions; $h_w a_w$ measures heat transfer between the descending emulsion and the wall. In deriving the above equations, the radial concentration distribution and the flow of the descending gas are neglected. Converting to dimensionless forms,

$$-(1 + \lambda m)(d\theta_u/dZ) + \tfrac{1}{2}\eta_c\epsilon_{ed}\Omega N_{Ar}e^{-\Gamma/\theta_u}C - N_{Hex}(\theta_u - \theta_d) = 0 \qquad (9\text{-}9)$$

$$\lambda m(d\theta_d/dZ) + \tfrac{1}{2}\eta_c\epsilon_{ed}\Omega N_{Ar}e^{-\Gamma/\theta_d}C$$
$$+ N_{Hex}(\theta_u - \theta_d) - N_{Hw}(\theta_d - \theta_w) = 0 \qquad (9\text{-}10)$$

where Ω [equal to $(-\Delta H)c^0/c_{pg}\rho_g T_{de}$] denotes the dimensionless heat of reaction and Γ (equal to E/RT_{de}) is the dimensionless activation energy. For an adiabatic condition, $N_{Hw} = 0$. Also, for $N_{Hex} \gg 1$, θ_u becomes equal to θ_d. Thus,

$$-(d\theta/dZ) + \eta_c\epsilon_{ed}\Omega N_{Ar}e^{-\Gamma/\theta}C = 0 \qquad (9\text{-}11)$$

For $N_{Hex} \approx 0$, the temperature of the ascending zone is given by

$$-(1 + \lambda m)(d\theta_u/dZ) + \tfrac{1}{2}\eta_c\epsilon_{ed}\Omega N_{Ar}e^{-\Gamma/\theta_u}C = 0 \qquad (9\text{-}12)$$

In the actual bed, solid exchange between the ascending and descending zone is significant, and the circulation is seen as a series combination of localized recirculation. If this scheme of solid mixing is more realistic, the temperature profile may be suitably expressed in terms of an effective thermal diffusivity, which will be used later for discussion of instability in the dilute phase.

In this section, calculations were carried out for the case of $\eta_c = 1$, $N_{Hex} \gg 1$, and $\bar{\epsilon}_{ed} = (Z - Z_t)\bar{\epsilon}_{e,de}/(1 - Z_t)$. The last relation means that catalyst density decreases linearly with bed height from the top of the dense phase to some point (Z_t) in the dilute phase. This Z_t is not the total bed height, but a hypothetical intermediate height where the amount of catalyst particles becomes negligible.

a. *Consecutive Reactions.* The problem of selectivity for B in consecutive reactions (A $\xrightarrow{1}$ B $\xrightarrow{2}$ C) is referred to as selectivity of type III by Wheeler (W11). Wheeler's analysis concerned the effect of pore diffusion on selectivity of catalysis. According to him, selectivity decreases along with the increase of diffusion resistance. In case of fluid bed reactors, the mass-transfer resistance between bubble and emulsion phases causes similar effects; the selectivity of B is greater for the case of larger N_{ob}. Also, in case of isothermal reactions it will be greater if e is chosen larger, because direct contact is assured in the dilute phase. For nonisothermal reactions the aforementioned expectation is not true because of the temperature distribution existing in the dilute phase.

For this reaction system, rates of increase of A, B, C are given by $R_A = -k_{r1}c_A$, $R_B = k_{r1}c_A - k_{r2}c_B$, and $R_c = k_{r2}c_B$, respectively. The k_r's may be expressed by the Arrhenius equation $k_r = A_r \exp(-E/RT)$. Calculations were made according to the equations given above. Details of the equations for this special case are shown elsewhere (M28). Results for the isothermal reactions are given in Fig. 83. Here, the rate of reaction 2 is assumed to be smaller than that of reaction 1 by a factor of 10. For $N_{ob} < 3$, the yield of B is higher for $Z_t = 2$ than that at the exit of the dense phase, i.e., $Z_t = 1$. If Z_t is larger, the amount of catalyst in the dilute phase is large, and the yield of B becomes large, as is obvious from Fig. 83. Thus the use of the dilute phase improves selectivity for isothermal consecutive reactions.

For nonisothermal reactions, the previous discussion does not hold. Figure 84 shows the yield of B with respect to the dimensionless heat of reaction Ω. If Ω is positive, the reaction is exothermic; if negative—endothermic; and if zero—isothermal. Values of Ω for real systems may sometimes be even larger, but because of solid mixing the effective Ω would be reduced. From the figure, it is seen that yield of B decreases significantly with increasing Ω, and especially so for large activation energy of reaction 2 (B $\rightarrow$ C) which proceeds more rapidly at higher temperatures in the dilute phase. Many oxidation reactions operated in fluid

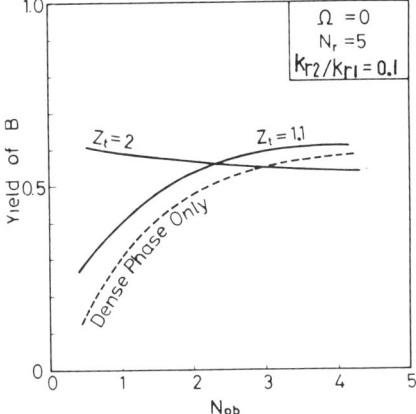

FIG. 83. Selectivity of isothermal consecutive reactions (A $\overset{1}{\rightarrow}$ B $\overset{2}{\rightarrow}$ C).

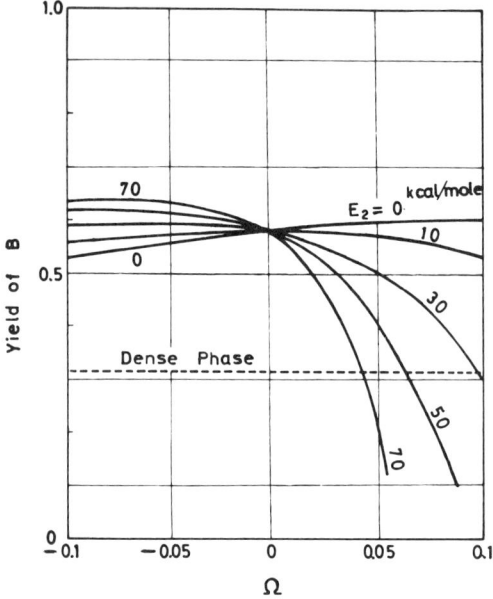

FIG. 84. Selectivity of nonisothermal consecutive reactions (A $\xrightarrow{1}$ B $\xrightarrow{2}$ C). $N_{ob} = 1$, $Z_t = 2$, $E_1 = 83.6$ kJ/mole, $N_{Ar1} = 2 \times 10^8$, $N_{Ar2} = N_{Ar1}/10$.

beds are exothermic with $E_2 > E_1$ and $E_2 \gg 10$ kcal/mole. In these reactions it is profitable to suppress the role of the catalyst in the dilute phase.

In case of endothermic reactions, the yield of B does not vary significantly from isothermal reactions. For $E_2 > 30$ kcal/mole, the use of the dilute phase even becomes profitable. Therefore, it may be desirable to use the dilute phase in case of endothermic reactions.

The case of solids mixing with consecutive reactions is shown in Fig. 85. The magnitude of solid mixing is considerable ($N_{Pe} = 0.1$), but the selectivity is affected strongly by the heat of reaction, especially for the case of highly exothermic reactions. The qualitative effect of dilute phase on selectivity is not affected by solid mixing; therefore the effect will be discussed for the case of negligible solid mixing, at smaller values of effective Ω.

b. *Denbigh's System with Optimal Temperature Distribution.* Denbigh (D11) proposed the following reaction system as a general case for organic reactions;

$$A \xrightarrow{1} X \xrightarrow{3} Y$$
$$\searrow^{2} \searrow^{4}$$
$$\phantom{A\xrightarrow{1}} P \phantom{\xrightarrow{3}} Q$$

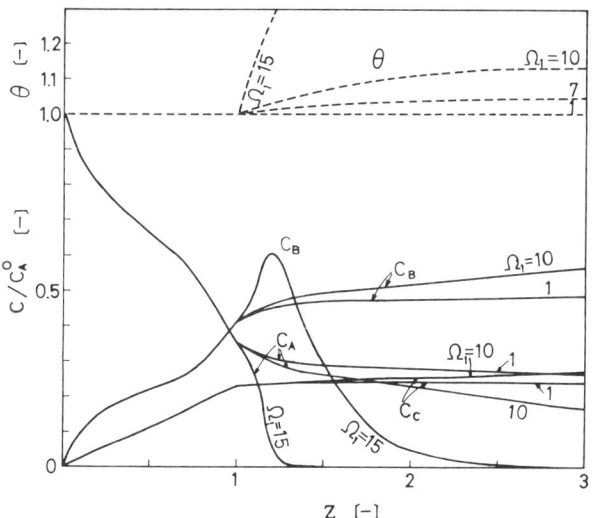

FIG. 85. Variation of selectivity of consecutive reaction $(A \xrightarrow{1} B \xrightarrow{2} C)$. $N_{Hw} = 0$, $N_{Ar1} = 2 \times 10^8$, $N_{Ar2} = 2 \times 10^7$, $\Omega_2 = \Omega_1$, $N_{Pe} = 0.1$, $E_1 = E_2 = 83.6$ kJ/mole.

This is a system with an optimal temperature distribution, while the system in the previous section is one with an optimal residence time. In this system, the cases of (1) $E_1 < E_2$ and $E_3 > E_4$ or (2) $E_1 > E_2$ and $E_3 < E_4$ are especially interesting. Several calculations were made for these cases; details are not shown here (cf. M28). Generally speaking, the use of the dilute phase is preferable for the case of the Denbigh system, except in the case of exothermic reactions for $E_1 > E_2$ and $E_3 < E_4$. In this case, high conversion in the dense phase is preferable, but increasing N_{ob} does not give significant improvement because the yield of P increases while that of Q decreases. Therefore, operation with a large U_f may possibly be the economically optimal step, even for the cases mentioned above. Care is needed if Y decomposes further to form by-products; in this case, for exothermic reaction of Y it is preferable to suppress the dilute phase for the same reason as stated above for consecutive reactions.

c. *Parallel Reactions with an Equilibrium.* The parallel reaction [A → X, A → Y] is another important reaction system. If the orders of the two reactions are equal, selectivity does not depend upon the type of reactor. An example of industrially important parallel reactions is [A $\underset{2}{\overset{1}{\rightleftharpoons}}$ X, A $\xrightarrow{3}$ Y] with X the required product (D12). Examples of this reaction system are hydrocarbon and methanol synthesis reactions, where thermodynamically favorable side reactions are suppressed by the use of appropriate catalyst. Packed-bed reactors with an internal cooling system have been used for this type of reaction. As expected, X should be removed from the reactor as soon as it is produced. In fluid beds, if reaction 1 is exothermic ($\Delta H_1 < 0$), the equilibrium is unfavorable for high temperatures. Thus, use of the dilute phase for exothermic reaction is unfavorable and a large N_{ob} is necessary in order for the reaction to proceed in the dense phase. On the other hand, if reaction 1 is endothermic ($\Delta H_1 > 0$) the equilibrium is more favorable at higher temperature ($\Omega > 0$). Assuming $\Omega_1 = \Omega_3$, the yield of the unwanted Y is plotted against ΔH_1 or N_{ob} in Figs. 86 and 87. Here, input values of $N_{Ar1} = 1.387 \times 10^{12}$ and $E_1 = 126$ kJ/mole give $k_{r1}L_f/U_f = 5.0$ at $T = 573°K$. If $\Omega = 0.1$ then $\Delta H \approx -17$ kJ/mole at $c^0 = 2.1 \times 10^{-7}$ mole/cm³ (~10%), and $T_{de} = 573°K$. It is shown that yield of X will decrease for exothermic reactions with small values of N_{ob}. Thus, in this reaction system, a large N_{ob} is preferred in order to obtain more desired product (X).

4. *Discussion of Selectivity*

In discussing the yield from a reactor, the temperature distribution inside the reactor must be investigated. In the dense phase of fluid beds, the heat-transfer coefficient between the bed and wall has been widely studied (L10, M14, M15, M16, T22, V5, W3, W5). Botterill (B12) has reviewed the recent literature; studies of heat transfer in the dilute phase are quite limited in number. Shirai (S9, S11), Furusaki (F15), and Morooka *et al.* (M50) studied the heat-transfer coefficient in the dilute phase as well as in the dense phase. They found that the heat-transfer coefficient between the bed and the wall decreases as the bed density decreases, which will cause an axial distribution of temperature in bed.

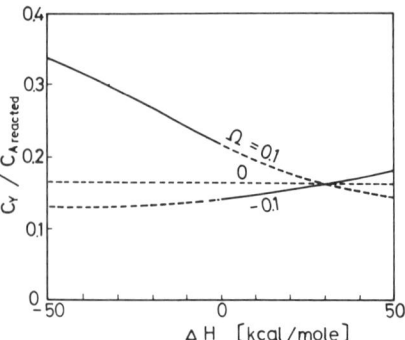

FIG. 86. Effect of heat of reaction on selectivity for X in competitive reactions with an equilibrium: $A \underset{2}{\overset{1}{\rightleftharpoons}} X$, $A \overset{3}{\rightarrow} Y$. Parameters are same as in Fig. 87, except $N_{ob} = 1$.

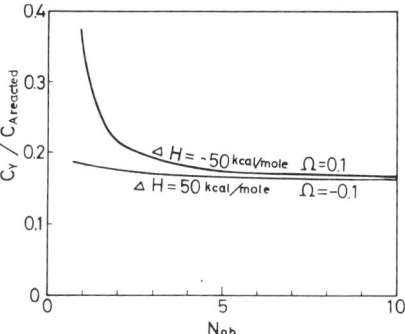

FIG. 87. Effect of N_{ob} on yield of byproduct Y in the competitive-reaction system $A \underset{2}{\overset{1}{\rightleftharpoons}} X$, $A \overset{3}{\rightarrow} Y$. $N_{Ar1} = 1.387 \times 10^{12}$, $N_{Ar3} = N_{Ar1}/10$, $E_1 = E_3 = 128$ kJ/mole, $Z_t = 3$, $K_R = 10$.

The influence of the temperature distribution on selectivity varies according to the reaction scheme. Among such schemes, the consecutive reaction (A → B → C) qualitatively represents many organic reactions with by-products. As shown in the previous section, the use of dilute phase is recommended for endothermic reactions, but prohibited for exothermic reactions. This conclusion agrees with the development of fluid bed reactors for partial oxidations (exothermic) and cracking (endothermic). This knowledge may help one to design or develop new fluid bed contactors.

B. Stability of Fluid Bed Reactors

In discussing the stability inside reactors, we must deal with several stability problems, such as multiplicity of steady states, local stability against perturbation, sensitivity to external conditions, oscillation, etc. (F19, S5). Froment (F8) and Endoh *et al.* (E11) discussed sensitivity in packed-bed reactors. Froment (F8) showed that multiple steady states and oscillation are rare for catalytic packed-bed reactors, when $(PeB)_f \gg 1$ and $N_{Pe} \gg 1$. In case of fluid bed reactors longitudinal mixing of solid particles is sometimes much larger in magnitude, so N_{Pe} becomes much smaller than $(PeB)_f$. This will introduce the possibility of multiple steady states and of instability. On the other hand, stability against perturbation is increased by solid mixing. Conversely, with less solid mixing, the steady state is essentially unstable and perturbations lead to random fluctuations (F15). Stability in the dense phase is discussed in other articles (B18, B19, E10).

The most important stability problem in fluid beds is that of runaway phenomena in the freeboard and also in the spaces under baffle plates inside the dense bed—the so called dilute-phase regions. Good contact and low heat-transfer rates are the major causes of runaway. The most significant temperature rise is considered to occur by a coupling effect of a medium degree of temperature rise due to good contact in the dilute phase (including the transition zone) and of noncatalytic reactions which are initiated by the temperature rise. Thus a model reaction to consider the possibility of runaway is:

$$A \xrightarrow{1} B \xrightarrow{2} C$$
$$\searrow^{3} \quad \downarrow^{4} \quad \swarrow^{5}$$
$$D$$

Here, A is the raw material, B is the required product, C and D are by-products, but D is produced by noncatalytic reactions (3–5), such as combustion or polymerization. The production of D is highly exothermic, and also the contributing reactions have large activation energies. Hence the production of D tends to give rise to violent reaction.

The material balances and enthalpy balance in the dilute phase are given by the following equations in dimensionless form:

$$(dC_A/dZ) + \nu N_{Ar1}e^{-\Gamma_1/\theta}C_A + N_{Ar3}e^{-\Gamma_3/\theta}C_A = 0 \quad (9\text{-}13)$$

$$(dC_B/dZ) + \nu(N_{Ar2}e^{-\Gamma_2/\theta}C_B - N_{Ar1}e^{-\Gamma_1/\theta}C_A) + N_{Ar4}e^{-\Gamma_4/\theta}C_B = 0 \quad (9\text{-}14)$$

$$(dC_C/dZ) - \nu N_{Ar2}e^{-\Gamma_2/\theta}C_B + N_{Ar5}e^{-\Gamma_5/\theta}C_C = 0 \quad (9\text{-}15)$$

$$(N_{Pe})^{-1}(d^2\theta/dZ^2) - (d\theta/dZ) + \nu(\Omega_1 N_{Ar1}e^{-\Gamma_1/\theta}C_A$$
$$+ \Omega_2 N_{Ar2}e^{-\Gamma_2/\theta}C_B) + (\Omega_3 N_{Ar3}e^{-\Gamma_3/\theta}C_A$$
$$+ \Omega_4 N_{Ar4}e^{-\Gamma_4/\theta}C_B + \Omega_5 N_{Ar5}e^{-\Gamma_5/\theta}C_C) - N_{Hw}(\theta - \theta_c) = 0 \quad (9\text{-}16)$$

The boundary conditions for these equations are (at $Z = 1$) $C_A = C_{A0}$, $C_B = C_{B0}$, $C_C = C_{C0}$, $\theta = 1$ and (at $Z = Z_t$) $d\theta/dZ = 0$. For simplicity, $N_{Ar3} = N_{Ar4} = N_{Ar5}$, $\Gamma_3 = \Gamma_4 = \Gamma_5$, and $\Gamma_1 = \Gamma_2$ are assumed. The Peclet number N_{Pe} accounts for the temperature diffusion due to solid mixing. If N_{Pe} is zero, the temperature in the dilute phase is completely uniform; N_{Hw} is the dimensionless heat-transfer coefficient; and sometimes the effect of radiative transfer must also be accounted for.

An example of concentration and temperature distribution is given in Fig. 88. Reaction 1 is an ordinary reaction with the activation energy of 83.6 kJ/mole (20 kcal/mole); the rate of reaction 2 is chosen as one-tenth of that for reaction 1. The noncatalytic reactions 3–5 are rather violent with the activation energy of 167.2 kJ/mole and heats of reaction (A → D) of 3×10^3 kJ/mole. This reaction does not occur at the temperature of the dense phase, but is quite active at higher temperatures because of the

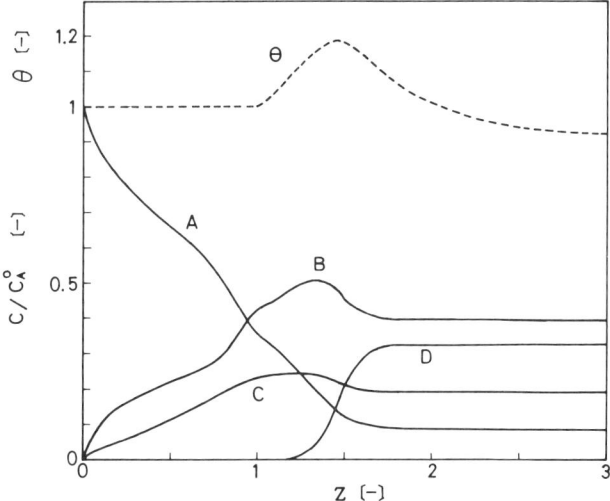

FIG. 88. Temperature rise by noncatalytic reaction with axial mixing of solid particles. $N_{Ar1} = 2 \times 10^8$, $N_{Ar2} = N_{Ar1}/10$, $N_{Ar3} = 1.08 \times 10^{13}$, $E_1 = E_2 = 83.6$ kJ/mole, $E_3 = 167$ kJ/mole, $N_{Pe} = 0.1$, $N_{Hw} = 50$, $\theta_c = 0.9$, $\Omega_1 = \Omega_2 = 1$, $\Omega_3 = 50$.

large activation energy. The value $N_{Hw} = 50$ is that obtained for the dilute phase with an effective diameter of about 5 cm. The values $N_{Ar1} = 2 \times 10^8$ and $E_1 = 83.6$ kJ/mole give $k_{r1}L_f/U_f = 5$; $\Omega_3 = 50$ means $-\Delta H_3 = 2800$ kJ/mole at $c_f^0 = 6.4 \times 10^{-7}$ mole/cm^3 (~30%) at 573°K and 1 atm. From Fig. 88 it is obvious that there is considerable temperature rise in the dilute phase, especially due to the noncatalytic reactions. The reaction system is very sensitive to N_{Pe} and N_{Hw}. The stabilizing effect of the dense phase, due to good thermal conductivity, is completely lost in the dilute phase. If N_{Pe} is larger, the temperature rise at the bottom of the dilute phase becomes so large as to be uncontrollable.

The case of large N_{Pe} may be studied by calculation for the case of unmixed solid particles. Calculation of this case shows that cooling of the dilute phase is essential for steady-state operation of the reactor for such unstable reactions. It is hard to recommend using the dilute phase, because side reactions should be avoided (Fig. 89).

The extreme case of small N_{Pe} is the case of the thermally completely mixed dilute phase. As shown above, multiple steady states are possible. The analyses by heat generation and rejection curves (V4) are shown in Figs. 90 and 91. Q is the dimensionless heat being generated or rejected, and θ is the dimensionless temperature in the dilute phase. The value of $N_{Hw}(L_t - L_f)/L_f$ is 100–200 for the transition zone and 30–50 for the dilute phase. There are two stable steady states and one unstable point for E_3 of 167 and 293 kJ/mole. If E_3 is reduced to 40 kJ/mole, only one stable point

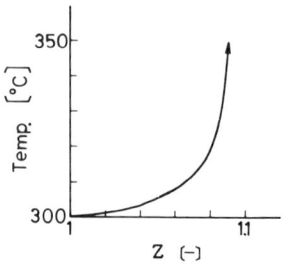

FIG. 89. Runaway in the dilute phase. $N_{Ar1} = 2 \times 10^8$, $N_{Ar2} = N_{Ar1}/10$, $N_{Ar3} = 2.92 \times 10^{24}$, $N_{pe} = \infty$, $\Omega_1 = \Omega_2 = 1$, $\Omega_3 = 50$, $T_c = 100°C$, $N_{Hw} = 1.17$ $E_1 = E_2 = 83.6$ kJ/mole, $E_3 = 293$ kJ/mole.

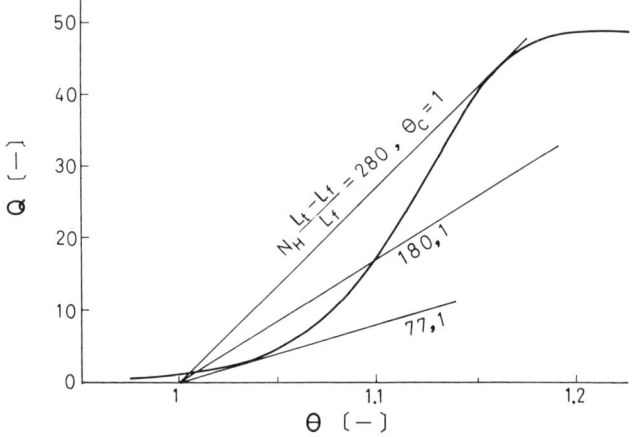

FIG. 90. Thermal effect in the dilute phase, with thermally complete mixing. $E_1 = E_2 = 83.6$ kJ/mole $E_3 = 167$ kJ/mole, $T_{de} = 300°C$, $\Omega_1 = \Omega_2 = 1$, $\Omega_3 = 50$, $N_{Ar1} = 2 \times 10^8$, $N_{Ar2} = N_{Ar1}/10$, $N_{Ar3} = 1 \times 10^{13}$, $L_t/L_f = 3.42$, $\theta_c = T/T_c$.

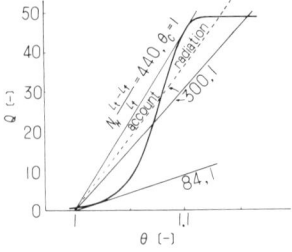

FIG. 91. Thermal effect in the dilute phase, thermally complete mixing. Parameters are same as those in Fig. 90 except $E_3 = 293$ kJ/mole $N_{Ar3} = 2.9 \times 10^{24}$.

is possible. Thus, spontaneous ignition is possible for large values of E_3, and instability is probable for these unstable reaction systems even in the case of thermal homogeneity.

In conclusion, the possibility of highly exothermic noncatalytic reactions must be carefully eliminated, for all values of solid mixing. This is done by the control of composition of the feed mixture so as, for example, to suppress an excess of oxygen. Also, appropriate cooling is necessary in the dilute phase.

X. Discussion and Summary

A. Applicability of Reactor Models

The main advantages of the reactor models are:

i. Reaction performance can be expressed by the minimum number of measurable parameters.
ii. The relationship between the parameters and variables, e.g., catalyst properties, reactor design, and operational variables, becomes clear and easy to apply.
iii. The equations obtained are simple and easy to use.

Most of the models for fluidized catalytic reactions have not devoted enough attention to flow properties of FCBs. The important features of the FCB are as follows: (a) Bubbles grow much more slowly than in teeter beds; (b) A circulating flow exists, centrally upward and peripherally downward; (c) Above the dense phase are the transition region and the dilute phase, where particle density decreases gradually with height.

These properties introduce another uncertainty in developing a large-scale FCB, which must be operated at high superficial velocity (about 50 cm/sec) and at a large bed height (5–10 m), with more than 30 percent of fines fraction. The successive contact mechanism (M25, M26) is the simplest theory dealing with FCB flow properties. The necessary parameters to calculate reactor performance are as follows: (a) bubble diameter; (b) particle distribution through the bed height, which may be obtained from static pressure distribution; (c) contact efficiency distribution in the transition region and the dilute phase. If relationships between these three parameters and the operational variables are obtained from FCB performance in more than two laboratory steps, it is possible to calculate the same reaction schemes in large-scale fluidized catalyst beds.

The direct contact model of Lewis *et al.* (L12) is a special case of the successive contact mechanism. If the effect of circulation of the emulsion on reaction can be ignored, and contact efficiency in the dilute phase is considered to be 100%, then the overall contact efficiencies by the two models approximately coincide.

The successive contact mechanism and simplified theory of recirculation provide much useful information concerning the FCB. For example, Furusaki et al. (F18) showed experimentally, in hydrogenation of ethylene, that the dilute phase plays an important role in the progress of the reactions, and affects conversion and selectivity significantly. They also showed that the fines fraction promoted the reaction in the dilute phase, as well as improved the fluidity. Variables such as fraction of catalyst in direct contact, improvement of contact efficiency, and temperature control in the dilute phase were found to affect significantly the extent of reaction in FCB. This result gives important perspective for reactor design, as to size distribution of catalyst particles, control of activity distribution with respect to particle sizes, selection of superficial velocity of gaseous flow, and improvement of the internals arrangement.

The effect of dilute phase on selectivity depends on the type of reaction. Generally, reaction systems with exothermic side reactions should avoid use of the dilute phase. On the other hand, it is advantageous to enhance the use of dilute phase for isothermal or endothermic reactions. Discussions on this subject by Miyauchi and Furusaki (M28) are pertinent in the design of FCBs.

According to the simplified theory of recirculation, the turbulent viscosity of the fluid bed determines its flow pattern. Turbulent viscosity for FCC particles may be given by a function of the bed diameter, as for bubble columns of low-viscosity liquids. This relation will provide useful information for practical design of FCBs, such as control of backmixing of particles and gas, gas distribution from the distributor, design of the sparger, control of aging of the catalyst, and prevention of erosion.

B. Development of Industrial Fluidized Catalyst Beds

1. Fluid Catalytic Cracking of Petroleum

The initial application of fluidized beds in the petroleum industry was the upflow dilute-phase reactor (M45). The obvious disadvantage of this design is that all of the flowing catalyst passes overhead and must be removed in dust-removal equipment. Later, the basic design which finds widest application is the downflow dense-bed reactor (M2), which has the following major advantages (K26):

i. Catalyst load to the cyclone decreases, and the cyclone may be made smaller. Loss of catalyst and rate of corrosion decrease.

ii. Catalyst inventory in the reactor is large enough to stabilize the operation, so that liquid oil may be introduced.
iii. Plant size becomes smaller, and the cost of the construction decreases.

Catalyst also serves as a heat-transfer medium between the reactor and the regenerator, circulating in a high flow rate between the two units.

Development of the microspherical catalyst in 1946, from Al_2O_3–SiO_2 sol by means of spray drying, contributed to major advances in fluidized catalyst-bed technology, especially for improvement of catalyst fluidity, decrease of attrition loss, and decrease of erosion in transfer lines.

Recently, new upflow operation (riser cracking) has become popular, through the development of highly active zeolite catalyst (V15).

2. *Fluid Catalytic Reforming*

Fluid catalytic reforming is a process of catalytic isomerization of a naphtha fraction, obtained by direct crude distillation, to a high-octane-number gasoline. The catalyst used is molybdena–alumina with a mean particle diameter ~60 μm. Superficially this process appears similar to FCC, but there are some important differences. This process is operated at a much higher pressure (about 14 atm); because the pressure drop in catalyst recirculation is small compared with total pressure, control of solid handling is sometimes difficult. Compared with catalytic cracking, reaction rates are slow. Consequently the reactor is considerably larger than the catalyst regenerator, and the resulting low catalyst recirculation rate gives poor thermal efficiency (G3).

3. *Production of Phthalic Anhydride*

Catalytic oxidation of naphthalene to phthalic anhydride is the second application of FCB which was initiated in 1945. Riley (R12) reported problems on afterburning and on the control of bed temperature, and introduced many improvements in the fluidized bed. However, the most important change to make scale-up easier and to achieve good performance was the use of microspherical catalyst, or moderately active German-type catalyst (B11, G14).

This was the first application of a fluidized bed to a process that necessitates removal of large heat of reaction (about 450 kcal/gm-mole naphthalene) and high yield. The fluidized process surpassed fixed-bed processes in safe operation at high concentration, in yield, in reduced pollution, and in plant cost—probably because the reaction was rather simple and the products were stable. Fluid bed catalysts currently in use consist principally of vanadium oxide on a silica gel base, and have a particle-size range of less than 300 μm (G14).

4. Acrilonitrile by the Sohio Process

The Sohio process is considered one of the most successful applications of FCB. Problems in industrial application of the reaction arose from the strong exothermicity of propylene ammoxidation and from the intermediate production of acrylonitrile in the consecutive reactions (V9). It is particularly noticeable that the catalyst gives high selectivity, and the reactor design aims at better fluidization and higher contact efficiency than in the FCC process.

The first plant was put on stream in 1960. Since then, catalysts have been modified, i.e., CAT-A (P · Mo · Bi), CAT-21 (U · Sb), and CAT-41 (P · Mo · Bi · Fe · X) have been used successively; yield of acrylonitrile based on propylene has also improved from about 57%, to 63%, and to 67% due to the modifications of catalyst (A8), all microspherical particles with the properties of typical fluid bed catalysts (S15).

5. Fischer–Tropsch Synthesis

The synthesis of hydrocarbons from H_2 and CO is strongly exothermic. The Hydrocol process mentioned in Section I is a modified Fischer–Tropsch synthesis, using powdered iron catalyst in a dense-phase fluidized bed. In spite of elaborate scale-up based on intensive research, many problems arose in commercial plants, and the operation was terminated in 1957 for economic reasons (Z5).

The major difficulty of this process arose from the physically and chemically unstable catalyst. The iron particles broke down in size very rapidly under the synthesis conditions. The fluffy nature of the product made fluidization difficult, if not impossible. Also, catalyst particles were "waxed up," i.e., coated with carbonaceous materials, and activity decreased significantly (S22).

Another difficulty of this process was low conversion in the commercial plant, which could not be predicted from the results of the pilot plants. Grekel et al. (G15) reported the effects of particle size distribution, gas inlet devices, and internals, on contact efficiency. Volk et al. (V12) also emphasized the effect of bed internals. These developmental studies became a very useful guide for applications of fluidized catalyst beds.

The Kellogg process has been successful for its dilute-phase transfer-line reactor in Sasol's South African plant. The void fraction in the reaction zone is more than 95%, and the superficial gas velocity ranges from 3 to 12 m/sec (G2). Careful consideration is thus necessary in applying FCB to processes where the catalyst can suffer physical and chemical attrition, and is so heavy that good fluidization is hard to attain.

6. *The Two Newest Applications in Japan*

Oxychlorination of ethylene to produce ethylene dichloride was put on stream in 1969 by Mitsui Toatsu Chemicals. Formerly, chlorine and/or alkyl chlorides had been produced in fluidized beds by the Shell process (A7, E12, F4). The previous catalyst had been made by dipping and calcining $CuCl_2$ onto various carriers. The present catalyst is made by spray-drying a gel of mixed $CuCl_2$ and Al_2O_3, followed by calcining, and has a particle size distribution suitable for fluid beds. It is very active, and has a long life and high attrition resistance (M19).

The reasons that fluid beds are applied to this process are: (a) The high heat-transfer capacity is appropriate for this highly exothermic system. (b) The reaction product is stable, and consecutive reactions do not occur. (c) Operational safety can be expected. Scale-up was successful with the use of baffles to prevent channeling and of a gas distributor to obtain high contact efficiency. Several steps of pilot-plant test were examined in developing the process. Performance of the commercial unit revealed the overall yield of 97%, which had been anticipated (M18).

Another example is the production of maleic anhydride by Mitubishi Chemical Industries, Ltd. in 1970, which is oxidation of BB-fraction. Synthesis of maleic anhydride from butylene is a typical consecutive reaction, and it was considered difficult to get high yield by the use of fluid beds. However, it is highly exothermic and application of fluid beds is preferable.

In scaling-up this process, appropriate reactor design was conducted elaborately by the results of several tests. The following two points are characteristic of the scale-up. One is the development of catalyst suitable for fluid beds. Selectivity should not decrease by the increase of hydrocarbon concentration. Also, reaction rate of the secondary reaction should be small. These problems were solved by the development of rather simple P-V catalyst (K2, K3). Another device is the instantaneous mixing inlet mechanism for hydrocarbon and air. By this, activity decline of the catalyst can be prevented (T12). Thus a process has been developed which is equally as profitable as the benzene process.

C. Recent Trends in Fluidized Catalyst Beds

Recently many inventions about fluidized catalyst beds, which are useful in industrial applications, have been reported (I6–8), initiating trends counter to traditional technology since FCC. Some of the important improvements are listed with patent references:

(a) Improvements in use of catalyst:
   i. Effective execution of consecutive reactions by mixed fluidization of catalysts of different activities (German Offen 2603770, Japan 75-30825).
   ii. Accomplishment of good fluidization by means of mixed fluidization of inactive particles of different size and of different specific gravity (Japan 73-31827, 74-27263).

(b) Improvements in feed systems of reactant gases:
   i. Prevention or decrease of decline of catalyst activity by instantaneous mixing of reactant gases (Japan 74-29166, 74-29167, 74-29168).
   ii. Simultaneous manufacture of plural main products by simultaneous feed of different main reactants (Britain 1238347, Japan 76-16615).
   iii. Establishment of uniform fluidization by introducing a reactant into the middle stage of fluidized beds (South Africa 7308033, Britain 1238347).
   iv. Increase of selectivity by means of split feed of reactants (U.S.A. 3546268; Britain 1208191).

(c) Improvements in internals:
   i. Establishment of good fluidization by a modified arrangement of vertical internals (Japan 75-15772, 77-7905).
   ii. Increase of contact efficiency by means of improved arrangement of sieve trays (Japan 74-47725, 74-47726).
   iii. Decrease of backmixing by providing horizontal ridges on the vertical surfaces (Britain 1359377).

(d). Prolonged catalyst activity:
   Providing an autoregeneration zone in fluid beds and improving its effect (Britain 1126617, Japan 76-29114).

(e) Modification to decrease the carry-over of catalyst particles:
   Curtailing carry-over by arranging the internals in the upper part of the dense phase (i.e., the transition region) and in the dilute phase (U.S.A. 3859405).

D. Technical Problems in FCB Design

Marshall (M2) presented the following six factors in the design of downflow fluidized solid reactors.

1. Required reaction volume: This is decided by experimental determination of the necessary space velocity. It will be influenced by the extent

of gas mixing and gas backmixing and by the ratio of length to diameter of the bed.

2. Desired diameter: This is determined by gas velocity, which will be in the range between the minimum fluidization velocity and the velocity determined by excessive entrainment. The most economic velocity will be set by balancing gas circulation cost with other costs.

3. Type of gas distributor: There are several types such as conical bottom, packed bed, and grid.

4. Freeboard above the reactor bed: Determined by the need for dust-removal equipment, and for free settling prior to entry to that equipment.

5. Disengaging-space diameter: The enlarged diameter will lower the gas velocity, aid settling of particles, and reduce carry-out.

6. Equipment required for heat exchange, if any.

Even now, the above considerations are the basis for FCB design. Elaborate devices are installed on the gas distributor and internals for high-performance reactions (Section I).

A panel discussion on fluid particle technology (Philadelphia, 1973) was significant in correctly recognizing the central problems in industrial application of fluidized beds to catalytic reactions (H2). Some of the comments are communicated in the following paragraphs.

Optimum size distribution is important for a fluid bed reactor (Bergougnou). Models based on bubbles are not yet capable of predicting the wall effect (Wen). Vertical baffles are most effective in breaking up large bubbles (Volk). The height of the bottom ends of vertical tube bundles above the grid will set the attainable bubble size at the bottom of the bundle. The bundles then essentially maintain the bubble size (Zenz). Horizontal perforated baffle plates reduce the mean residence time of elutriable fine particles in a fluidized bed (Buckham). Observations on attrition in cyclones indicate that it is an exponential function of velocity (Tenney).

In general, industrial application of FCBs is possible from pilot-plant tests (including cold runs) of appropriate scale, with the 35-year experience of industrial plants and research since the initiation of FCC. However, the following conditions are necessary. First, the catalyst should have such properties and activity as to be suitable for FCB. Second, the heat of reaction must be appropriate for use in FCB, considering heat transfer and the heat capacity of fluidized beds. Third, the reaction products must be stable, and high selectivity must not be required if consecutive reactions are involved. Generally, the catalyst is in an equilibrium state in industrial reactors due to physical and chemical impacts, and artificial controls from outside corresponding to those impacts.

Problems on the optimization of industrial fluidized catalyst beds are still left unsolved. Certainly optimization in a rigorous sense is impossible.

However, approximate treatment is possible. Ikeda and Tashiro (19) report an optimization of catalytic reactions in fluid beds. They find that the maximum yield of the intermediate product decreases, and that the optimum contact time increases for first-order consecutive and parallel reaction systems if contact efficiency in the reactor decreases. They also showed the most economical "equilibrium activity" and the optimal size distribution of catalyst.

Modern technology on fluidized catalyst beds started late in the 1930s with the pioneer work by W. K. Lewis and E. R. Gilliland, and has found a place in some large-scale industrial applications. It will find further applications, as its phenomenological and theoretical background become clearer. Specific flow properties which distinguish the fluidized catalyst beds from the usual teeter beds have been highlighted. The study of bed properties has been advanced by the concept of successive contact mechanisms. However, the science of fluid catalyst beds is still young, and much is still unknown about their physical and chemical behavior. The technology, following the trends reported in this review, has benefited from constantly improving technical knowledge. In particular, better fluidity of the beds has been found to lead to higher reactor efficiency.

Acknowledgements

The authors wish to express their thanks to Professor Y. Kato, Kyushu University, for stimulating discussions on this article, and to Professor T. Vermeulen for his valuable comments and help in editing the manuscript. Thanks are also to Mr. T. Kikuchi of the University of Tokyo for drawing figures, and to Misses K. Sekiguchi and T. Hongo, University of Tokyo, for typing manuscripts.

Nomenclature

A_r Free area of baffles; also, frequency factor in the Arrhenius plot
A_T Cross-sectional area of fluid bed
a Fraction of catalyst in direct contact with gas phase, Eq. (7-35)
a_b Contact area of bubbles per unit volume of bed
a_c Specific surface area of cluster phase
a_{ex} Area for the exchange of material between ascending and descending emulsion
a_w Area of heat transfer per unit volume of bed
B Width of bed
C, C_A, C_B
C_C, C_R Concentration in dilute phase (dimensionless)
C_b Concentration of free gas in bubble phase (dimensionless)
C_e Concentration of free gas in emulsion phase (dimensionless)
c Molar concentration; also concentration in freely flowing gas in dilute phase
c_b Concentration in bubble phase

c_c Concentration in cloud-overlap phase; also concentration in cluster phase (Section VIII)
c_e Concentration in emulsion phase
c_{ed} Concentration in descending emulsion
c_{eu} Concentration in ascending emulsion
c_{fe} Concentration in fluid phase in emulsion
c_{Lf} Concentration at L_f
c^0 Concentration in inlet gas
c_0 Concentration in outlet gas
c_{pe} Apparent heat capacity of emulsion
c_{pf} Heat capacity of fluid
c_{pg} Heat capacity of gas
c_{pl} Heat capacity of liquid
c_{ps} Heat capacity of solid
c_{se} Concentration in solid phase in emulsion
D_e Effective gas diffusivity in particle
D_{eq} Equivalent bed diameter
D_{eff} Effective gas diffusivity in emulsion
D_{eH} Equivalent bed diameter on horizontal surfaces
D_{eV} Equivalent bed diameter on vertical surfaces
D_G Gas diffusivity
D_N Diameter of nozzle
D_T Diameter of bed
d_b Diameter of bubble
$(d_b)_{eff}$ Effective bubble diameter
d_{bi} Initial bubble diameter
d_{bM} Maximum bubble diameter
d_{bms} Maximum stable bubble diameter given by Davidson and Harrison (D3)
d_{bs} Steady bubble diameter in turbulent bubble flow
d_p Diameter of particle
$\bar{d}_b$ Number-averaged mean bubble diameter
d_{32} Volume-surface mean bubble diameter
E Activation energy

E_e Apparent longitudinal dispersion coefficient of gas in emulsion phase based on empty vessel
E_{ef} $\epsilon_{fe}E_{zs}$ in case of fine particles
E_{es} $\epsilon_{se}E_{zs}$ in case of fine particles
E_0 Eötvos number, $d_b^2 g\rho_l/\sigma$ (dimensionless)
E_r Radial diffusivity
$\bar{E}_r$ Cross-sectionally averaged radial diffusivity
E_{sr} Lateral dispersion coefficient of solids based on empty vessel
E_z Axial diffusivity defined by Eq. (6-25)
E_{zr} Axial diffusivity by the Taylor dispersion
E_{zT} Total axial diffusivity Eq. (4-2)
E_{zL} E_{zT} of liquid in bubble column
E_{zs} E_{zT} of emulsion in fluidized bed
$\bar{E}_z$ Cross-sectionally averaged axial diffusivity
e Fraction of catalyst in dilute phase with respect to volume of catalyst in emulsion phase, Eq. (7-25)
e_b Refer to Eq. (3-20) (dimensionless)
F Gas exchange rate between the bubble and emulsion phase based on the volume of the emulsion after Lewis et al. (L12)
$f(c)$ $k_{or}(z)/r(c)$
$F(d_b)$ Distribution function of bubble sizes existing in bed
f_w Ratio of wake volume to the total bubble volume
Gr_c Grashof number, $\bar{\epsilon}_b g D_T^3/\nu_M^2$ (dimensionless)
g Gravitational acceleration
H Height of freeboard
$(H_{ob})_{app}$ Apparent overall height of a transfer unit refer to Eq. (6-3)

H_{oR}	Overall height of a transfer unit for whole bed		k_{or}	$[(k_{ob}a_b)^{-1} + (\xi\bar{\epsilon}_e k_r)^{-1}]^{-1}$
ΔH	Heat of reaction		k_{oR}	refer to Eq. (7-4)
h_b	Bubble-side heat-transfer coefficient		k_{oRc}	refer to Eq. (8-14)
h_{ex}	Heat-exchange coefficient between ascending and descending regions (c.g.s. units)		k_r	Rate constant for first-order irreversible reaction based on volume of emulsion phase

h_{ob} Overall heat-transfer coefficient from bubble to emulsion phase

h_w Wall-to-bed heat-transfer coefficient

J Parameter defined in Eq. (6-15)

K_{oR} Overall reaction rate constant based on settled bed

K_R Equilibrium constant (dimensionless)

k Numerical coefficient (dimensionless); also wave number

k Wave number

k_b Mass-transfer coefficient in bubble phase

k_{bc} Mass-transfer coefficient defined by Eq. (6-19)

k_{ce} Mass-transfer coefficient defined by Eq. (6-20)

k_e Mass-transfer coefficient in emulsion phase

k_{eff} Effective thermal conductivity in emulsion phase

k_{ex} Mass-exchange coefficient between ascending and descending emulsion

k_f Mass transfer coefficient between gas and particle

k_g Thermal conductivity of gas

k_0 Numerical constant, Eq. (4-15) (dimensionless)

k_{ob} Overall mass-transfer coefficient between bubble and emulsion phase

k_{oc} Mass-transfer coefficient to or from cluster phase

k_{oJ} Mass-transfer coefficient between jet and emulsion

$k_{r1}, k_{r2}\ldots$ Reaction rate constant for reactions 1, 2 . . .

k_s Effective thermal diffusivity, $c_{ps}\rho_s E_{zS}\epsilon_{se}$

L Height of mixed phases

L_e Equivalent height of emulsion phase

L_f Height of dense-phase fluid bed

L_m Probe position measured from the distributor

L_{mf} Height of fluid bed at U_{mf}

L_q Height of settled bed

L_t Total bed height

ΔL Distance between stages

l Distance between observation points; also length of bubble signals

$\bar{l}_{10}$ Arithmetical mean length of bubble signals

$M(l)$ Distribution function of bubble signals measured with a point-shaped probe

m Adsorption equilibrium constant for emulsion, equal to $\epsilon_{fe} + m\epsilon_{se}$; also, $c_{ps}\rho_s/c_{pg}\rho_g$

m_H $\sqrt{k_r D_{eff}}/k_e$

m_s Adsorption equilibrium constant for catalyst particles (c_{se}/c_{fe})

N Number of horizontal baffle plates

N_{Ar} $A_r L_f/U_f$

N_b $k_b a_b L_f/U_G$

N_{ex} $mk_{ex}a_{ex}L_f/U_f$

N_{Hw} $h_w a_w L_f/U_f c_{pg}\rho_g$

N_{Hex} $L_f h_{ex}a_{ex}/U_f c_{pg}\rho_g$

N_{ob} $k_{ob}a_b L_f/U_f$

N_{Pe} $L_f U_G c_{pg}\rho_g/k_s$

N_r Radial mass flux in Eq. (4-4)

N_r $k_r L_f/U_f$

N_{rb} $\nu_b k_r L_f/U_f$

N_r	mU_e/U_G
n	Growth factor of disturbances; also constant defined by Eq. (2-13)
n_d	Number of holes of perforated plate
$\bar{P}$	Static pressure
Pe	Peclet number, $m\,\delta^2 \bar{u}_s/4d_b D_{eff}$
$(PeB)_f$	$U_G L_t/E_z$
$(PeB)_s$	$U_G L_t/mE_{zs}\epsilon_e$
ΔP	Pressure drop of bed
Q	Specific converting power; also net transport defined by Eq. (4-7); also dimensionless heat generated or rejected in Figs. 90–91
q_f	Flow rate of gas in gas cloud phase
q_s	Flow rate of solids in gas cloud phase
R	Radius of bed; also gas constant; also column radius
R_{A-C}	Reaction rates
r	Coordinate of radial position
r_*	r at $u = 0$
r_b	Radius of bubble
$r(c)$	reaction rate
r_c	Radius of gas cloud
r_p	Radius of particle
T	Temperature; also radially averaged temperature in dilute phase
T_c	Temperature of coolant
T_d	Temperature in descending zone of dilute phase
T_{de}	Temperature in dense phase
T_u	Temperature in ascending zone of dilute phase
T_w	Temperature at heat-transfer well
$\hat{t}$	Characteristic contact time of pockets
$\bar{U}$	Superficial gas velocity as defined by Eq. (3-26)
U^*	Velocity as defined by Eq. (5-14)
U_e	Superficial circulation velocity of emulsion $\bar{\epsilon}_{eu}c_{eu}$
U_f, U_G	Superficial gas velocity
U_{fe}	Superficial circulation velocity of gas in emulsion phase
U_L	Superficial net liquid velocity
U_{mb}	Minimum bubbling velocity
U_{mf}	Minimum fluidization velocity
U'_s	Superficial intermixing velocity of solids through baffle plate
U_{se}	Superficial circulation velocity of solids in emulsion phase
U_ψ	Tangential velocity along the bubble–emulsion interface
u	Time-averaged interstitial velocity of liquid
$\bar{u}$	Interstitial mean liquid velocity, Eq. (4-15)
u_b	Ascending velocity of bubbles
$\bar{u}_b$	$U_G/\bar{\epsilon}_b$, mean bubble velocity
u_{b0}	Bubble velocity along the column axis
$\bar{u}_{b0}$	Rising velocity of finite-size bubble swarm
u_e	Circulation velocity of emulsion
u_{e0}	Circulation velocity of emulsion along column axis
u_j	Velocity of jet
u_l	Circulation velocity of liquid
u_{l0}	Circulation velocity of liquid along column axis of bed
$\bar{u}_l$	Area-averaged circulation velocity of liquid
$\bar{u}_{lu}$	Mean interstitial velocity of upflow, Eq. (5-9)
u_{mf}	Interstitial velocity of gas in emulsion phase
u_0	u along the column axis
u_s	Slip velocity of bubble relative to liquid
$\bar{u}_s$	Mean slip velocity of bubble
$\bar{u}_{s0}$	Free-rising velocity of single bubble
u_t	Terminal velocity of a particle
$\|u_w\|$	Absolute liquid velocity at column wall

u'_r	Radial velocity fluctuation of liquid	ϵ_{eu}	Volume fraction of ascending emulsion, $\epsilon_{eu} + \epsilon_{ed} = \epsilon_e$
u'_z	Axial velocity fluctuation of liquid	ϵ_f	Volume fraction of gas, $\epsilon_f + \epsilon_s = 1$
u_δ	Velocity at $y = \delta$, Eq. (3-7)	ϵ_{fe}	Volume fraction of gas in emulsion, $\epsilon_{fe} + \epsilon_{se} = 1$
V_b	Volume of a bubble	ϵ_L	Local liquid holdup (dimensionless), $\epsilon_b + \epsilon_L = 1$
v_0	$gD_T^2/192\nu_t$, characteristic velocity	$\bar{\epsilon}_L$	Averaged value of ϵ_L
v^*	Frictional velocity	ϵ_m	Rate of energy dissipation per unit mass
We	Weber number, $d_b \rho_l u_s^2/\sigma$ (dimensionless)	ϵ_{mf}	Volume fraction ϵ_f at U_{mf}
x	Coordinate of position	ϵ_s	Volume fraction of solids, $\epsilon_f + \epsilon_s = 1$
y	$R - r$	ϵ_{se}	Volume fraction of solids in emulsion, $\epsilon_{fe} + \epsilon_{se} = 1$
Z	z/L_f	η	Correction factor defined by Eq. (6-22a)
Z_t	z_t/L_f	η_c	Contact efficiency
z	Coordinate of bed height	θ	Polar coordinate with origin at bubble center, also dimensionless temperature (T/T_{de}), also time
z'	Length of space existing directly under horizontal baffle plate		
z_t	Total bed height		

GREEK LETTERS

α	$\bar{u}_s/u_{mf}$; also numerical coefficient of order unity (dimensionless)	λ	Wavelength, $2\pi/k$; also ratio of superficial flow of solids to that of gas in dilute phase
β	Hatta number; also numerical coefficient of order unity (dimensionless)	λ_{min}	Minimum wavelength, Eq. (5-18)
β_r	Modified Hatta number for the emulsion	λ_0	Minimum eddy size
β_r^*	refer to Eq. (8-16)	μ	Volume ratio of catalyst in bubble phase
Γ	E/RT_{de}	μ_b	Apparent viscosity of fluid bed
δ	Thickness of gas cloud; also thickness of laminar sublayer	μ_g	Viscosity of gas
ϵ	Entrainment rate	μ_L	Viscosity of liquid
ϵ_b	Local gas-bubble holdup	μ_p	Effective Newtonian shear viscosity of particulate fluid
$\bar{\epsilon}_b$	Averaged value of ϵ_b	μ_t	Turbulent viscosity
ϵ_e	Volume fraction of emulsion, $\epsilon_e + \epsilon_b = 1$	ν	Fraction of catalyst particles contacting directly with gas
$\bar{\epsilon}_e$	Averaged value of ϵ_e	ν_b	Fraction of catalyst in direct contact with the bubble gas
ϵ_{ed}	Volume fraction of descending emulsion; also ϵ_e in dilute phase	ν_d	ν in the dilute phase
$\bar{\epsilon}_{ed}$	Averaged ϵ_e in dilute phase; also averaged value of ϵ_{ed}	ν_m	Turbulent diffusivity of mass, $\nu_m = \alpha \nu_t$
$\bar{\epsilon}_{e,de}$	Averaged ϵ_e in dense phase		

ν_M	Molecular kinematic viscosity	φ	Parameter defined by Eq. (7-18)
ν_t	Turbulent kinematic viscosity	χ	Tortuosity
ξ	Defined by Eq. (7-28)	ψ	Angle measured away from the vertical (radians)
ρ_b	Bed density	Ω	$(-\Delta H)c_f^0/c_{pg}\rho_g T_{de}$
ρ_e	Density of emulsion phase in fluid bed	ω	Frequency of wave
ρ_f	Density of fluid		
ρ_g	Density of gas		
ρ_l	Density of liquid		
ρ_p	Density of particles including inner pore volume		
ρ_q	Density of settled bed		
ρ_s	Density of solids		
σ	Interfacial tension		
σ^2	Variance		
τ	Shearing stress		
τ_a	Time available for growth		
τ_{am}	Maximum value for τ_a		
τ_b	Local surface renewal time		
τ_e	Growth time of disturbances		
τ_p	Time constant for particle		
τ_w	Shear stress at column wall		
$\Phi(c)$	Rate of reaction or mass transfer		
ϕ	Radial coordinates r/R (dimensionless)		
ϕ_*	Radial coordinates r_*/R (dimensionless)		

SUBSCRIPTS

b	bubble
d	descending flow
de	dense phase
di	dilute phase
e	emulsion
ex	exchange
f	fluid phase
g	gas phase
H	horizontal, also heat
L, l	liquid
M	molecular
o	overall
p	particle
r	radial
s	solid, or slip
t	turbulent
u	ascending flow
V	vertical
w	wall
z	axial

References

A1. Ahlborn, F., *Z. Tech. Phys.* **12**, 482 (1931).
A2. Akehata, T., Shirai, T., Sugita, M., Yoshino, K., and Shirai, K., *Hokkaido Meet. Soc. Chem. Eng. Jpn. Preprint* No. B8 (1968).
A3. Akita, K., and Yoshida, F., *Ind. Eng. Chem. Process Des. Dev.* **12**, 76 (1973).
A4. Akita, K., and Yoshida, F., *Ind. Eng. Chem. Process Des. Dev.* **13**, 84 (1974).
A5. Alexander, B. F., and Shah, Y. T., *Chem. Eng. J.* **11**, 153 (1976).
A6. Anderson, T. T., *AIChE J.* **10**, 776 (1964).
A7. Anonymous, *Chem. Eng.*, Oct., 392 (1953).
A8. Anonymous, *Hydrocarbon Process.* **46**, 141 (1967); **48**, 146 (1969); **52**, 99 (1973).
A9. Aoyama, Y., Ogushi, K., Koide, K., and Kubota, H., *J. Chem. Eng. Jpn.* **1**, 158 (1968).
A10. Argo, W. B., and Cova, D. R., *Ind. Eng. Chem. Process Des. Dev.* **4**, 352 (1965).
A11. Aris, R., *Proc. R. Soc.* **A235**, 67 (1956).
A12. Askins, J. W., Hinds, G. P., Jr., and Kunreuther, F., *Chem. Eng. Progr.* **47**, 401 (1951).

B1. Baeyens, J., and Geldart, D., "Fluidization and Its Applications," p. 263. Toulouse, France, 1973.
B2. Bailie, R. C., Chung, D. S., and Fan, L. T., *Ind. Eng. Chem. Fundam.* **2**, 245 (1963).
B3. Baird, M. H. I., and Rice, R. G., *Chem. Eng. J.* **9**, 171 (1975).
B4. Bakker, P. J., and Heertjes, P. M., *Chem. Eng. Sci.* **12**, 260 (1960).
B5. Bartholomew, R. N., and Casagrande, R. M., *Ind. Eng. Chem.* **49**, 428 (1957).
B6. Basov, V. A., Makhevka, V. I., Malik-Akhnozarov, T. Kh., and Orochko, D. I., *Int. Chem. Eng.* **2**, 263 (1962).
B7. Bedura, R., Deckwer, W. D., Warnecke, H. J., and Langemann, H., *Chem. Ing. Tech.* **46**, 399 (1974).
B8. Behie, L. A., Doctoral dissertation, Western Ontario University, 1972.
B9. Behie, L. A., and Kehoe, P., *AIChE J.* **19**, 1070 (1973).
B10. Bellman, R., and Pennington, R. H., *Q. Appl. Math.* **12**, 151 (1954).
B11. Betts, W. D., *Ind. Chem.* July, 370 (1963).
B12. Botterill, J. S. M., "Fluid-Bed Heat Transfer." Academic Press, New York, 1975.
B13. Botton, R. J., *Chem. Eng. Prog. Symp. Ser.* **66** (101), 8 (1970).
B14. Bridge, A. G., Lapidus, L., and Elgin, J. C., *AIChE J.* **10**, 819 (1964).
B15. Broadhurst, T. E., and Becker, H. A., *AIChE J.* **21**, 238 (1975).
B16. Brötz, W., *Chem. Ing. Tech.* **28**, 165 (1956).
B17. Burgess, J. M., and Calderbank, P. H., *Chem. Eng. Sci.* **30**, 1511 (1975).
B18. Burkur, D., and Amundson, N. R., *Chem. Eng. Sci.* **30**, 847, 1159 (1975).
B19. Burkur, D., Wittmann, C. V., and Amundson, N. R., *Chem. Eng. Sci.* **29**, 173 (1974).
B20. Bart, R., Sc.D. Thesis, Mass. Inst. of Tech, Cambridge, Mass., 1950, cited by G. H. Reman (R4).
B21. Bohle, W., and Swaay. W. P. M., "Fluidization," Proc. 2nd. Eng. Foundation Conf. 1978, p. 167, Cambridge Univ. Press, London and New York (1978).
B22. Bukur, D. B., *Ind. Eng. Chem. Fundam.* **17**, 120 (1978).
C1. Callahan, J. L., Grasselli, R. K., Milberger, E. C., and Strecker, H. A., *Ind. Eng. Chem. Prod. Res. Dev.* **9**, 134 (1970).
C2. Carlsmith, L. E., and Johnson, F. B., *Ind. Eng. Chem.* **37**, 451 (1945).
C3. Chiba, T., and Kobayashi, H., *Chem. Eng. Sci.* **25**, 1375 (1970).
C4. Chiba, T., Terashima, K., and Kobayashi, H., *Chem. Eng. Sci.* **27**, 965 (1972).
C5. Chiba, T., Terashima, K., and Kobayashi, H., *J. Chem. Eng. Jpn.* **6**, 78 (1973).
C6. Clift, R., Grace, J. R., and Weber, M. E., *Ind. Eng. Chem. Fundam.* **13**, 45 (1974).
C7. Clift, R., and Grace, J. R., *Chem. Eng. Prog. Symp. Ser. 116* **67**, 23 (1971).
C7a. Chavarie, C., and Grace, J. R., *Ind. Eng. Chem. Fundam.* **14**, 75, 79, 86 (1976).
C7b. Chavarie, C., and Grace, J. R., *Chem. Eng. Sci.* **31**, 741 (1976).
C8. Cleland, F. A., and Wilhelm, R. H., *AIChE J.* **2**, 489 (1956).
C9. Calderbank, P. H., and Toor, F. D., *in* "Fluidization" (J. F. Davidson and D. Harrison, eds.), p. 383. Academic Press, New York, 1971.
D1. Danckwerts, P. V., *Trans. Faraday Soc.* **46**, 300 (1952).
D2. Danckwerts, P. V., *Chem. Eng. Sci.* **2**, 1 (1953).
D3. Davidson, J. F., and Harrison, D., "Fluidized Particles." Cambridge Univ. Press, London and New York, 1963.
D4. Davidson, J. F., and Harrison, D., *Chem. Eng. Sci.* **21**, 731 (1966).
D5. Davidson, J. F., and Harrison, D. (eds.), "Fluidization." Academic Press, New York, 1971.
D6. Davies, L., and Richardson, J. F., *Trans. Inst. Chem. Eng.* **44**, T293 (1966).
D7. de Groot, J. H., *Proc. Int. Symp. Fluidization, Eindhoven*, p. 348.

D8. de Vries, R. J., van Swaaij, W. P. M., Mantovani, C., and Heijkoop, A., *Proc. 5th Eur., 2nd Int. Symp. Chem. React. Eng.* B9-59 (1972).
D9. Deckwer, W. D., Burckhart, R., and Soll, G., *Chem. Eng. Sci.* **29**, 2177 (1974).
D10. DeMaria, F., and Longfield, J. E., *Chem. Eng. Prog. Symp. Ser. 38* **58**, 16 (1962).
D11. Denbigh, K. G., *Chem. Eng. Sci.* **8**, 125 (1958).
D12. Denbigh, K. G., "Chemical Reactor Theory." Cambridge Univ. Press, London and New York, 1965.
D13. Dimotakis, P. E., and Brown, G. L., *J. Fluid Mech.* **78**, 535 (1976).
D14. Deckwer, W., Graeser, U., Langemann, H., and Serpemen, Y., *Chem. Eng. Sci.* **28**, 1223 (1973).
D14a. Donsi, G., and Massimilla, L., "Fluidization and Its Applications," p. 41. Capadues, Toulouse, 1973.
D15. Donsi, G., and Massimilla, L., *AIChE J.* **19**, 1104 (1973).
D16. Drinkenburg, A. A. H., *Proc. Int. Symp. Fluidization, Eindhoven*, p. 468 (1967).
D17. Drinkenburg, A. A. H., and Rietema, K., *Chem. Eng. Sci.* **27**, 1765 (1972).
D18. Drinkenburg, A. A. H., and Rietema, K., *Chem. Eng. Sci.* **28**, 259 (1973).
D19. Darton, R. C., *Trans. Inst. Chem. Eng.* **57**, 134 (1979).
D20. De Lasa, H. I. and Grace, J. R., *AIChE J.* **25**, 984 (1979).
E1. Eberly, P. E., *J. Phys. Chem.* **65**, 68 (1961).
E2. Eberly, P. E., *J. Phys. Chem.* **66**, 812 (1962).
E3. Eberly, P. E., and Kimberlin, C. N., *Trans. Faraday Soc.* **57**, 1169 (1961).
E4. Echigoya, E., Toyoda, K., and Morikawa, K., *Kagaku Kogaku* **32**, 364 (1968).
E5. Echigoya, E., Iwasaki, M., Kanemoto, T., and Niiyama, H., *Kagaku Kogaku* **32**, 571 (1968).
E6. Eissa, S. H., El-Halwagi, M. M., and Saleh, M. A., *Ind. Eng. Chem. Process Des. Dev.* **10**, 31 (1971).
E7. Eissa, S. H., and Schügerl, K., *Chem. Eng. Sci.* **30**, 1251 (1975).
E8. El Halwagi, M. M., and Gomezplata, A., *AIChE J.* **13**, 503 (1967).
E9. El Nashaie, S., *Chem. Eng. Sci.* **32**, 295 (1977).
E10. El Nashaie, S., and Yates, J. G., *Chem. Eng. Sci.* **28**, 515 (1973).
E11. Endoh, I., Furusawa, T., and Matsuyama, H., *Kagaku Kogaku* **41**, 232 (1977).
E12. Engel, W. F., and Waale, M. J., *Chem. Ind.* Jan. 76 (1962).
F1. Fair, J. R., Lambright, A. J., and Anderson, J. W., *Ind. Eng. Chem. Process Des. Dev.* **1**, 33 (1962).
F2. Fan, L. T., Lee, C. J., and Bailie, R. C., *AIChE J.* **8**, 239 (1962).
F3. Finnerty, R. G., Maa, J. R., Vossler, A. M., Yeh, H. S., Crouse, W. W., and Rice, W. J., *Ind. Eng. Chem. Fundam.* **8**, 271 (1969).
F4. Fleurke, K. H., *Chem. Eng.* March, CE41 (1968).
F5. Fournol, A. B., Bergougnou, M. A., and Baker, C. G. J., *Can. J. Chem. Eng.* **51**, 401 (1973).
F6. Frantz, J. F., *Chem. Eng.* **69**, Sept., 161 (1962).
F7. Freedman, W., and J. F. Davidson, *Trans. Inst. Chem. Eng.* **47**, T251 (1969).
F8. Froment, G. F., *Adv. Chem. Ser. 109* p. 1 (1972).
F9. Fryer, C., Ph.D. thesis, Monash University, Australia, 1974.
F10. Fryer, C., and Potter, O. E., *Ind. Eng. Chem. Fundam.* **11**, 338 (1972).
F11. Fryer, C., and Potter, O. E., "Fluidization and Its Applications," p. 440. Capadues, Toulouse, 1973.
F12. Fryer, C., and O. E. Potter, *AIChE J.* **22**, 38 (1976).
F13. Furusaki, S., *Kagaku Kogaku* **32**, 1033 (1968).

F14. Furusaki, S., *AIChE J.* **19**, 1009 (1973).
F15. Furusaki, S., Dr. Eng. dissertation, University of Tokyo, 1976.
F16. Furusaki, S., and Miyauchi, T., *Annu. Meet. Soc. Chem. Eng. Jpn. 42nd,* Preprint (1977).
F17. Furusaki, S., Kikuchi, T., and Miyauchi, T., *Autumn Meet., 9th, Soc. Chem. Eng. Jpn.* Preprint (1975).
F18. Furusaki, S., Kikuchi, T., and Miyauchi, T., *AIChE J.* **22**, 354 (1976).
F19. Furusawa, T., Endoh, I., and Matsuyama, H., *Kagaku Kogaku* **33**, 949 (1969).
F20. Furukawa, J., and Ohmae, T., *Ind. Eng. Chem.* **50**, 821 (1958).
G1. Gabor, J. D., *AIChE J.* **10**, 345 (1964).
G2. Garrett, L. W., *Chem. Eng. Prog.* **56**, (4), 39 (1960).
G3. Geldart, D., *Chem. Ind.* 1474 (1967).
G4. Geldart, D., *Powder Technol.* **6**, 201 (1972).
G5. Geldart, D., *Powder Technol.* **7**, 285 (1973).
G6. Gill, W. N., and Sankarasubramanian, R., *Proc. R. Soc. (London) Ser. A* **316**, 314 (1970).
G7. Gilliland, E. R., and Knudsen, C. W., *Chem. Eng. Prog. Symp. Ser. 116* **67**, 168 (1971).
G8. Gilliland, E. R., and Mason, E. A., *Ind. Eng. Chem.* **41**, 1191 (1949).
G9. Gilliland, E. R., and Mason, E. A., *Ind. Eng. Chem.* **44**, 218 (1952).
G9a. Gilliland, E. R., Mason, E. A., and Oliver, R. C., *Ind. Eng. Chem.* **45**, 1177 (1953).
G9b. Gilliland, E. R., *N. J. Section Meet.* AIChE May (1953); cited by W. F. Pansing (P1).
G10. Gondo, S., Tanaka, S., Kazikuri, K., and Kusunoki, K., *Chem. Eng. Sci.* **28**, 1437 (1973).
G11. Gordon, A. L., and Amundson, N. R., *Chem. Eng. Sci.* **31**, 1163 (1976).
G12. Goto, Y., Okamoto, T., and Terahata, T., *Annu. Meet. Soc. Chem. Eng. Jpn. 34th,* April (1969).
G13. Grace, J. R., *Can. J. Chem. Eng.* **48**, 30 (1970).
G14. Graham, J. J., and Way, P. F., *Chem. Eng. Prog.* **58**, (1), 96 (1962).
G15. Grekel, H., Hujsak, K. L., and Mungen, R., *Chem. Eng. Prog.* **60**, (1), 56 (1964).
G16. Guigon, P., Bergougnou, M. A., and Baker, C. G. J., *AIChE Symp. Ser. 141* **70**, 63 (1974).
G17. Gunn, D. J., *Chem. Eng. (London)* CE 153 (1968).
G18. Gometzplata, A., and Shuster, W. W., *AIChE J.* **6**, 454 (1960).
G19. Griffith P., and Wallis, G. B., *Trans. Am. Soc. Mech. Eng.* **83C**, 307 (1961).
G20. Grace, J. R., and De Lasa, H. I., *AIChE J.* **24**, 364 (1978).
G21. George, S. E., and Grace, J. R., *AIChE Symp. Ser. No. 176,* **74**, 67 (1978).
H1. Hall, C. C., and Taylor, A. H., *J. Inst. Petrol.* **41** (376), 102 (1955).
H2. Halow, J. S., *AIChE Symp. Ser. 141* **70** (141), 1 (1974).
H3. Hanada, K., B. Eng. thesis, Dept. of Applied Chem., Kyushu University, 1975.
H4. Harberman, W. L., and R. K. Morton, *Trans. Am. Soc. Civil Eng.* **121**, 227 (1956).
H5. Harmathy, T. Z., *AIChE J.* **6**, 281 (1960).
H6. Harrison, D., and Leung, L. S., *Trans. Inst. Chem. Eng.* **39**, 409 (1961).
H7. Henriksen, H. K., and Østergaard, K., *Chem. Eng. Sci.* **29**, 626 (1974).
H8. Hetzler, R., and Williams, M. C., *Ind. Eng. Chem. Fundam.* **8**, 668 (1969).
H9. Hills, J. H., *Trans. Inst. Chem. Eng.* **52**, 1 (1974).
H10. Hills, J. H., and Darton, R. C., *Trans. Inst. Chem. Eng.* **54**, 258 (1976).
H11. Hikita, H., and Kikukawa, H., *Chem. Eng. J.* **8**, 191 (1974).
H12. Hinze, J. O., *AIChE J.* **1**, 289 (1955).
H13. Hiraki, I., and Kunii, D., *Kagaku Kogaku* **33**, 680 (1969).

H14. Hiraki, I., Yoshida, K., and Kunii, D., *Kagaku Kogaku* **29**, 846 (1965).
H15. Hirama, T., Ishida, M., and Shirai, T., *Kagaku Kogaku Ronbunshu* **1**, 272 (1975).
H16. Hovmand, S., and Davidson, J. F., *Trans. Inst. Chem. Eng.* **46**, T190 (1968).
H17. Hovmand, S., and Davidson, J. F., "Fluidization" (J. F. Davidson and D. Harrison, eds.), p. 193. Academic Press, 1971.
H18. Hurt, D. M., *Ind. Eng. Chem.* **35**, 522 (1943).
H19. Handlos, A. E., Kunstman, R. S., and Schissler, D. O., *Ind. Eng. Chem.* **49**, 25 (1957).
H20. Hagyard, T., and Sacerdote, A. M., *Ind. Eng. Chem. Fundam.* **5**, 500 (1966).
H21. Horio, M., and Wen, C. Y., *AIChE Symp. Ser.* No. 161, **73**, 9 (1977).
H22. Hoebink, J. H. B. J., and Rietema, K., "Fluidization," Proc. 2nd. Eng. Foundation Conf. 1978, p. 327, Cambridge Univ. Press, London and New York (1978).
I1. Ikeda, Y., *Kagaku Kogaku* **27**, 667 (1963).
I2. Ikeda, Y., *Kagaku Kogaku* **29**, 57 (1965).
I3. Ikeda, Y., *Kagaku Kojo* **9**, (12), 63 (1965).
I4. Ikeda, Y., *Kagaku Kikai Gijutsu* **18**, 191 (1965).
I5. Ikeda, Y., *Kagaku Kogaku* **34**, 1013 (1970).
I6. Ikeda, Y., *Kemikaru Enjiniyaringu* 973 (1977).
I7. Ikeda, Y., *J. Jpn Petrol. Inst.* **20**, 619 (1977).
I8. Ikeda, Y., and Tashiro, S., *Kagaku Kogaku* **29**, 956 (1965).
I9. Ikeda, Y., and Tashiro, S., "Saikin no Kagaku Kogaku," p. 195. Maruzen, Tokyo, 1967.
I10. Ikeda, Y., Tashiro, S., Yamaguchi, T., and Kawai, M., *Annu. Meet. Soc. Chem. Eng. Jpn. 35th,* Preprint D202 (1970).
I11. Ishii, T., and Osberg, G. L., *AIChE J.* **11**, 279 (1965).
I12. Ivanov, M. E., and Bykov, V. P., *Theor. Found. Chem. Eng.* **4**, 119 (1970).
I13. Iwasaki, M., Furuoya, I., Sueyoshi, H., Shirasaki, H., and Echigoya, E., *Kagaku Kogaku* **29**, 892 (1965).
J1. Jackson, R., *Trans. Inst. Chem. Eng.* **41**, 21 (1963).
J2. Johnstone, H. F., Batchelor, J. D., and Shen, C. Y., *AIChE J.* **1**, 318 (1955).
K1. Kaji, H., M.S. thesis, Dept. of Chem. Eng., University of Tokyo, 1966.
K2. Kamimura, S., *J. Jpn. Petrol. Inst.* **20**, 627 (1977).
K3. Kamimura, S., *Kagaku Kogaku* **33**, 1082 (1973).
K4. Kato, K., and Wen, C. Y., *Chem. Eng. Sci.* **24**, 1351 (1969).
K5. Kato, Y., *Kagaku Kogaku* **27**, 366 (1963).
K6. Kato, Y., and Nishiwaki, A., *Kagaku Kogaku* **35**, 912 (1971), *Int. Chem. Eng.* **5**, 182 (1972).
K7. Kato, Y., Nishinaka, M., and Morooka, S., *Kagaku Kogaku Ronbunshu* **1**, 530 (1975).
K8. Kehoe, P. W. K., and Davidson, J. F., *AIChE Symp. Ser. 128* **69**, 34 (1973).
K9. Kehoe, P. W. K., and Davidson, J. F., *AIChE Symp. Ser. 128* **69**, 41 (1973).
K10. Kimura, T., M.S. thesis, Dept. of Chem. Eng., University of Tokyo, March, 1972.
K11. Klinkenberg, A., Jr., *Trans. Inst. Chem. Eng.* **43**, T141 (1965).
K12. Kobayashi, H., "Saikin no Kagaku Kogaku," p. 17. Soc. Chem. Eng., Japan, 1966.
K13. Kobayashi, H., Arai, F., and Chiba, T., *Kagaku Kogaku* **29**, 858 (1965).
K14. Kobayashi, H., and Arai, F., *Kagaku Kogaku* **29**, 885 (1965).
K15. Koide, K., Morooka, S., Ueyama, K., Matsuura, A., Yamashita, F., Iwamoto, S., Kato, Y., Inoue, H., Suzuki, S., and Akehata, T., *J. Chem. Eng. Jpn.* **12**, 98 (1979).
K16. Kojima, E., Akehata, T., and Shirai, T., *J. Chem. Eng. Jpn.* **8**, 108 (1975).
K17. Kölbel, H., Beinhauer, R., and Langemann, H., *Chem. Ing. Tech.* **44**, 697 (1972).

K18. Kölbel, H., Borchers, E., and Muller, K., *Chem. Ing. Tech.* **30**, 729 (1958).
K19. Kölbel, H., Borcher, E., and Martins, J., *Chem. Ing. Tech.* **32**, 84 (1960).
K20. Kumar, A., Degaleesan, T. E., Laddha, G. S., and Hoelscher, H. E., *Can. J. Chem. Eng.* **54**, 503 (1976).
K21. Kumar, R., and Kuloor, N. R., "Advances in Chemical Engineering," Vol. 8, p. 256. Academic Press, New York, 1970.
K22. Kunii, D., and Levenspiel, O., *Ind. Eng. Chem. Fundam.* **7**, 446 (1968).
K23. Kunii, D., and Levenspiel, O., *Ind. Eng. Chem. Process Des. Dev.* **7**, 481 (1968).
K24. Kunii, D., and Levenspiel, O., "Fluidization Engineering." Wiley, New York, 1969.
K25. Kunii, D., and Levenspiel, O., *J. Chem. Eng. Jpn.* **2**, 84 (1969).
K26. Kunii, D., Akiyama, T., and Takagi, K., "Shin Kagaku Kogaku Koza," IV-5, Fluidization. Nikkan-Kogyo-Shimbun, Tokyo, 1957.
K27. Kunii, D., Yoshida, Y., and Hiraki, I., *Proc. Int. Symp. Fluidization, Eindhoven*, p. 243 (1967).
K28. Kono, T., B. Eng. thesis, Dept. of Appl. Chem., Kyushu University, 1976.
K29. Khan, A. R., Richardson, J. F., and Shakiri, K. J., "Fluidization," Proc. 2nd. Eng. Foundation Conf. 1978, p. 351, Cambridge Univ. Press, London and New York (1978).
K30. Kikuchi, T., private communication (1977).
L1. LaNauze, R. D., *Powder Technol.* **15**, 117 (1976).
L2. Lanneau, K. P., *Trans. Inst. Chem. Eng.* **38**, 125 (1960).
L3. Larroux, G. J., Kim, Y. G., and Bankoff, S. G., *Chem. Eng. Sci.* **27**, 447 (1972).
L4. Latham, R. L., and Potter, O. E., *Chem. Eng. J.* **1**, 152 (1970).
L5. Latham, R. L., Hamilton, C., and Potter, O. E., *Br. Chem. Eng.* **13**, 666 (1968).
L6. Leva, M., *Chem. Eng. Prog.* **47**, 39 (1951).
L7. Leva, M., "Fluidization." McGraw-Hill, New York, 1959.
L8. Leva, M., and Grummer, M., *Chem. Eng. Prog.* **48**, (6), 307 (1952).
L9. Levenspiel, O., and Bischoff, K. B., "Advances in Chemical Engineering," Vol. 4, p. 95. Academic Press, New York, 1963.
L10. Levenspiel, O., and Walton, J. S., *Chem. Eng. Prog. Symp. Ser.* **50**, 1 (1954).
L11. Lewis, W. K., Gilliland, E. R., and Girouard, H., *Chem. Eng. Prog. Symp. Ser. 38* **58**, 87 (1962).
L12. Lewis, W. K., Gilliland, E. R., and Glass, W., *AIChE J.* **5**, 419 (1959).
L13. Lewis, W. K., Gilliland, E. R., and Lang, P. M., *Chem. Eng. Prog. Symp. Ser. 38* **58**, 65 (1962).
M1. MIT, A Bibliography of Fluidization Research at the Massachusetts Institute of Technology, 1939-1955.
M2. Marshall, S., *Chem. Eng.* May, 219 (1953).
M3. Massimilla, L., *AIChE Symp. Ser. 128* **69**, 11 (1973).
M4. Massimilla, L., and Johnstone, H. F., *Chem. Eng. Sci.* **16**, 105 (1961).
M5. Massimilla, L., and Westwater, J. W., *AIChE J.* **6**, 137 (1960).
M6. Massimilla, L., Donsi, G., and Zucchini, C., *Chem. Eng. Sci.* **27**, 2005 (1972).
M7. Matheson, G. L., Herbst, W. A., and Holt, P. H., *Ind. Eng. Chem.* **41**, 1099 (1949).
M8. Mathis, J. F., and Watson, C. C., *AIChE J.* **2**, 518 (1956).
M9. Matsen, J. M., *Chem. Eng. Prog. Symp. Ser. 101* **66**, 47 (1970).
M10. Matsen, J. M., *AIChE Symp. Ser. 128* **69**, 30 (1973).
M11. Matsen, J. M., and Tarmy, B. L., *Chem. Eng. Prog. Symp. Ser. 101* **66**, 1 (1970).
M12. May, W. G., *Chem. Eng. Prog.* **55**, (12), 49 (1959).
M13. Merrick, D., and Highley, J., *AIChE Symp. Ser. 137* **70**, 336 (1974).
M14. Mickley, H. S., and Fairbanks, D. F., *AIChE J.* **1**, 374 (1955).

M15. Mickley, H. S., and Trilling, C. A., *Ind. Eng. Chem.* **41,** 1135 (1949).
M16. Mickley, H. S., Fairbanks, D. F., and Hawthorn, R. D., *Chem. Eng. Prog. Symp. Ser.* **32 57,** 51 (1961).
M17. Miwa, K., Mori, S., Kato, T., and Muchi, I., *Kagaku Kogaku* **35,** 770 (1971); *Int. Chem. Eng.* **12,** 181 (1972).
M18. Miyauchi, Takeshi, and Oyamada, N., *Jpn. Petrol. Inst.* **20,** 624 (1977).
M19. Miyauchi, T., Sato, Y., Michiki, H., and Fujimoto, K., Japan Patent 45-39616, 1970.
M20. Miyauchi, Terukatsu, "Ryukeisosa-to-Kongotokusei." Nikkan-Kogyo-Sinbun, Tokyo, 1960.
M21. Miyauchi, T., *Annu. Meet. Soc. Chem. Eng. Jpn. 30th,* Preprint No. 6110 (1965).
M22. Miyauchi, T., *Annu. Meet. Soc. Chem. Eng. Jpn. 31st,* Preprint No. 244 (1966).
M23. Miyauchi, T., *J. Chem. Eng. Jpn* **4,** 238 (1971).
M24. Miyauchi, T., *CHISA Congr., 4th, Prague* Sept. (1972).
M25. Miyauchi, T., *J. Chem. Eng. Jpn.* **7,** 201 (1974).
M26. Miyauchi, T., *J. Chem. Eng. Jpn.* **7,** 207 (1974).
M27. Miyauchi, T., *Fall Meet. Soc. Chem. Eng. Jpn. 11th,* Invited Lecture, Oct. (1977).
M27a. Miyauchi, T., and Kikuchi, T., *Annu. Meet. Soc. Chem. Eng. Jpn., 42nd, Hiroshima* April (1977).
M28. Miyauchi, T., and Furusaki, S., *AIChE J.* **20,** 1087 (1974).
M29. Miyauchi, T., and Morooka, S., *Kagaku Kogaku* **33,** 369 (1969); *Int. Chem. Eng.* **9,** 713 (1969).
M30. Miyauchi, T., and Morooka, S., *Kagaku Kogaku* **33,** 880 (1969).
M31. Miyauchi, T., and Shyu, C. N., *Kagaku Kogaku* **34,** 958 (1970).
M32. Miyauchi, T., and Yamada, K., *Kagaku Kogaku* **35,** 547 (1971).
M33. Miyauchi, T., Kaji, H., and Saito, K., *J. Chem. Eng. Jpn.* **1,** 72 (1968).
M34. Mori, S., and Muchi, I., *Kagaku Kogaku* **34,** 510 (1970).
M35. Mori, S., and Muchi, I., *J. Chem. Eng. Jpn.* **5,** 251 (1972).
M36. Mori, S., and Wen, C. Y., *AIChE J.* **21,** 109 (1975).
M37. Mori, Y., and Nakamura, K., *Kagaku Kogaku* **29,** 868 (1965).
M38. Morooka, S., Dr. Eng. dissertation, University of Tokyo, 1969.
M39. Morooka, S., and Miyauchi, T., *Kagaku Kogaku* **33,** 569 (1969).
M40. Morooka, S., Kato, Y., and Miyauchi, T., *J. Chem. Eng. Jpn.* **5,** 161 (1972).
M41. Morooka, S., Nishinaka, M., and Kato, Y., *Kagaku Kogaku* **37,** 485 (1973).
M42. Morooka, S., Nishinaka, M., and Kato, Y., *Kagaku Kogaku Ronbunshu* **2,** 71 (1976); *Int. Chem. Eng.* **17,** 254 (1977).
M43. Morooka, S., Tajima, K., and Miyauchi, T., *Kagaku Kogaku* **35,** 680 (1971); *Int. Chem. Eng.* **12,** 168 (1972).
M44. Morse, R. D., and Ballow, C. O., *Chem. Eng. Prog.* **47,** (9), 199 (1951).
M45. Murphree, E. V., Brown, C. L., Fischer, H. G. M., Gohr, E. J., and Sweeney, W. J., *Ind. Eng. Chem.* **35,** 768 (1943); *Chem. Tech.* **6,** 523 (1976).
M46. Murray, J. D., *J. Fluid Mech.* **21,** 465 (1965).
M47. Murray, J. D., *J. Fluid Mech.* **22,** 57 (1965).
M48. Muchi, I., *Mem. Fac. Eng. Nagoya Univ.* **17,** (1), 79 (1965).
M49. Muchi, I., Mamuro, T., and Sasaki, K., *Kagaku Kogaku* **25,** 757 (1961).
M50. Morooka, S., Maruyama, Y., Kawazuishi, K., Higashi, S., and Kato, Y., *Kagaku Kogaku Ronbunshu* **5,** 275 (1979).
M51. Morooka, S., Kawazuishi, K. and Kato, Y., *Powder Technol.* **26,** 75 (1980).
N1. Nguyen, H. V., and Potter, O. E., *Adv. Chem. Ser.* **133,** 290 (1974).
N2. Nguyen, H. V., and Potter, O. E., "Fluidization Technology," Vol. 2, p. 193. Hemisphere, Washington, D.C., 1976.

N3. Nguyen, X. T., Bergougnou, M. A., and Baker, C. G. J., *Can. J. Chem. Eng.* **51**, 573 (1973).
N4. Nicklin, D. J., *Chem. Eng. Sci.* **17**, 693 (1962).
N5. Nicklin, D. J., Wilkes, J. O., and Davidson, J. F., *Trans. Inst. Chem. Eng.* **40**, 61 (1962).
N6. Nishinaka, M., Morooka, S., and Kato, Y., *Powder Technol.* **9**, 1 (1974).
N7. Nishinaka, M., Morooka, S., and Kato, Y., *Kagaku Kogaku Ronbunshu* **1**, 81 (1975).
N8. Nishinaka, M., Morooka, S., and Kato, Y., *Kagaku Kogaku Ronbunshu* **1**, 605 (1975).
N9. Nishinaka, M., Morooka, S., and Kato, Y., *Kagaku Kogaku Ronbunshu* **2**, 96 (1976).
N10. Nguyen, H. V., Whitehead, A. B., and Potter, O. E., "Fluidization," Proc. 2nd. Eng. Foundation Conf. 1978, p. 140, Cambridge Univ. Press, London and New York (1978).
O1. Ogura, Y., "Taiki-Ranryu-Ron" (Atmospheric Turbulence), 5th. ed., pp. 2–7, 68. Chijin-Shokan, Tokyo, 1959.
O2. Ohki, Y., and Inoue, H., *Chem. Eng. Sci.* **25**, 1 (1970).
O3. Oki, K., and Shirai, T., "Fluidization Technology," Vol. 1, p. 95. Hemisphere, Washington, D.C., 1976.
O4. Olin, H. L., and Dean, O. C., *Petrol. Eng.* March, C-23 (1953).
O5. Oltrogge, R. D., Preliminary outline of thesis, Technical University of Eindhoven and University of Michigan, 1971.
O6. Orcutt, J. C., Davidson, J. F., and Pigford, R. L., *Chem. Eng. Prog. Symp. Ser. 38* **58**, 1 (1962).
O7. Ormiston, R. M., Mitchell, F. R. G., and Davidson, J. F., *Trans. Inst. Chem. Eng.* **43**, T209 (1965).
O8. Osberg, G. L., and Charlesworth, D. H., *Chem. Eng. Prog.* **47**, 566 (1951).
O9. Østergaard, K., "Advances in Chemical Engineering," Vol. 7, p. 71. Academic Press, New York, 1968.
O10. Othmer, D. F., "Fluidization," Reinhold, New York, 1956.
O11. Overcashier, R. H., Todd, D. B., and Olney, R. B., *AIChE J.* **5**, 54 (1959).
O12. Ogasawara, S., Kihara, M., Nishiyama, M., Shirai, T., and Morikawa, K., *Kagaku Kogaku* **28**, 59 (1964).
P1. Pansing, W. F., *AIChE J.* **2**, 71 (1956).
P2. Partridge, B. A., and Rowe, P. N., *Trans. Inst. Chem. Eng.* **44**, T335 (1966).
P3. Pavlov, V. P., *Khim. Prom.* (9), 698 (1965).
P4. Perkins, T. K., and Johnston, O. C., *Soc. Petrol. Eng. J.* 70 (March 1963).
P5. Potter, O. E., and Thiel, W. "Fluidization Technology," Vol. 2, p. 185. Hemisphere, Washington, D.C., 1976.
P6. Pozin, L. S., Aérov, M. E., and Bystrova, T. A., *Theor. Found. Chem. Eng.* **3**, 714 (1969), (English trans.).
P7. Prandtl, L., *ZAMM* **22**, 241 (1942); see H. Schlichting, "Boundary Layer Theory," Chap. 19, 4th English ed. McGraw-Hill, New York, 1960.
P8. Pyle, D. L., and Harrison, D., *Chem. Eng. Sci.* **22**, 1199 (1967).
P9. Pyle, D. L., and Rose, P. L., *Chem. Eng. Sci.* **20**, 25 (1965).
P10. Pyle, D. L., and Stewart, P. S. B., *Chem. Eng. Sci.* **19**, 842 (1964).
P11. Pyle, D. L., and Harrison, D., *Chem. Eng. Sci.* **22**, 531 (1967).
R1. Raghuraman, J., and Verma, Y. B. G., *Chem. Eng. Sci.* **30**, 145 (1975).
R2. Reichart, H., *VDI Forschungsh.* (*Berlin*) 414 (1942).
R3. Reith, T., Renken, S., and Israel, B. A., *Chem. Eng. Sci.* **23**, 619 (1968).
R4. Reman, G. H., *Chem. Ind.* Jan., 46 (1955).

R5. Rice, W. J., and Wilhelm, R. H., *AIChE J.* **4**, 423 (1958).
R6. Richardson, L. F., *Proc. R. Soc. (London) Ser. A* **110**, 709 (1926).
R7. Richardson, J. F., and Zaki, W. N., *Trans. Inst. Chem. Eng.* **32**, 35 (1954).
R8. Rieke, R. D., and Pigford, R. L., *AIChE J.* **17**, 1096 (1971).
R9. Rietema, K., *Proc. Int. Symp. Fluidization Eindhoven* p. 154 (1967).
R10. Rietema, K., *Proc. Int. Symp. Fluidization Eindhoven* p. 176 (1967).
R11. Rietema, K., and Hoebink, J., "Fluidization Technology," Vol. 1, p. 279. Hemisphere, Washington, D.C., 1976.
R12. Riley, H. L., *Trans. Inst. Chem. Eng.* **37**, 305 (1957).
R13. Rowe, P. N., *Chem. Eng. Sci.* **31**, 285 (1976).
R14. Rowe, P. N., and Everett, D. J., *Trans. Inst. Chem. Eng.* **50**, 55 (1972).
R15. Rowe, P. N., and Partridge, B. A., *Chem. Eng. Sci.* **18**, 511 (1963).
R16. Rowe, P. N., and Partridge, B. A., *Trans. Inst. Chem. Eng.* **43**, T157 (1965).
R17. Rowe, P. N., Partridge, B. A., and Lyall, E., *Chem. Eng. Sci.* **19**, 973 (1964).
R18. Rietema, K., *Chem. Eng. Sci.* **34**, 571 (1979).
R19. Rowe, P. N., MacGillivray, H. J., and Cheesman, D. J., *Trans. Inst. Chem. Eng.* **57**, 194 (1979).
R20. Rowe, P. N., Santoro, L., and Yates, J. G., *Chem. Eng. Sci.* **33**, 133 (1978).
S1. Sada, E., "Kiho-Ekiteki-Kagaku." Soc. Chem. Eng. Japan, 1969.
S2. Sankarasubramarian, R., and Gill, W. N., *Proc. R. Soc. (London) Ser. A* **333**, 115 (1973).
S3. Saxton, J. A., Fitton, J. B., and Vermeulen, T., *AIChE J.* **16**, 120 (1970).
S4. Schlichting, H., "Boundary Layer Theory," 4th ed. McGraw-Hill, New York, 1960.
S5. Schmitz, R. A., "Chem. Reaction Eng. Reviews" (H. H. Hulburt, ed.), *Adv. Chem. Series* **148**, 156 (1975).
S6. Schügerl, K., *Proc. Int. Symp. Fluidization, Eindhoven*, p. 782. (1967).
S7. Shen, C. Y., and Johnstone, H. F., *AIChE J.* **1**, 349 (1955).
S8. Sherwood, T. K., Pigford, R. L., and Wilke, C. R., "Mass Transfer." McGraw-Hill, New York, 1975.
S9. Shirai, T., *Kagaku Kogaku* **26**, 637 (1962).
S10. Shirai, T., "Ryudoso," 2nd ed. Kagaku-gijutsusha, Kanazawa, 1973.
S11. Shirai, T., Yoshitome, H., Shoji, Y., Tanaka, S., Hojo, K., and Yoshida, S., *Kagaku Kogaku* **29**, 880 (1965).
S12. Shrikhande, K. Y., *J. Sci. Ind. Res.* **14B**, 457 (1955).
S13. Shyu, C. N., and Miyauchi, T., *Kagaku Kogaku* **35**, 663 (1971).
S14. Squires, A. M., *Chem. Eng. Prog.* **58**, (4), 66 (1962).
S15. Standard Oil Co. (Br.) **1**, 126, 617 (1968).
S16. Steiner, H. H., *Ind. Eng. Chem.* **36**, 618, 840 (1944).
S17. Stemerding, S., see Reman, G. H., *Chem. Ind.* Jan., 46 (1955).
S18. Stephens, G. K., Sinclair, R. J., and Potter, O. E., *Powder Technol.* **1**, 157 (1967).
S19. Stewart, P. S. B., Ph.D. dissertation, Cambridge University, 1965.
S20. Stewart, P. S. B., *Trans. Inst. Chem. Eng.* **46**, T60 (1968).
S21. Stewart, P. S. B., and Davidson, J. F., *Powder Technol.* **1**, 61 (1967).
S22. Storch, H. H., Glumbic, N., and Anderson, R. B., "The Fischer–Tropsch and Related Syntheses." Wiley, New York, 1951.
S23. Sit, S. P. and Grace, J. R., *Chem. Eng. Sci.* **33**, 1115 (1978).
T1. Tadaki, T., and S. Maeda, *Kagaku Kogaku* **25**, 254 (1961).
T2. Takamura, T., B. Eng. thesis, Dept. Chem. Eng., University of Tokyo, 1970.
T3. Talmor, E., and Benerati, R. F., *AIChE J.* **9**, 536 (1963).
T4. Tam, Le Van, and Miyauchi, T., *Kagaku Kogaku* **35**, 650 (1971).

T5. Takahashi, M., M.S. thesis, Dept. Chem. Eng., University of Tokyo, 1977.
T6. Tanaka, I., and Shinohara, H., *J. Chem. Eng. Jpn.* **5**, 57 (1972).
T7. Tanaka, I., Shinohara, H., Hirose, H., and Tanaka, Y., *J. Chem. Eng. Jpn.* **5**, 51 (1972).
T8. Taylor, G. I., *Proc. R. Soc. (London) Ser.* **A201**, 192 (1950).
T9. Taylor, G. I., *Proc. R. Soc. (London) Ser.* **A219**, 186 (1953).
T10. Taylor, G. I., *Proc. R. Soc. (London) Ser.* **A223**, 446 (1954).
T11. Taylor, G. I., *Proc. R. Soc. (London) Ser.* **A225**, 473 (1954).
T12. Terahata, T., Tazawa, S., Kakizaki, T., Minoda, N., Miyazima, S., Ito, M., Imai, H., Kawazu, T., Japan patent No. 49-29166, 1974.
T13. Tett, H. C., "Fluidization," p. 1. Soc. Chem. Ind., London, 1964.
T14. Tichacek, L. J., Barkelew, C. H., and Baron, T., *AIChE J.* **3**, 439 (1957).
T15. Todt, J., Lücke, J., Schügerl, K., and Renken, A., *Chem. Eng. Sci.* **32**, 369 (1977).
T16. Toei, R., and R. Matsuno, *Proc. Int. Symp. Fluidization, Eindhoven*, p. 271 (1967).
T17. Toei, R., Matsuno, R., Oichi, M., and Yamamoto, K., *J. Chem. Eng. Jpn.* **7**, 447 (1974).
T18. Toei, R., Matsuno, R., Hotta, H., Oichi, M., and Fujine, Y., *J. Chem. Eng. Jpn.* **5**, 273 (1972).
T19. Toei, R., Matsuno, R., Kojima, H., Nagai, Y., Nakagawa, K., and Yu, S., *Kagaku Kogaku* **29**, 851 (1965).
T20. Toei, R., Matsuno, R., Nishitani, K., Hayashi, H., and Imamoto, T., *Kagaku Kogaku* **33**, 668 (1969).
T21. Tone, S., Seko, H., and Otake, T., *J. Chem. Eng. Jpn.* **7**, 44 (1974).
T22. Toomy, R. D., and Johnstone, H. F., *AIChE Symp. Ser. 5* **49**, 51 (1953).
T23. Towell, G. D., Strand, C. P., and Ackerman, G. H., Paper presented at Am. Inst. Chem. Eng. and Inst. Chem. Eng. Joint Meeting, p. 1272, London (1965).
T24. Towell, G. D., and Ackerman, G. H., *Proc. 5th Eur., 2nd Int. Symp. Chem. Reaction Eng. Amsterdam* B3-1 (1972).
T25. Trawinski, H., *Chem. Ing. Tech.* **23**, 416 (1951).
T26. Trawinski, H., *Chem. Ing. Tech.* **25**, 201 (1953).
T27. Tsutsui, T., and Miyauchi, T., *Kagaksu Kogaku Ronbunshu* **5**, 40 (1979).
T28. Turner, J. C. R., *Chem. Eng. Sci.* **21**, 971 (1966).
T29. Tsutsui, T., M. S. thesis, Dept. of Chem. Eng., University of Tokyo, 1974.
T30. Tellis, C. B., and Hulburt, H. M., *Proc. Pacific Chem. Eng. Conf. 1st Soc. Chem. Eng. Jpn. AIChE Kyoto* Session 8, p. 178 (1972).
U1. Ueyama, K., and Miyauchi, T., *Kagaku Kogaku Ronbunshu* **2**, 430 (1976).
U2. Ueyama, K., and Miyauchi, T., *Kagaku Kogaku Ronbunshu* **2**, 595 (1976).
U3. Ueyama, K., and Miyauchi, T., *Kagaku Kogaku Ronbunshu* **3**, 19 (1977).
U4. Ueyama, K., and Miyauchi, T., *Kagaku Kogaku Ronbunshu* **3**, 115 (1977).
U5. Ueyama, K., and Miyauchi, T., *AIChE J.* **25**, 258 (1979).
V1. Van Deemter, J. J., *Chem. Eng. Sci.* **13**, 143 (1961).
V2. Van Deemter, J. J., *Proc. Int. Symp. Fluidization Eindhoven* p. 334 (1967).
V3. Van der Laan, *Chem. Eng. Sci.* **7**, 187 (1958).
V4. Van Heerden, C., *Ind. Eng. Chem.* **45**, 1242 (1953).
V5. Van Heerden, C., Nobel, P., and van Krevelen, D. W., *Chem. Eng. Sci.* **1**, 51 (1951).
V6. Van Krevelen, D. W., and Hoftijzer, P. J., *Chem. Eng. Prog.* **46**, (1), 29 (1950).
V7. Van Swaaij, W. P. M., and Zuiderweg, F. J., *Proc. 5th Eur. 2nd Int. Symp. Chem. Reaction Eng.* B9-25 (1972).
V8. Van Swaay, W. P. M., and Zuiderweg, F. J., "Fluidization and Its Applications," p. 454. Cepadues, Toulouse, 1973.

V8a. Van Swaaij, W. P. M., "Chem. Reaction Eng. Reviews," (D. Luss and V. W. Weekman Jr., eds.), *Am. Chem. Soc. Symp, Ser.,* **72,** 193 (1978).
V9. Veatch, F., Callaham, J. L., Idol, J. D., and Milberger, E. C., *Perol. Refiner* **41,** (11), 187 (1962).
V10. Verma, Y. B. G., *Powder Technol.* **12,** 167 (1975).
V11. Vermeulen, T., Moon, J. S., Hennico, A., and Miyauchi T., *Chem. Eng. Prog.* **62,** (9), 95 (1966).
V12. Volk, W., Johnson, C. A., and Stotler, H. H., *Chem. Eng. Prog.* **58,** (3), 44 (1962).
V13. Vreedenberg, H. A., *Chem. Eng. Sci.* **9,** 52 (1958).
V14. Vreedenberg, H. A., *Chem. Eng. Sci.* **11,** 274 (1960).
V15. Venuto, P. B. and Habib, E. T. Jr., "Fluid Catalytic Cracking with Zeolite Catalysts." Marcel Dekker, 1979.
W1. Wakabayashi, T., and Kunii, D., *J. Chem. Eng. Jpn.* **4,** 226 (1971).
W2. Wen, C. Y., and Hashinger, R. F., *AIChE J.* **6,** 220 (1960).
W3. Wen, C. Y., and Leva, M., *AIChE J.* **2,** 482 (1956).
W4. Wen, C. Y., and Yu, Y. H., *Chem. Eng. Prog. Symp. Ser. 62* **62,** 100 (1966).
W5. Wender, L., and Cooper, G. T., *AIChE J.* **4,** 15 (1958).
W6. Werther, J., *Trans. Inst. Chem. Eng.* **52,** 149, 160 (1974).
W7. Werther, J., *AIChE Symp. Ser. 141* **70,** 53 (1974).
W8. Werther, J., *Powder Technol.* **15,** 155 (1976).
W9. Werther, J., "Fluidization Technology," Vol. 1, p. 215. Hemisphere, Washington, D.C., 1976.
W10. Werther, J., *Int. J. Multiphase Flow* **3,** 367 (1977).
W10a. Werther, J., *Chem. Ing. Tech.* **50,** 850 (1978).
W11. Wheeler, A., "Catalysis II" (P. H. Emmett, ed.), p. 105. Reinhold, New York, 1958.
W12. Whitehead, A. B., Gartside, G., and Dent, D. C., *Powder Technol.* **14,** 61 (1976).
W13. Wilhelm, R. H., and Kwauk, M., *Chem. Eng. Prog.* **44,** 201 (1948).
W14. Wollard, J. M., and Potter, O. E., *AIChE J.* **14,** 388 (1968).
W15. Wan, C. G., and Ziegler, E. N., *Ind. Eng. Chem. Fundam.* **12,** 55 (1973).
W16. Wijffels, Ir. J-B., and Rietema, K., *Trans. Inst. Chem. Eng.* **50,** 224 (1972).
W17. Wijffels, Ir. J-B., and Rietema, K., *Trans. Inst. Chem. Eng.* **50,** 233 (1972).
X1. Xavier, A. M., Lewis, D. A. and Davidson, J. F., *Trans. Inst. Chem. Eng.* **56,** 274 (1978).
Y1. Yagi, S., and Miyauchi, T., *Kagaku Kogaku* **17,** 382 (1953).
Y2. Yamagoshi, T., B.S. thesis, Dept. Chem. Eng., University of Tokyo, 1969.
Y3. Yamazaki, M., *J. Chem. Eng. Jpn.* **8,** 420 (1975).
Y3a. Yamazaki, M., *Annu. Meet. Soc. Chem. Eng. Jpn., 40th,* Nagoya (1975).
Y4. Yamazaki, M., and Miyauchi, T., *J. Chem. Eng. Jpn.* **4,** 324 (1971).
Y5. Yamazaki, M., and Miyauchi, T., *Kagaku Kogaku Ronbunshu* **3,** 261 (1977).
Y6. Yamazaki, M., Ito, N., and Jimbo, G., *Kagaku Kogaku Ronbunshu* **3,** 272 (1977).
Y7. Yang, W. C., and Keairns, D. L., *AIChE Symp. Ser. 141* **70,** 27 (1974).
Y8. Yang, W. C., and Keairns, D. L., "Fluidization Technology," Vol. 2, p. 51. Hemisphere, 1976.
Y9. Yates, J. G., and Constans, J. A. P., *Chem. Eng. Sci.* **28,** 1341 (1973).
Y10. Yates, J. G., and Rowe, P. N., *Trans. Inst. Chem. Eng.* **55,** 137 (1977).
Y11. Yerushalmi, J., Turner, D. H., and Squires, A. M., *Ing. Eng. Chem. Process Des. Dev.* **15,** 47 (1976).
Y12. Yerushalmi, J., Gluckman, M. J., Graff, R. A., Dobner, S., and Squires, A. M., "Fluidization Technology," Vol. 2, p. 437. Hemisphere, 1976.
Y13. Yokota, T., Hidaka, Y., and Yasutomi, T., *Kagaku Kogaku Ronbunshu* **1,** 399 (1975).

Y14. Yoshida, A., B.S. thesis, Dept. Chem. Eng., University of Tokyo, 1971.
Y15. Yoshida, K., and Kunii, D., *J. Chem. Eng. Jpn.* **1,** 11 (1968).
Y16. Yoshida, K., and Kunii, D., *J. Chem. Eng. Jpn.* **7,** 34 (1974).
Y17. Yoshida, K., Kunii, D., and Levenspiel, O., *Ind. Eng. Chem. Fundam.* **8,** 402 (1969).
Y18. Yoshida, K., Kunii, D., and Levenspiel, O., *Int. J. Heat Mass Transfer* **12,** 529 (1969).
Y19. Yoshida, K., Ueno, T., and Kunii, D., *Chem. Eng. Sci.* **29,** 77 (1974).
Y20. Yoshitome, H., *Kagaku Kogaku* **27,** 27 (1963).
Y21. Yoshitome, H., Dr. Eng. dissertation, Tokyo Inst. Tech., 1967.
Y22. Yoshitome, H., and Shirai, T., *J. Chem. Eng. Jpn.* **3,** 29 (1970).
Y23. Yoshitome, H., Mannami, Y., Mukai, K., Yoshikoshi, N., and Kanazawa, T., *Kagaku Kogaku* **29,** 19 (1965).
Y24. Yacono, C., Rowe, P. N. and Angelino, H., *Chem. Eng. Sci.* **34,** 789 (1979).
Y25. Yerushalmi, J. and Cankurt, N. T., *Chem. Tech.* **8,** 564 (1978).
Y26. Yoshida, K., Nakajima, K., Hamatani, N. and Shimizu, F., "Fluidization," Proc. 2nd. Eng. Foundation Conf. 1978, p. 13, Cambridge Univ. Press, London and New York (1978).
Z1. Zabrodsky, S. S., "Hydrodynamics and Heat Transfer in Fluidized Bed." Gosenergoizdat, Moscow, 1963; English trans., MIT Press, Cambridge, Massachusetts, 1966.
Z2. Zabrodsky, S. S., Antonishin, N. V., and Parnas, A. L., *Can. J. Chem. Eng.* **54,** 52 (1976).
Z3. Zalewski, W. Ċ., and Hanesian, D., *AIChE Symp. Ser.* *128* **69,** 58 (1973).
Z4. Zenz, F. A., *Petrol. Refiner* **36** (5), 261 (1957).
Z5. Zenz, F. A., and Othmer, D. F., "Fluidization and Fluid-Particle Systems." Reinhold, New York, 1960.
Z6. Zenz, F. A., and Weil, N. A., *AIChE J.* **4,** 472 (1958).
Z7. Zuber, N., *Chem. Eng. Sci.* **19,** 897 (1964).

Note Added in Proof

Some related literature that appeared subsequent to the preparation of this article may be of interest to the reader and is therefore listed here:

Topic	Reference
Flow in fluid beds	B22, R20, X1, Y24
Bubble phenomena	D19, R18, R19, Y26
Heat and mass transfer	B21, H22, K29, S23
Axial dispersion	N10
Elutriation or dilute phase phenomena	G21, M50, M51, Y25
Reactions in freeboard or grid region	D20, G20
Models	V8a, W10a

INDEX

A

Absorption columns
 axial dispersion in, 114
 laboratory simulation of, 114–123
  practical selection, 116–119
 mass-transfer coefficients of, 1–133
 scale-up problems in, 112–114
Acrilonitrile, by Sohio process, 428
Aluminum industry, in India, 175–177
Andhra Pradesh (India), scientific and technical research in, organization chart for, 153
Atomic mineral industry, in India, 179

B

Biological systems, gas–liquid reactions in, 2
Birla Institute of Technology and Science (BITS) education program at, 190–192
Bubble(s)
 formation of, in fluidized catalyst beds, 290–297
  effect on performance, 340–360
 size, measurement of, 37–38
Bubble-cap column, mass-transfer coefficients of, 87–88
Bubble column
 longitudinal dispersion in, 335–338
 mass-transfer coefficients of, 75–76, 90–93
 turbulent-flow phenomena in, 310–330
  theory compared to experiments, 317–326
Bubble reactors, mechanically agitated, mass-transfer coefficients of, 97–107
Bulk flow, in fluidized catalyst beds, 297–304

C

Catalytic reactions, in fluidized catalyst beds, 391
Caustic soda industry, in India, 164–165
Chemical industry, in India, 135–197
Chlor-alkali industry, in India, 163–166
Chlorine industry, in India, 165–166
Coalescence frequency, in liquid–liquid dispersions, 228–233
Cocurrent packed columns, mass-transfer coefficients of, 76–87
Copper industry, in India, 177–178
Council of Scientific and Industrial Research (India), organization of, 152
Countercurrent columns, mass-transfer coefficients of, 69–75

D

Drop(s)
 breakage rates, phenomenological models, 209–215, 244–249
  daughter droplets, 214–221
  frequency, 210–214
 collision rates between, 216–221
 size
  in liquid–liquid dispersions, 223–228
  in turbulent dispersions, 207–209

E

Ejector reactor, mass-transfer coefficients of, 109–110
Electrolytes
 diffusivity of, 32–35
 solubility of, 26–28

F

Fertilizer industry, in India, 160–163
Film model, for gas–liquid reactors, 4
Fischer–Tropsch synthesis, on fluidized catalyst beds, 428–429
Flow field, in agitated dispersions, 200–206
Fluid catalytic reforming, description of, 427
Fluidization, various states of, 278
Fluidized catalyst beds
  axial distribution to bed structure in, 305–307
  axial distribution of reactivity in, 402–413
  bed internals effects, 307–310
  bubbles in
    bed performance and, 340–360
    dynamics, 360–363
    formation, 290–297
    holdup, 299–301
    mass transfer to, 365
    splitting, 350–360
    velocity, 342–345
    wake, 346–350
  bulk flow within, 297–304
  catalytic reactions in, 391
  desired catalyst particle properties, 284
  emulsion-phase expansion, 286–290
  emulsion viscosity in, 302–304
  fines effect on, 285–286
  flow features of, 280
  flow properties of, 285–310
  fluidization quality related to particle properties, 283
  heat and mass transfer in, 360–381
    axial and radial mixing, 373–379
    wall heat transfer, 379–381
  industrial, 426–429
  longitudinal dispersion phenomena in, 330–340
    concept for, 338–339
    radial and axial diffusivity in, 333–335
  nonisothermal effect on performance of, 413–425
    steady reaction, 413–421
  operation of, 297
  overall rate constant for reaction in, 382–383
  particle capacitance effect in, 371–373, 375–379
  particle properties in relation to fluidity, 285–289
    desirable properties, 288
  properties of, 277–281
  reactor models for applicability, 425–426
  recent trends of, 429–430
  scale-up ratio and reaction in, 282
  stability of, 421–425
  successive contact mechanisms in, 381–413
    axial distribution, 407–413
    reactor models, 383–390
  technical problems in design of, 430–432
  transport phenomena in, 275–448
  turbulent-flow phenomena in, 310–330
  turbulent kinematic viscosity in, 326

G

Gas holdup, measurement of, 36–37
Gas–liquid reactors
  mass-transfer rates in, 1–133
    with chemical reaction, 7–18
    physical absorption, 3–7
Gas scrubbing, gas–liquid reactions in, 2
Gases, in liquids, solubility and diffusivity of, 20–35

H

Hatta number, significance of, 19–20
Heat transfer, in fluidized catalyst beds, 360–381
Heavy chemical industry, in India, 157–166
Henry constant, 21

I

Indian chemical industry, 135–197
  awards for progress in, 184–187
  control of, 148
  development from 1800 to 1947
  development to 1800, 137–142
  geographical locations of, 155
  for heavy chemicals, 157–166
  for iron and steel, 172–175
  licensing and financing of, 149

INDEX 451

needs of, 189–195
for nonferrous metals, 175–180
for organic chemicals, 166–168
for pesticides, 170
for pharmaceuticals, 180–184
for plastics, 170
professional societies and education in, 187–189
scientific and technological research organization in, 151
since independence (1947), 154–184
for soaps and detergents, 168–170
structure of, 147–154
summary of, 136–137
Interfacial area(s)
in absorber scale-up, 67–114
Interfacial area
of liquid-in-liquid dispersions, 221–223
mathematical models, 234–236
in various reactors, 113
measurement of, 35–67
chemical methods, 39–49
physical methods, 36–40
Interphase reactions, in liquid–liquid dispersions, 199–273
Iron and steel industry, in India, 172–175

J

Japan, applications of fluidized catalyst beds in, 429
Jet reactors, mass-transfer coefficients of, 107–112

K

Kinetic regimes, identification of, 52

L

Longitudinal dispersion phenomena, in fluidized catalyst beds, 330–340
Laminar/transitional flow dispersions, bheavior in, 205–206
Lead industry, in India, 178
Lewis–Gilliland–Glass plot, 409
Liquids, gasses in, solubility and diffusivity of, 20–35

Liquid-in-liquid dispersions
agitated, flow field in, 200–206
behavior of, 207–221
dispersed phase interaction models, 237–238
drop size distribution in, 223–228
mathematical models, 236–237
interaction models for, 238–239
Monte Carlo simulation models, 253–259
interfacial area measurements, 221–223, 234–236
interphase reactions and mass transfers in, 199–273
laminar/transitional flow dispersions, 205–206
macro- and micromixing concepts of, 259–262
mass transfer analysis, 249–253
property measurement and analysis, 221–233
Liquid-phase processes, gas–liquid reactions in, 2

M

Mass transfer
in fluidized catalyst beds, 360–381
with reaction in liquid–liquid dispersion, 233–262
Mass-transfer coefficients
in absorber scale-up, 67–114
of agitated bubble reactors, 97–107
of bubble columns, 75–76, 90–93
with chemical and purely physical processes, 65–67
of ejector reactors, 109–110
of jet reactors, 107–112
measurement of, 35–67
apparatus for, 49, 53
chemical methods, 39–49
physical methods, 36–40
of plate columns, 87–90
scale-up problems and, 112–114
of simulative laboratory models, 116–123
of spray towers, 94–97
of tube reactors, 93–94
of Venturi scrubbers, 110–112

Mass-transfer rates, in gas–liquid absorbers and reactors, 1–133
Monte Carlo simulation models, for interaction in liquid–liquid dispersions, 253–259

N

Nonelectrolytes, diffusivity of, in liquids, 28–32
Nonferrous metal industry, in India, 175–180

O

Organic chemical industry, in India, 166–168
Ostwald coefficient, 21

P

Packed columns
  mass-transfer coefficients of, 67–87
    bubble columns, 75–76
    countercurrent columns, 69–75
    cocurrent packed columns, 76–87
Particle capacitance, in fluidized catalyst beds, 371–373
Pesticides industry, in India, 170
Petroleum cracking, by fluid catalysis, 426–427
Pharmaceutical industry, in India, 180–184
Phthalic anhydride, production on fluid catalyst beds, 427–428
Plastics industry, in India, 170–171
Plate columns, mass-transfer coefficients of, 87–90
Plunging-jet reactors, mass-transfer coefficients of, 107–109
Population balance equation, mass transfer analysis by, 249
Prandtl's hypothesis, turbulent viscosity based on, 326–327
Pure-product manufacture, gas–liquid reactions in, 2

S

Salting coefficients
  for gases, 27
  for inorganic ions, 27
Sieve-plate columns, mass-transfer coefficients of, 88–90
Soap and detergent industry, in India, 168–70
Sodium sulfite oxidation, use in interfacial area measurement, 56–65
Spray towers, mass-transfer coefficients of, 94–97
Stirred cells and vessels
  hydrodynamics in, 204–205
  simulating a packed column, 119–120
Submerged-jet reactors, mass-transfer coefficients of, 107–109
Sulfuric acid industry, in India, 157–161
Surface-renewal models, for gas–liquid reactors, 4–5, 15–16

T

Teeter bed, properties of, 279
Transport phenomena
  in fluidized catalyst beds, 275–448
Tube reactors, mass transfer coefficients of, 93–94
Turbulent dispersions, drop size in, 207–209
Turbulent flow, in bubble columns and fluidized catalyst beds, 310–330

V

Venturi scrubbers, mass-transfer coefficients of, 110–112

Z

Zinc industry, in India, 178–179

RAYMOND H. FOGLER LIBRARY